T0093222

Smart Cities

This book discusses the basic principles of sustainable development in a smart city ecosystem to better serve the life of citizens. It examines smart city systems driven by emerging IoT-powered technologies and the other dependent platforms.

Smart Cities: AI, IoT Technologies, Big Data Solutions, Cloud Platforms, and Cybersecurity Techniques discusses the design and implementation of the core components of the smart city ecosystem. The editors discuss the effective management and development of smart city infrastructures, starting with planning and integrating complex models and diverse frameworks into an ecosystem. Specifically, the chapters examine the core infrastructure elements, including activities of the public and private services as well as innovative ICT solutions, computer vision, IoT technologies, data tools, cloud services, AR/VR technologies, cybersecurity techniques, treatment solution of the environmental water pollution, and other intelligent devices for supporting sustainable living in the smart environment.

The chapters also discuss machine vision models and implementation as well as real-time robotic applications. Upon reading the book, users will be able to handle the challenges and improvements of security for smart systems and will have the know-how to analyze and visualize data using big data tools and visualization applications. The book will provide the technologies, solutions as well as designs of smart cities with advanced tools and techniques for students, researchers, engineers, and academics.

Smart Cities

IoT Technologies, Big Data Solutions, Cloud Platforms, and Cybersecurity Techniques

Edited by
Alex Khang, Shashi Kant Gupta, Sita Rani, and Dimitrios A. Karras

Cover image: © Shutterstock

First edition published 2024
by CRC Press
2385 NW Executive Center Drive, Suite 320, Boca Raton FL 33431

and by CRC Press
4 Park Square, Milton Park, Abingdon, Oxon, OX14 4RN

CRC Press is an imprint of Taylor & Francis Group, LLC

© 2024 selection and editorial matter, Alex Khang, Shashi Kant Gupta, Sita Rani, and Dimitrios A. Karras; individual chapters, the contributors

Reasonable efforts have been made to publish reliable data and information, but the author and publisher cannot assume responsibility for the validity of all materials or the consequences of their use. The authors and publishers have attempted to trace the copyright holders of all material reproduced in this publication and apologize to copyright holders if permission to publish in this form has not been obtained. If any copyright material has not been acknowledged please write and let us know so we may rectify in any future reprint.

Except as permitted under U.S. Copyright Law, no part of this book may be reprinted, reproduced, transmitted, or utilized in any form by any electronic, mechanical, or other means, now known or hereafter invented, including photocopying, microfilming, and recording, or in any information storage or retrieval system, without written permission from the publishers.

For permission to photocopy or use material electronically from this work, access www.copyright.com or contact the Copyright Clearance Center, Inc. (CCC), 222 Rosewood Drive, Danvers, MA 01923, 978-750-8400. For works that are not available on CCC please contact mpkbookspermissions@tandf.co.uk

Trademark notice: Product or corporate names may be trademarks or registered trademarks and are used only for identification and explanation without intent to infringe.

ISBN: 978-1-032-45111-4 (hbk)
ISBN: 978-1-032-45230-2 (pbk)
ISBN: 978-1-003-37606-4 (ebk)

DOI: 10.1201/9781003376064

Typeset in Times
by KnowledgeWorks Global Ltd.

Contents

Biography of Editors

Dr. Alex Khang is a Professor in Information Technology (IT), AI and data scientist, software industry expert, and the chief of technology officer (AI and Data Science Research Center) at the Global Research Institute of Technology and Engineering, North Carolina, United States. He has over 28 years of teaching and research experience in information technology at the Universities of Science and Technology in Vietnam, India, and USA. He has published 52 authored books (in computer science 2000-2010), 2 authored books (software development), and 20 book chapters. He has published 10 edited books, and 11 edited books (calling for book chapters) in the fields of AI ecosystem (AI, ML, DL, IoT, Robotics, Data science, Big data, and Quantum computing), smart city ecosystem, healthcare ecosystem, Fintech technology, and blockchain technology (since 2020). He has over 28 years of working experience as a software product manager, data engineer, AI engineer, cloud computing architect, solution architect, software architect, database expert in the foreign corporations of Germany, Sweden, the United States, Singapore, and multinationals (former CEO, former CTO, former Engineering Director, Product Manager, and Senior Software Production Consultant).

Dr. Shashi Kant Gupta is a researcher and director, independent academic scholar at research department, CREP, Lucknow, Uttar Pradesh, India. He has completed his Ph.D. in CSE from Integral University, Lucknow, Uttar Pradesh, India, and worked as assistant professor in the department of computer science and engineering, PSIT, Kanpur, Uttar Pradesh, and India. He has published more than 17 Indian patents, one patent is under grant approval. He has already granted some Germany patents as well. He has more than 28 years of teaching experience and 2 years of industrial experience. He has been publishing 10 book chapters for Taylor and Francis Group – CRC Press, and 04 edited books (calling for book chapters) in the fields of AI Ecosystem (AI, Robotics, data science, Big Data, and IoT), smart city ecosystem, healthcare ecosystem, and Blockchain technology. He has published many research papers in reputed international journals with SCOPUS and ESCI indexed journals and published many papers in national and international conferences and as well as in Seminars. He is currently founder and CEO of CREP, Lucknow, Uttar Pradesh, India. He is a member of Spectrum IEEE & Potentials Magazine IEEE since 2019 and many more international organization for research activities. He has published many research papers in reputed international journals with SCOPUS and ESCI indexed journals and published many papers in national and international conferences and in seminars. He has organized various faculty development programs, seminars, workshops, and short-term courses at University level. His main research work focuses on performance enhancement through cloud computing, Big data analytics, IoT and computational intelligence-based education. He is currently working as a reviewer in various international journals like BJIT and many more. He has published many Indian and Australian patents in the field of information technology, computer science, and management.

Dr. Sita Rani is a Faculty of Computer Science and Engineering at Guru Nanak Dev Engineering College, Ludhiana. Previously, she has served as a Professor – Computer Science & Engineering and Deputy Dean (Research) at Gulzar Group of Institutions, Khanna (Punjab). She has completed her B.Tech and M.Tech degrees in the faculty of Computer Science and Engineering from Guru Nanak Dev Engineering College, Ludhiana. She obtained her Ph.D. in Computer Science and Engineering from I.K. Gujral Punjab Technical University, Kapurthala, Punjab. At present, she is also Postdoctoral Fellow at Data Mining and Virtualization Laboratory, South Ural State University, Russia. She has more than 19 years of teaching experience. She is an active member of IEEE, ISTE, and IAEngg. She is the receiver of ISTE Section Best Teacher Award and International Young Scientist Award. She has contributed to the various research activities while publishing articles in the renowned SCI journals and conference proceedings. She has published five international and one Indian patents. She has delivered many expert talks in A.I.C.T.E. sponsored Faculty Development Programs and organized many International Conferences during her 19 years of teaching experience. She is a member of Editorial Board of four international journals of repute. Her research interest includes Parallel and Distributed Computing, Machine Learning, and Internet of Things (IoT).

Dr. Dimitrios A. Karras, Associate Professor, he received Diploma and M.Sc. Degree in Electrical and Electronic Engineering from the National Technical University of Athens (NTUA), Greece in 1985 and the Ph.D. in Electrical Engineering, NTUA, Greece in 1995, with honors. During 1990–2004, he collaborated as visiting professor and researcher with several universities and research institutes in Greece and Heidelberg, Germany (DKFZ). During 2004–2018, he has been with the Sterea Hellas Institute of Technology, Automation Dept., Greece as assoc. prof. in Digital Systems and Signal Processing, as well as visiting prof. of Hellenic Open University, Dept. Informatics in Communication Systems (2002–2010). Since 1/2019, she has remained an Associate Prof. in Digital and Intelligent Systems & Signal Processing, in National and Kapodistrian University of Athens, Greece, School of Science, Dept. General (dakarras@uoa.gr) as well as adjunct Assoc. Prof. Dr. with the School of Basic Sciences, BIHER University, Chennai, India, as well as with the GLA university, Mathura, India and EPOKA & CIT universities, Computer Engineering Dept., Tirana. He has published more than 80 research refereed journal papers in intelligent and distributed/multi-agent systems, pattern recognition, and image/signal processing and neural networks as well as in bioinformatics and more than 185 research papers in international refereed scientific conferences. His research interests span the fields of intelligent and distributed systems, multi-agent systems, pattern recognition and computational intelligence, image and signal processing/systems, biomedical systems, communications and networking, as well as security and sustainability applications. He has served as program committee member as well as program/general chair at several international workshops and conferences in signal, image, communication, and automation systems. He is, also, former editor in chief (2008–2016) of the International Journal in Signal and Imaging Systems Engineering (IJSISE), Academic editor in the Applied Computational Intelligence and Soft Computing, TWSJ, ISRN Communications and the Applied Mathematics Hindawi journals, as well as associate editor in various scientific journals, including CAAI, IET, Bentham Science Recent Patents in Engineering Section Editor, Sustainable Solutions and Society journal (Editorial Team | Sustainable Solutions and Society (spast.org)), IJANA and Academic Publishing, Information System and Smart City. He has been cited in more than 2500 research papers, his H/G-indices are 21/51 (Google Scholar) and his Erdos number is 5. His RG score is 31.94 in the highest 10% of scientific researchers. Apart from his scientific research, Assoc. Prof. Dr. D.A. Karras is involved in humanitarian projects as Sustainability, Human Rights and Peace passionate being Director of Research and Documentation at AdiAfrica NGO.

well as associate editor in various scientific journals, including [illegible], IET, De Plante Science, Recent Patents in Engineering, [illegible], Sustainable [illegible] Society Journal [illegible] Sustainable [illegible] and [illegible] [illegible] [illegible] and Academic Publishing: Information System and Smart City, [illegible] [illegible] in more than 2[illegible] [illegible] His [illegible] [illegible] (Google Scholar) and his [illegible] number [illegible] His RG score is [illegible] [illegible] [illegible] [illegible] [illegible] [illegible] [illegible] [illegible] [illegible] [illegible] [illegible] [illegible] [illegible] [illegible] [illegible]

Contributors

Ahmad Al Yakin
Universitas Al Asyariah Mandar
Sulawesi Barat
Indonesia

Faris S. Alghareb
CIE Department, Ninevah University
Mosul, Iraq

Abuzarova Vusala Alyar
Azerbaijan State Oil and Industry University
Baku, Azerbaijan

Shweta Bansal
Centre of Excellence
Department of Computer Science
School of Engineering and Technology
K.R. Mangalam University
Gurugram, Haryana, India

Khushi Bhoj
Thakur College of Engineering and Technology
Mumbai, Maharashtra, India

Geetha C.
Satyabama Institute of Science and Technology Business Incubator
Semmancheri, Chennai, Tamil Nadu, India

Vadivelraju Chandrasekar
Department of Computer science and Engineering
Saveetha School of Engineering, SIMATS
Chennai, Tamil Nadu, India

NL Sowjanya Cherukupalli
Department of Computer Science and Engineering
Koneru Lakshmaiah Education Foundation
Guntur, Andhra Pradesh, India

Ankur Goel
Meerut Institute of Technology
MIET Group
Meerut, Uttar Pradesh, India

Swati Gupta
Centre of Excellence
Department of Computer Science
School of Engineering and Technology
K.R. Mangalam University
Gurugram, Haryana, India

Abdullayev Vugar Hajimahmud
Azerbaijan State Oil and Industry University
Baku, Azerbaijan

Balqees Talal Hasan
CIE Department
Ninevah University
Mosul, Iraq

Lopamudra Hota
Department of Computer Science and Engineering
National Institute of Technology
Rourkela, Odisha, India

Olena Hrybiuk
Department of Computer Science and Software Engineering (CSSE)
International Scientific and Technical University
Kyiv, Ukraine

Luke Jebaraj
Electrical and Electronics Engineering
P.S.R. Engineering College
Sivakasi, Tamil Nadu, India

Shobhna Jeet
School of Legal Studies
K.R. Mangalam University
Gurgaon, Haryana, India

Sushma S. Kamlu
Department of Electrical and Electronics Engineering
Birla Institute of Technology
Mesra, Ranchi, Jharkhand, India

Alex Khang
Professor in Information Technology
Universities and Institutions of Science and Technology in Vietnam, India and United States
AI and Data Scientist, Department of AI and Data Science
Global Research Institute of Technology and Engineering
Fort Raleigh, North Carolina

Ritu Kothiwal
Vishwa Vishwani Institute of Sytems
Management, Thumkunta (V)
Hyderabad, Telangana, India

Arun Kumar
Department of Computer Science and Engineering
National Institute of Technology
Bhubaneswar, Odisha, India

Praveen Kumar
Department of Computer Science and Engineering
National Institute of Technology
Rourkela, Odisha, India

Anand Kumar M.
Department of Information Technology
National Institute of Technology Karnataka
Surathkal, Mangalore, Karnataka, India

Jayashree Mahale
School of Computer Science and Engineering
Sandip University
Nashik, Maharashtra, India

Ritvik Mahesh
Department of Information Technology
National Institute of Technology Karnataka
Mangalore, Karnataka, India

Abdul Mukit
Sekolah Tinggi Agama Islam Darul Ulum Banyuanyar Pamekasan
East Java, Indonesia

Gadirova Elmina Musrat
Baku State University
Baku, Azerbaijan

Muthmainnah
Universitas Al Asyariah Mandar Sulawesi Barat
West Sulawesi, Indonesia

Arpita Nayak
KIIT School of Management
KIIT University
Bhubaneswar, Odisha, India

Biraja Prasad Nayak
Department of Computer Science and Engineering
National Institute of Technology
Bhubaneswar, Odisha, India

Sanjeev Patel
Department of Computer Science and Engineering
National Institute of Technology
Rourkela, Odisha, India

Atmika Patnaik
King's College
London

B. C. M. Patnaik
School of Management
KIIT University
Bhubaneswar, Odisha, India

Antony Richard Pravin
Department of Electrical and Electronics Engineering
S.R.M. Institute of Science and Technology
Chennai, Tamil Nadu, India

Prashasti Pritiprada
KIIT School of Management
KIIT University
Bhubaneswar, Odisha, India

Parul Priya
Department of Electrical and Electronics Engineering
Birla Institute of Technology
Ranchi, Jharkhand, India

Sohanraj R.
Department of Information Technology
National Institute of Technology Karnataka
Surathkal. Mangalore, Karnataka, India

Subhashini R.
Cambridge Institute of Technology Bangalore
Bengaluru, Karnataka, India

Kumar Ratnesh
Dewan VSGI
Meerut, Uttar Pradesh, India

Bholanath Roy
Department of Computer Science and Engineering
Maulana Azad National Institute of Technology
Bhopal, Madhya Pradesh, India

Dillip Rout
School of Computer Science and Engineering
Sandip University
Nashik, Maharashtra, India

Jason Krithik Kumar S.
Department of Information Technology
National Institute of Technology Karnataka
Mangalore, Karnataka, India

Neduncheliyan S.
Satyabama Institute of Science and Technology Business Incubator
Chennai, Tamil Nadu, India

Amar Saraswat
Department of Computer Science
School of Engineering and Technology
K.R. Mangalam University
Gurugram, Haryana, India

Ipseeta Satpathy
School of Management
KIIT University
Bhubaneswar, Odisha, India
Lynchburg College
Lynchburg, Virginia,

Ankita Sharma
Chandigarh University
Sahibzada Ajit Singh Nagar, Punjab, India

Deepak Sharma
Dewan Institute of Management Studies, Meerut, Uttar Pradesh, India

Yash Sharma
Thakur College of Engineering and Technology
Mumbai, Maharashtra, India

Arvind Kumar Shukla
School of Computer Science & Applications
IFTM University
Moradabad, Uttar Pradesh, India

Nupur Soni
School of Computer Applications
Babu Banarasi Das University
Lucknow, Uttar Pradesh, India

Kumar Sriram
Department of Electrical and Electronics Engineering
St. Anne's College of Engineering and Technology
Panruti, Tamil Nadu, India

Aman Syed
Thakur College of Engineering and Technology
Mumbai, Maharashtra, India

Akanksha Tandon
Department of Computer Science and Engineering
National Institute of Technology
Rourkela, India

Rashmi Thakur
Thakur College of Engineering and Technology
Mumbai, Maharashtra, India

Vicky Tyagi
Department of Computer Science
School of Engineering and Technology
K. R. Mangalam University
Gurugram, Haryana, India

Anil Vasoya
Thakur College of Engineering and Technology
Mumbai, Maharashtra, India

Satish Chandra Velpula
IIM Sambalpur
Burla, Sambalpur, Odisha, India
New Castle, USA

Meenu Vijarania
Centre of Excellence
Department of Computer Science
School of Engineering and Technology
K.R. Mangalam University
Gurugram, Haryana, India

Joseph Wheeder
School of Computer Science and Engineering
Sandip University
Nashik, Maharashtra, India

Preface

Nowadays, most of contemporary cities in the world are operating and managing a lot of complex systems in areas of socio-economic life such as public services, shopping, banking, energy, transportation, construction, education, logistic, and healthcare system. Planning the maintenance and development for them requires more smart Internet of Things (IoT) technologies, new innovative ideas for IoT technologies, data solutions, cloud platforms, cybersecurity techniques, and another advanced intelligent technologies. The goals of using smart solutions are to constantly improve the quality of services and implement the basic principles of sustainable development in a smart city ecosystem to serve better for the life of citizens in the era of society 5.0.

To get the success of designing and implementing the core components of the smart city ecosystem is effective management of the development of smart city infrastructure, starting with planning and integrating the complex models and diversity of frameworks into an ecosystem, especially the core infrastructure elements, are including activities of the public and private services as well as innovative modern solutions, advanced technologies, cloud services, security techniques, and other intelligent devices for supporting sustainable living in the smart environment.

The book brings insight into the smart city systems and offers the implementation of emerging IoT technologies and data-driven applications. It includes current developments and future directions and covers the concept of the smart city systems along with its ecosystem. It focuses on the Information and Communications Technology (ICT) solutions along with IoT technologies, Big data Solutions, cybersecurity techniques, cloud services, and intelligent prediction with visualization simulation for the smart city ecosystems. This book is also useful to architects, engineers, companies, experts, students, scholars, and researchers involved in artificial intelligence engineering, data science, IoT technologies, cloud platforms, and cybersecurity techniques in the smart city ecosystem that they can be a reference for future research.

Thank everyone!
Happy reading!

Editorial team: Alex Khang, Shashi Kant Gupta,
Sita Rani, Dimitrios A. Karras

Preface

Acknowledgments

This book is based on the design and implementation of Internet of Things (IoT), Artificial Intelligence (AI), data science, big data solutions, cloud platforms, cybersecurity technology, and emerging technologies in the smart city ecosystem.

Preparing and designing a book outline to introduce to readers across the globe are the passion and noble goal of the editorial team. To be able to make ideas to a reality and the success of this book, the biggest reward belongs to the efforts, experiences, enthusiasm, and trust of the contributors.

To all the reviewers with whom we have had the opportunity to collaborate and monitor their hard work remotely, we acknowledge their tremendous support and valuable comments not only for the book but also for future book projects.

We also express our deep gratitude for all the pieces of advice, support, motivation, sharing, collaboration, and inspiration we received from our faculty, contributors, educators, professors, scientists, scholars, engineers, and academic colleagues.

Last but not least, we are really grateful to our publisher CRC Press (Taylor & Francis Group) for the wonderful support in making sure the timely processing of the manuscript and bringing out this book to the readers soonest.

Thank you, everyone.

Editorial team: Alex Khang, Shashi Kant Gupta, Sita Rani, Dimitrios A. Karras

1 Smart City: Concepts, Models, Technologies and Applications

Luke Jebaraj, Alex Khang, Vadivelraju Chandrasekar, Antony Richard Pravin, and Kumar Sriram

1.1 INTRODUCTION

A smart city is a metropolis, which employs the strategy of using Internet of Things (IoT), combined with Artificial Intelligence techniques, to improve the operational efficacy of the city. Information and communication technology (ICT) plays a crucial role in the construction of smart cities (Khang & Rani et al., 2024).

Around 66–70% of global population is expected to be in urban zones, by next few decades. Therefore, it is essential to construct the infrastructure, to meet swift growth of urban population, through technological tools like ICT, big data, IoT, data fusion and mining.

This section proposes to examine the components, architecture, challenges and safety objectives of the planned smart city (Khang & Gupta et al., 2023).

1.1.1 Importance

The ease of living is a major demand by communities and individuals in an urban setting to enhance their lifestyle with the help of technology. This trend necessitates the importance of new technology-based, urban lifestyle for citizens. Therefore, the smart city design should be based on technology-based contemporary and eco-friendly infrastructure, well-organized and accessible surroundings and appropriate transportation.

1.1.2 Features

The following features are the substantial factors, to be considered for upgrading a city into a smart city, in terms of infrastructure, health, transportation, governance, security etc.

- To expand the local economy.
- To generate additional opportunities for employment.
- Well-organized transportation system.
- Technology-enabled education model.

DOI: 10.1201/9781003376064-1

- To cater to housing, infrastructure and health care necessitates of society.
- To ensure effectual resource management.
- Cost-effective and reliable key services to the community.
- Smart plan for housing.
- To evolve a strategy for meeting the challenges of climate change and growth of population.
- To provide smart technology utilization for community requirements.

1.1.3 Key Platforms, Areas and Components

Basic framework of a smart city is divided into three platforms like smart corporeal infrastructure, smart civilization infrastructure and smart digital infrastructure. Smart corporeal infrastructure platform comprises the smart living areas, smart environment, smart usefulness and smart mobility (Rani & Khang et al., 2022).

Smart lifestyle, citizen, economy and governance are deemed as the domain of smart civilization infrastructure. The area of smart digital infrastructure comprises smart sensor, smart data execution and smart networks (Hahanov & Khang et al., 2022). Each area is constructed with multiple key components. The entire key structure is depicted in Figure 1.1 and described in the following sections.

1.1.3.1 Smart Corporeal Infrastructure

The principal role of smart physical infrastructure is to create the basic framework of smart cities (Albino & Berardi et al., 2015). This platform consists of a variety of critical areas like smart living, environment, utilities and mobility, with numerous key components.

1.1.3.1.1 Smart Living

Smart living helps to provide the state of the art facilities to the communities living in urban areas (Vinod Kumar, 2020). Smart living implies enriched public facilities, enhanced quality of life and security for public clients. Smart living also provides scope for its creative expressions. Smart living encompasses various components like housing, electronic connection, cultural amenities, security and protection against emergency.

i. **Housing**: Twenty-first-century citizens tend to migrate from traditional housing to advanced quality housing and smart living meets the challenges of lack of resources, housing crisis and security and social threats (Bass & Sutherland et al., 2018).
ii. **Electronic connection**: The modern digital technology is one of the significant components for smart living because only it can provide improved quality services, devolution, security and multiple e-payment processes through sustainable electronic communication (Vrushank & Vidhi et al., 2023).
iii. **Cultural amenities**: Smart living retains the traditional and spiritual backdrops and promotes countrywide cultivation while enhancing the

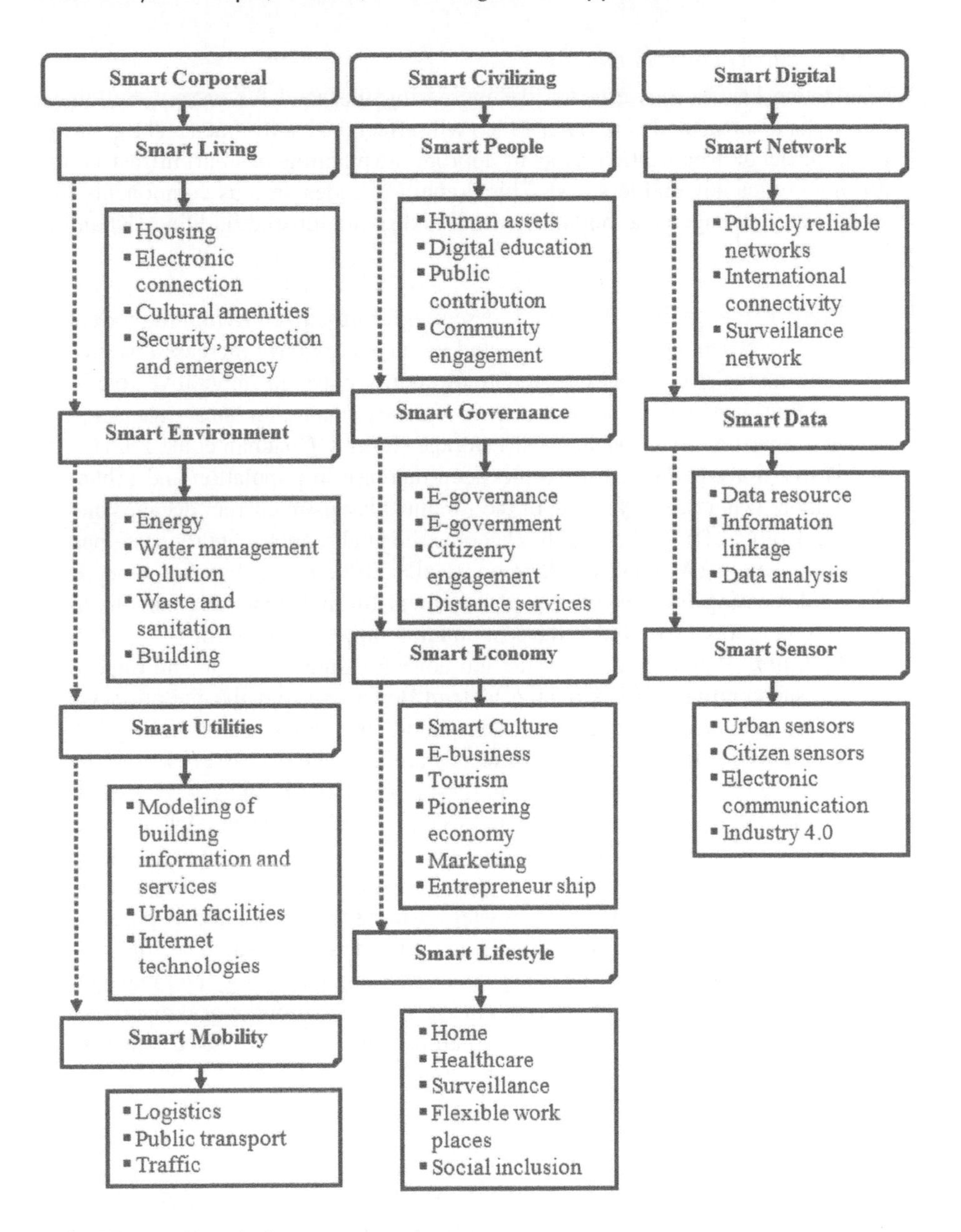

FIGURE 1.1 Key platforms, areas and components.

quality of city life. In other words, the wisdom of the cultural past is not ignored in the process of creating smart living.

iv. **Security, protection and emergency**: Smart living ensures security and protection against emergency for citizens, to help them to tide over unexpected accidents and crisis. For establishing such as circle of protection, proper planning is necessary.

1.1.3.1.2 Smart Environment

Urbanization has various effects, like increasing the need for energy, buildings, waste water disposal and pure water which will affect the environment. The environmental impact of urbanization is met by appropriate planning of smart infrastructure with more efficiency and less cost. This domain includes various components like energy, water management, pollution, waste and sanitation and building (Khanh & Khang, 2021).

i. **Energy**: Energy infrastructure is essential for urbanized living. In modern days, electrical energy utilization and green energy have increased against conventional fossil fuel energy. Smart energy plays an imperative role in meeting energy demand in the form of smart generation, smart transmission, smart distribution and smart storage (Rani & Chauhan et al., 2021).
ii. **Water management**: Due to the steep increase in population and urbanization, water scarcity is one of the foremost issues in current decades and availability of potable water has become the vital requirement for sustained urbanization (Achmad & Nugroho et al., 2018). It can be overcome by implementing improved water management through innovative technologies and tools to minimize the cost and maximize the reliability.
iii. **Pollution**: Pollution of air, water and noise are anthropogenic and imperil the surroundings. It is essential to trim down or avoid the pollution via pollution control management system, integrated with smart technologies.
iv. **Waste and sanitation**: Waste materials are unwanted substances from the activities of human, and it can be in form of solid, liquid and gas. Waste disposal is a critical issue for communities due to the increasing population. In smart cities, digital infrastructure has been employed to overcome this issue, through smart waste management, monitoring the collection and separation of waste, recycling and reuse of waste, waste transportation and disposal.
v. **Building**: The efficiency, optimal water and energy consumption and customer satisfaction are achieved by smart building technology and management system. Such buildings consume less energy from the grid due to the installation of renewable sources in their premises (Jucevičiusa & Patašienoʊ et al., 2014).

1.1.3.1.3 Smart Utilities

There are several smart utilities like water and energy, which are necessary for the operation and maintenance of support services. This domain comprises various components like modeling of building information and services, urban facilities and internet technologies.

i. **Modeling of building information and services**: It is essential to create corporeal infrastructure for the planning of providing the uppermost quality of service to citizens. Operation and maintenance of the modeling of smart building information technology would optimize the use of resources. Information about the infrastructure should be accessible when we require it (Rana & Khang et al., 2021).

ii. **Urban facilities**: There are numerous urban facilities, proffered to citizens by the city services management. In order to provide more efficient services, every smart citizen should be connected with these online facilities through information and communication technologies.
iii. **Internet technologies**: Many internet technologies like big data, cyber physical systems, cloud computing and IoT are used extensively, in various urban domains, like construction, planning, sustainable development and management (Khang & Hahanov et al., 2022).

1.1.3.1.4 Smart Mobility

Smart mobility signifies the movement of citizen, from one place to another, using different modes of transport, in a smart way. It incorporates the complete and optimal use of various types of vehicles, through authentic data and precise information, provided by ICT.

Smart mobility reduces the usage of personal cars, improves the quality of performance of public transport, protects the environment and saves the energy consumption. This domain contains components like logistics, public transport and traffic.

i. **Logistics**: In smart logistics, the information technology connects the logistics organization and distribution centers for stage points, stations and freight to allow any mobility as required. Mobile applications are also used for this application.
ii. **Public transport**: The public transport communication is performed through IoT technologies via mobile applications, to obtain real-time data about the route status and travel information, integrated with smart transport management (Khang et al., IoT, 2023).
iii. **Traffic**: The smart traffic management system operates with the help of sensors, IoT and radio frequency identification, to facilitate more efficient traffic flow. The smart traffic management system consists of various traffic subsystems like vehicle information, public transport information, active traffic management, fire service vehicles, ambulances, public parking, travel guides and real-time road navigation (Khang et al., Advanced IoT, 2024).

1.1.3.2 Smart Civilization Infrastructure

This platform comprises various essential domains like smart people, governance, economy and lifestyle with several key components (Chakravorti et al., 2017).

1.1.3.2.1 Smart People

Smart people are the citizens of smart cities and they apply their individual knowledge to give smart solutions to city issues. This field includes different components like human assets, digital education, public contribution and community engagement (Khang & Gupta et al., 2023).

i. **Human assets**: Human assets include citizens and their societies, with their related infrastructures. Human resource is a deciding factor to form the smart cities, due to their ability, skill level, inspiration and authority.

The outlining and shaping of smart cities, without human assets, are not conceivable (Allam & Newman, 2018).

ii. **Digital education**: The digital education for smart city development is essential for older and middle-aged citizens, who may be averse to technology. Hence, the need is to educate them.
iii. **Public contribution**: Public contribution means people should be engaged in governance and become urban collaborators in an extensive range of activities, related to public management like housing, safety, education and health (Bednarska-Olejniczak & Olejniczak et al., 2019).
iv. **Community engagement**: A well-organized community engagement system is essential for a supportive network because creating a smart society is not possible without community engagement.

1.1.3.2.2 Smart Governance

Smart civilization infrastructure is not possible without smart governance. This domain contains components like e-governance, e-government, distance services and citizenry engagement.

i. **E-governance**: The continued existence of e-governance is essential for smart governance, through transparent and innovative governance networks, using updated applications and tools. The e-governance ensures the implementation of its policies efficiently and effectively. The success of e-governance depends upon public contribution to every governance choice.
ii. **E-government**: The prime objective of e-government is to maximize information accessible to business and citizens and help them to avail better public services. A good-quality administration, people-friendly organization and the high-quality policy are the essential factors to construct a good e-government.
iii. **Citizenry engagement**: Key infrastructures, like transparency of open information technologies, self-governing requirements, participatory policy-making and urban hall services, are essential for citizenry engagement in smart governance. The urban government encourages the capability of stakeholders and it becomes a considerable factor in developing citizenry engagement (Khang & Gupta et al., 2023).
iv. **Distance services**: Integrated electronic services enable distance services, which is a fine tool for promoting social equality. It is also called a supplementary service, along with additional communication ports, and props up mutual communication between stakeholder and governance.

1.1.3.2.3 Smart Economy

The smart economy plays a principal role in the growth of the urban zone and enables pioneering and inspired business. This field includes components like smart culture, e-business, tourism, pioneering economy, marketing and entrepreneurship.

i. **Smart culture**: Smart culture means the culture of the smart city. It is based on ICT and motivates the growth of urban ability. It is an economic constituent of smart city infrastructure.

ii. **E-business**: Financial communication, purchase and after-sale, purchaser communication and financial transactions are made easy through financial-based e-business. It maintains a mutual relationship between purchasers and employers.
iii. **Tourism**: Smart tourism is related to every facet of the smart city like smart governance, resource management, method of interaction through communication technology, corporeal infrastructure and living capability (Dabeedooal & Dindoyal et al., 2019). Tourists are attracted through smart infrastructure.
iv. **Pioneering economy**: Pioneering economy is groundbreaking because it differs radically from traditional economic components. Smart economy transcends political borders and yields economic affluence by depending on smart digital and social infrastructure.
v. **Marketing**: Economic development of the society necessitates marketing to be customized to suit the smart economic components and tools. Smart marketing is development of economic activities through connecting with worldwide markets.
vi. **Entrepreneurship**: The digital infrastructure-based entrepreneurship amplifies economic transactions. Smart entrepreneurship eventually leads to economic development (Komninos, 2011).

1.1.3.2.4 Smart Lifestyle

The basic issues of lifestyle of the people should be resolved through smart social infrastructure so that their requirements are achieved in a smart way. It is important to develop the smart society infrastructure through required changes in lifestyle and maximizing the advantages of contributors. This domain includes components like home, health care, surveillance, flexible workplaces, social inclusion and social services.

i. **Home**: Smart homes permit occupants to capitalize on the natural environment and energy efficiency. It provides the security for occupants and helps to enhance the psychological and physical health.
ii. **Health care**: The technological innovations in the health sector have enabled smart health care. The digital smart infrastructure helps to keep the electronic records of patients (Khang & Rana et al., 2023).
iii. **Surveillance**: Surveillance is implemented by image-based network tools for face recognition. E-surveillance helps to enhance the security of the community, during periods of social or crime crisis.
iv. **Flexible workplaces**: The employees can select their work under any circumstances, in a flexible manner, and it is called flexible workplaces. It overcomes drawbacks associated with traditional way of work.
v. **Social inclusion**: The dignity and aptitude of poor people are ensured by social inclusion. Smart city enables the achievement of social inclusion, through innovative and creative community integration.
vi. **Social services**: Smart social services provide public services, recreation and health, to all types of people, including children, women and senior citizens and help them to improve their lifestyle.

1.1.3.3 Smart Digital Infrastructure

The smart digital infrastructure builds interaction between residents and government employees and enables functional activities to operate horizontally (Dustdar & Nastić et al., 2017). This platform includes vital domains like smart network, smart data and smart sensor with numerous key components.

1.1.3.3.1 Smart Network

Smart networks play the role of gathering, amassing and delivering information and data, in either directions or locations. The transmission and reception of concurrent data are energetically synchronized by smart networks.

The active structure change of socio-economic activities necessitates the high adoptive smart networks (Angelidou, 2017). This domain comprises components like publicly reliable networks, international connectivity, and surveillance network and dashboard.

i. **Publicly reliable networks**: Publicly reliable networks can be used by the majority of the citizens of smart city, who can communicate through priority-based mutual communication.
ii. **International connectivity**: International connectivity provides smart communication, through the smart grid channel between every major stakeholder so as to take away the sequential and spatial connectivity constraints, which plagued traditional mode. International connectivity helps to gather and process information through the digital infrastructure (Chaturvedi & Matheus et al., 2019).
iii. **Surveillance network**: Surveillance network enables the data processing in a multidimensional and additional qualitative way, with less quantity of data than the total of the separate parts. It is also known as tangential network.
iv. **Dashboard**: The stakeholders from smart cities need to communicate and record the information flatly and perpendicularly, with equipped program interaction. An interactive dashboard can be created to concentrate on the necessity of flow of information horizontally.

1.1.3.3.2 Smart Data

Smart data are collected digitally by the smart grids, via various sensors, and suitable data can be processed qualitatively, which helps the processing of information in a speedy way. This domain encompasses components like data resource, information linkage and data analysis.

i. **Data resource**: The data sources should be from the standard compilation of data strategy to facilitate the storage, repossession and processing in a multimodal way, along with necessitates of individual stakeholders. This framework is of processing, distributing and storing data, in a secure way, under a united equipped structure (Rani & Khang et al., 2023).
ii. **Information linkage**: Information can be linked through various smart cities' digital and physical connectivity components, with relevant technologies, and realize extremely qualitative, effectual and demand-driven capacity.

iii. **Data analysis**: The data analysis is an expensive and time-consuming process, since large-scale data have to be handled. Therefore, it needs the contemporary smart data analysis methods, with adequate processing ability.

1.1.3.3.3 Smart Sensor

Smart sensor examines the input data, using preset internal algorithms, and eliminates the superfluous, processed and unrelated data and sends it to the data centers. This key component of smart networks records the precise and automatic information (Alharbi & Soh, 2019). This field covers components like urban sensors, citizen sensors, electronic communication and Industry 4.0.

i. **Urban sensors**: Urban sensors give the essential information about city environment to the stakeholders, through processing centers. They check the complete urban situation so as to change the lifestyle and life quality.
ii. **Citizen sensors**: Citizen sensors are executed in two different ways. These are used to observe the usual behavior and health position of citizens initially. Later, these are used to share the information and help citizens to experience it through the electronic gadgets.
iii. **Electronic communication**: The high-speed electronic communication between the main components of smart city is executed through ICT tools like wireless networks, system instrumentation, fiber-optic and Wi-Fi networks. It enables qualitative, easier, safer and faster communication with the rest of the society.
iv. **Industry 4.0**: Industry 4.0 formulates machine to machine correspondence and human to machine correspondence via IoT components and gadgets. It looks for optimizing every smart city parameters, by creating smart connections between all stakeholders in the smart city.

1.1.4 Solution Architecture

The following common and specific smart city architectures are generally used in majority of smart city projects.

1.1.4.1 Common Architecture

The design of smart city and its implementation is based on the evolving situation and different necessities. Though these are architecture models like cloud computing, edge computing and fog computing, they could be customized to meet the special needs of a smart city.

1.1.4.2 Cloud Computing Architecture

Usually, this architecture is sectionalized in numerous layers. Each layer uses the services given with its individual and from other layers. The bases of architecture are being alienated into services and they are in the computation and decentralization of the information (Bhambri & Khang et al., 2022).

1.1.4.3 Edge Computing Architecture

The model of edge computing architecture generates data at the network edge and it is competent to process that data on the rim of it. The edge architecture is not forever the least sensor in the network and it bridges the cloud and local network.

1.1.4.4 Fog Computing Architecture

It is an extension of cloud computing architecture, which is done by the responsibilities and performance of the end nodes of the network. This type of architecture is used for geographically scattered and low-latency-required applications.

1.1.5 Challenges

Smart cities ensure that various functions like electronic governing, energy, transportation, health care, commerce and law implementation are executed, by using smart computing and electronic technologies in an effective way. Smart cities are likely to face challenges, depending upon the level of their functions.

- Deficiency of appropriate infrastructure for smart cities.
- Lack of knowledge and technical skills of residents in smart cities.
- Lack of information while planning for a smart city.
- Residents in smart city report short-range state of mind.
- Data privacy and transparency.
- Problems in private and public sector coordination.
- Lack of capacity to implement smart city plans.
- Lack of political will and its differences.
- Communal inclusivity of smart city plans.

1.1.6 Safety Objectives

All the associates of bionetworks like ventures, software suppliers, energy suppliers, network service suppliers, device makers and government have to do their part and incorporate solutions that stand by four main safety objectives, as given below:

i. **Accountability**: Responsible interactions are essential for client's access to a network. These accesses ought to be hard to falsify and have dependable truth security. The solutions of user ID management, having strong authentication, should ensure that the data are shared only with approved parties, to attain these safety core intentions.
ii. **Confidentiality**: It is necessary to avert the illegal revelation of responses, collected for analyzing the perceptive details of the customers.
iii. **Integrity**: Smart cities rely on precise and consistent information. It is necessary to take appropriate measures to ensure that information or data are exact and free from manipulation.
iv. **Availability**: The smart city cannot succeed without reliable and real-time access to data. The amassing, distillation and sharing of information are decisive, and solutions of security ought to avert unfavorable outcomes on accessibility.

1.2 SMART CITY TECHNOLOGIES

The technology-enabled smart cities yield a multitude of advantages. It facilitates well-organized communication between the government and citizens and introduces the public choice approach. For healthier management, it adopts some of the technologies to enhance its better functioning.

The entire smart city-related imperative technologies, like IoT technology, ICT, blockchain technology (BT), sensor technology, geospatial technology (GT) and artificial intelligence technology (AIT), are discussed here and depicted in Figure 1.2.

1.2.1 Internet of Things (IoT) Technology

The IoT refers to the software-entrenched collective network, sensors and other imperative technologies, linking and swapping data with some additional devices and networks over the internet.

Systems and networks from conventional background networking, are added to the IoT. It links and spreads crosswise each and every point like veins. Every smart city device requires being associated with each other, to get decisions for them, through sharing of resources.

IoT gives the ideal template structure of communicating devices, to give smart solutions to daily problems. IoT devices are formed for the use of customers, including house automation, linked health, associated vehicles and appliances with distant observing capabilities.

Habitual energy savings are the most significant long-term advantage in the smart home technology. It gives the superior quality of life, by ensuring the additional liberty of clients (Khang & Vrushank et al., 2023).

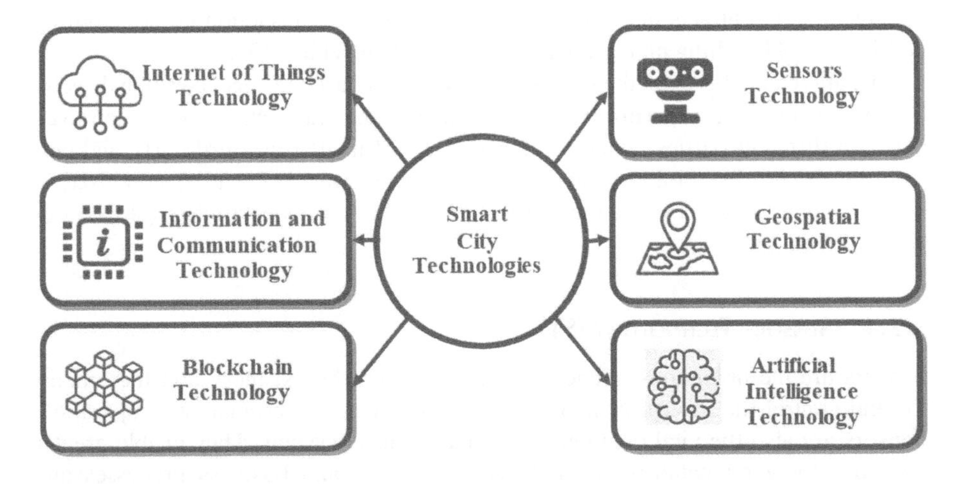

FIGURE 1.2 Smart city technologies.

1.2.2 Information and Communication Technology (ICT)

The role of ICT tools is to make the smarter two-way communication between the citizens and it builds a viaduct between the government and citizens of a smart city.

ICT is an extended edition of information technology, to incorporate the computers, telecommunication networks and signals that enable clients to access the manipulated, transmitted and stored information.

ICT helps to create a resource pool, to analyze the state of the demand for government intervention and to address them online. It refers to the audiovisual convergence of the telephone network systems, combined with the cyber network system, through a united cable link system, distribution of signal and management.

ICT encompasses any kind of component that will receive, recover and transmit the digital form of information. ICT helps to create a collective intelligence by means of communication through electronic medium, which can be organized for optimizing resources by using deep learning and analytics.

It includes some of the antiquated components like radio, television sets, land line telephones, from past decades, and continued with some new cutting-edge components like robotics, smart phones, smart TVs and artificial intelligence technologies.

1.2.3 Blockchain Technology (BT)

The application of BT is new to the concept of smart city. It is a high-tech, dispersed, pervasive and groundbreaking technology, which ensures confidentiality and accessibility of every agreement and piece of information. It is a contributed, accessible and unchallengeable digital ledger, which facilitates the procedure of recording transactions in a trade system (Khang & Chowdhury et al., Blockchain, 2022).

BT also means an arrangement of records, kept in a disseminated database, with incessantly increasing call blocks, which can be protected from tampering and emendation. The transparency and security are enhanced, with the incorporation of BT into all connected services of smart cities. BT is predicted to influence cities through smart contracts in some ways, which assist the transaction process, billing and facility of handling management (Tailor & Khang et al., 2022).

The agreement between the sellers and buyers in smart contract is written directly by coding lines. They permit trusted agreements and transactions to be approved between different parties, without the necessary of mediating third party and performing the process in a quicker, cheaper and safer way. BT can be now used in smart grids to make possible the energy sharing between smart cities (Hussain & Khang et al., 2022).

1.2.4 Sensors Technology (ST)

Sensors are the ubiquitous, hidden elements of the urban scenery and they monitor and control the enterprise management to enhance the efficiency of operation. Sensors are also the vital component of smart control system. They enable greater visibility for work patterns of employee, workflows and business processes and determine the conditions of environmental facilities.

Generally, sensors convert the physical nature parameters into a digital electronic signal and it can be applied in a self-governing system. Majority of smart building sensors have ability to give both automation and data information. The improvement of process is based on the awareness of the control system environment, where the sensors are fixed to collect the necessary information or data.

Sensors regulate their operations due to the use of suitable variables to typify its situation. A massive number of different sensors are obtainable to develop the technology continuously, which enables applications, not possible in the past owing to the restricted availability and high cost.

1.2.5 Geospatial Technology (GT)

GT has turned into a vital part of smart cities. GT enables the earth-orientated data or information, obtained by users, for creating modeling, visualization, simulations and analysis.

The GT offers an essential structure for transforming observation and data collection, to enable computer-oriented solution about smart infrastructure. It facilitates significant knowledgeable decisions, based on the resources precedence.

Smart city is built with a sustainable correct plan, which requires precise and significant data from GT, which gives the final and original foundation. A smart city project requires the digital system development for successful execution that can envision and supervise the geospatial data in an accessible atmosphere.

1.2.6 Artificial Intelligence Technology (AIT)

AIT refers to the human intelligence simulation, programmed machines to reflect human intelligence and similar to humans and help human action.

AIT reports attributes, connected with a mind of human-like problem-solving and learning. Smart city generates a large-scale data in a digital form and they are processed and information is produced in return. Such enormous quantity of generated data is processed by AIT and by that can create sense out of that information.

AIT coordinates with IoT, to manage the challenges, caused by huge population in urban areas, which demand energy management, traffic management, health care management etc. AIT can improve the quality of business and life of citizens, in a smart city.

1.3 SMART CITY MODELS

Smart city models help the developer and planner of smart city, to simulate models, related to their personal, socio-financial conditions and strategic objectives. Four distinct smart city models are planned and used worldwide.

1.3.1 Smart Transportation Model

The smart transportation model is designed for moving goods and citizens, in a densely inhabited area, within the urban limits. It plans to regulate the traffic congestion in urban zones through necessary technologies, like communication and information technology, public transportation, freight trucks and private or rented taxies. This model is already in practice in Dubai and Singapore.

1.3.2 Necessary Services Model

The essential services model is designed for various services like digital health care, mobile networks and digital communication, through emergency management systems. It plans to set up good communication infrastructures and place their capital into a small number of appropriate programs of smart city. This category of the model is being carried out in Copenhagen and Tokyo.

1.3.3 Business Bionetwork (Ecosystem) Model

The business bionetwork model is designed to jumpstart economic activity, by means of using appropriate information and communication technologies. It comprises setting up the specific digital skills training as an essential accessory, to make qualified personnel and to promote ultra-modern trades. This category of the model is evident in Amsterdam, Cape Town and Edinburgh.

1.3.4 Wide Spectrum Model

The wide spectrum model is designed to meet the challenges of services like water management, waste and sewage management and to evolve technical resolutions intended for control of pollution, of a problem in megacities. This category of model is already in practice in Barcelona, Beijing and Vancouver.

1.4 SMART CITY APPLICATIONS

The principal goal of smart cities is to improve the quality of life of the citizens and to stimulate the economic growth, without compromising on sustainability and efficiency, by means of decision-making improvement by citizens. There are two major classifications of smart city applications like precise applications and decisive applications. The application areas are displayed in Figure 1.3.

1.4.1 Precise Applications

The precise applications of smart city are relevant in various areas like buildings, surroundings, mobility, utilities, governance, public services, health care, economy and citizens (Khang & Hajimahmud et al., 2022).

1.4.1.1 Buildings

The key objective of the smart buildings is to construct both commercial and residential buildings, for comfort and energy proficient for living and working. A substantial number of sensors and actuators are interconnected, to monitor their own everyday tasks (Apanaviciene & Vanagas et al., 2020).

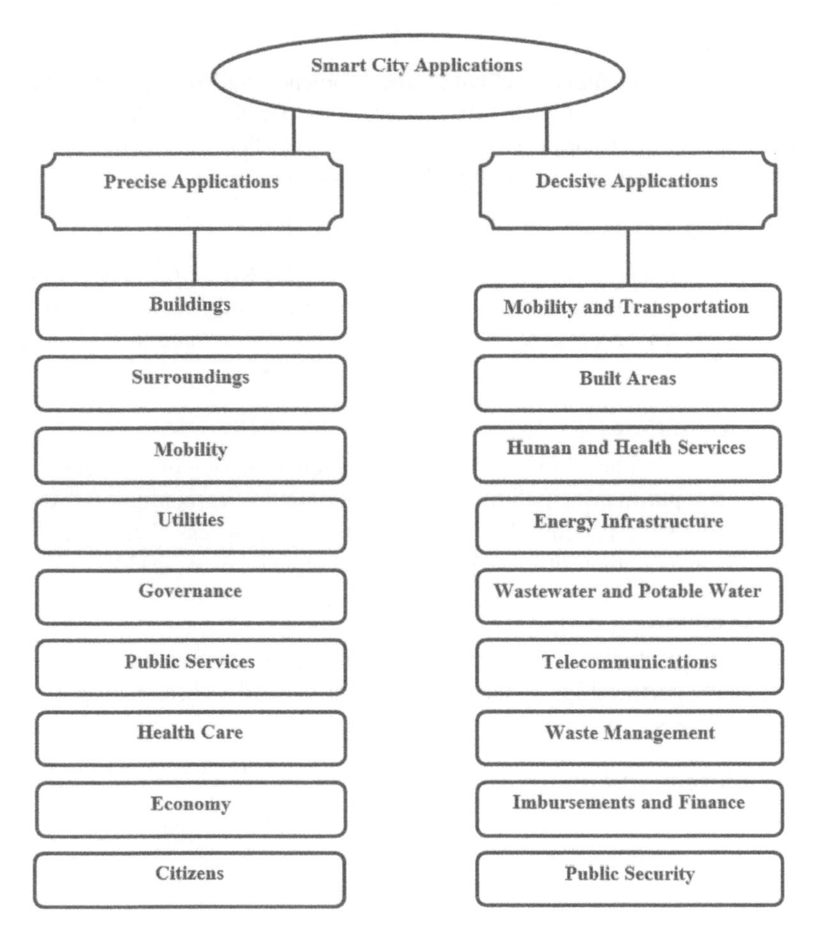

FIGURE 1.3 Smart city applications.

1.4.1.2 Surroundings

Smart surroundings are necessary to enhance the quality, safety and sustainability of cities. Smart technologies would help to detect the risk of unhealthy conditions around its surroundings and remedy them.

1.4.1.3 Mobility

Smart mobility is associated with public and private transportation system, which improves the safety with efficiency and decreases the unwanted transit times of the citizens. Recently, advanced intelligent vehicles are introduced, with the help of sensors, to enhance the quality of mobility.

1.4.1.4 Utilities

Smart utilities monitor the use of necessary resources like water, energy and gas, for efficiency, economic enlargement and sustainability, with the employment of some recent data-driven software.

1.4.1.5 Governance

Smart governance attempts to enhance the efficiency and transparency of home government activities and benefits its citizens to cross-reference and access various sources of data or information. Contribution of citizens helps to engage in city planning and growth processes, using digital electronic applications.

1.4.1.6 Public Services

The purpose of smart community services is to organize the resources, associated with the public, in an efficient and effective manner. Audio or visual monitoring systems are also used for appropriateness of safety services related to public (Chen & Wei et al., 2020).

1.4.1.7 Health Care

Smart health care implies the effective and competent provision of health care. Smart medical centers play a principal role in improving medical care, aligned with digital health records of the patients. It authorizes patients to access their digital health records or data about their fitness (Ahmad & Khujamatov et al., 2022).

1.4.1.8 Economy

Smart economy promotes economic progress through modern business models, private and public partnerships. It supports the joint networks of the entrepreneur and establishes office space, through broadband connectivity.

1.4.1.9 Citizens

Smart Citizens are developed by executing lifelong digital educational programs for employability and interactive information poles, to access various citizen services. Broadband connectivity can also support citizens in remote areas.

1.4.2 Decisive Applications

The decisive applications of smart city are employed for various purposes like mobility and transportation, built areas, health services, energy infrastructure, management of waste water and potable water, telecommunication, imbursements and finance and public security.

1.4.2.1 Mobility and Transportation

Mobility and transportation are very perilous today due to the traffic congestion. It is necessary to keep public transports, vehicles, bike lanes, streets and roads in an organized healthier way. Automated and connected vehicle services technology plays a vital role in reducing the majority of accidents with incredible efficiency.

1.4.2.2 Built Areas

The built areas refer to all kinds of buildings and their environments. It includes schools, colleges, hospitals, police stations, shopping malls and public administrative offices. The IoT-based sensors are connected to observe the daily tasks.

1.4.2.3 Human and Health Services

Educational virtualization and telemedicine services are provided to urban communities. It also includes online services, including community outreach.

1.4.2.4 Energy Infrastructure

Energy infrastructure generates and supplies electricity, to meet all requirements, in an efficient manner. It comprises generation, transmission, substation and distribution assets, with metering arrangements, controlled by urban and rural power authorities. The smart grid, micro-grid, solar and wind generation and energy storage have made tremendous progress in this area (Bedi & Goyal et al., 2022).

1.4.2.5 Waste Water and Potable Water

The infrastructure of water includes the distribution, compilation, purity, hygiene, metering and repossession. The waste water infrastructure comprises the recycle and treated process of waste water, which could be used for irrigation purpose.

1.4.2.6 Telecommunications

Telecommunication faces a challenging task, in the context of safety and comfortability of smart city citizens. High capacity broadband network is essential for all kinds of business activities, from server farms, call centers and distribution hubs (Manzanilla-Salazar & Malandra et al., 2020)

1.4.2.7 Waste Management

The modern-day waste management is based on smart collection, distribution and recycling of waste materials in urban zones. IoT-enabled solid waste management plays a major role in processing urban waste stream.

1.4.2.8 Imbursements and Finance

The revenue rose through a payment subscription model, which eliminates payment upfront contribution from the smart city. Unified payment portals can generate the revenue, with single domain registration, in a scientific manner.

1.4.2.9 Public Security

The employees and agencies of public safety infrastructure ensure the safety of citizens. It includes the IoT-enabled fire services, police, courts and agencies from emergency and disaster management.

1.5 RESULTS AND DISCUSSION

This chapter examined the basics, technologies, models and key applications of smart city, related to various fields. Recently, the various infrastructures, areas, components, technologies, models and applications have received growing attention from the smart city researchers.

As a result, by way of responsibility, smart city technologies, models and applications have become more competent, scalable and durable. These technological structures and their models would be well appreciated for a long period. Hence, the need to build a smart city based on various infrastructures, technologies, models and applications.

1.6 CONCLUSION

This chapter has elucidated various infrastructures, technologies, models and applications of smart city, integrated with different platforms and architectures. These platforms and architectures have been devised, to manipulate various aspects and constraints, in a customized manner.

The extensibility of the smart city technology-based solutions has been substantiated to be better than the other notable urban architectures, counting the effort and handling. As the application of smart cities, it could be employed in several other fields.

1.7 SUMMARY

In this chapter, various smart city infrastructures like corporeal infrastructure, civilization's infrastructure and digital infrastructure have been presented. The major smart city technologies like IoT, ICT, BT, ST, GT and AIT have also been focused.

In addition, various key models and applications of smart city are presented, based on earlier research. Smart cities and its different adaptations have also been explained.

REFERENCES

Achmad, K.A., Nugroho, L.E., Djunaedi, A., Widyawan. "Smart city for development: towards a conceptual framework", *4th International Conference on Science and Technology (ICST)*, pp. 1–6 (2018). https://ieeexplore.ieee.org/abstract/document/8528677/

Ahmad, K.A.B., Khujamatov, H., Akhmedov, N., Bajuri, M.Y., Ahmad, M.N., Ahmadian, A. "Emerging trends and evolutions for smart city healthcare systems", *Sustainable Cities and Society*, vol. 80 (2022). https://doi.org/10.1016/j.scs.2022.103695

Albino, V., Berardi, U., Dangelico, R.M. "Smart cities: definitions, dimensions, performance, and initiatives", *Journal of Urban Technology*, vol. 22, no. 1, pp. 3–21 (2015). https://www.tandfonline.com/doi/abs/10.1080/10630732.2014.942092

Alharbi, N., Soh, B. "Roles and challenges of network sensors in smart cities", *International Conference on Smart Power & Internet Energy Systems*, vol. 322, pp. 1–8 (2019). https://iopscience.iop.org/article/10.1088/1755-1315/322/1/012002/meta

Allam, Z., Newman, P. "Redefining the smart city: culture, metabolism and governance", *Smart Cities*, vol. 1, no. 1, pp. 1–22 (2018). https://espace.curtin.edu.au/handle/20.500.11937/70707

Angelidou, M. "The role of smart city characteristics in the plans of fifteen cities", *Journal of Urban Technology*, vol. 24, no. 4, pp. 3–28 (2017). https://www.tandfonline.com/doi/abs/10.1080/10630732.2017.1348880

Apanaviciene, R., Vanagas, A., Fokaides, P.A. "Smart building integration into a smart city (SBISC): development of a new evaluation framework". *Energies*, vol. 13, pp. 1–19 (2020). https://ieeexplore.ieee.org/abstract/document/8528677/

Bass, T., Sutherland, E., Symons, T. "*Reclaiming the Smart City Personal Data, Trust and the New Commons – A Report*", July 23, 2018. https://media.nesta.org.uk/documents/DECODE-2018_report-smart-cities.pdf

Bedi, P., Goyal, S.B., Rajawat, A.S., Shaw, R.N., Ghosh, A. "Application of AI/IoT for Smart Renewable Energy Management in Smart Cities". *In:* Piuri, V., Shaw, R.N., Ghosh, A., Islam, R. (eds.) *AI and IoT for Smart City Applications. Studies in Computational Intelligence*, vol. 1002, pp. 115–138 (2022). https://doi.org/10.1007/978-981-16-7498-3_8

Bednarska-Olejniczak, D., Olejniczak, J., Svobodova, L. "Towards a smart and sustainable city with the involvement of public participation—the case of Wroclaw", *Sustainability*, vol. 11, no. 2, pp. 1–33 (2019). https://www.mdpi.com/393270

Bhambri, P., Rani, S., Gupta, G., Khang, A. "*Cloud and Fog Computing Platforms for Internet of Things*" (2022). CRC Press. ISBN: 978-1-032-101507. https://doi.org/10.1201/9781003213888

Chakravorti, B., et al. "Building smart societies—a blueprint for action", *The Fletcher School, Tufts University* (2017). https://sites.tufts.edu/digitalplanet/files/2020/06/Building-Smart-Societies.pdf

Chaturvedi, K., Matheus, A., Nguyen, S.H., Kolbe, T.H. "Securing spatial data infrastructures for distributed smart city applications and services", *Future Generation Computer Systems*, vol. 101, pp. 726–736 (2019). https://www.sciencedirect.com/science/article/pii/S0167739X18330024

Chen, M., Wei, X., Chen, J., Wang, L., Zhou, L. "Integration and provision for city public service in smart city cloud union: architecture and analysis", *IEEE Wireless Communications*, vol. 27, no. 2, pp. 148–154 (2020). https://ieeexplore.ieee.org/abstract/document/9003307/

Dabeedooal, Y.J., Dindoyal, V., Allam, Z., Jones, D.S. "Smart tourism as a pillar for sustainable urban development: an alternate smart city strategy from Mauritius", *Smart Cities*, vol. 2, no. 2, pp. 1–10 (2019). https://www.mdpi.com/456916

Dustdar, S., Nastić, S., Šćekić, O. "Introduction to Smart Cities and a Vision of Cyber-Human Cities", *Smart Cities*, pp. 3–15 (2017). Springer. https://doi.org/10.1007/978-3-319-60030-7_1

Hahanov, V., Khang, A., Litvinova, E., Chumachenko, S., Hajimahmud, V.A., Alyar, A.V. "The Key Assistant of Smart City – Sensors and Tools," *AI-Centric Smart City Ecosystems: Technologies, Design and Implementation* (1st Ed.) (2022). CRC Press. https://doi.org/10.1201/9781003252542-17

Hussain, S.H., Sivakumar, T.B., Khang, A. "Cryptocurrency Methodologies and Techniques," *The Data-Driven Blockchain Ecosystem: Fundamentals, Applications, and Emerging Technologies* (1st Ed.), pp. 149–164 (2022). CRC Press. https://doi.org/10.1201/9781003269281-2

Jucevičiusa, R., Patašienoʼ, I., Patašiusc, M. "Digital dimension of smart city: critical analysis", *Procedia – Social and Behavioral Sciences*, vol. 156, pp. 146–150 (2014). https://www.sciencedirect.com/science/article/pii/S1877042814059576

Khang, A., Chowdhury, S., Sharma, S. "*The Data-Driven Blockchain Ecosystem: Fundamentals, Applications, and Emerging Technologies*" (2022). CRC Press. https://doi.org/10.1201/9781003269281

Khang, A., Gupta, S.K., Dixit, C.K., Somani, P. "Data-Driven Application of Human Capital Management Databases, Big Data, and Data Mining," *Designing Workforce Management Systems for Industry 4.0: Data-Centric and AI-Enabled Approaches* (1st Ed.), pp. 113–133 (2023). CRC Press. https://doi.org/10.1201/9781003357070-7

Khang, A., Gupta, S.K., Hajimahmud, V.A., Babasaheb, J., Morris, G. "*AI-Centric Modelling and Analytics: Concepts, Designs, Technologies, and Applications*" (1st Ed.) (2023). CRC Press. https://doi.org/10.1201/9781003400110

Khang, A., Gupta, S.K., Rani, S., Karras, D.A. "*Smart Cities: IoT Technologies, Big Data Solutions, Cloud Platforms, and Cybersecurity Techniques*" (1st Ed.) (2023). CRC Press. https://doi.org/10.1201/9781003376064

Khang, A., Gupta, S.K., Shah, V., Misra, A. "*AI-aided IoT Technologies and Applications in the Smart Business and Production*" (1st Ed.) (2023). CRC Press. https://doi.org/10.1201/9781003392224

Khang, A., Hahanov, V., Abbas, G.L., Hajimahmud, V.A. "Cyber-Physical-Social System and Incident Management," *AI-Centric Smart City Ecosystems: Technologies, Design and Implementation* (1st Ed.) (2022). CRC Press. https://doi.org/10.1201/9781003252542-2

Khang, A., Ragimova, N.A., Hajimahmud, V.A., Alyar, A.V. "Advanced Technologies and Data Management in the Smart Healthcare System," *AI-Centric Smart City Ecosystems: Technologies, Design and Implementation* (1st Ed.) (2022). CRC Press. https://doi.org/10.1201/9781003252542-16

Khang, A., Rana, G., Tailor, R.K., Hajimahmud, V.A. "*Data-Centric AI Solutions and Emerging Technologies in the Healthcare Ecosystem*" (1st Ed.) (2023). CRC Press. https://doi.org/10.1201/9781003356189

Khang, A., Vrushank, S., Rani, S. "*AI-Based Technologies and Applications in the Era of the Metaverse*" (1st Ed.) (2024). IGI Global Press. https://doi.org/10.4018/9781668488515

Khanh, H.H., Khang, A. "The Role of Artificial Intelligence in Blockchain Applications," *Reinventing Manufacturing and Business Processes through Artificial Intelligence*, pp. 20–40 (2021). CRC Press. https://doi.org/10.1201/9781003145011-2

Komninos, N. "Intelligent cities: variable geometries of spatial intelligence", *Intelligent Buildings International*, vol. 3, no. 3, pp. 172–188 (2011). https://www.tandfonline.com/doi/abs/10.1080/17508975.2011.579339

Manzanilla-Salazar, G., Malandra, F., Mellah, H., Wetté, C., Sansò, B. "A machine learning framework for sleeping cell detection in a smart-city IoT telecommunications infrastructure," *IEEE Access*, vol. 8, pp. 61213–61225 (2020). doi: 10.1109/ACCESS.2020.2983383

Rana, G., Khang, A., Sharma, R., Goel, A.K., Dubey, A.K. "*Reinventing Manufacturing and Business Processes through Artificial Intelligence*" (2021). CRC Press. https://doi.org/10.1201/9781003145011

Rani, S., Bhambri, P., Kataria, A., Khang, A. "Smart City Ecosystem: Concept, Sustainability, Design Principles and Technologies," *AI-Centric Smart City Ecosystems: Technologies, Design and Implementation* (1st Ed.) (2022). CRC Press. https://doi.org/10.1201/9781003252542-1

Rani, S., Bhambri, P., Kataria, A., Khang, A., Sivaraman, A.K. "*Big Data, Cloud Computing and IoT: Tools and Applications*" (1st Ed.) (2023). Chapman and Hall/CRC. https://doi.org/10.1201/9781003298335

Rani, S., Chauhan, M., Kataria, A., Khang, A. "IoT Equipped Intelligent Distributed Framework for Smart Healthcare Systems," *Networking and Internet Architecture* (2021). Vol.2, p: 30. https://doi.org/10.48550/arXiv.2110.04997

Tailor, R.K., Pareek, R., Khang, A. "Robot Process Automation in Blockchain," *The Data-Driven Blockchain Ecosystem: Fundamentals, Applications, and Emerging Technologies* (1st Ed.), pp. 149–164 (2022). CRC Press. https://doi.org/10.1201/9781003269281-8

Vinod Kumar, T.M. "Smart Environment for Smart Cities", *Advances in 21st Century Human Settlements*, pp. 1–59 (2020). Springer. https://doi.org/10.1007/978-981-13-6822-6_1

Vrushank, S., Vidhi, T., Khang, A. "Electronic Health Records Security and Privacy Enhancement Using Blockchain Technology," *Data-Centric AI Solutions and Emerging Technologies in the Healthcare Ecosystem* (1st Ed.), p. 1 (2023). CRC Press. https://doi.org/10.1201/9781003356189-1

2 Benchmarking the Collaborative and Integrated Smart City Model with Industry 5.0

A Way Forward

Ankur Goel, Ritu Kothiwal, Satish Chandra Velpula, Kumar Ratnesh, and Deepak Sharma

2.1 INTRODUCTION

The conceptualization and representation of 'smart city' encompass infinite variations in terms of inherent characteristics so as to define it in a more aligned, concrete and structured manner. The paradigm shift in intelligence capabilities and the consistent transition in technologies have carved a path for overall smart thinking of human brain leading to the development of 'smart cities'.

However, the composition of six basic features like smart living, smart mobility, smart environment, smart people, smart infrastructure and smart governance is one of the most optimized and best fit domains for defining the 'smart city' till date (Khang & Rani et al., 2022).

In current era, 'smart city' is a blend of creativity and innovation, learning and knowledge based, network and information oriented, IoT enabled (Internet of Things), big data driven, environment and ecofriendly, decision support facilitated, 360 degree digitally transformed, intelligently molded, people accustomed, well equipped with safety and security mechanisms, concerned toward socioeconomic attributes etc. which collectively framed into three unmatched root level systems of living, operational and infrastructural (Rani & Khang et al., 2022).

The 'smart city' implies the collective controlling and efforts of smart architects, smart public policy makers, smart leaders, smart design thinkers, smart social system, smart data analysis, smart process engineering etc. (Jebaraj & Khang et al., 2024).

Furthermore, the collaboration and integration of such efforts with housing utilities, transport, infrastructure, municipal services, primary healthcare, recycling of goods (waste management), parking, communication, consumer facilities, academic and educational institutions and water supply are overall essentially required making them more accessible, sustainable and resilient (Khang & Hajimahmud et al., 2022).

DOI: 10.1201/9781003376064-2

However, the applicability of Industry 4.0 tools (composed of digitalization, automation, big data, IoT, block chain technologies, machine learning algorithms, etc.) has already resolved this issue up to certain extent. But even then, there are more explicit challenges and an observatory obstacle in implementing the uniform 'Smart City Model' to all the urban economies altogether (Hajimahmud & Khang et al., 2022).

The determined complexity and undergoing challenge in the development of uniform 'Smart City Model' encourage the researchers for an intensive literature review and systematic study of progression in worldwide smart cities till date (Jebaraj & Khang et al., 2024).

The enormous data and variety of models are available in this regard, but the detailed, insightful and mini-micro analysis with a paired comparative study of at least best smart cities like Singapore, London, Amsterdam, Barcelona, New York of the world in different economic scenario is required for proposing the desired model (Hahanov & Khang et al., 2022).

The ideology is to determine and predict the future-oriented 'Smart City Model' which is pandemic enabled and manageable in severe crisis situations – even more drastic than Covid-19 as well (Khang & Hahanov et al., 2022).

Entire world has witnessed that in recent Covid-19, even the best super smart models excluding digitalization concepts related to any field have demonstrated standstill situations leveraging zero productivity. As a resultant, in such drastic conditions, all smart resources are of no value and inherent advantages being they are not used by any means, e.g. lockdown situations (Khang & Gupta et al., 2023).

2.2 RELATED WORK

Industry 5.0 is an upcoming way forward revolution which is expected and predicted to retransform the professional and personal livelihood patterns through interaction of human intelligence and cognitive computing capabilities with resourcefulness in collaborative and integrated systems all around. It refers to the synergetic compliances of all smart domains (machines, architecture, infrastructure, thinking, planning, etc.) to be pooled with robotic and manmade (artificial) intelligence facilities (Hajimahmud & Khang et al., 2022).

It assures enhanced version of increased profitability, utmost viability, exact feasibility, agile productivity and improved efficiencies/adaptabilities in all core dimensions. It functions as a catalyst for more responsive, responsible and ever-sustainable environment in near future which is also required for predicted 'Smart City' model of this proposed study (Tailor & Khang et al., 2023).

This proposed Industry 5.0-enabled 'Smart City Model' is supposedly to perform even in the worst drastic situations of pandemic like lockdown in near future if so.

2.3 RESEARCH METHODOLOGY

In accordance with the proposal of this research study, there is an intense need for the formulation of super 'Smart City Model' enabling the forthcoming structure of Industry 5.0 characteristics.

The model is thoroughly collaborative with the current smart city models (digitalization and automation based) and could be uniformly benchmarked for all urban economies. Furthermore, it is entirely equipped with integrated root level system (living, operational and infrastructural) and serves as a mechanism to fight with disastrous conditions.

The study ascertains to answer the research questions as follows:

- What should be the proposed and benchmarked super smart city model enabled with Industry 5.0?
- Whether the proposed model is fully collaborative and integrated with Industry 5.0 characteristics to overcome the drastic situations in near future.

Finally, the research questions lead to the specific research objectives of the study as follows:

- To identify, propose and benchmark the Industry 5.0 super smart city model for future.
- To collaborate and integrate the proposed model with all basic characteristics of Industry 5.0.

The study uses the secondary data (materials) for which an analysis has been conducted through systematic literature review and intensive research was conducted based on available sources.

2.4 REVIEW OF LITERATURE

Villegas-Ch and Palacios-Pacheco et al. (2019) noted that, currently, the integration of technologies such as the IoT and big data seeks to cover the needs of an increasingly demanding society that consumes more resources.

The massification of these technologies fosters the transformation of cities into smart cities. Smart cities improve the comfort of people in areas such as security, mobility, energy consumption and so forth.

However, this transformation requires a high investment in both socioeconomic and technical resources. To make the most of the resources, it is important to make prototypes capable of simulating urban environments and for the results to set the standard for implementation in real environments.

Gil and Cortes et al. (2019), mentioned that e-platforms represent the use of information and communication technologies with the aim of encouraging citizen participation in decision-making processes, improving information and service delivery, reinforcing transparency, accountability as well as credibility.

Juceviius and Patašien et al. (2014), specified that a deeper analysis of the concept of smart social systems shows that many social systems can be smart without necessarily basing their activities on information and communication technologies (ICT).

Boykova and Ilina et al. (2016), maintained that in the light of the increasingly complex socioeconomic processes and changes, today's cities as complex systems will not be able to respond to numerous challenges unless they possess a governance model that can flexibly adjust to shifting external conditions.

Vasile Baltac (2019), defined smart city projects are considered real challenges to the development of cities everywhere. The concept itself has many definitions, but a smart city should be defined less based on implemented IT solutions, and more based on optimization of its basic functions using new technologies. There are societal aspects of smart city implementations, similar to eGovernment early projects, and aspects of the use of digital technology that raise concerns.

Anita Kokx and Ronald van Kempen (2010), stated that many accounts of urban governance emphasize municipal and neighborhood scales, featuring local participation, social cohesion and the relationship between local government and residents.

The key result is that Dutch urban policy incorporates dominant neo-liberal multi-scalar meta-governance, owing to the simultaneously strong market orientation and state regulation.

Joao Seixas and Abel Albet i Mas (2010), highlighted about how the concept of governance has been evolving into not only one of the most important but also dubious concepts in urban politics.

The enlightening perspectives of cooperation, participation and collective construction are accompanied by shadowed fears of public demission, oligarchic regimes and less local democracy.

These lights and shadows and the dilemmas they bring along are particularly relevant when observing the cities of the south of Europe, the socio-cultural specificities of which very much structure local political and policy materialization.

2.5 PROPOSED 'SUPER SMART CITY MODEL'

As an outlay of this chapter, Industry 5.0-enabled (automatic) 'Super Smart City Model' which is completely based on 'Artificial Intelligence and Robotic Transformation' has been proposed with the following distinguished characteristics:

- This model is Industry 5.0 technologies and resources enabled, fulfilling all the modernized living standards and criterions (Rana & Khang et al., 2021).
- This model is predicted to serve as a sound mechanism in drastic situations and for disaster management.
- This model is above the level of 'Data Driven and IoT Enabled' smart cities of the world (Rani & Chauhan et al., 2021).
- This model is best applied to the worst conditions of recent pandemic like Covid-19 as far as smart city is concerned.
- This model is assumed to be the best fit model in VUCA (volatility, uncertainty, complex and ambiguous) situations.

However, the uniform implementation and further management of such model are highly unpredictable to huge investments and enormous infrastructural requirements.

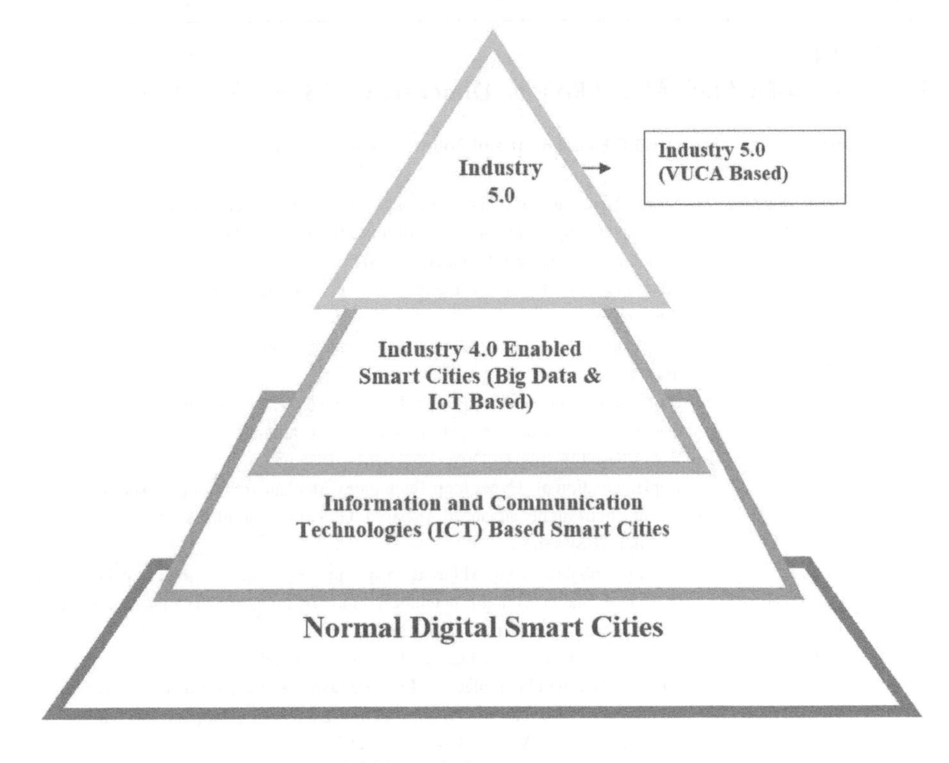

FIGURE 2.1 Smart cities model hierarchal structure.

Furthermore, the controlling dimension of such 'Robotic Transformation' at the smart city level is an exceptional challenge as shown in Figure 2.1.

Industry 5.0-based 'Super Smart City Model' as a part of this study has been eventually proposed after the in-depth analysis of the most predominant data-driven smart city models of 'London' and 'Barcelona'.

The dimensions presented in the 'IoT and big data'-enabled models of these cities have been augmented one step further to integrate and collaborate them with 'AI and Robotic Transformation' supposedly to assist the well-being and living of citizens in drastic conditions as well like pandemic (Rani & Khang et al., 2023). Tabular representation of the proposed model is as shown in Table 2.1.

2.6 ISSUES AND CHALLENGES

Issues, challenges and concerns for implementing the proposed model

- Increased responsibility for citizens in local initiatives resulting in increased the risks of mistakes due to lacking skills or incorrect information. Need of verification and validation of information, and stronger educational programs.
- Unemployment issue related to the enormous implementation of robots is one of the core concerns.

TABLE 2.1
Industry 5.0-Enabled (AI and Robots) Dimensions of Super Smart City

S. No	Basic Dimensions	Industry 5.0-Enabled (AI and Robots) Super Smart City
1	Transport, traffic and mobility	• More efficient use of transport through 'Automatic Autonomous Vehicles'. • AI-based transport infrastructure significantly reduces the demand for parking area in cities (Khang & Gupta, et al., 2023). • AI and real-life sensor information used for 'Automated Traffic Control Systems'. • 'Unmanned Aerial Vehicles' exploit underused urban airspace (Khang & Gupta, et al., 2023). • AI-based prediction of congestion for enhancing accelerated mobility. • Use of 'Robotic Bike Garages' for super smart mobility. • Robot-based 'Advanced Driver Assistance' systems. • Implementation of 'Hyper loop Technology Mechanisms' for pool sharing. • Placement of highly advanced AI-based 'Global Positioning & Location Monitoring Systems'.
2	Healthcare system and medical facilities	• Robotic ambulances and AI based on spot process of diagnosing critical diseases like cancer and providing immediate solution by integration of distant medical facilities (Khang & Rana et al., 2023). • Availability of 'Automated Caring Providers' for patients. • AI and robots will be in place to diagnose diseases and administer medical treatment, monitor patients' condition, transplant organs, implant medical devices and so on (Vrushank & Khang, 2023). • Integration of human medical capacities and technologies for having neuro-interfaces and digital medical assistants.
3	Environment and ecosystem concerns	• Availability of 'Robotic Disruptive Mobility'-based ecosystem. • AI-based 'Smart Environmental Monitoring' (Khang & Hajimahmud et al., 2023). • 'Robotic Actions and Responses' to environmental stimuli. • Automation and AI provide the climate control needed to manage advances in controlled internal environments for food growing and leisure (Khanh & Khang, 2021). • 'Robotic Monitoring Network' for air, water, energy and power quality.
4	Civic security and safety	• Robotic monitoring replacing the use of sensors. Application and implementation of 'Drones' for the same purpose. • AI-based public safety and surveillance, e.g. facial recognition.
5	Urban planning and infrastructure	• Automation enables buildings and infrastructure to respond to climate change (e.g. regulating energy use and comfort, air quality). • Sensors and AI can underpin the development and management of green infrastructure. • More efficient monitoring, repair and control of infrastructure through robotics, especially in contexts where human accessibility is difficult or unpleasant as in the case of recent pandemic. • AI-based monitoring system conveys information regarding power, heat, gas, water supply, weather and ecological aspects of the city. • AI integrated 'Multi Commodity Grids'. AI enables prediction of renewable energy generation from intermittent sources such as the sun and wind, and other decentralized sources such as geothermal and aqua-thermal energy. • AI can be used to optimize sustainable urban waste processing and all other municipal services.

(Continued)

TABLE 2.1 (*Continued*)
Industry 5.0-Enabled (AI and Robots) Dimensions of Super Smart City

S. No	Basic Dimensions	Industry 5.0-Enabled (AI and Robots) Super Smart City
6	Citizen reach, well-being and participation	• 'Robotic Emotional Intelligence' is well placed and fully implemented for the citizens. • Automated and robotic health and social care support assisted living. Scope to extend age-friendly urban environments. • Application of 'chatbots' and 'robo-advisors (Robo-advisors are automated portfolio managers)' to provide best public engagement services. • AI-based 'Swarm Intelligence' for tasks optimization and spatial monitoring. • Implementation of 'Smart Amazon Alexa' in homes of citizens for smart acccssing of phonc, answcring calls, monitoring and security through doorbells. • Disappearance of physical work in homes by tasking through robotic automation as a part of intelligent capitalism. • Implementation of UDPs (Urban Data Platforms) to create AI for improved or new value adding public services.
7	Training programs and educational institutes	• Robotic assistants for solving problems and tedious assignments. • Robotic mechanisms to teach subjects like mathematics to primary and junior school students. • High-tech AI and robots enable educational institutions equipped with exemplary facilities for online education in pandemic-like situations. • AI-based 'Indigenous Excellence after Schools Programs' and 'Experiential Learning Libraries' for HEI's of proposed model of smart cities. • Trained AI mechanisms with images and video segments of students engaging in malpractices, tracking their activity and behavior to check for any deviations from the normal pattern, which can then be rejected or reviewed by the invigilator.
8	Innovative labs and research centers	• Unemployment issues ascertained through the application of 'AI and Robotics' to be tackled through deployment of human resources in researching innovative technologies such as robotics, AI, 3D simulation, scanning and print, the internet of things, augmented and virtual reality, big data, machine learning, block chain etc. and many more (Bhambri & Khang et al., 2022). • Addressing viable future careers in 'Research & Training' in the field of updated technology for each and every citizen in a variety of disciplines related to their research interest area. Super smart city must be equipped with multiple innovative labs and research centers for this implementation.
9	Disaster management centers and ethical practices in AI and robotics	• Commercialization of AI that prevents harm and advances humanity increases societal and environmental concerns and respects humanity. • Contribution of AI in damage detection, damage prediction, damage classification, damage localization, condition assessment and life-time prediction during disastrous situations. • AI-based proactive evaluative exposure by simulating impact of events like storm, droughts, floods and tsunami on critical facilities.

Source: Author's own formation on the basis of 'Systematic Literature Review' and 'content Analysis'.

- Competitive AI-based expertise and skill set are highly necessary for its implementation.
- Necessity of high-performance-oriented computing requirements for AI is one of the most significant challenges.
- AI urban system's biggest challenge is missing or incomplete data, accuracy and availability of data.
- Increased dependency on private parties as developers and implementers of AI and Robotic Transformation is one of the prominent challenges.
- Huge financial outlay and budgetary constraints related to the implementation of AI and Robotic Transformation.
- Ethical challenges of AI related to the conflicts of interest, biasness in decision-making, economic pressures, inequalities, trust and transparency (Khang & Vrushank, et al., Metaverse, 2023).

2.7 CONCLUSION

The phenomenon of 'Smart City' is completely dynamic and depends on the emergence of specific requirements of the urban economies all around. Gradually, these fundamental features invariably branched out to numerous sub-features and multi-dimensional aspects to present various synchronized, persistent and demonstrative parameters related to 'Smart City Models' from time to time.

Research and development issues pertaining to AI and robotic technology are to be addressed (Khang & Abdullayev et al., 2024). AI system will require open, dynamic, hyper-connected and unregulated environment to work in more efficient and better way.

AI and Robotic Transformation (ART) assist the well-being and living of citizens in drastic conditions but need collaboration and integration at various managerial levels and dimensions of smart city.

REFERENCES

Bhambri, P., Rani, S., Gupta, G., Khang, A., *Cloud and Fog Computing Platforms for Internet of Things* (2022). CRC Press. ISBN: 978-1-032-101507. https://doi.org/10.1201/9781003213888

Boykova, M., Ilina, I., Salazkin, M., "The smart City approach as a response to emerging challenges for urban development", *Foresight and STI Governance* vol. 10, no 3 (2016). https://cyberleninka.ru/article/n/the-smart-city-approach-as-a-response-to-emerging-challenges-for-urban-development

Gil, O., Cortes, M.E., Cantador, I., *International Journal of E- Planning and Research*, vol. 8, no. 1 (2019). https://www.igi-global.com/article/citizen-participation-and-the-rise-of-digital-media-platforms-in-smart-governance-and-smart-cities/217705

Hahanov, V., Khang, A., Litvinova, E., Chumachenko, S., Hajimahmud, V.A., Alyar, A.V., "The Key Assistant of Smart City – Sensors and Tools", *AI-Centric Smart City Ecosystems: Technologies, Design and Implementation* (1st Ed.) (2022). CRC Press. https://doi.org/10.1201/9781003252542-17

Hajimahmud, V.A., Khang, A., Hahanov, V., Litvinova, E., Chumachenko, S., Alyar, A.V., "Autonomous Robots for Smart City: Closer to Augmented Humanity", *AI-Centric Smart City Ecosystems: Technologies, Design and Implementation* (1st Ed.) (2022). CRC Press. https://doi.org/10.1201/9781003252542-7

Jebaraj, L., Khang, A., Chandrasekar, V., Pravin, A.R., Sriram, K., "Smart City Concepts, Models, Technologies and Applications," *Smart Cities: IoT Technologies, Big Data Solutions, Cloud Platforms, and Cybersecurity Techniques* (1st Ed.) (2024) CRC Press. https://doi.org/10.1201/9781003376064-1

Juceviius, R., Patašien, I., Patasius, M., "Digital dimension of smart city: critical analysis 19th International Scientific Conference;" *Economics and Management, ICEM* 2014, April 2014. https://www.sciencedirect.com/science/article/pii/S1877042814059576

Khang, A., Abdullayev, V., Hahanov, V., Shah, V., *Advanced IoT Technologies and Applications in the Industry 4.0 Digital Economy* (1st Ed.) (2024). CRC Press. https://doi.org/10.1201/978-1-003-43426-9

Khang, A., Gupta, S.K., Hajimahmud, V.A., Babasaheb, J., Morris, G., *AI-Centric Modelling and Analytics: Concepts, Designs, Technologies, and Applications* (1st Ed.) (2023) CRC Press. https://doi.org/10.1201/9781003400110

Khang, A., Gupta, S.K., Rani, S., Karras, D.A., *Smart Cities: IoT Technologies, Big Data Solutions, Cloud Platforms, and Cybersecurity Techniques* (1st Ed.) (2023). CRC Press. https://doi.org/10.1201/9781003376064

Khang, A., Gupta, S.K., Shah, V., Misra, A., *AI-Aided IoT Technologies and Applications in the Smart Business and Production* (1st Ed.) (2023). CRC Press. https://doi.org/10.1201/9781003392224

Khang, A., Hahanov, V., Abbas, G.L., Hajimahmud, V.A., "Cyber-Physical-Social System and Incident Management", *AI-Centric Smart City Ecosystems: Technologies, Design and Implementation* (1st Ed.) (2022). CRC Press. https://doi.org/10.1201/9781003252542-2

Khang, A., Hahanov, V., Litvinova, E., Chumachenko, S., Triwiyanto, V.A., Hajimahmud, R.N., Ali, A.V., Alyar, Anh, P.T.N., "The Analytics of Hospitality of Hospitals in Healthcare Ecosystem", *Data-Centric AI Solutions and Emerging Technologies in the Healthcare Ecosystem*, p. 4 (1st Ed.) (2023). CRC Press. https://doi.org/10.1201/9781003356189-4

Khang, A., Ragimova, N.A., Hajimahmud, V.A., Alyar, A.V., "Advanced Technologies and Data Management in the Smart Healthcare System", *AI-Centric Smart City Ecosystems: Technologies, Design and Implementation* (1st Ed.) (2022). CRC Press. https://doi.org/10.1201/9781003252542-16

Khang, A., Rana, G., Tailor, R.K., Hajimahmud, V.A., *Data-Centric AI Solutions and Emerging Technologies in the Healthcare Ecosystem* (1st Ed.) (2023) CRC Press. https://doi.org/10.1201/9781003356189

Khang, A., Rani, S., Sivaraman, A.K., *AI-Centric Smart City Ecosystems: Technologies, Design and Implementation* (1st Ed.) (2022). CRC Press. https://doi.org/10.1201/9781003252542

Khang, A., Vrushank, S., Rani, S., *AI-Based Technologies and Applications in the Era of the Metaverse* (1st Ed.) (2023), IGI Global Press. https://doi.org/10.4018/9781668488515

Khanh, H.H., Khang, A., "The Role of Artificial Intelligence in Blockchain Applications", *Reinventing Manufacturing and Business Processes through Artificial Intelligence*, pp. 20–40 (2021). CRC Press. https://doi.org/10.1201/9781003145011-2

Kokx, A., van Kempen, R., "Dutch urban governance: Multi-level or multi-scalar," *European Urban and Regional Studies*, 2010. https://journals.sagepub.com/doi/pdf/10.1177/0969776409350691

Lee, S.K., Kwon, H.R., Cho, H.A., Kim, J., Lee, D.; IDB, Jun 2016; https://dx.doi.org/10.18235/0000409

Rana, G., Khang, A., Sharma, R., Goel, A.K., Dubey, A.K., *Reinventing Manufacturing and Business Processes through Artificial Intelligence* (2021). CRC Press. https://doi.org/10.1201/9781003145011

Rani, S., Bhambri, P., Kataria, A., Khang, A., "Smart City Ecosystem: Concept, Sustainability, Design Principles and Technologies", *AI-Centric Smart City Ecosystems: Technologies, Design and Implementation* (1st Ed.) (2022). CRC Press. https://doi.org/10.1201/9781003252542-1

Rani, S., Bhambri, P., Kataria, A., Khang, A., Sivaraman, A.K., *Big Data, Cloud Computing and IoT: Tools and Applications* (1st Ed.) (2023). Chapman and Hall/CRC. https://doi.org/10.1201/9781003298335

Rani, S., Chauhan, M., Kataria, A., Khang, A., "IoT Equipped Intelligent Distributed Framework for Smart Healthcare Systems", *Networking and Internet Architecture* (2021). CRC Press. https://doi.org/10.48550/arXiv.2110.04997

Seixas, J., Mas, A.A., "Urban governance in the South of Europe: cultural identities and global dilemmas," *Southern Europe*, 2010. https://www.taylorfrancis.com/chapters/edit/10.4324/9781315548852-12/integrated-urban-interventions-greece-local-relational-realities-unsettled

Tailor, R.K., Pareek, R., Khang, A., "Robot Process Automation in Blockchain", *The Data-Driven Blockchain Ecosystem: Fundamentals, Applications, and Emerging Technologies*, pp. 149–164 (1st Ed.) (2023). CRC Press. https://doi.org/10.1201/9781003269281-8

Vasile, B., Smart Cities—A View of Societal Aspects. Smart Cities, vol. 2, no. 4, pp. 538-548, (2019). https://doi.org/10.3390/smartcities2040033

Villegas-Ch, W., Palacios-Pacheco, X., Luján-Mora, S., *Sustainability*, vol. 11, p. 2857 (2019). doi: 10.3390/su11102857. www.mdpi.com/journal/sustainability.

Vrushank, S., Khang, A., "Internet of Medical Things (IoMT) Driving the Digital Transformation of the Healthcare Sector", *Data-Centric AI Solutions and Emerging Technologies in the Healthcare Ecosystem*, p. 1. (1st Ed.) (2023). CRC Press. https://doi.org/10.1201/9781003356189-2

3 The Role of Internet of Things (IoT) in Smart City Framework

Subhashini R and Alex Khang

3.1 INTRODUCTION

The Internet of Things (IoT) depicts the organization of actual items called things that are installed with sensors, programming, and different advancements to associate and trading information with different gadgets and frameworks over the web. These gadgets range from common family objects to modern apparatuses. Within excess of 7 billion associated IoT gadgets today, specialists are anticipating that this number should develop to 10 billion by 2020 and 22 billion by 2025.

3.1.1 Importance of IoT

Throughout the course of recent years, IoT has become one of the main advancements of the 21st century. Now that we can associate ordinary articles—kitchen apparatuses, vehicles, indoor regulators, child screens—to the web through installed gadgets, consistent correspondence is conceivable between individuals, cycles, and things (Khang & Gupta et al., 2023).

Through minimal expense registering, the cloud, large information, investigation, and versatile innovations, actual things can share and gather information with negligible human intercession. In this hyper connected world, advanced frameworks can record, screen, and change every communication between associated things. The actual world meets the computerized world—and they participate (Luke & Khang et al., 2024).

3.1.2 Technologies behind IoT

While the possibility of IoT has been in presence for quite a while, an assortment of late advances in various advances has made it functional.

- Admittance to minimal expense, low-power sensor innovation. Reasonable and dependable sensors are making IoT innovation feasible for additional producers.
- Availability. A large group of organization conventions for the web has made it simple to interface sensors to the cloud and to other "things" for proficient information move.
- Distributed computing stages. The expansion in the accessibility of cloud stages empowers the two organizations and shoppers to get to the framework they need to increase without really overseeing everything.

DOI: 10.1201/9781003376064-3

- AI and investigation. With propels in AI and examination, alongside admittance to differed and tremendous measures of information put away in the cloud, organizations can assemble bits of knowledge quicker and all the more without any problem. The rise of these united advances keeps on pushing the limits of IoT and the information delivered by IoT additionally takes care of these innovations (Rana & Khang et al., 2021).
- Conversational man-made consciousness (AI). Progresses in brain networks have brought normal language handling (NLP) to IoT gadgets (like computerized individual collaborators Alexa, Cortana, and Siri) and made them engaging, reasonable, and practical for home use (Khanh & Khang, 2021).

3.1.3 IoT Applications

IoT applications run on IoT gadgets and can be made to be intended for pretty much every industry and vertical, including medical services, modern mechanization, shrewd homes and structures, car, and wearable innovation. Progressively, IoT applications are utilizing AI and AI to add insight to gadgets. IoT Intelligent Applications are prebuilt software-as-a-service (SaaS) applications that can investigate and introduce caught IoT sensor information to business clients by means of dashboards. We have a full arrangement of IoT Intelligent Applications (Rani & Chauhan et al., 2021).

IoT applications use AI calculations to dissect gigantic measures of associated sensor information in the cloud. Utilizing continuous IoT dashboards and alarms, you gain perceivability into key execution markers, insights for mean time among disappointments, and other data (Bhambri & Khang et al., 2022). AI-based calculations can distinguish hardware oddities and send alarms to clients and, surprisingly, trigger mechanized fixes or proactive counter measures. With cloud-based IoT applications, business clients can rapidly improve existing cycles for supply chains, client assistance, HR, and monetary administrations. There is compelling reason need to reproduce whole business processes.

3.2 REAL-WORLD APPLICATIONS OF IoT

The number of real-world IoT applications will increase as technology advances in the next years. IoT and AI are anticipated to be combined in the near future to provide intelligent solutions for all current technological problems. Here, we'll go through ten significant real-world instances of IoT applications (Rani & Khang et al., 2022).

3.2.1. Smart City

A smart city, often known as the "City of the Future," is a well-known IoT application idea that places technology in the category of being responsible for enhancing urban infrastructure to make urban centers more effective, less expensive, and more pleasant to live in. Additionally, it seeks to enhance economic development and contribute to environmental sustainability (Hahanov & Khang et al., 2022). Planning and public administration are addressed by the "Smart City" idea through the automation of services in an innovative and sustainable manner. Government services, transportation and traffic management, electricity, healthcare, water, creative urban agriculture, and waste management are just a few of the industries that smart cities enhance and modernize (Khang & Rani et al., 2022). (Figure 3.1).

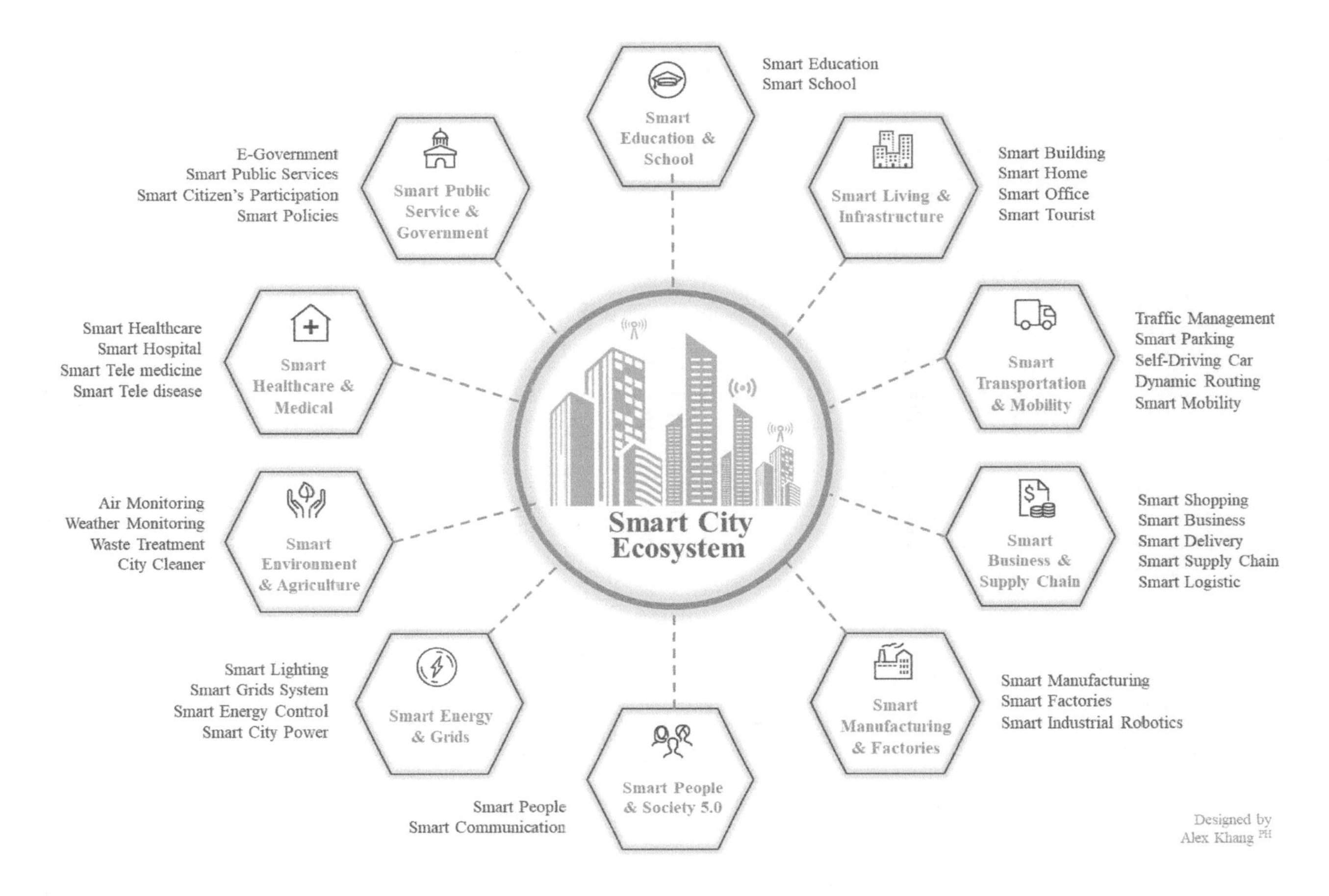
Smart Education & School
Smart Education
Smart School
Smart Public Service & Government
E-Government
Smart Public Services
Smart Citizen's Participation
Smart Policies
Smart Living & Infrastructure
Smart Building
Smart Home
Smart Office
Smart Tourist
Smart Healthcare & Medical
Smart Healthcare
Smart Hospital
Smart Tele medicine
Smart Tele disease
Smart Transportation & Mobility
Traffic Management
Smart Parking
Self-Driving Car
Dynamic Routing
Smart Mobility
Smart City Ecosystem
Smart Environment & Agriculture
Air Monitoring
Weather Monitoring
Waste Treatment
City Cleaner
Smart Business & Supply Chain
Smart Shopping
Smart Business
Smart Delivery
Smart Supply Chain
Smart Logistic
Smart Energy & Grids
Smart Lighting
Smart Grids System
Smart Energy Control
Smart City Power
Smart Manufacturing & Factories
Smart Manufacturing
Smart Factories
Smart Industrial Robotics
Smart People & Society 5.0
Smart People
Smart Communication
Designed by
Alex Khang PH

FIGURE 3.1 Illustration of Smart City (Khang, 2021).

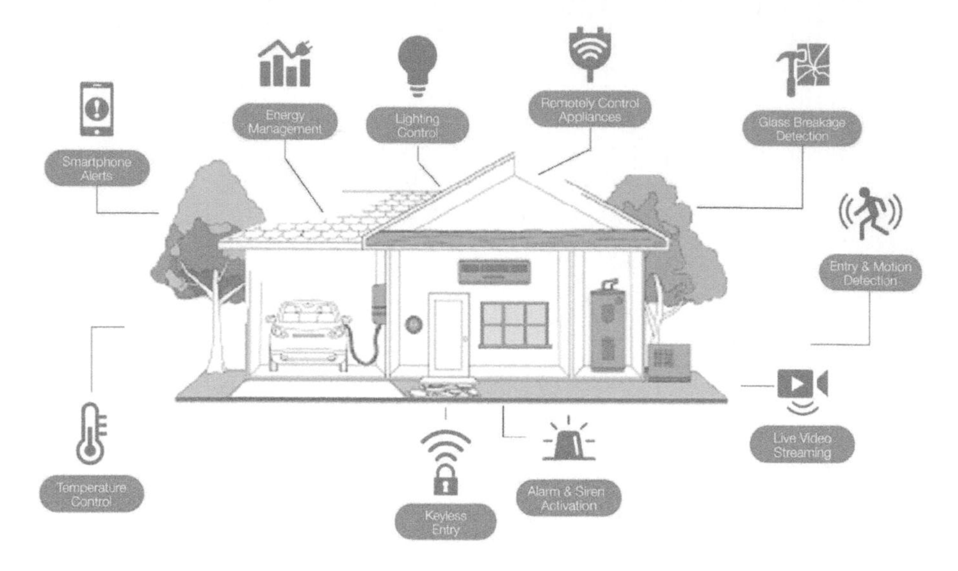

FIGURE 3.2 Illustration of smart home.

3.2.2 Smart Home

The most popular IoT application on this list is smart houses. Smart houses include intricate, sophisticated security systems that help to ensure home security and safety while controlling household appliances like lights, alarms, and water flow from faucets. Through their smartphones, tablets, and computers, homeowners can monitor and manage house operations in smart homes. If you have forgotten to switch off your oven, you might be able to do so easily using your phone. You can control all of your house appliances from one location with a smart home (Khang & Gupta et al., 2023). (Figure 3.2).

3.2.3 Smart Self-Driving Cars

Formerly a thing of the future, self-driving and operating automobiles are now a reality. Certain functions can be managed by smart device through the use of smart car technology. To calculate things like engine oil level and radiator water temperature, central computers installed in the automobile use data from sensors placed all around the vehicle. Even better, you can use a phone app to keep tabs on the condition of the vehicle, including its position, fuel level, and oil level. (Figure 3.3).

3.2.4 IoT in Farming

Farmers have the opportunity to change the agricultural business by using sophisticated IoT farming applications to optimize several labor-intensive farm processes. You may use IoT to develop soil-chemistry-based fertilizer profiles, monitor soil nutrient and moisture concentrations, and assist you choose the optimal time to harvest crops. In order to track the health of the cattle, a number of sensors are deployed

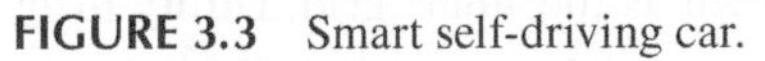

FIGURE 3.3 Smart self-driving car.

across the agricultural sector in smart farming. One illustration of a smart farm device that aids farmers in weather data forecasting and crop and livestock status monitoring is Pycno Agriculture Sensors.

3.2.5 Fitness Trackers

You can track your progress and optimize your fitness objectives using IoT-connected products. Fitness trackers keep tabs on a variety of everyday activities, including sleeping habits, heart rate, movement patterns, workout data, calories burnt, and more. These gadgets perform all of this using sensors that gather information from your skin.

3.2.6 IoT-Connected Factories

IoT technology, sometimes referred to as the industrial IoT, is used in smart factories to collect data on industrial equipment and processes in order to develop strategic goals and increase productivity. Sensors are connected to production equipment and machine tools to enhance analytics. To safeguard revenues and boost supply chain productivity, such technology may assist manufacturers in reducing energy usage, enhancing asset monitoring, and identifying equipment concerns early.

3.2.7 IoT Hospitality and Tourism

Operations in the hotel and tourist sectors can be greatly improved by the IoT. For hotels and motels in the hospitality sector, manpower is a significant expenditure, but IoT may automate some exchanges to lighten staffing obligations. For instance, mobile electronic keys enable hotel visitors to enter and check into their rooms without contacting a member of staff. Customers may order room service and report any problems with their rooms, such as a shortage of towels, through their smartphones, which helps hotels get vital information more promptly.

3.2.8 Retail IoT

The retail industry may benefit from savings, efficiency, and innovation with IoT technologies, much as the industrial sector. A lot can be achieved with retail IoT, including better customer experiences, precise and real-time product tracking, smarter staffing plans, and all-around effective inventory management. With the use of the IoT, retailers can also keep track of consumer buying habits, identifying their purchase history, trends, and location data to better guide customer initiatives.

3.2.9 Smart Grid

The IoT that takes care of energy systems is known as the smart grid. Utility firms employ smart grid technology to identify energy savings by tracking energy use, forecasting energy shortages and power outages, and gathering information on how diverse people and businesses use energy. The typical person may monitor their personal energy usage and identify efficiencies in their home by using the smart grid's data (Rani & Khang et al., 2023).

3.2.10 IoT Applications for Health Monitoring

The COVID-19 pandemic made it clear that the whole healthcare industry needs substantial modernization, efficiency, and flexibility improvements. How else may a patient get care if they are unable to make it to a doctor's appointment (Vrushank & Vidhi et al., 2023). The businesses that would profit from utilizing sensor devices in their operational procedures are those that are most suited for IoT.

3.2.11 Manufacturing

By integrating production-line monitoring to enable preventive repair on equipment when sensors indicate an approaching malfunction, manufacturers may obtain a competitive edge. In fact, sensors are able to detect when industrial output is being disrupted. Manufacturers can immediately verify equipment for correctness or take it out of production while it is being repaired with the use of sensor warnings. This enables businesses to lower operational expenses, increase uptime, and enhance asset performance management.

3.2.12 Automotive

The deployment of IoT applications has the potential to provide the automobile sector with major benefits. Sensors can identify approaching equipment failure in cars that are already on the road and can warn the driver with facts and advice, in addition to the advantages of applying IoT to manufacturing processes. Because of the combined data, IoT-based apps have been collected by car manufacturers (Hajimahmud & Khang et al., 2022).

3.2.13 Transportation and Logistics

Different IoT applications have positive effects on logistics and transportation networks. Thanks to IoT sensor data, fleets of cars, trucks, ships, and trains that transport merchandise may be redirected based on the weather, the availability of available vehicles, or the availability of drivers. Additionally, sensors for temperature monitoring and track-and-trace might be included within the inventory itself. IoT monitoring systems that give warnings when temperatures increase or decrease to a level that threatens the product would be extremely helpful to the food and beverage, floral, and pharmaceutical sectors, which frequently carry inventory that is temperature-sensitive.

3.2.14 Retail

Retail businesses may manage inventory, enhance customer service, streamline the supply chain, and cut costs by using IoT apps. For instance, smart shelves with weight sensors may gather RFID-based data and transmit it to the IoT.

3.2.15 Public Sector

The advantages of IoT in the public sector and other areas where services are provided are also many. For instance, IoT-based apps may be used by government-owned utilities to alert its customers to both major and minor delays in the supply of water, electricity, or sewer services. Applications built for the IoT can gather information about the extent of an outage and allocate resources to help utilities recover from outages more quickly.

3.2.16 Healthcare

The healthcare sector benefits greatly from IoT asset monitoring. The precise position of patient support equipment, such as wheelchairs, is frequently necessary information for doctors, nurses, and orderlies. When wheelchairs at a hospital are fitted with IoT sensors, they can be tracked via the IoT asset-monitoring application, allowing anyone seeking for one to identify the closest wheelchair right away, various hospitals (Vrushank & Khang, 2023).

3.2.17 General Safety across All Industries

The IoT may be used to increase worker safety in addition to tracking physical assets. Workers in hazardous workplaces, such as mines, oil and gas fields, chemical facilities, and power plants, for instance, need to be aware of any hazardous events that might have an impact on them. They can be alerted to accidents or rescued as soon as feasible when they are connected to IoT sensor-based apps. Wearables that can track environmental factors and human health also leverage IoT applications. These programmers allow doctors to remotely monitor patients in addition to assisting consumers in understanding their own health.

3.3 IoT APPLICATION DEPLOYMENT

The capacity of IoT to give sensor data as well as empower gadget-to-gadget correspondence is driving an expansive arrangement of utilizations. Coming up next are probably the most well-known applications and what they do. Make new efficiencies in assembling through machine observing and item quality checking. Machines can be ceaselessly observed and dissected to ensure they are performing inside required resistances. Items can likewise be checked progressively to distinguish and address quality deformities. Following empowers organizations to decide resource area rapidly. Ring-fencing permits them to ensure that high-esteem resources are shielded from burglary and evacuation.

3.3.1 Use Wearables to Screen Human Wellbeing Investigation and Natural Circumstances

IoT wearables empower individuals to all the more likely figure out their own wellbeing and permit doctors to screen patients from a distance. This innovation likewise empowers organizations to follow the wellbeing and security of their representatives, which is particularly valuable for laborers utilized in unsafe circumstances.

3.3.2 Drive Efficiencies and Additional Opportunities in Existing Cycles

One illustration of this is the utilization of IoT to build effectiveness and security in associated coordinated operations for armada the board. Organizations can utilize IoT armada checking to coordinate trucks, progressively, to further develop effectiveness.

3.3.3 Empower Business Process Changes

An illustration of this is the utilization of IoT gadgets for associated resources for screen the strength of remote machines and trigger assistance calls for preventive upkeep. The capacity to remotely screen machines is likewise empowering new item as-a-administration plans of action, where clients never again need to purchase an item yet rather pay for its use (Khang & Rana et al., 2023).

3.4 RELATED WORK

3.4.1 IoT Changing the World

By allowing linked automobiles, IoT is completely redefining the automotive. With IoT, automobile owners may remotely control their vehicles—for instance, by preheating the vehicle before the driver gets in it or by remotely calling for a vehicle through phone. Cars will even be able to schedule their own servicing appointments when necessary thanks to IoT's capacity to facilitate device-to-device connectivity.

The linked automobile gives automakers and dealers the ability to completely alter the car ownership model. In the past, producers and individual consumers have not interacted (or none at all). Once the automobile was delivered to the dealer, the manufacturer's involvement with it essentially came to an end. Automobile manufacturers or dealers may maintain a constant interaction with their consumers thanks to linked automobiles. They can impose fees in instead of selling automobiles.

This is so that they can be classified across many application domains and geographical areas, which is how IoT systems are made. As a result, it generates a lot of dependencies across platforms, domains, and services. Due to the interdependence of IoT systems and devices, a framework that is intelligent and connection-aware is now required. This is where IoT architecture comes into play.

Imagine having one "brain" in charge of all of your smart IoT devices, from sensors and actuators to Internet gateways and data acquisition systems! The IoT architecture may be thought of as the brain in this scenario, and the caliber of its constituent parts directly affects the architecture's efficacy and applicability. There are many methods to IoT architecture depending on how a system interacts and the various tasks an IoT device completes since we can call.

3.4.2 IoT Architecture: A Deeper Analysis of the "Brain"

In its simplest form, an IoT architecture is a collection of various components, such as sensors, protocols, actuators, cloud services, and layers. The IoT architectural layers are differentiated in order to track a system's consistency across protocols and gateways in addition to devices and sensors. Researchers have put out a variety of designs, and it is clear that no architecture has gained widespread acceptance for the IoT. A three-layer architecture is the most fundamental kind.

3.4.3 The Different Layers of IoT Architecture

The perception, network, and application layers are the three separate layers.

The physical layer known as the perception layer includes sensors for detecting and acquiring environmental data. It detects certain physical factors or other intelligent items in the surrounding environment. Correlation with servers, network equipment, and other smart objects is the responsibility of the network layer. Additionally, the features are employed to transmit and analyze sensor data. The application layer effectively provides the user with application-specific services. It outlines a variety of use cases for IoT services, including smart homes, smart cities, and smart health. The IoT's central idea is defined by a three-layer architecture, although it is seen to be insufficient for. (Figure 3.4).

- Business Layers
- Application Layers
- Processing Layers
- Transport Layers
- Perception Layers

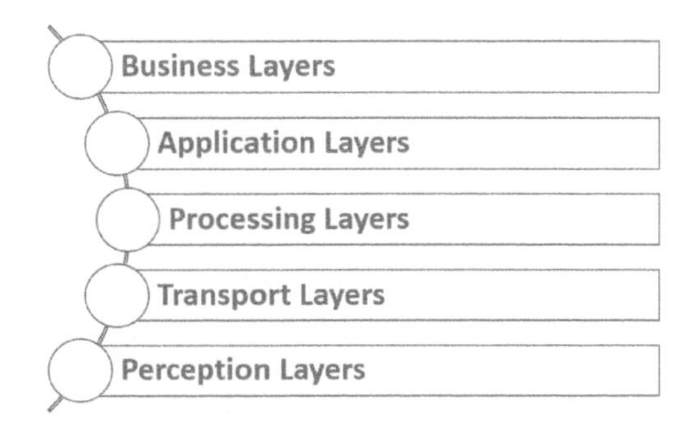

FIGURE 3.4 Five-layer architecture of IoT.

The transport layer primarily moves sensor data from the perception layer to the processing layer and vice versa, as its name suggests. Networks, including SDN/NFV, 5G, wireless local area networks, RFID, Bluetooth, and, NFC are mostly used for this transportation.

Massive volumes of data from the transport layer are stored, analyzed, and processed by the processing layer, also referred to as the middleware layer. It is capable of managing and offering various sets of services to the lower tiers. It uses a variety of technologies, including big data analytics, cloud computing, and databases. The IoT system as a whole, comprising business models, apps, and companies, is managed by the business layer.

3.4.4 Enterprise Benefit from IoT Architecture

Now that we are aware of the various IoT architecture levels, how do they benefit organizations and how can they make the most of IoT? Isolated, siloed, and fragmented data from the devices are what is meant when the term "Internet of Things" (IoT) is used to describe linked devices and protocols. These fragmented insights may not alone give enough data to support investing significant resources in an IoT strategy. The interactions between the devices must be open, and more device and system synergies must be made feasible if businesses are to benefit fully from IoT as Figure 3.5.

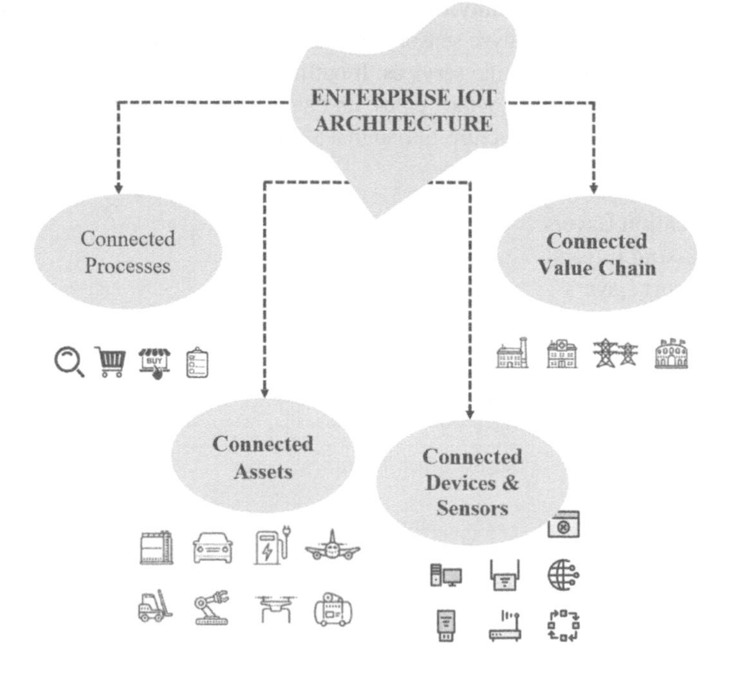

FIGURE 3.5 Enterprise Architecture of IoT (Khang A., 2023).

Enterprises may automate business operations by utilizing a variety of technologies, including cloud platforms, Internet-based communication, and embedded devices with actuators and sensors. Additionally, these enterprise IoT apps may give devices control instructions based on certain business principles.

The insights gleaned from IoT data sets will become important for businesses to make choices as big data analytics advances. IoT architecture serves as a single node that monitors all operations, including connecting IoT devices, mining sensor data sets, and operating software platforms. These are Internet gateways, Edge IoT, and Datacenter and cloud.

- Internet Gateways: This part, which is based on standard IoT Gateways, serves as a go-between for business data centers that are often cloud-based and the world of field devices.
- Edge IoT: The word edge derives from edge computing, where data is processed as near to the data's source as feasible at the network's edge. Making data processing and decision-making as near to real-time as possible is the primary element that makes edge processing so important.
- Datacenter and cloud: In an IoT architecture, virtualization enables efficient use of hardware resources and IoT devices. The application servers for the various cloud service models are also introduced, and they include both HTTP and MQTT servers. The MQTT servers enable a high number of device connections and real-time communication among devices, whilst the HTTP servers can implement services for devices and end users. The supporting databases are also presented for functionality, availability, and performance, among other important factors.

3.5 SECURITY RISKS IN IoT

Some IoT video cameras had weak passwords because IoT manufacturers didn't follow regulations, which in turn allowed the Mirai software, one of the most destructive botnet assaults, to spread. We will focus on the most significant IoT security concerns out of the numerous that exist. The following IoT security concerns can be categorized as causes or effects.

3.5.1 IoT Manufacturers' Noncompliance

Nearly every day, new IoT devices are released, and every single one of them has weaknesses. The main cause of the majority of IoT security problems is that manufacturers do not devote enough time and money to security. For instance, most Bluetooth fitness trackers stay visible after the first connecting, and a smart refrigerator can reveal a Gmail login. One of the major security risks with IoT is exactly this. Manufacturers will keep making devices with weak security as long as there are no global IoT security standards.

Manufacturers that began integrating Internet connectivity into their products may not necessarily prioritize "security" as a design consideration when creating new products. IoT device makers run the following security concerns with their products:

- Poor, easy-to-guess, or hard-coded passwords
- Hardware problems
- Absence of a reliable updating system
- Outdated and unpatched embedded software and operating systems
- Unsecure data storage and transport

3.5.2 Lack of User Knowledge and Awareness

Internet users have become accustomed to avoiding spam and phishing emails, running virus checks on their computers, and setting strong passwords to protect their Wi-Fi networks. However, IoT is a new technology, and few people are familiar with it. Although manufacturing continues to provide the majority of the IoT security hazards, consumers and business procedures might still pose more serious dangers.

The user's ignorance and lack of understanding of IoT functioning is one of the main IoT security dangers and difficulties. Everyone is placed at danger as a result. The quickest approach to enter a network is typically via tricking a person. Social engineering is a sort of IoT security issue that is frequently disregarded.

The 2010 Stuxnet assault against an Iranian nuclear plant employed social engineering. Industrial programmable logic controllers (PLCs), another type of IoT device, were the target of the assault. The assault caused the factory to burst and corrupt 1000 centrifuges. The internal network was supposedly separated from the public network to prevent assaults, yet all it took was for a staff member to insert a USB flash drive into one of the internal PCs to compromise the system.

3.5.3 IoT Security Problems in Device Update Management

Insecure software or firmware is a different source of IoT security issues. Even if a vendor sells a product with the most recent software update, new vulnerabilities will inevitably appear. Updates are essential for keeping IoT devices secure. As soon as new vulnerabilities are found, they should be updated. Still, some IoT devices continue to be used despite the need for upgrades, unlike smartphones or laptops that receive updates automatically. Another danger is that a gadget updating itself can transmit its backup to the cloud, causing a brief outage. A hacker might steal confidential data if the connection is not encrypted and the update files are not secured.

3.5.4 Lack of Physical Hardening

Security problems with IoT might also result from a lack of physical hardening. Even while certain IoT devices should be able to function independently without any user input, they nevertheless need to be physically protected from external hazards. These devices may occasionally be left in isolated areas for extended periods of time, when they may be physically tampered with, for instance, by inserting a USB flash drive that contains malware.

An IoT device's manufacturer is the first step in ensuring its physical security. The challenge of producers is made more difficult by the need to incorporate secure sensors and transmitters into the already affordable equipment. IoT device security must also be maintained physically by users. If not protected, a smart motion sensor or a video camera placed outside a home might be altered.

3.5.5 Botnet Attacks

Malware on one IoT device by itself does not really threaten anything; it takes a large number of devices to seriously compromise anything. A hacker builds a bot army by infecting them with malware and then instructs them to make thousands of requests per second to the target in order to bring it down.

After the Mirai bot attack in 2016, there was a lot of commotion regarding IoT security. Numerous DDoS (Distributed Denial of Service) assaults were launched against the DNS that supported websites like GitHub, Twitter, Reddit, Netflix, and Airbnb by infecting hundreds of thousands of IP cameras, NAS, and home routers. IoT devices have the drawback of being extremely susceptible to malware assaults. No, they don't

3.5.6 Industrial Espionage and Eavesdropping

Spying could not be the only option if hackers seize control of surveillance at a spot by infecting IoT devices. Such attacks might also be carried out in order to seek ransom money. Therefore, violating privacy is a significant IoT security risk. IoT device hacking and spying are serious issues since a variety of sensitive data might be hacked and exploited against the owner.

A hacker could only wish to control a camera and utilize it for espionage. Nevertheless, it is important to remember that many IoT gadgets, including wearables, smart toys, and medical equipment, collect user information. On an industrial scale, hackers may gather huge data from a firm to reveal crucial corporate information.

Specific IoT devices with security issues are beginning to be banned in several jurisdictions. For instance, the interactive IoT doll with a Bluetooth pin lets anybody within a 25- to 30-meter radius to use the toy's microphone and speaker. Germany outlawed the doll after classifying it as an espionage item.

3.5.7 High Jacking Your IoT Devices

One of the worst virus kinds to have ever existed has been called ransomware. Your private files are not destroyed by ransomware; rather, it encrypts them to prevent access and prevents that access. The hacker who infected the device will then demand a ransom payment in exchange for the decryption key that will release the contents. Although ransomware infections of IoT devices are uncommon, the idea is gradually gaining popularity among black hat hackers. However, in the future, wearables, medical technology, smart homes, and other intelligent ecosystems may be in danger.

Both positive and terrible news are presented here. Because most IoT data is hosted in the cloud, this virus might not have important data to lock down, but it might disable the complete operation of the device. Imagine that you are required to pay a ransom in order for your car to start, or that your home is secured and the thermostat is set to the highest setting.

3.5.8 Data Integrity Risks of IoT Security in Healthcare

With IoT, data is always on the move. It is being transmitted, stored, and processed.

The majority of IoT devices extract and gather data from the surrounding environment.

A smart thermostat, HVAC, TV, or medical equipment are some examples. But occasionally, these gadgets transfer the data they've gathered unencrypted to the cloud.

As a result, a hacker may be able to access a medical IoT device, take control of it, and modify the data it gathers. A controlled IoT medical gadget may be used to give misleading signals, leading medical professionals to take measures that might be harmful to their patients' health. When a battery is actually ready to die, a medical IoT device that has been compromised may indicate to the maintenance station that it is completely charged. IoT security threats are even worse for medical equipment like pacemakers and insulin injectors. Hackers were able to access the implanted cardiac device made by St. Jude Medical, modify the pacing or shocks, or even worse, drain the battery.

3.5.9 Rogue IoT Devices

We may already be aware of the IoT's explosive expansion, with Ericsson forecasting that there will be 18 billion connected devices worldwide by 2022. The issue with this many devices is not limited to the BYOD (Bring Your Own Devices) strategy in businesses, but also in residential networks.

Being able to control all of our devices and secure the perimeter is one of the biggest threats to IoT security and problems. However, rogue or fake harmful IoT devices are starting to be uninvitedly deployed in guarded networks. To gather or manipulate sensitive information, a malicious device substitutes a legitimate one or joins a group as a member. These gadgets breach the network's defenses.

The Raspberry Pi and Wi-Fi Pineapple are two examples of rogue IoT devices. These can secretly collect incoming data transmissions as a rogue AP (Access Point), thermostat, video camera, or MITM (Man in the Middle). Future rogue device emergence may potentially include other variants. Interestingly, the idea served as inspiration for the horror film "Child's Play," which offers an intriguing illustration. In the film, Chucky is an IoT gadget that has gone rogue and started commanding other devices in a smart home system, posing a serious threat to people's lives (Khang & Gujrati et al., 2023).

3.5.10 Crypto Mining with IoT Bots

Bitcoin mining requires massive amounts of CPU and GPU resources, and as a result of this prerequisite, another IoT security risk has emerged: IoT bots that mine cryptocurrency. This kind of assault targets IoT devices with infected botnets with the intention of mining cryptocurrencies rather than doing any physical harm (Khang et al., Blockchain, 2022).

One of the first cryptocurrencies to be mined via hacked IoT devices, including video cameras, is the open-source cryptocurrency Monero. An army of video cameras has the resources to mine bitcoin, even though one camera does not. IoT botnet miners are a serious danger to the cryptocurrency industry since they may overwhelm and disrupt the whole market with a single attack (Hussain & Khang et al., 2022).

3.6 IMPROVE SECURITY REQUIREMENTS

Necessity to impose security requirements on IoT devices and their manufacturers, the following are the justifications for imposing security standards on makers of IoT devices.

3.6.1 Inadequate Access Controls

The only individuals who should have access to an IoT device's services are the owner and the people they trust to be in their near vicinity. However, a device's security mechanism frequently fails to adequately enforce this. IoT devices may have a high enough level of confidence in the local network that no additional authentication or authorization is needed. Any additional hardware attached to the same network is also trusted. This is especially a problem when the device is connected to the Internet: everyone in the world can now potentially access the functionality offered by the device.

The fact that all devices of the same type are sent with the same default password (such as "admin" or "password123") is a typical concern. For devices of the same model, the firmware and default settings are often the same. The credentials for the device may be used to access all devices in that series because they are known to the public, supposing that they are not changed by the user, which happens often. IoT devices frequently have a single account or privilege level that is both externally and internally accessible. This indicates that there is no additional access control after obtaining this permission. Multiple vulnerabilities are not covered by this one degree of security.

3.6.2 Overly Large Attack Surface

Every possible connection to a system offers a fresh set of chances for an attacker to identify and take advantage of weaknesses. A device can be attacked more often the more services it provides over the Internet. Attack surface is the term for this. One of the first stages in the process of securing a system is lowering the attack surface. A device can execute services on open ports that aren't necessarily necessary for functioning. By not exposing the service, an attack against such an unneeded service might be easily avoided. While seldom required in production, services like Telnet, SSH, or a debug interface may be crucial during development.

3.6.3 Outdated Software

It's crucial to share the latest version of software as soon as a vulnerability is found and fixed in order to be protected from it. This implies that IoT devices must have updated software that is free of known vulnerabilities when they are shipped, as well as update functionality to fix any vulnerabilities that are discovered after the device is deployed. Take the malware Linux as an example. Darlloz was originally identified in the latter part of 2013, and it took use of a defect that had been reported and repaired earlier in the year.

3.6.4 Lack of Encryption

A "Man-in-the-Middle" attacker can get any data being transferred with a client device or backend service when a device interacts in plain text (MitM). Anyone with the ability to get access to the network channel between a device and its endpoint can examine the network traffic and perhaps gather sensitive information like login passwords.

Using a plain text version of a protocol (like HTTP) when an encrypted version is available is a common issue in this category (HTTPS). A Man-in-the-Middle assault occurs when an attacker sneakily intercepts communications, transmits them, and then modifies them without the target parties' knowledge. Even if data is encrypted, flaws could still exist if the encryption is incomplete or set up improperly. For instance, a gadget might malfunction.

Encryption must also safeguard sensitive data that is kept on a device (at rest). Lack of encryption and storing passwords or API tokens in plain text on a device are typical security flaws. Other issues include the application of weak cryptographic methods or the unauthorized use of cryptographic algorithms. Encryption must also safeguard sensitive data that is kept on a device (at rest). Lack of encryption and storing passwords or API tokens in plain text on a device are typical security flaws. Other issues include the application of weak cryptographic methods or the unauthorized use of cryptographic algorithms.

3.6.5 Application Vulnerabilities

The first step in safeguarding IoT devices is to acknowledge that software includes vulnerabilities. It may be possible to cause functionality in the gadget that the designers had not anticipated due to software flaws. In few circumstances, this might lead to the attacker running their own code on the system, making it feasible to get private data or launch an attack on another person. It is difficult to totally prevent security vulnerabilities while building software. This is true of all software flaws. However, there are methods to avoid well-known vulnerabilities or reduce the possibility of vulnerabilities. This involves using recommended procedures to prevent application flaws, including consistently validating input.

3.6.6 Lack of Trusted Execution Environment

A critical first step in safeguarding IoT devices is admitting that software includes vulnerabilities. Device functionality that was not intended by the creators may be activated via software flaws. In some circumstances, this might lead to the hacker executing their own code on the system, making it feasible to harvest sensitive data or target other parties.

Security flaws are inescapable while creating software, just like any other type of issue. There are ways to avoid well-known vulnerabilities or lower the likelihood of vulnerabilities, though. For example, continuously carrying out input validation is one of the best practices to prevent application vulnerabilities. Programs signing must be done during the boot process with the aid of hardware in order to completely restrict the code that is allowed to run on the device. It might be challenging to execute this properly. Errors in the implementation of trusted execution environments lead to so-called jailbreaks in gadgets like the Apple iPhone, Microsoft Xbox, and Nintendo Switch.

3.6.7 Vendor Security Posture

When security flaws are discovered, the vendor's response heavily influences the outcome. The vendor's responsibilities include gathering information on potential vulnerabilities, creating mitigation, and updating deployed devices. Whether a vendor has a mechanism in place to effectively manage security concerns is frequently what determines the vendor's security posture.

The customer mostly interprets the vendor's security posture as enhanced security-related contact with the vendor. It will probably not assist to reduce the problem if a vendor does not give contact information or advice on what to do in the event of reporting a security concern. End users will continue to use the equipment as intended if they are unaware of any limits. The environment might become less secure as a result. Vendors may simplify things for consumers by providing information on how frequently security upgrades are released for devices and how to safely dispose of or resell the device so that sensitive data is not transferred.

3.6.8 Insufficient Privacy Protection

Sensitive data is routinely stored on consumer electronics. The password for a wireless network is stored on devices connected to that network. Cameras can record audio and video of the house where they are installed. A serious privacy violation would occur if attackers were able to acquire this information.

IoT devices and associated services must handle sensitive data appropriately, securely, and only with the end user's permission. This is true for both the distribution and storage of private data. The vendor is crucial in terms of privacy protection. In addition to an external attacker, the seller or a connected party may be in charge of a privacy violation. Without express consent, the manufacturer or service provider of an IoT device may collect data on user behavior for uses like market research. There are known instances when IoT gadgets, such smart televisions, may be listening in on family talks.

3.6.9 Intrusion Ignorance

From the user's perspective, a corrupted device frequently continues to operate properly. Most of the time, no extra power or bandwidth utilization is noticed. To advise the user of any security issues, the majority of devices lack logging or alerting features.

If they have, when the device is compromised, these can be erased or deactivated. As a result, users are frequently unaware that their device is being attacked or has been hacked, making it difficult for them to take preventative action.

3.6.10 Insufficient Physical Security

Attackers can open a gadget and assault the hardware if they have physical access to it. Any protection software can be disregarded, for instance, by simply accessing the contents of the memory components.

Debugging contacts that are available only after opening the device may also be present, giving an attacker greater options. Attacks that involve physical contact have an effect on a single device. We do not consider this to be one of the largest security issues because it is impossible to carry out these assaults in bulk via the Internet, but it is still mentioned.

A physical attack can have a significant effect if it discovers a device key that is shared by all devices of the same model, compromising a large number of devices. However, in that situation, we believe that the key sharing issue, rather than physical security, is the more pressing issue.

3.6.11 User Interaction

By making it simple to set up their devices securely, vendors may encourage the secure deployment of their products. Users can be encouraged to configure secure settings by paying careful attention to usability, design, and documentation. This category and the others mentioned above do not entirely overlap. For instance, the issue with improper access control noted above involves the use of risky or default passwords. Making it very simple or even required for users to set a safe password on the device is one method to address this.

It might be challenging for a non-technical user to determine whether a device complies with the need for the majority of the aforementioned security categories. By definition, the end user can observe user interaction; therefore, the consumer can assess how effectively a gadget handles user interaction. In order to ensure that installed security measures are engaged and properly applied, user contact is a crucial element. If changing the default password is feasible but the user is unaware of it or unable to use it, it is pointless.

3.7 BEST PRACTICES OF SECURITY

Best practices for ensuring the security of IoT systems, IoT security best practices can help you increase the protection of three main components of IoT systems: devices, networks, and data. Let's start by discussing ways to secure smart devices.

3.7.1 Secure Smart Devices

Make sure the hardware is tamper-proof. Attackers may steal IoT devices to mess with them or access private data. Make sure your product is tamper-proof to protect

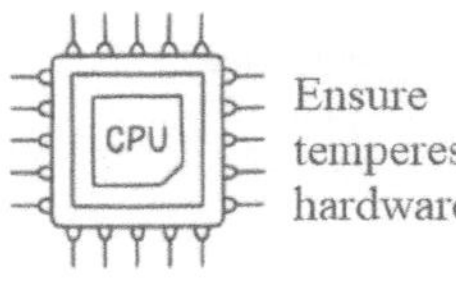

FIGURE 3.6 How to secure smart devices.

device data. By implementing port locks, camera covers, strong boot-level passwords, and other measures that will render the device inoperable in the event of tampering, you may assure physical security (Khang & Gupta et al., 2023). (Figure 3.6).

Offer updates and fixes. Continuous gadget upkeep requires extra expenses. However, regular updates and patches are the only way to guarantee effective product security. It is ideal to implement automated and required security updates that don't need end users to take any activity. Customers should be made aware of the length of the product's support term as well as what to do after it expires. After your system has been published, be sure to monitor future vulnerabilities and create updates as necessary. Run thorough testing. Penetration testing is your main tool for finding vulnerabilities in IoT firmware and software and reducing the attack surface as much as possible. You can use static code analysis to find the most obvious flaws, and you can use dynamic testing to dig up well-hidden vulnerabilities.

Put data safeguards on your devices. The security of data should be ensured by IoT devices both during and after exploitation. Ensure that the nonvolatile device memory is where cryptographic keys are kept. You may also offer to discard unwanted goods or provide them with a means to do so without disclosing sensitive information.

Comply with performance standards for components. To ensure good usability, IoT device hardware must adhere to a set of performance standards. Hardware for the IoT, for instance, should have great computing capability while using low electricity. Devices also need to provide wireless connectivity, strong data encryption, and permission. Additionally, it is ideal for your IoT solution to continue operating even if its Internet connection is momentarily lost.

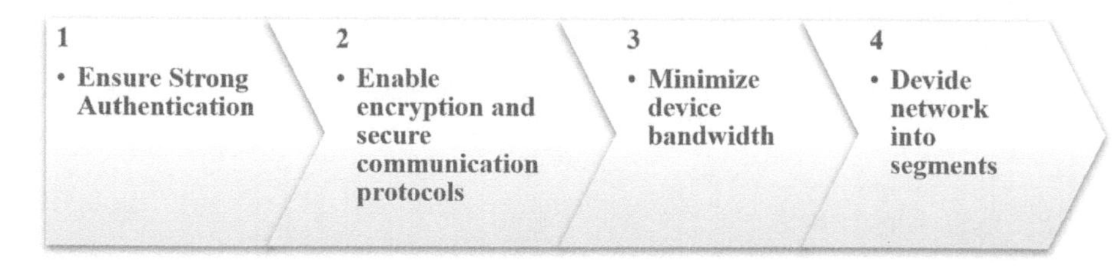

FIGURE 3.7 How to secure IoT networks.

3.7.2 Secure Networks

Make sure authentication is strong. By utilizing distinctive default credentials, this is possible. Use the most recent conventions when identifying or addressing your items to ensure their continued use. Give your product multi-factor authentication if at all possible. (Figure 3.7).

Enable secure communication methods and encryption. Security protection is also necessary for device communication. However, given the IoT devices' constrained capabilities, cryptographic methods need to be modified. You can use Lightweight Cryptography or Transport Layer Security for these applications (Khang & Hahanov et al., 2022).

You may employ wireless or wired technologies, including RFID, Bluetooth, Cellular, ZigBee, Z-Wave, Thread, and Ethernet, with an IoT architecture. Additionally, you may guarantee network security by using enhanced protocols like IPsec and Secure Sockets Layer. Reduce the device bandwidth. Limit network traffic to what is required for the IoT device to operate. If at all feasible, configure the hardware and kernel-level bandwidth limits and suspicious traffic detection on the device.

By doing this, you can defend your product from potential denial of service (DoS) assaults. Because malware has the ability to take control of the product and use it as a botnet to launch distributed denial-of-service attacks, the product should also be built to reboot and clean code if malware is found. Create segments in your networks. Create next-generation firewall security by slicing up large networks into multiple smaller ones. Utilize VLANs or IP address ranges for this. Use a VPN in your IoT system to provide safe Internet access (Hajimahmud & Khang et al., 2023).

3.7.3 Secure Data

Guard confidential data. Install distinct default passwords for every product, or demand password changes right once when a device is used. To guarantee that only authorized users have access to data, use rigorous authentication. Additionally, if the user chooses to return or resell the device, a reset option can be added to further improve privacy protection by enabling the erasure of private information and the wiping of configuration settings.

Only gather the information you need. Ensure that just the data required for your IoT product's functionality is collected. This would lessen the chance of data leakage, safeguard the privacy of customers, and remove the dangers of breaking numerous data protection standards, laws, and regulations (Khang & Abdullayev et al., 2023).

Communications across a secure network. Limit unneeded IoT network connectivity for your product for increased security. By making your product invisible through inbound connections by default, you may assure secure communication and not just rely on the network firewall.

Use encryption techniques that have been specifically designed to meet the requirements of IoT systems, such as the Advanced Encryption Standard, Triple Data Encryption Standard (DES), RSA (Rivest-Shamir-Adleman), and Digital Signature Algorithm. Apart from the practices mentioned above, make sure to follow recommendations like the NIST guide on IoT device cybersecurity, released to address challenges raised in the IoT Cybersecurity Improvement Act of 2020.

3.8 INTERNET OF THINGS SECURITY CHALLENGES

There were 1.51 billion breaches of IoT devices from January to June of 2021, compared to 639 million breaches recorded by Kaspersky for the entire year of 2020. It is unacceptable to undervalue the significance of cybersecurity while creating IoT systems (Khang & Khang et al., 2023).

Exploring possible cybersecurity concerns is crucial before learning how to safeguard IoT systems as Figure 3.8.

3.8.1 Software and Firmware Vulnerabilities

Because many smart devices are resource-constrained and have little processing capacity, it can be challenging to ensure the security of IoT systems. They are therefore more vulnerable than non-IoT devices since they are unable to execute robust, resource-intensive security measures.

For the following reasons, many IoT systems have security flaws:

- A lack of processing power to provide effective built-in security.
- Inadequate IoT system access control.
- Insufficient funding for properly evaluating and enhancing firmware security.

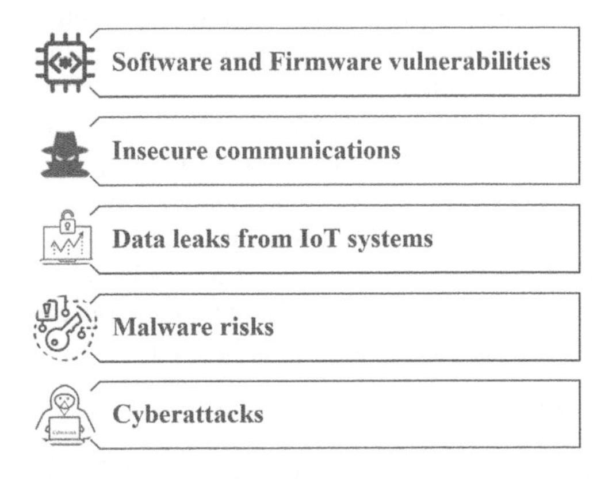

FIGURE 3.8 IoT cybersecurity challenges.

- Lack of frequent patches and upgrades because to IoT devices' technological limitations and restricted budgets. Users may not update their devices, thus restricting vulnerability patching.
- Over time, older devices might not be able to receive software upgrades.
- Inadequate defense against physical assaults: a perpetrator only has to get near enough to insert a chip or use radio waves to compromise the device.

An IoT system is a target for malicious actors that want to infiltrate its communications, introduce malware, and steal sensitive data. Hackers were able to access Ring smart cameras, for instance, by using weak, recycled, and default credentials. Using the microphone and speakers on the webcam, they were even able to contact with victims remotely.

3.8.2 Insecure Communications

Since IoT devices have limited resources, it is challenging to deploy the majority of existing security techniques. Traditional security methods are therefore less effective at safeguarding the communication of IoT devices. The potential for a man-in-the-middle (MitM) assault is one of the most harmful dangers brought on by unsecured communications. If your device doesn't employ secure encryption and authentication protocols, hackers can simply carry out MitM attacks to compromise an update process and take control of your device.

Even virus installation and functional changes are possible by attackers. Even if your device is not the target of a MitM attack, hackers may still be able to intercept the data it transfers via clear text conversations with other devices and systems. Connected devices are susceptible to attacks from other devices. For instance, if attackers gain access to just one device in a home network, they can easily compromise all other un-isolated devices in it.

3.8.3 Data Leaks from IoT Systems

We've consistently reported that hackers may access the data that your IoT system processes by intercepting unencrypted messages. This may even contain private information like your location, financial information, and medical history. Attackers can also obtain useful information by leveraging inadequately protected communications, albeit this is not the sole method.

All data are transported through and kept in the cloud, and services housed in the cloud are likewise susceptible to outside threats. As a result, both the devices themselves and the cloud environments to which they are attached might leak data. Another potential cause of a data leak in your IoT systems are third-party services. For instance, it was discovered that Ring smart doorbells were improperly transferring user data to Facebook and Google. This incident appeared because of third-party tracking services enabled in the Ring mobile app.

3.8.4 Malware Risks

Set-top boxes, smart TVs, and smartwatches were determined to be the gadgets most susceptible to malware attacks, according to a recent study by Zscaler.

An IoT device's functionality might be altered, personal information could be collected, and other attacks could be launched if attackers manage to introduce malware into the system. In addition, certain gadgets may come pre-infected with viruses if their producers don't take proper software security precautions.

The most well-known IoT-targeted malware has already been dealt with by several firms in creative ways. A Microsoft tutorial on how to proactively defend your systems against the Mozi IoT botnet is available, and an FBI agent recently discussed how the agency stopped the Mirai botnet assaults. But hackers continue to develop new techniques for abusing IoT networks and devices. BotenaGo, malware created in Golang in 2021, was found by researchers to be capable of exploiting over 30 distinct vulnerabilities in smart devices.

3.8.5 Cyberattacks

Apart from the malware and MITM attacks discussed above, IoT systems can also be susceptible to various cyberattacks. Here's a list of the most common types of attacks on IoT devices.

3.8.5.1 Attacks that Cause a Denial of Service (DoS)

Due of their low computing power, IoT devices are extremely susceptible to denial-of-service assaults. A device's capacity to react to legitimate requests is jeopardized by a deluge of bogus traffic during a DoS attack.

3.8.5.2 Attacks that Deny Users Sleep (DoSL)

Since sensors linked to a wireless network should continually monitor their surroundings, they frequently run on batteries that don't need to be charged frequently. By leaving the gadget in sleep mode the majority of the time, battery life is maintained. According to the communication requirements of various protocols, such as medium access control, sleep and waking modes are managed (MAC).

3.8.5.3 Attackers Could Use the MAC Protocol's Flaws to Launch a DoSL Attack

Because of the battery depletion caused by this form of assault, the Device spoofing. This attack is possible when a device has improperly implemented digital signatures and encryption. For instance, a poor public key infrastructure (PKI) may be exploited by hackers to "spoof" a network device and disrupt IoT deployments.

3.8.5.4 Physical Intrusion

Though most attacks are performed remotely, physical intrusion of a device is also possible if it's stolen. Attackers can tamper with device components to make them operate in an unintended way.

3.8.5.5 Application-Based Attacks

These types of attacks are possible when there are security vulnerabilities in device firmware or software used on embedded systems or weaknesses in cloud servers or backend applications.

3.9 CONCLUSION

IoT influence security, because of several properties of the underlying technology, threats against IoT systems and devices translate to greater security concerns. IoT settings are useful and effective because of these qualities, but threat actors may take advantage of them (Nayak & Satpathy et al., 2023). These qualities consist of the processes depicted in the following sections.

3.9.1 Accumulation of a Lot of Data

IoT sensors and equipment collect a wealth of information from people and their surroundings. The IoT settings require this data to operate effectively. However, if not protected or if it is stolen or somehow hacked, this data might have a number of undesirable cascade implications.

3.9.2 Interconnection between the Physical and Virtual Worlds

Many IoT devices may operate using the data they collect from their respective contexts. It reduces the separation between virtual and physical systems because to this capability. Although it may be easy for consumers, it may also make it easier for cyber threats to manifest as physical repercussions, having a higher impact.

3.9.3 Construction of Intricate Settings

Thanks to the increasing variety and availability of devices, complex IoT settings are now possible. In the context of the IoT, the term "complex" refers to an environment where there are sufficient IoT devices operating that dynamic interactions between them are feasible. Although this complexity increases an IoT environment's capabilities, it also increases the attack surface.

3.9.4 Architecture Centralization

Applying a conventional, centralized design to IoT applications can compromise security. The information obtained by each device and sensor will be sent to a base station thanks to a centralized design. In an organization, hundreds of devices that collect an incredible quantity of data may all use the same core database. Although this could be less expensive than maintaining separate databases, it runs the danger of creating a larger attack surface that is closely linked to a single root.

REFERENCES

Bhambri, P., Rani, S., Gupta, G., Khang, A., *Cloud and Fog Computing Platforms for Internet of Things* (2022). CRC Press. https://doi.org/10.1201/9781032101507

Hahanov, V., Khang, A., Litvinova, E., Chumachenko, S., Hajimahmud, V.A., Alyar, A.V., "The Key Assistant of Smart City – Sensors and Tools", *AI-Centric Smart City Ecosystems: Technologies, Design and Implementation* (1st Ed.) (2022). CRC Press. https://doi.org/10.1201/9781003252542-17

Hajimahmud, V.A., Khang, A., Hahanov, V., Litvinova, E., Chumachenko, S., Alyar, A.V., "Autonomous Robots for Smart City: Closer to Augmented Humanity," *AI-Centric Smart City Ecosystems: Technologies, Design and Implementation* (1st Ed.) (2022). CRC Press. https://doi.org/10.1201/9781003252542-7

Hajimahmud, V.A., Khang, A., Gupta, S.K., Babasaheb, J., Morris, G., *AI-Centric Modelling and Analytics: Concepts, Designs, Technologies, and Applications* (1st Ed.) (2023). CRC Press. https://doi.org/10.1201/9781003400110

Hussain, S.H., Sivakumar, T.B., Khang, A., "Cryptocurrency Methodologies and Techniques", *The Data-Driven Blockchain Ecosystem: Fundamentals, Applications, and Emerging Technologies*, pp. 149–164 (2022). CRC Press. https://doi.org/10.1201/9781003269281-2

Khang, A., Abdullayev, V., Hahanov, V., Shah, V., *Advanced IoT Technologies and Applications in the Industry 4.0 Digital Economy* (1st Ed.) (2023). CRC Press. https://doi.org/10.1201/978-1-003-43426-9

Khang, A., Chowdhury, S., Sharma, S., *The Data-Driven Blockchain Ecosystem: Fundamentals, Applications, and Emerging Technologies* (2022). CRC Press. https://doi.org/10.1201/9781003269281

Khang, A., Gupta, S.K., Rani, S., Karras, D.A., *Smart Cities: IoT Technologies, Big Data Solutions, Cloud Platforms, and Cybersecurity Techniques* (2023). CRC Press. https://doi.org/10.1201/9781003376064

Khang, A., Hahanov, V., Abbas, G.L., Hajimahmud, V.A., "Cyber-Physical-Social System and İncident Management", *AI-Centric Smart City Ecosystems: Technologies, Design and Implementation* (1st Ed.) (2022). CRC Press. https://doi.org/10.1201/9781003252542-2

Khang A. (2021). Material4Studies, *Material of Computer Science, Artificial Intelligence, Data Science, IoT, Blockchain, Cloud, Metaverse, Cybersecurity for Studies.* Retrieved from https://www.researchgate.net/publication/370156102_Material4Studies

Khang, A., Rana, G., Tailor, R.K., Hajimahmud, V.A., *Data-Centric AI Solutions and Emerging Technologies in the Healthcare Ecosystem* (1st Ed.) (2023). CRC Press. https://doi.org/10.1201/9781003356189

Khang, A., Rani, S., Gujrati, R., Uygun, H., Gupta, S.K., (Eds.) *Designing Workforce Management Systems for Industry 4.0: Data-Centric and AI-Enabled Approaches* (1st Ed.) (2023). CRC Press. https://doi.org/10.1201/99781003357070

Khang, A., Rani, S., Sivaraman, A.K., *AI-Centric Smart City Ecosystems: Technologies, Design and Implementation* (1st Ed.) (2022). CRC Press. https://doi.org/10.1201/9781003252542

Khanh, H.H., Khang, A., "The Role of Artificial Intelligence in Blockchain Applications", *Reinventing Manufacturing and Business Processes through Artificial Intelligence*, pp. 20–40 (2021). CRC Press. https://doi.org/10.1201/9781003145011-2

Luke, J., Khang, A., Chandrasekar, V., Pravin, A.R., Sriram, K. (Eds.). "Smart City Concepts, Models, Technologies and Applications", *Smart Cities: IoT Technologies, Big Data Solutions, Cloud Platforms, and Cybersecurity Techniques* (1st Ed.) (2024). CRC Press. https://doi.org/10.1201/9781003376064-1

Nayak, A., Satpathy, I., Patnaik, B.C.M., Baral, S.K., Khang, A., "Impact of Artificial Intelligence (AI) on Talent Management (TM): A Futuristic Overview", *Designing Workforce Management Systems for Industry 4.0: Data-Centric and AI-Enabled Approaches*, pp. 32–50. (1st Ed.) (2023). CRC Press. https://doi.org/10.1201/9781003357070-9

Rana, G., Khang, A., Sharma, R., Goel, A.K., Dubey, A.K., *Reinventing Manufacturing and Business Processes through Artificial Intelligence* (2021). CRC Press. https://doi.org/10.1201/9781003145011

Rani, S., Bhambri, P., Kataria, A., Khang, A., "Smart City Ecosystem: Concept, Sustainability, Design Principles and Technologies", *AI-Centric Smart City Ecosystems: Technologies, Design and Implementation* (1st Ed.) (2022). CRC Press. https://doi.org/10.1201/9781003252542-1

Rani, S., Bhambri, P., Kataria, A., Khang, A., Sivaraman, A.K., *Big Data, Cloud Computing and IoT: Tools and Applications* (2023). Chapman and Hall/CRC. https://doi.org/10.1201/9781003298335

Rani, S., Chauhan, M., Kataria, A., Khang, A., "IoT Equipped Intelligent Distributed Framework for Smart Healthcare Systems", *Networking and Internet Architecture* (2021). CRC Press. https://doi.org/10.48550/arXiv.2110.04997

Vrushank, S., Khang, A., "Internet of Medical Things (IoMT) Driving the Digital Transformation of the Healthcare Sector", *Data-Centric AI Solutions and Emerging Technologies in the Healthcare Ecosystem* (1st Ed.), p. 1 (2023). CRC Press. https://doi.org/10.1201/9781003356189-2

Vrushank, S., Vidhi, T., Khang, A., "Electronic Health Records Security and Privacy Enhancement Using Blockchain Technology", *Data-Centric AI Solutions and Emerging Technologies in the Healthcare Ecosystem* (1st Ed.), p. 1 (2023). CRC Press. https://doi.org/10.1201/9781003356189-1

4 Applications of the Internet of Things (IoT) in Smart Cities

A Step toward Innovation

Arpita Nayak, Atmika Patnaik, Ipseeta Satpathy, Alex Khang, and B.C.M. Patnaik

4.1 INTRODUCTION

International Business Machines Corporation (IBM) is the organization that originated the phrase "smart city." With its smarter city challenge initiative, the firm has built a global vision of urbanization based on data centralization and a heavy emphasis on security.

The phrase "smart city" refers to the use of technology, namely data collecting, to improve the running of cities. The concept behind smart cities is that the more information local governments have about their residents, the better the services they can provide (Khang & Gupta et al., 2023a). However, the phrase signifies different things to different parties, ranging from corporations to governments. Two measures of smart cities are proposed by the World Bank.

The first is a "technology-intensive metropolis, with sensors everywhere and extremely efficient public services, due to real-time data generated by thousands of networked devices."

The second is "a city that cultivates a stronger interaction between residents and governments – leveraged by accessible technologies. They rely on public feedback to assist enhance service delivery and are developing systems to collect this information." The diverse priorities, traits, and phrases used in each description, in this case, provided by the same organization, illustrate the sector's variety (Rana & Khang et al., 2021).

The phrase "smart city" alludes to emerging sectors that integrate communication and information technologies with urban functions and the environment (Sterbenz, 2017). In its most basic in this context, the phrase refers to the integration and combination of ICT and urban services. Smart cities, on the other extreme, can be defined as the combination of ICT, the surrounding habitat, clean technology, and vital infrastructure inside urban and residential regions (Park et al., 2018).

A vast majority of new employees are migrating to cities, as well as a huge number of rural families are increasing the population of cities by relocating. Sensing devices established techniques, and background settings are necessary to construct the essential infrastructure of a smart city (Jebaraj & Khang et al., 2023).

DOI: 10.1201/9781003376064-4

The Internet of Things (IoT) is among the most crucial components in the general deployment of such a city (IoT) (Hajimahmud & Khang et al., 2023). Tokyo, the metropolis with the highest population density in the world, continues to develop and has the most residents of any city on the planet. Japan's capital is the world's largest urban area, with a population of over 38 million individuals (38,050,000 people).

Furthermore, Jakarta, Indonesia has a population of over 31 million (32,275,000), while Delhi, India has a population of approximately 26 million. According to predictions, by 2030, large cities would contain 60% of the global population (Mao, 2019).

The term "Internet of Things" refers to the global interconnection of billions of physical objects that have been linked to the internet and are now acquiring and exchanging data. Sensing devices have also made their way into our daily life. You may not see these, yet they are all around you: in school, smart urban with IoT, and buildings (Rani & Chauhan et al., 2021).

Sensors can sense heat, pressure, water, and motion and communicate data with other related equipment and management systems. Sensors are increasingly being used in IoT applications (Tanwar & Tyagi et al., 2017; Rani & Kashyap et al., 2022).

Smart cities' key aims are to enhance policy efficiency, decrease waste and annoyance, increase socioeconomic and economic sustainability, and optimize social participation. The IoT is by far the most crucial aspect of this issue as smart cities rely on it to give connected solutions to the public (Subhashini & Khang, 2024).

The IoT refers to a network of interconnected things that communicate and exchange data, such as connected sensors, lights, and meters for data collecting and analysis. This information is then used by these communities to improve infrastructure, utility services, and services, among many other things (Bhambri & Rani et al., 2022).

Furthermore, a security system is being created to secure, monitor, and regulate data flow from the smart city network in order to maintain the network infrastructure's integrity (Shahrour & Xie, 2021).

Sensor smart cities offer common person management assistance such as surveillance devices, home automation, smart parking, vehicular traffic, and climatic change systems by interconnecting the virtual and physical worlds via electrical gadgets in roads, dwellings, vehicles, and buildings, environmental pollution, energy management systems, and smart grids. Several obstacles today stand in the way of the growth of the network ecosystems in intelligent systems (Khang & Hahanov et al., 2022).

As a corollary, sensing devices in smart city architecture are linked to the cloud in order to analyze data and make decisions. The IoT and cloud computing paradigms work together to deliver inputs and information to activities performed by integrated sensors, cars, humans, and mobile devices (Qian & Wu et al., 2019).

Because of the internet's power to transform the world, IoT is becoming one of the most important forms of infrastructure in smart cities. Because of the proper application of IoT technology, smart cities may now make much better use of public funds by enhancing service quality while lowering prices (Rani & Khang et al., 2023).

The fundamental purpose of IoT in smart cities is to provide individuals with immediate and unique access to their sources through enhanced public transit monitoring, water, energy, and maintenance (Ghazal & Hasan et al., 2021; Kumari & Gupta et al., 2021) as shown in Figure 4.1.

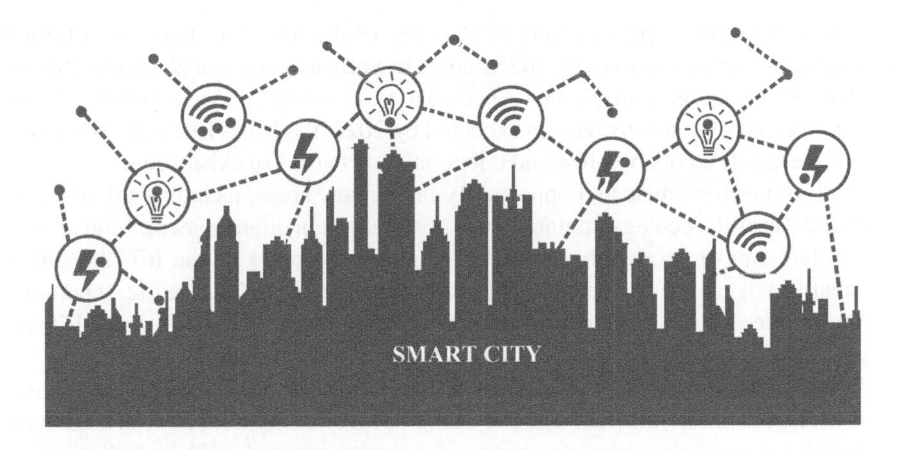

FIGURE 4.1 Illustration of smart city.

The notion of smart cities promotes more openness and action. A few IoT solutions are applied within smart cities, as seen in the graphic beneath:

- Gather information from the surrounding environment; based on the sensor type, for instance, collect data from numerous sources. Temperature sensors collect information from heat or cold. Humidity sensors get information from the atmosphere. Data is collected from the air through gas sensors. SCADA monitoring of water supply networks using a pressure sensor (Hajimahmud & Khang et al., 2022).
- Authentic alerts may be delivered by configuring the required software components to be a typical value that informs us if the value is more or lower than the standard. Some technologies, such as sensing devices, can automatically signal, allowing them to water plants at the right times by analyzing and acting on info concerning soil humidity. In smart cities, this may be utilized in the following ways: smart meter; an automated meter reading system that sends real-time data to customers and any appropriate departments for online monitoring and billing. Intelligent traffic control is a sort of adaptable traffic management that prevents traffic congestion by using actual information from sensing devices and CCTV.
- To go through data management activities, the system's software follows the process of utilizing data obtained from sensors – transforming data into critical information that may be valuable to a machine or device. Motion health screening tool for smart health to enhance primary care diagnosis. Waste collection; tracking waste container/bin fill levels to optimize pickup schedules and procedures. Intelligent lighting control – reducing street lamps; smart lighting as needed via brightness sensors and motion sensors.

Since the concept of a smart city was first introduced, IoT has been an important determinant of smart city growth. IoT technology will increase and have a bigger influence on how we live as technology advances and more countries adopt next-generation connectivity.

According to the report Enhancing Security of the IoT More than 75.44 billion connected IoT units will exist by 2025 thanks to Software-Defined Networks (SDN). With over 7.33 billion smartphones expected to be in use by 2023 and over 1.105 million connected wearable tech users expected by 2022, the IoT is likely to evolve into one of the greatest collaborative and cooperative networks in existence.

With so much promise and opportunity in so many areas, including urban transportation, security, ecology, maintenance, healthcare, and management, smart towns necessitate cities to comprehend the benefits and prospects of the IoT. Advanced connectivity is one of the fundamental building elements of the future generation of smart cities. People and organizations will be connected in unique ways. IoT has huge potential and advantages for smart cities (Khang & Rana et al., 2023).

The IoT is essential in every city. Tokyo has 38 million people, Delhi has 29 million, Shanghai has 26 million, and Sao Paolo has 21 million (Alam, 2021). Because of their huge populations, these megacities are noteworthy today. There will be many more of them in the future as populations get denser.

Cities are predicted to contain more than 60% of the worldwide people by 2030. It's a dangerous prediction that might boomerang if the proper measures aren't taken. Large populations need large quantities of resources. Residents will need water, efficient and environmentally friendly transit, clean air, and practical pollution and hygiene control.

Cities of the future will be able to serve their citizens' requirements more effectively and economically thanks to the unique use of smart city methods and IoT technology characteristic plans that can be formed utilizing connected technologies and enormous volumes of data. These solutions have the potential to ease problems, enhance the quality of life for city people, and minimize resource use. For a genuinely intelligent city, IoT is a helping hand (Appleton, 2022).

An IoT for data collecting and analysis, smart city employs a variety of IoT devices. IoT devices gather information from a variety of sources, including lights, sensors, and meters. As a result, smart cities are making significant progress in terms of public utilities, frameworks, and other devices.

A smart city is all about linking technologies, data, and people to increase efficiency and operate smarter. Internal integration and cooperation are critical to making this happen. Axis, in collaboration with our partners, provides flexible, innovative camera and IoT systems, leveraging our industry-leading experience and knowledge to bring your city together.

Making cities more habitable should thus be the overall goal. To do this, local officials, investors, and technology suppliers must collaborate to identify and build future-proof, scalable, and adaptable solutions (Cvar & Trilar et al., 2020).

4.2 RELATED WORK

4.2.1 How Did It Begin? The History of Smart City

Smart cities might have originated in the 1960s and 1970s, when the International Research Bureau had begun gathering data, releasing studies, and allocating resources to the region's need in order to avoid future calamities and alleviate penury. Since then smart cities have progressed in three stages.

Smart City 1.0 is led by technology suppliers. This age was concerned with implementing technology in cities despite the municipality's inability to fully appreciate the technology's possible repercussions or impact on daily life. Cities, on the other hand, pushed Smart City 2.0 to move forward.

Future-oriented municipal authorities contributed to describing the city's history and how new technologies and other advances may be used to attain that goal in this second generation.

Smart Local 3.0 is a new wave in which neither technology companies nor city governments have authority; instead, a method of community co-creation is used. Issues over inequality, as well as a desire to construct a centralized system of social integration, appear to be driving this most current invention.

Vienna, Austria, is one of the world's first cities to use this slashing third-generation tech. Cooperation with Wien Electricity, a local energy business, has been formed in Vienna. Vienna engaged residents to participate in local solar efforts as part of this partnership.

Likewise, in addressing issues such as gender equality and low-cost housing, Vienna has encouraged public participation. Vancouver, Canada, has also adopted the Smart City 3.0 plan, with 30,000 people contributing to the development of the Vancouver Greenest City 2020 Plan of Action (Shea & Burns, 2020).

4.2.2 Smart Traffic Management Through the Use of the Internet of Things

Cities have houses, businesses, workplaces, restaurants, and entertainment, among other things. However, having all of these places gathered together in a metropolitan region makes sense only if you can get to them.

What's the purpose of having a city full of great sites to see if you can't travel from point A to point B? This is why transportation technology exists. It is critical to have a mechanism to move people from place to place, whether it is public or private, in a vehicle, bus, or on a micro-mobility scooter.

However, the advent of various means of transportation has created as many difficulties as it has addressed. Pollution, traffic, and accidents are all typical occurrences that range in intensity from minor annoyances to life-threatening hazards.

Fortunately, the Internet of Things or IoT can provide some alternatives. IoT sensors and devices may be installed on roads and highways to observe, analyze, and exchange data to improve specific functions in smart cities, which employ smart technology to enhance their infrastructure and operations.

As a result, there may be less pollution, traffic, and accidents (Murphy, 2022; Atta & Abbas et al., 2020). It is critical to plan for potential technology issues with rapid progress of societies, as shown in Figure 4.2.

The next necessity is the IoT, which might include autos, home gadgets (fans, lights, doors, controller of air traffic), and smart city efforts are centered on the IoT; It is the potent catalyst that has enabled widespread digitization, spawning the concept of smart cities.

The IoT relates to the ubiquitous internet connection of commodities, allowing them to communicate data to the cloud and possibly receive orders for completing activities comprising data collection and data analytics to get information that can

FIGURE 4.2 Illustration of smart traffic management.

be utilized to enhance results and policy. And over 75 billion gadgets are expected to be connected to the internet by 2025 (Ashraf, 2021).

Traffic optimization is an issue for large smart cities. Los Angeles, California, as one of the world's biggest cities, has employed advanced transportation systems to regulate traffic flow.

Embedded sensors in the pavement send real-time traffic flow information to a centralized traffic control software which processes the information and modifies signal lights in seconds according to the traffic scenario. It predicts where traffic can flow based on past data, with no human intervention.

The IoT is making it much easier to optimize traffic patterns. Using sensors to detect and provide real-time traffic flow changes to a management solution, the system can interpret and modify traffic signals and other devices to the present situation in seconds, with little to no proper oversight (Dhingra & Madda et al., 2021).

About 68% of the world's population lives in cities, and an increasing number of individuals are focusing on urban development. As a result, there is a growing need for IoT-enabled smart cities. Smart cities and the IoT will assist to increase living quality and enhance what was before characteristics. The primary connecting control system on Cloud is a typical notion. They are equally beneficial to the environment as solar energy (Kadar Muhammad Masum & Kalim Amzad Chy, et al. 2018).

Sensors embedded in the pavements continue sending traffic flow information to a central traffic control platform. The IoT is already being utilized to assist wealthy nations and smart cities in reducing traffic congestion. Traffic planning is one of the most serious infrastructural concerns that rising countries confront today.

The IoT is already being used to help affluent countries and smart cities reduce traffic congestion. Smart traffic management, in contrast, to hand, incorporates aspects such as intelligent parking sensors, lighting controls, and smart roadways (Anand, 2021).

The use of IoT in daily life is vast, including switching TV channels using a mobile remote app, driving autonomous vehicles, and so on. Modern smart city projects make use of IoT technology by putting various sensors and cameras across cities (Hassankashi, 2018).

Based on the author's study, these IoT gadgets (smart traffic indicators) interface with linked automobile sensors and offer critical information regarding traffic light situations, such as color or time till green.

By boosting real-time surveillance, the goal is to prevent accidents and violations near traffic lights. There are several possible applications for IoT in traffic control. Consider the following examples as mentioned below (Saxena, 2021):

a. **Intelligent traffic signals** – Smart traffic signals resemble traditional stoplights, but they employ a network of sensors to monitor traffic in real time. The objective is frequently to minimize the amount of time automobiles sit idle. The numerous signs interact with one another and adjust to changing traffic circumstances in real time, thanks to IoT technology. This not only saves time in traffic but also reduces carbon emissions that automobiles emit into the atmosphere. Carnegie Mellon University is a participant in a pilot study in Pittsburgh that is evaluating this sort of technology, and the preliminary results are quite encouraging. The experimental run resulted in a 40% decrease in vehicle wait time, a 26% decrease in travel time, and a 21% decrease in projected vehicular emissions.

b. **IoT-enabled emergency help** – Road traffic accidents are the main cause of mortality among persons aged 1–54 in the United States. Getting into a car and driving to work or school is a relatively dangerous action that many people must engage in every day. IoT technology has the potential to make it a bit safer. This might take the form of real-time collision avoidance and notification, therefore decreasing the critical period an injured person sits untreated. In New Orleans, IoT solutions have been deployed across various emergency response professionals, including the fire department, police officers, and EMTs. 911 dispatchers may use IoT to expedite communications to these rescuers, enabling them to access data and make more educated choices faster.

c. **Commutes are made easier with applications like Waze** – Every driver using Waze or comparable software on the road serves as an IoT sensor. Waze collects and analyzes data for every road traveled. It may offer suggestions, identify ideal routes, warn of accidents or traffic congestion, and even propose the best time to depart using a strong algorithm. Waze cut travel time by 19% in a European pilot test. Less time on the road, like smart traffic signals, equals less pollution and is usually more convenient for everyone.

d. **Tolls and tickets have been improved** – In California, a pilot program is investigating how IoT may be used to replace the gas tax that the state now uses to pay for road maintenance and construction. Drivers would owe money under this system based on how much they drove. Sensors would detect this data and bill the user automatically based on it.

Getting into cars and driving to work or school is a statistically dangerous habit that many people must engage in each day. Manual traffic management by police officers is not effective. To improve traffic flow, an intelligent traffic management system is being developed obtained from the sensor's data, connectivity, and automated algorithms (Khang & Gujrati et al., 2023).

The purpose is to regulate the length of a green or red light for a given traffic signal at a junction as effectively as feasible. The traffic lights should not always flash the same color of green or red, but should instead fluctuate based on the number of cars on the route.

Green lights should be enabled for a longer period of time while traffic is flowing in one direction; when traffic is light, red signals should be activated for a more prolonged period of time. The period is expected to improve junction efficiency while simultaneously cutting travel expenses and emissions (Singh & Singh et al., 2019).

Every 500 meters, low-cost vehicular devices are shown in the road's middle. At least five sensors are joined together and communicate with a single IoT kit. The kits are all web connected. A range of parameters is considered when assessing each sensor intensity and adding one another sensor input, as well as leaving vehicle history road capacity (Abuga & Raghava, 2021).

4.2.3 Application of the Internet of Things in Parking: Smart Parking

In recent times, the smart city concept has risen in favor. The concept of a smart city appears to be becoming more practical as the IoT evolves. In the subject of IoT, ongoing efforts are being undertaken to improve the efficiency and dependability of urban infrastructure (Rani & Khang et al., 2022).

The IoT is tackling challenges such as traffic jams, a lack of parking spots, and road safety (Praveen & Harini, 2019; Khanna & Anand, 2016). The number of automobiles in many cities nowadays is fast consuming the available parking spaces in public locations.

Today, parking is a major concern in practically every city. Cities are countries. This issue can be remedied by using smart parking whenever possible. Smart parking directs each individual or car driver to available parking places, allowing them to make the most use of available parking space while reducing administrative expenses and traffic congestion. Living in cities may be difficult. Urbanization is constantly increasing.

Moreover, metropolitan transportation is becoming increasingly crowded. By the United Nations Department of Economic and Social Affairs (UN DESA), cities will account for 100% of the global population increase. By 2050, this increase is predicted to reach 68%. Population increase is not just a significant issue for the government, but it also affects the majority of citizens daily as shown in Figure 4.3.

Parking problems are uncommon, especially in major cities. According to IoT Analytics, market expenditure on smart parking services and products is expected to hit $3.8 billion by 2023, at a CAGR of 14%. The rise in market investment is great news since it will push people to work hard to fix these traffic problems instead of doing nothing (Peng & Clough et al., 2021).

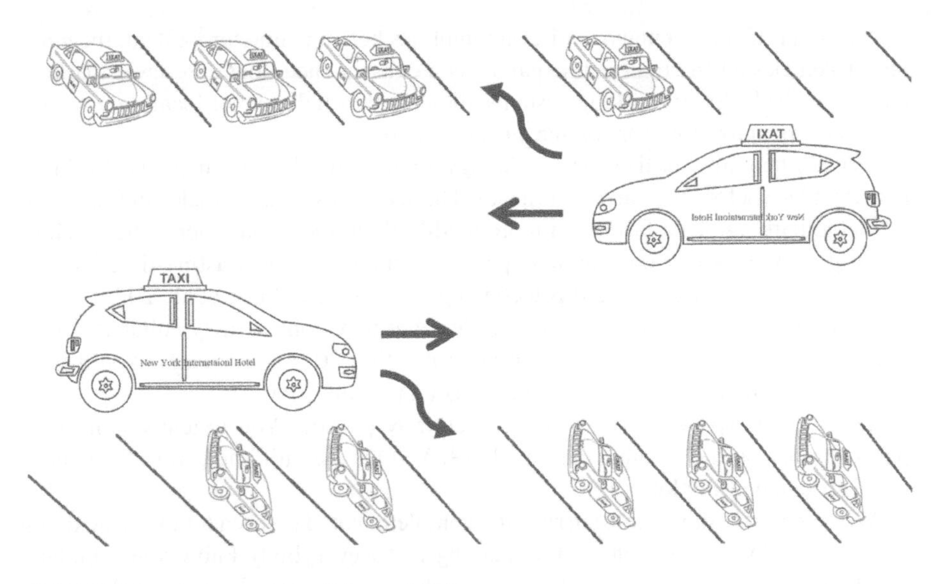

FIGURE 4.3 Smart parking (Khang, 2021).

The user gets real-time data on the accessibility of all available parking spots and chooses the best one. We conducted an internal study of the smart parking system's technology. The basic concept was to provide smart parking utilizing the IoT and infrared cameras are used, and empty parking bays are shown via a web service (Koba, 2019).

An IoT-based technology, also known as a connected parking system, is a centralized management system that allows autos to use a smartphone app to look for and reserve parking spots. The system's hardware consists of sensors that identify available parking spaces and send this information to all vehicles in the neighborhood. This information is often updated, so drivers never have to fear not finding a parking spot (Snehal & Babasaheb et al., 2023).

In addition to assisting cars in finding a parking place, the system gives notifications regarding peak periods and peak costs. These warnings are intended to help drivers save money while simultaneously minimizing traffic congestion (Barriga & Sulca et al., 2019).

Parking systems can be placed on the exterior or inside of structures. When a vehicle enters the parking area, receptors sense it and calculate the number of vacant parking places. This data is subsequently transmitted to the driver's phone through an app.

The parking guidance system also provides real-time occupancy statistics, which can be accessed via the app. This information is gathered from each sensor and updated every 5 minutes. One significant disadvantage of automated parking systems is that they have greater competition for parking places in metropolitan areas with little available space.

However, while these systems are beneficial to drivers, there are significant disadvantages to this sort of effort. Drivers who depend on public transit may not be able to utilize this app since they do not own or operate a car. These systems also need a significant amount of maintenance since numerous sensors must be changed regularly due to wear and tear or burglary (Lookmuang & Nambut et al., 2018; Kodali & Borra et al., 2018).

Parking is in high demand, and as automobiles become more propellant, the number of vehicles on the road will expand. As a consequence, there are fewer parking places available. Smart parking systems alleviate the challenge of locating an available place by informing cars of vacant spots nearby.

Drivers may also utilize the technology remote via their smartphone to find a parking place before they arrive at the lot. These devices assist vehicles in finding an available parking place faster and more readily than traditional approaches such as circling or waiting for someone to depart. Drivers may avoid wasting time hunting for a parking spot by utilizing this technology (Perkovic & Šolić et al., 2020).

Smart parking is a useful IoT application that may considerably enhance everyone's life. Assume you arrive for a crucial meeting 20 minutes early. You'll have more than enough time assuming you can get a parking spot.

The parking space at the building is entirely packed. You urgently hunt for a parking place on the road but can't find one. You try the building across the street's underground parking lot.

While driving in, you must arrive at a complete halt. There's a lot of traffic doing what you're doing. You contact the meeting to let everybody know you'll be late, but the parking garage complex has no cell phone coverage. The average American motorist wastes $345 each year, totaling more than $70 billion yearly. Furthermore, 40% of drivers polled stated they avoid going to brick-and-mortar stores because it is difficult to obtain a parking place.

According to recent studies, big metropolitan areas will house up to 68% of the world's population by 2050. This might have an immediate impact on how people park their automobiles in cities. Integrated new parking solutions paired with IoT connectivity aid in the resolution of this issue. Unoccupied parking spots are detected via installed IoT sensors. This IoT data is instantly sent to a distant server.

All parking facilities' data is gathered and processed in order to provide a map of available parking spots for anyone seeking a parking spot (Floris & Porcu et al., 2022).

4.2.4 How Do IoT Sensors Detect Available Parking Spaces?

Ultrasonic waves are used by IoT sensors to calculate the proximity to nearly anything. If a parking place is occupied, each detector buried in the ground detects the cars beneath. The three primary detection situations are shown in Figure 4.4.

- **Occupied space** – The sensor detects distances to objects ranging from 10 to 50 cm (approximately 4–20 inches).
- **Free space** – The response proximity of more than 50 cm to an item (about 20 inches).
- **Space is filthy** – The sensor detects a distance to an item of less than 10 cm (about 4 inches).

The application, which is powered by AWS IoT and AWS Lambda, displays free spaces for drivers. Sensor failures are shown in yellow, whereas filled areas are shown in green. The availability of parking spots in a parking lot has an impact on

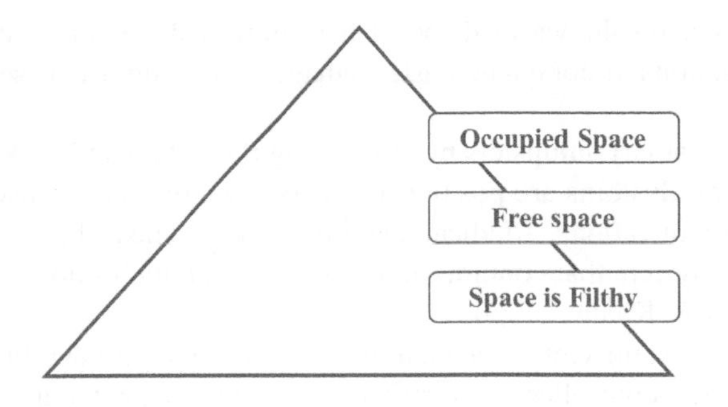

FIGURE 4.4 Extended model of three possible observable conditions.

the hardware and software requirements for IoT installation or system design. It is suggested that sensors in big parking lots utilize a router and the LPWAN interface is required.

LoRaWAN protocol implementation is a contemporary IoT trend and method for extending the working hours of automated driving by reducing power consumption. This reduces the need to replace batteries, according to LoRa Alliance guidelines. Battery life can be extended by up to five years before it has to be replaced.

Smart parking sensors with it are possibly used for ultrasonic, magnetic field monitoring, and IR sensing (Ultrasonic: Using ultrasonic to detect measurements enhances sensing accuracy. The downside of employing this sort of sensor is that dirt can accumulate).

In the process of electromagnetic field sensing, while metal objects come close to the sensor, even tiny changes in the magnetosphere can be detected through infrared sensor, which can sense motion and monitor changes in the surrounding temperatures. This type of sensor detects movement and monitors temperature changes in the surrounding environment (Gubochkin, 2022).

Smart parking systems appear to have a bright future. This solution is powered by IoT, AI, Machine Learning, and Advanced Analytics same technology that is propelling digitalization for businesses under the "Industry 4.0" banner. Parking 4.0 will boost the efficiency of parking systems by addressing urbanization concerns (Shlyakhetko & Braibant et al., 2022).

4.2.5 Smart Waste Management System Based on IoT

In today's society, we commonly see garbage cans or dustbins in public places around cities spilling due to the daily increase in waste. It generates an unfavorable environment for people as well as a foul stench in the surrounding region, which aids in the spread of numerous deadly illnesses and human ailments. We want to build "IoT Enabled Waste Management for Smart Cities" to prevent this problem" (Memon & Karim Shaikh et al., 2019).

The term "Internet of Things" refers to objects (embedded electronics) that are connected to the internet and may be remotely piloted. To collect real-time information from the smart trashcans in our system, the smart dustbins in our system are connected to the internet. In recent years, there has been significant population growth, and as a result, waste disposal has increased. To prevent the spread of hazardous infections, a complete waste management system is essential (Sohag & Podder, 2020).

For managing smart dumpsters by monitoring their state and making appropriate decisions, several dustbins are positioned across town or on campuses (Educational Institutions, Organizations, Medical Facilities, etc.). These bins are linked to a microcontroller system that contains an infrared sensor and radio frequency components (Vrushank & Khang, 2023).

When the IR system senses the quantity of dust in the garbage bin and sends the signals to the microcontroller, the very same output is encrypted and transferred via RF transmitter and receiver, and the data is decrypted by the receiver at the central system (Intel Galileo), and internet connection is facilitated via a network connection from the modem.

The data was collected, examined, and analyzed on the cloud, and the web-based GUI indicates the state of the garbage in the bin (Khedikar & Khobragade et al., 2017). We live in an ever-expanding technological environment. Industry 4.0 advancements have an impact on a wide range of markets, enterprises, and company owners.

While sectors are expanding, so is the demand for modern technologies. The IoT is gaining traction, notably in the waste management industry. Consider the following: from common household items to industrial vehicles, IoT is virtually everywhere. This variety has resulted in groundbreaking innovations in the waste management industry, which deals with rubbish in a variety of forms and locations.

However, most regions continue to handle trash in conventional methods. Municipalities, recycling centers, MRFs, and other waste management organizations are already reaping the benefits of IoT. Municipalities, recycling centers, and other waste management companies are beginning to use IoT technology into their operations (Ballal & Patil et al., 2019; Praveena & Sharanya et al., 2022).

Our trash production is constantly rising, culminating in a global garbage disaster. Even as we make sacrifices to build a more sustainable and greener planet by establishing 2050 climate goals, we keep failing to recycle or regulate our garbage creation before it's too late.

The only route out of this mess is to combine technology assistance with a vision of societal, economic, and sustainable development. Of course, solutions for smart cities use IoT technology to make it easier for humans to sense items and communicate.

Every day, governments, regions, towns, and municipalities incorporate "smart" technologies and solutions into their operations. As a result, significant waste management firms are already using digital solutions. As a result, incorporating IoT technology into your operations is a critical step.

There are several instances of IoT-based trash management for smart cities throughout the world. These solutions do more than just improve your operating strategies, they may also assist you in eliminating superfluous expenses and ensuring a more prudent budget.

Furthermore, they offer an example of environmentally sustainable waste management, and the additional resources generated by modern recycling processes contribute greatly to an ultimately, IoT-enabled smart trash management that has the potential to capture a sustainable society.

Overall, IoT-enabled smart waste management can catch: a consistent operating procedure and less managerial time, using limited resources wisely, and revenue generation at its peak (Goel & Goyal et al., 2021). A study in Evreka (2022) stated following are some instances of how a smart waste management IoT cooperation may benefit your firm:

a. **Sensors for smart bins** – Waste bins are an important part of waste management operations since they initiate the waste cycle. Smart sensors based on IoT enable you to use smart bin sensor technology from the start. This Fill Level Sensor, one of the greatest kinds of smart bin detectors backed by IoT technology, allows you to: with real-time data, you can monitor the location. View fullness levels to create daily optimum collection routes. Keep an eye on the temperatures of your smart bins. Receive immediate warnings in the event of an emergency, with location monitoring, you can minimize the number of lost containers and achieve enhanced inventory management. Temperature control will aid in the prevention of unanticipated consequences such as explosions and fires. One of the most significant parts of a multi-route makes system will be checking its fullness level.
b. **Dumpster rental** – A good illustration of an IoT-based waste management system is dumpster rental technology. It can utilize a sensor system, similar to smart bin sensors, to track and arrange unique and primarily on-demand waste collection activities.
c. **Management of fleet** – Fleet operations may become more productive using IoT. Evreka's Fleet Management Program is a great example of an Internet of Things-enabled Fleet Management Platform. Furthermore, this system will provide a structured atmosphere for managers and staff. Using data from smart bin sensors, to save operating costs, the Fleet Management Software may produce a route plan. Furthermore, drivers follow a pre-planned and environmentally beneficial path, which managers may track using a live map. Managers can also make fleet modifications and do performance analyses on customizable reports based on the past. There are no empty bins during garbage pickup thanks to the connection between Fill Level Sensor and the Evreka Fleet Management Solution. Most importantly, the best route optimization software informs management if a safer route will be established since the trucks are in residential districts.

IoT plays a crucial role in enhancing smart city applications by providing real-time monitoring and management of municipal processes. Conversely, with cities expected to host over two-thirds of the global population by 2030, rubbish disposal will be one of the most serious challenges cities will face. The world produces 2.01 billion tons of solid waste per year, with urban dwellers on track to create 3.40 billion tons by 2050 (Khanh & Khang, 2021).

Waste disposal prices are rising as well, with the World Bank forecasting that worldwide rubbish collection expenditures would exceed $375 billion over the next

FIGURE 4.5 Smart waste management.

Source: **181870631.JPEG, ©iStock.**

five decades. Fortunately, smart city programs are pushing trash management innovation as shown in Figure 4.5.

The smart management market was valued at slightly under $1.5 billion in 2018, and the estimation of smart management projects has gone up to $5 billion by 2025 (Bhuvaneswari & Hossen et al., 2020; Shamin & Fathimal et al., 2019).

IoT-enabled smart trash sensors enable cities to improve garbage collection, minimize the number of uncollected rubbishes, and manage resources, making smart waste management a vital part of the smart city ecology (Khang & Rani et al., 2022).

IoT in trash management has the potential to reduce unnecessary expenses caused by bottlenecks in rubbish collection operations. According to Berg Insight, the number of intelligent bins is expected to reach 2.4 million by 2025, with a 29.8% growth through 2025 if smart waste sensors are deployed quickly (Sosunova & Porras, 2022; Ravanan & Subasri et al., 2021).

Kevin Ashton, director of MIT's Auto-ID Center, developed the concept "The Internet of Things" in 1999. The current IoT enables sensing, actuation, collecting data, storage, and linking actual or virtual items to the Internet for computation.

Environmental monitoring, item tracking, traffic management, and smart home technologies are among the services provided by IoT apps that execute these duties (Badve & Chaudhari et al., 2020; Hong & Park et al., 2014).

In 2014, based on previous data, the city of Amsterdam installed trash collection vans with weighing equipment that instantaneously assesses how much a barrel weighs and helps anticipate fill rates with an accuracy of 80–90%.

In addition, 12,500 Enevo fill-level monitors were installed in garbage cans, and the system was evaluated using plastic debris. Amsterdam plans to save €3 million

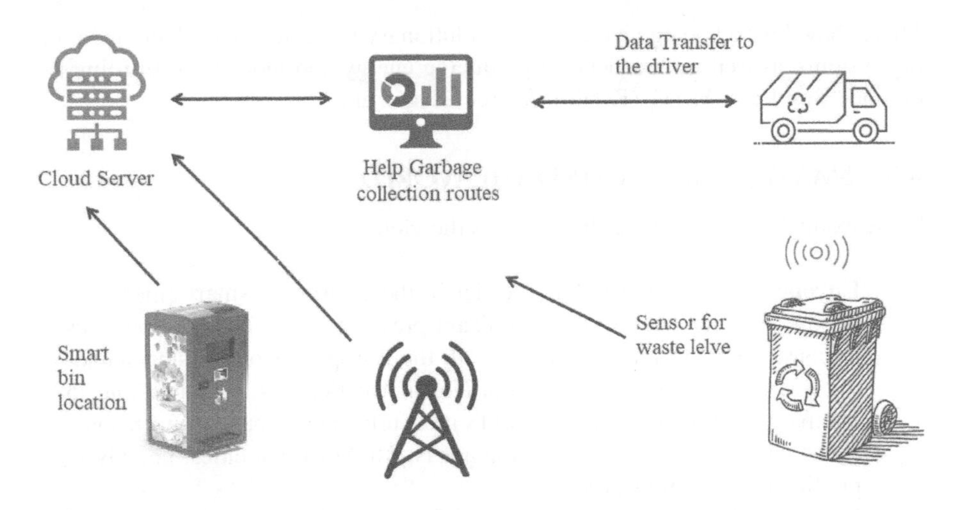

FIGURE 4.6 Extended model of the mechanism of how IoT works in waste management (Klubnikin, 2020).

in yearly garbage collection expenses by extending the IoT system from testing to city-wide usage. The image below depicts the method of IoT operation in trash management as shown in Figure 4.6.

The data is subsequently transmitted to the gateway through a low-power or mobile link. Wearable sensors can be intrinsically linked to the portal (via network topologies) or relay data to nearby nodes (mesh networks).

The site publishes detected data on the internet. Those kinds that process critical data locally may be included in modern waste disposal systems. The cloud technology converts raw sensor data into relevant data, which is then shown via interfaces. Plant workers and trash truckers use PCs and portable devices to scan data for empty bags and boxes and adjust their routes accordingly (Hahanov & Khang et al., 2022).

Sensor information is sent to the cloud via the gateway. Those kinds that process critical data locally may be integrated into sophisticated waste management systems in the future. The cloud platform converts raw sensors' collected data into meaningful information, which is then shown via dashboards.

Using PCs and portable devices, plant dispatchers and garbage truck drivers examine data for unoccupied bins and containers and change their routes accordingly. These technologies assist cities in reducing traffic congestion, CO_2 emission, and waste management expenses, which may account for up to 50% of budgetary allocation in most developing countries (Klubnikin, 2020).

In one of India's smart urban, an IoT-based garbage monitoring and management solution project was successfully tested. Ahmedabad has been called "GIFT City" (Gujarat International Finance Tec-City).

A complete solid waste collection system based on Swiss technology is established, in which garbage is disposed of away with little human participation. The rubbish will be recycled to provide organic manure and power for GIFT City.

Sensors may be used to manage landfills by detecting and ensuring that nearby regions fulfill Environmental Protection Agency groundwater contamination standards.

This is how IoT technology has proven revolutionary in waste disposal and monitoring systems, as well as beneficial in producing energy and money, so fulfilling the tagline "Best from Waste" (Karthik & Sreevidya et al., 2021).

4.3 SMART CITIES AROUND THE WORLD

Some examples of smart cities from across the globe:

- **Europe** – Europe is the global leader in the creation of smart cities. The European Union has exerted significant pressure on its member countries to construct smart city initiatives, with the European Commission spending 365 million euros on that aim. North America has trailed behind in comparison, despite being the world's most urbanized area. However, some smart city efforts are ongoing in major North American cities, notably in public safety and transportation.
- **Paris** – A French company, Vincent Callebaut Architectures, presented several high-rise constructions with positive energy generation (BEZOS). This program is consistent with the Paris Climate Energy Plan aims to cut greenhouse gas emissions by 75% by 2050.
- **London** – Westminster, London, installed Smart Park in 2014, a smart parking system that allows automobiles to rapidly identify parking spots and removes the need for lengthy scans for an open position. Therefore, traffic congestion in cities is lessened.
- **Copenhagen** – Copenhagen aspires to be the first fuel-smart city by 2025. Its Nordhavn area highlights how heating and intelligent connectivity may be used to combine electricity and heat, ecological structures, and electric vehicles into a single energy system.
- **New York City** – New York City is experimenting with connected vehicles (CTV) to assist minimize traffic deaths and injuries as well as car damage and infrastructure damage. The CTV infrastructure is largely geared at safety applications that depend on automotive, motorist-to-vehicle, and facility interconnections.
- **San Francisco** – Smart Traffic Signals in San Francisco The pilot will investigate the use of new technology such as Multimodal Smart Signalized Intersection Processes and then Communication, Trans - shipment Signal Prioritized by Trans Signal Priority, and Emergency Service Regulating interstate commerce to improve safety, start reducing collisions, and shorten emergency vehicle response times.
- **Amsterdam** – Amsterdam is an example of a smart city that understands the need of unlocking the data vault. The smart city initiative has evolved to include over 170 events since its beginning in 2009. It also makes transportation and traffic data accessible to relevant parties, such as developers, who use the information to construct mapping apps for the city's mass transit.

Everyone may now traverse the city with ease. There's more to it than that. To keep things flowing promptly, the city created "robots," which are autonomous delivery boats.

It also advocated a floating town of dwellings, which would provide a sustainable, energy-efficient solution to the city's overpopulation problem. Power is created within communities, and residents collect river water and filter it in their tanks.

4.4 CONCLUSION

The IoT is a game-changing technical innovation to transfigure the way we live, work, or entertain. A smart city, for example, may manage infrastructure and maintenance more effectively, lowering operating costs and enhancing inhabitants' lives (Khang & Gupta et al., 2023b).

The more cities throughout the world integrate IoT technology, the more cities, their inhabitants, and the world can profit from its utilization.

Cities are always changing and evolving. Smart cities are intended to be responsive and adaptable to their people's ever-changing requirements.

City planners and administrators may access essential data and insights to better the lives of their inhabitants and address their most urgent concerns by implementing the newest IoT-based smart city technology (Khang & Hajimahmud et al., 2022).

REFERENCES

Abuga, D., Raghava, N. *Smart Traffic Management and Control System Based on IoT.* (2021). https://doi.org/10.21203/rs.3.rs-314224/v1

Alam, T. "Cloud-based IoT applications and their roles in smart cities", *Smart Cities*, vol. 4, no. 3, pp. 1196–1219 (2021). https://doi.org/10.3390/smartcities4030064

Anand, A. "4 uses of IOT in traffic management. Analytics steps." Retrieved October 12, 2021, from https://www.analyticssteps.com/blogs/4-uses-iot-traffic-management

Appleton, J. "What is IOT and why is it important for smart cities? The global smart city knowledge center." Retrieved November 8, 2022, from https://www.beesmart.city/en/solutions/what-is-iot-and-why-is-it-important-for-smart-cities

Ashraf, S. "A proactive role of IOT devices in building smart cities", *Internet of Things and Cyber-Physical Systems*, vol. 1, pp. 8–13 (2021). https://doi.org/10.1016/j.iotcps.2021.08.001

Atta, A., Abbas, S., Khan, M.A., Ahmed, G., Farooq, U. "An adaptive approach: smart traffic congestion control system", *Journal of King Saud University - Computer and Information Sciences*, vol. 32, no. 9, pp. 1012–1019 (2020). https://doi.org/10.1016/j.jksuci.2018.10.011

Badve, M., Chaudhari, A., Davda, P., Bagaria, V., Kalbande, D. "Garbage collection system using IOT for smart city", *2020 Fourth International Conference on I-SMAC (IoT in Social, Mobile, Analytics and Cloud) (I-SMAC)* (2020). https://doi.org/10.1109/i-smac49090.2020.9243387

Ballal, V.S., Patil, S.S., Dange, N.P. "Smart city waste management system using IoT SERVER", *Management*, vol. 6, pp. 955–958 (2019). https://www.academia.edu/download/59804350/IRJET-V6I420720190620-49608-3qinkm.pdf

Barriga, J.J., Sulca, J., León, J.L., Ulloa, A., Portero, D., Andrade, R., Yoo, S.G. "Smart parking: a literature review from the technological perspective", *Applied Sciences*, vol. 9, no. 21, p. 4569 (2019). https://doi.org/10.3390/app9214569

Bhambri, P., Rani, S., Gupta, G., Khang, A. *Cloud and Fog Computing Platforms for Internet of Things* (2022), ISBN: 978-1-032-101507. CRC Press. https://doi.org/10.1201/9781032101507

Bhuvaneswari, T., Hossen, J., Asyiqinbt, N., Hamzah, A., Velrajkumar, P., Jack, O.H. "Internet of things (IoT) based smart garbage monitoring system", *Indonesian Journal of Electrical Engineering and Computer Science*, vol. 20, no. 2, pp. 736–743 (2020). https://pdfs.semanticscholar.org/95d8/56b15f8bab61b3f533729c8da3b7049e4ffb.pdf

Cvar, N., Trilar, J., Kos, A., Volk, M., Stojmenova Duh, E. "The use of IOT technology in smart cities and smart villages: similarities, differences, and future prospects", *Sensors*, vol. 20, no. 14, p. 3897 (2020). https://doi.org/10.3390/s20143897

Dhingra, S., Madda, R.B., Patan, R., Jiao, P., Barri, K., Alavi, A.H. "Internet of things-based fog and cloud computing technology for smart traffic monitoring", *Internet of Things*, vol. 14, p. 100175 (2021). https://doi.org/10.1016/j.iot.2020.100175

Evreka, "IoT-based smart waste management systems for revolutionary changes 'Eureka.'" RetrievedJuly26,2022,fromhttps://evreka.co/blog/iot-based-smart-waste-management-systems/

Floris, A., Porcu, S., Atzori, L., Girau, R. "A social IoT-based platform for the deployment of a smart parking solution", *Computer Networks*, vol. 205, p. 108756 (2022). https://www.sciencedirect.com/science/article/pii/S1389128621005995

Ghazal, T.M., Hasan, M.K., Alshurideh, M.T., Alzoubi, H.M., Ahmad, M., Akbar, S.S., Al Kurdi, B., Akour, I.A. "IoT for smart cities: machine learning approaches in smart healthcare—A review", *Future Internet*, vol. 13, no. 8, p. 218 (2021). https://doi.org/10.3390/fi13080218

Goel, M., Goyal, A.H., Dhiman, P., Deep, V., Sharma, P., Shukla, V.K. "Smart garbage segregator and IoT based waste collection system", In *2021 International Conference on Advance Computing and Innovative Technologies in Engineering* (ICACITE), pp. 149–153 (2021, March). IEEE. https://ieeexplore.ieee.org/abstract/document/9404692/

Gubochkin, V. "How to use IoT for smart parking solution development." *IoT for All*. Retrieved May 16, 2022, from https://www.iotforall.com/how-to-use-iot-for-smart-parking-solution-development

Hahanov, V., Khang, A., Litvinova, E., Chumachenko, S., Hajimahmud, V.A., Alyar, A.V. "The key assistant of smart city – Sensors and tools", *AI-Centric Smart City Ecosystems: Technologies, Design and Implementation* (1st Ed.) (2022). CRC Press. https://doi.org/10.1201/9781003252542-17

Hajimahmud, V.A., Khang, A., Gupta, S.K., Babasaheb, J., Morris, G. *AI-Centric Modelling and Analytics: Concepts, Designs, Technologies, and Applications* (1st Ed.) (2023). CRC Press. https://doi.org/10.1201/9781003400110

Hajimahmud, V.A., Khang, A., Hahanov, V., Litvinova, E., Chumachenko, S., Alyar, A.V. "Autonomous robots for smart city: closer to augmented humanity", *AI-Centric Smart City Ecosystems: Technologies, Design and Implementation* (1st Ed.) (2022). CRC Press. https://doi.org/10.1201/9781003252542-7

Hassankashi, M. "WhatisIoTandwhyweneedIOT?Codeproject."RetrievedMay27,2018,from https://www.codeproject.com/Articles/833251/What-is-IoT-and-Why-we-need-IoT

Hong, I., Park, S., Lee, B., Lee, J., Jeong, D., Park, S. "IoT-based smart garbage system for efficient food waste management", *The Scientific World Journal*, pp. 1–13 (2014). https://doi.org/10.1155/2014/646953

Jebaraj, L., Khang, A., Chandrasekar, V., Pravin, A.R., Sriram, K. "Smart City Concepts, Models, Technologies and Applications", *Smart Cities: IoT Technologies, Big Data Solutions, Cloud Platforms, and Cybersecurity Techniques* (1st Ed.) (2023). CRC Press. https://doi.org/10.1201/9781003376064-1

Kadar Muhammad Masum, A., Kalim Amzad Chy, M., Rahman, I., Nazim Uddin, M., Islam Azam, K. "An internet of things (IOT) based Smart Traffic Management System: a context of Bangladesh", In 2018 *International Conference on Innovations in Science, Engineering and Technology (ICISET)*. https://doi.org/10.1109/iciset.2018.8745611

Karthik, M., Sreevidya, L., Devi, R.N., Thangaraj, M., Hemalatha, G., Yamini, R. "An efficient waste management technique with IoT based smart garbage system", *Materials Today: Proceedings Physical Review* vol. 47, pp. 777–780 (2021). https://www.sciencedirect.com/science/article/pii/S2214785321050380

Khang, A., Gupta, S.K., Rani, S., Karras, D.A. *Smart Cities: IoT Technologies, Big Data Solutions, Cloud Platforms, and Cybersecurity Techniques* (1st Ed.) (2023a). CRC Press. https://doi.org/10.1201/9781003376064

Khang, A., Gupta, S.K., Shah, V., Misra, A. *AI-Aided IoT Technologies and Applications in the Smart Business and Production* (1st Ed.) (2023b). CRC Press. https://doi.org/10.1201/9781003392224

Khang, A., Hahanov, V., Abbas, G.L., Hajimahmud, V.A. "Cyber-Physical-Social System and İncident Management", *AI-Centric Smart City Ecosystems: Technologies, Design and Implementation* (1st Ed.) (2022). CRC Press. https://doi.org/10.1201/9781003252542-2

Khang, A. (2021). *Material of Computer Science, AI, Data Science, IoT, Blockchain, Cloud, Cybersecurity for Studies.* Retrieved from https://www.researchgate.net/publication/370156102_Material4Studies

Khang, A., Rana, G., Tailor, R.K., Hajimahmud, V.A. *Data-Centric AI Solutions and Emerging Technologies in the Healthcare Ecosystem* (1st Ed.) (2023). CRC Press. https://doi.org/10.1201/9781003356189

Khang, A., Rani, S., Gujrati, R., Uygun, H., Gupta, S.K. *Designing Workforce Management Systems for Industry 4.0: Data-Centric and AI-Enabled Approaches* (2023). CRC Press. https://doi.org/10.1201/99781003357070

Khang, A., Rani, S., Sivaraman, A.K. *AI-Centric Smart City Ecosystems: Technologies, Design and Implementation* (1st Ed.) (2022). CRC Press. https://doi.org/10.1201/9781003252542

Khanh, H.H., Khang, A. "The Role of Artificial Intelligence in Blockchain Applications", *Reinventing Manufacturing and Business Processes through Artificial Intelligence*, pp. 20–40 (2021). CRC Press. https://doi.org/10.1201/9781003145011-2

Khanna, A., Anand, R. "IoT based smart parking system", In *2016 International Conference on Internet of Things and Applications (IOTA)*, pp. 266–270 (2016, January). IEEE. https://ieeexplore.ieee.org/abstract/document/7562735/

Khedikar, M.A., Khobragade, M.M., Sawarkar, M.N., Nikita, M. "Garbage management of smart city using IoT", *International Journal for Research in Science Engineering*, pp. 35–38 (2017). https://www.academia.edu/download/52558415/17March2.pdf

Klubnikin, A. *Smart City Waste Management with IOT: Top Use Cases. Custom Software Development Company.* Retrieved November 3, 2020, from https://www.softeq.com/blog/how-smart-cities-are-leveraging-iot-for-waste-management

Koba, S. *IoT-Based Smart Parking System Development. MobiDev.* Retrieved October 9, 2019, from https://mobidev.biz/blog/iot-based-smart-parking-system

Kodali, R.K., Borra, K.Y., G. N., S.S., Domma, H.J. "An IoT based smart parking system using LoRa", In *2018 International Conference on Cyber-Enabled Distributed Computing and Knowledge Discovery (CyberC)*, pp. 151–1513 (2018, October). IEEE. https://ieeexplore.ieee.org/abstract/document/8644697/

Kumari, A., Gupta, R., Tanwar, S. "Amalgamation of blockchain and IOT for smart cities underlying 6G communication: a comprehensive review", *Computer Communications*, vol. 172, pp. 102–118 (2021). https://doi.org/10.1016/j.comcom.2021.03.005

Lookmuang, R., Nambut, K., Usanavasin, S. "Smart parking using IoT technology", In *2018 5th International Conference on Business and Industrial Research (ICBIR)*, pp. 1–6 (2018, May). IEEE. https://ieeexplore.ieee.org/abstract/document/8391155/

Mao, Y.M. *What Is the Role of IOT in Smart Cities? Finextra Research.* Retrieved September 26, 2019, from https://www.finextra.com/blogposting/17931/what-is-the-role-of-iot-in-smart-cities

Memon, S.K., Karim Shaikh, F., Mahoto, N.A., Aziz Memon, A. "IoT based Smart garbage monitoring & collection system using WEMOS & ultrasonic sensors", In *2019 2nd International Conference on Computing, Mathematics and Engineering Technologies (ICoMET)* (2019). https://doi.org/10.1109/icomet.2019.8673526

Murphy, M. 7 ways IoT can improve traffic management. *Cellular Connectivity for the Internet of Things*. Retrieved February 28, 2022, from https://www.hologram.io/blog/7-ways-iot-can-improve-traffic-management/

Park, E., del Pobil, A., Kwon, S. "The role of the internet of things (IOT) in smart cities: technology roadmap-oriented approaches", *Sustainability*, vol. 10, no. 5, p. 1388 (2018). https://doi.org/10.3390/su10051388

Peng, G., Clough, P.D., Madden, A., Xing, F., Zhang, B. "Investigating the usage of IOT-based smart parking services in the borough of Westminster", *Journal of Global Information Management*, vol. 29, no. 6, pp. 1–19 (2021). https://doi.org/10.4018/jgim.20211101.oa25

Perkovic, T., Šolić, P., Zargariasl, H., Čoko, D., Rodrigues, J.J. "Smart parking sensors: state of the art and performance evaluation", *Journal of Cleaner Production*, vol. 262, p. 121181 (2020). https://www.sciencedirect.com/science/article/pii/S0959652620312282

Praveena, P., Sharanya, K.K., Srinithi, S., Suvatha, B., "IoT based solar-powered smart waste disposal system", In *Proceedings of the International Conference on Intelligent Technologies in Security and Privacy for Wireless Communication, ITSPWC* 2022, (2022, August). 14–15 May 2022, Karur, Tamil Nadu, India. https://eudl.eu/doi/10.4108/eai.14-5-2022.2318874

Praveen, M., Harini, V. "NB-IOT based smart car parking system", In *2019 International Conference on Smart Structures and Systems (ICSSS)* (2019). https://doi.org/10.1109/icsss.2019.8882847

Qian, Y., Wu, D., Bao, W., Lorenz, P. "The internet of things for smart cities: technologies and applications", *IEEE Network*, vol. 33, no. 2, pp. 4–5 (2019). https://ieeexplore.ieee.org/abstract/document/8675165/

Rana, G., Khang, A., Sharma, R., Goel, A.K., Dubey, A.K. *Reinventing Manufacturing and Business Processes through Artificial Intelligence* (2021). CRC Press. https://doi.org/10.1201/9781003145011

Rani, S., Bhambri, P., Kataria, A., Khang, A. "Smart city ecosystem: concept, sustainability, design principles and technologies", *AI-Centric Smart City Ecosystems: Technologies, Design and Implementation* (1st Ed.) (2022). CRC Press. https://doi.org/10.1201/9781003252542-1

Rani, S., Bhambri, P., Kataria, A., Khang, A., Sivaraman, A.K. *Big Data, Cloud Computing and IoT: Tools and Applications* (2023). Chapman and Hall/CRC. https://doi.org/10.1201/9781003298335

Rani, S., Chauhan, M., Kataria, A., Khang, A. "IoT equipped intelligent distributed framework for smart healthcare systems", *Networking and Internet Architecture* (2021). CRC Press. https://doi.org/10.48550/arXiv.2110.04997

Rani, R., Kashyap, V., Khurana, M. "Role of IOT-cloud ecosystem in smart cities: review and challenges", *Materials Today: Proceedings*, vol. 49, pp. 2994–2998 (2022). https://doi.org/10.1016/j.matpr.2020.10.054

Ravanan, V., Subasri, R., Kumar, M.G.V., Dhivya, K.T., Kumar, P.S., Roobini, K. "Next generation smart garbage level indication and monitoring system using IoT", In *2021 Smart Technologies, Communication and Robotics (STCR)*, pp. 1–4 (2021, October). IEEE. https://ieeexplore.ieee.org/abstract/document/9588961/

Saxena, P. Role of IOT in road safety and traffic management. *The National AI Portal of India*. Retrieved March 23, 2021, from https://indiaai.gov.in/article/role-of-iot-in-road-safety-and-traffic-management

Shahrour, I., Xie, X. "Role of internet of things (IoT) and crowdsourcing in smart city projects", *Smart Cities*, vol. 4, no. 4, pp. 1276–1292 (2021). https://doi.org/10.3390/smartcities4040068

Shamin, N., Fathimal, P.M., Raghavendran, R., Prakash, K. "Smart garbage segregation & management system using Internet of Things (IoT) & Machine Learning (ML)", In *2019 1st International Conference on Innovations in Information and Communication Technology (ICIICT)*, pp. 1–6 (2019, April). IEEE. https://ieeexplore.ieee.org/abstract/document/8741443/

Shea, S., Burns, E. What is a smart city? Definition from whatis.com. *IoT Agenda.* Retrieved July 16, 2020, from https://www.techtarget.com/iotagenda/definition/smart-city

Shlyakhetko, O., Braibant, A., Czechowska, E., Fryczka, M., Hadrich, R. "IoT Project: Smart Parking", *Developments in Information & Knowledge Management for Business Applications*, pp. 37–58. (2022). Springer, Cham. https://link.springer.com/chapter/10.1007/978-3-030-95813-8_2

Singh, S., Singh, B., Ramandeep, Singh, B., Das, A. "Automatic vehicle counting for IoT based smart traffic management system for Indian urban settings", In *2019 4th International Conference on Internet of Things: Smart Innovation and Usages (IoT-SIU)* (2019). https://doi.org/10.1109/iot-siu.2019.8777722

Snehal, M., Babasaheb, J., Khang, A. "Workforce management system: concepts, definitions, principles, and implementation", *Designing Workforce Management Systems for Industry 4.0: Data-Centric and AI-Enabled Approaches*, pp. 1–13 (1st Ed.) (2023). CRC Press. https://doi.org/10.1201/9781003357070-1

Sohag, M.U., Podder, A.K. "Smart garbage management system for a sustainable urban life: an IoT based application", *Internet of Things*, vol. 11, p. 100255 (2020). https://www.sciencedirect.com/science/article/pii/S2542660520300901

Sosunova, I., Porras, J. "IoT-enabled smart waste management systems for smart cities: a systematic review", *IEEE Access* (2022) https://ieeexplore.ieee.org/abstract/document/9815071/

Sterbenz, J.P.G. "Smart city and IoT resilience, survivability, and disruption tolerance: challenges, modeling, and a survey of research opportunities", In *2017 9th International Workshop on Resilient Networks Design and Modeling (RNDM)* (2017). https://doi.org/10.1109/rndm.2017.8093025

Subhashini, R., Khang, A. "The Role of Internet of Things (IoT) in Smart City Framework", *Smart Cities: IoT Technologies, Big Data Solutions, Cloud Platforms, and Cybersecurity Techniques* (2024). CRC Press. https://doi.org/10.1201/9781003376064-3

Tanwar, S., Tyagi, S., Kumar, S. "The role of internet of things and smart grid for the development of a smart city", *Intelligent Communication and Computational Technologies*, pp. 23–33 (2017). https://doi.org/10.1007/978-981-10-5523-2_3

Vrushank, S., Khang, A. "Internet of medical things (IoMT) driving the digital transformation of the healthcare sector", *Data-Centric AI Solutions and Emerging Technologies in the Healthcare Ecosystem*, p. 1 (1st Ed.) (2023). CRC Press. https://doi.org/10.1201/9781003356189-2

5 Internet of Things (IoT)-Based Implementation of Smart Cities Using Emerging Technologies and Its Challenges

Shweta Bansal, Amar Saraswat, Meenu Vijarania, and Swati Gupta

5.1 INTRODUCTION

With growing global inhabitants and expansion of urbanization (which is expected to increase by a further 10% in the next 30 years), resulting in 70% of people living in cities by 2050, countries worldwide are envisioned to equip towns to handle the colossal number of people as well as the aggravation it will bring to existing urban systems (Ahvenniemi & Huovila et al., 2017; Kurdachenko & Semko et al., 2017).

Intelligent cities have emerged as a critical push by many governments to make cities more accessible and inviting to the predicted population rise and offer city people a better living experience, as proven by the numerous public and commercial initiatives now underway (Janssen & Luthra et al., 2019).

We evaluate Internet of Things (IoT) usage in smart cities in this study and highlight how it facilitates such endeavors. However, because this issue is of great interest to academics, various surveys were discovered in the literature review (Hollands, 2018).

Anthopoulos and Reddick (2016) explore the problems that IoT system implementation in smart cities encounters and prioritize them depending on the specialists accessible to them. Our work differs from Silva and Khan et al. (2018) in that we cover the challenges without regard to expert opinion and thus are not restricted to their application area alone; instead, we provide an all-encompassing discussion of the essential facets that IoT architects in smart cities must consider and go over best practices for acknowledging each of those facets.

Sánchez (2018) emphasizes the design and use of IoT in smart cities and a brief description of the technologies employed. The investigations of Atat and Liu et al. (2018) give an application-oriented study on specific systems that have been designed

DOI: 10.1201/9781003376064-5

for different bright city aspects, along with the case studies of cities, including ongoing projects on smart cities.

Anthopoulos and Reddick (2016) examine the use of large datasets for cognitive computing, including data creation methods, collecting, storage, analytics, security, and measures to make such systems more environmentally friendly. They also discuss the many applications that such systems may be used for. Unlike Anthopoulos and Reddick (2016), we present a full review of the many fundamental components and technologies utilized in smart city deployments and a discussion of their current stage of use/deployment. The survey's structure is shown in Figure 5.1.

In this chapter, we describe smart cities as the use of various communication and information technology to enhance the quality of life for a city's residents. This includes implementing these technologies in all previously mentioned areas, such as public governance, transportation, healthcare, commerce, sustainable living, social participation, and opportunity provision.

In an idealistic world, the concept of a smart city extends beyond the traditionally defined borders of a conventional city's administrative and social structures by allowing interaction between them, allowing it to function more cohesively and efficiently. Smart cities have various benefits compared to a traditional city ecosystem (in value terms). Smart cities are at the frontline of cutting-edge technology that will assist governments in meeting their climate change goals (Khang & Rani et al., 2022).

Smart cities are concerned with power generation, intelligent transportation infrastructure, intelligent smart local administrators to reduce cities' carbon footprints, and enabling the implementation of emerging sustainable technologies for cleaner and healthier living.

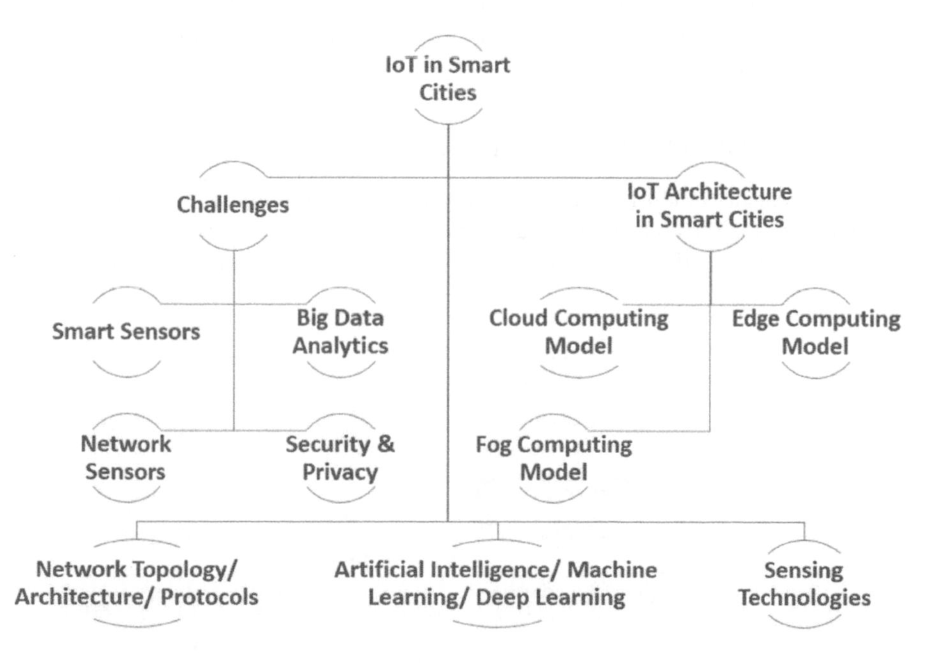

FIGURE 5.1 Overall structure of the survey.

Intelligent innovative city enterprises will be worth USD 1 trillion (Koubaa & Aldawood et al., 2020), providing a significant financial incentive for government firms to effectively contribute to the improvement of technologies that promote innovative development of the city.

The goal of an intelligent city project aims to raise the standard of living for residents of the city and the development of an equitable society in which all viewpoints are respected, and equal opportunities are offered. In the framework of smart cities, information systems are a critical component of providing public services through improvements by improving citizen interactions with the city antimere (Rani & Khang et al., 2022).

The following are the chapter's key contributions:

- It discusses the IoTs' different applications, architectures, and components, in the context of smart cities.
- It presents a thorough examination of IoT technologies utilized at differing stages of the IoT architecture.
- It discusses the technical difficulties associated with the adoption of IoT in smart cities and provides its solutions.
- It examines the present status of IoT utilization and covers how artificial intelligence (AI) has been employed in IoT for smart cities utilizing clustering, classification, and other techniques. Furthermore, the many applications, solutions, and data that is used to achieve the overarching architecture of smart cities are thoroughly described. The data sources, algorithms, activities done, and methods of deployment leveraged by these suggested techniques are also discussed (Khang & Gupta et al., 2023).
- It makes recommendations for the future on critical areas of IoT deployment in smart city applications.

5.2 COMPONENTS OF SMART CITIES

An intelligent city comprises several elements, as shown in Figure 5.2. The first part of smart cities is the collection of data. The next step is the transmission and reception of the data; the third step is storing the data, and the last is the analysis of data.

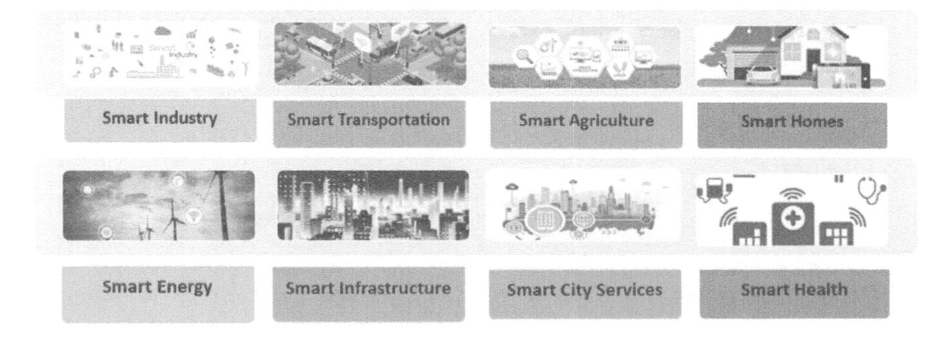

FIGURE 5.2 Various elements of the smart cities.

Data gathering is application-based, and it has acted as a significant driver of sensor development in various fields. The second phase is data interchange, which entails sending data from data gathering equipment to the cloud for storage and processing.

Smart agriculture uses various sensors in the fields to measure numerous factors to assist in decision-making and the prevention of multiple diseases, vermin, etc. (Rojek & Studzinski, 2019).

Meticulousness agriculture, which entails placing sensors in plants to provide specific dimensions and therefore allowing for tailored care mechanisms to be implemented, is one aspect of the intelligent agriculture paradigm (Khang, 2023a).

Precision agriculture will be required for food protection and security (Pardini & Rodrigues et al., 2019) and is thus a critical component of the struggle for long-term food production. Crop surveillance and disease detection, crop care, and decision-making are the most common AI applications in IoT for agriculture (Rana & Khang et al., 2021).

The operations that support a city's population and provide the services are referred to as innovative city services. They include municipal chores, such as the provision of water, trash management, and ecological control and monitoring.

Water quality sensors can offer continuous updates on the water quality utilized in the city and detect leakage (Dutta & Chowdhury et al., 2017). Smart Energy Efficiency: A typical electrical system has a one-way energy flow from a significant generating source, commonly a fossil fuel or hydroelectric power plant.

Detecting errors and taking appropriate measures in such scenarios is also a time-consuming procedure. Furthermore, as renewable energy technologies become more affordable, today's users receive a supply from the primary company and generate their energy.

The phrase "smart health" is one of the information and communication technology (ICT) methods used to ensure the availability and quality of health treatment. With an aging population and rising healthcare expenses, academics and healthcare professionals have been concentrating their efforts in this field.

Current healthcare systems are overloaded, and as a result, they cannot meet rising public demand. In this sense, innovative health strives to make healthcare accessible to as many people as feasible through video conferencing (Trencher & Karvonen, 2019) and AI-assisted diagnosis aids for clinicians (Haverkort & Zimmermann, 2017).

The smart home is a critical component of smart cities since it is at the heart of the residents' lives. The usage of sensing devices deployed through a person's house to offer information about the home and its occupants is known as smart homes.

The sensors like environmental sensors, motion trackers, and power/energy usage sensors are some examples of these sensors used for the monitoring of the user activity.

Industries worldwide are constantly striving to improve their efficiency and productivity while lowering costs. The concept of an industry 4.0 plant is one in which all of its intermediaries' functionaries are fully integrated and operate in sync with one another. Due to the apparent use of this technology (Vijarania & Dahiya et al., 2021), this is conceivable.

Application of this technology in production along with manufacturing mechanisms, as well as cyber-physical processes that merge workers as well as machineries, has provided the marketplace with many benefits, including better and faster innovation, better manufacturing schemes (resources and operations), higher product quality, along with improved factory worker safety.

Working with a collection of different devices and machines, on the other hand, has its own set of challenges and necessitates cyber-physical procedures that are flexible in configuration, connection, and deployment for IoT applications in Smart Industry (Trakadas & Simoens et al., 2020). AI and the IoT are being used to build and implement Industry 4.0 services.

A city's infrastructure is critical to its residents' quality of life; local governments must develop new bridges, roads, and buildings and execute upkeep to ensure continued use.

By applying sensors for detecting structural, constructional conditions for monitoring using accelerometers (Farag, 2019) and intelligent materials (Wang & Ram et al., 2016), intelligent infrastructure assists cities in maintaining their infrastructures are in good shape and useable.

The data obtained by these sensors enables condition monitoring of this critical equipment to keep the city running smoothly. The various sensor technologies for IoT in smart cities are shown in Figure 5.3.

Congestion, pollution, public transportation scheduling, and cost-cutting are all challenges that have taken root in many metropolitan regions. Deployment of the same for merging Vehicle Infrastructure for the usage by Pedestrian communication is one of the prominent features of this technology.

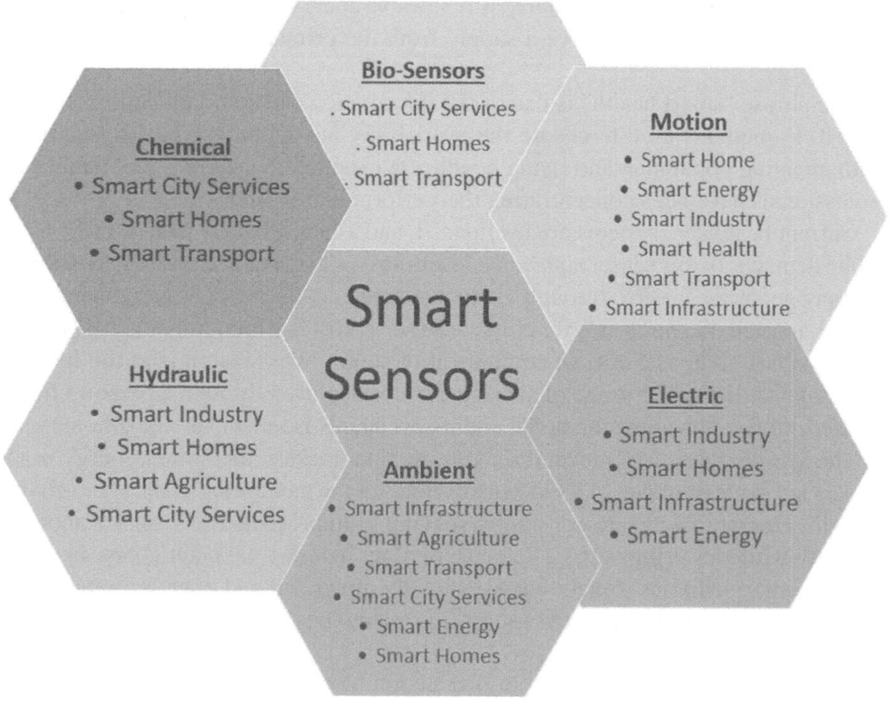

FIGURE 5.3 Smart sensors.

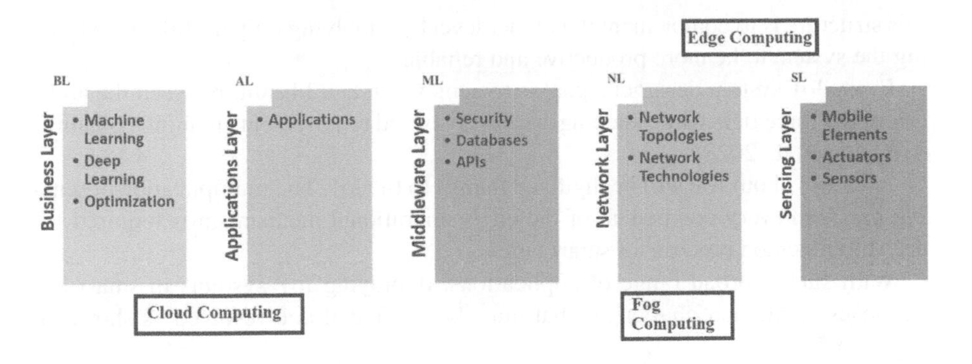

FIGURE 5.4 Generic design of IoT.

Several systems rely on GPS data to detect driving behavior and traffic patterns now that every car has a tracking device as well as every driver has a cellphone. This real-time data is already being used in apps like Waze and Google Maps for route planning, as well as for public transportation trip booking and other analyses.

Sensors in smart cities use the IoT to collect and transport data about the city to a cloud server, where it is further processed and analyzed for decision-making as well as pattern extraction.

By utilizing cloud services, the IoT integrates sensing of the data, its transmission and reception, processing, and storage functions. As illustrated in Figure 5.4, a general IoT design is made up of five levels, each of which operates on the information from the preceding layer. It also depicts the three distinct designs for IoT systems (Rani & Chauhan et al., 2021).

Table 5.1 lists the features of each of the three tiers of the IoT system. The three IoT topologies discussed are not really mutually exclusive; rather, the objective of

TABLE 5.1
Cloud, Fog, and Edge Model

Cloud Model	Fog Model	Edge Model
On a global scale, contextual awareness encompasses all components of the application.	The Fog layer is aware of the context of the localized sensing circumstance.	Typically, devices based on edge only have information about their state. Although an exchange approach is viable, it is confined to the local vicinity.
Heterogeneous information from a variety of sensing devices is often used.	It uses a variety of data, but only within a limited area.	Limited access to certain type of data.
The expense of networking is substantial.	Medium networking cost.	Least networking cost.
Raw data may be transferred to the Cloud, posing a potential privacy issue.	When compared to Cloud computing, there is a higher level of privacy.	Even more, privacy protection than the fog computing approach is conceivable.
As decision-making is centralized, it is the least robust.	More reliable than the cloud model	As decision-making is distributed, it is the most robust.

this structure is to complement the higher level by supplying only useful data, allowing the system to be more productive and reliable.

Every IoT system designer's goal is to achieve an equilibrium between the abilities of the three tiers while keeping system costs and requirements in mind (Bhambri & Khang et al., 2022).

All areas of our life will be digitized thanks to the IoT. The multiplication of sensing nodes in every component of the city's operational mechanism is required for this digitalization process for smart cities.

With such a broad range of applications, deploying IoT systems in smart cities poses significant challenges that must be evaluated and resolved as shown in Figure 5.5.

Data security and privacy are among the most important considerations in smart cities. Any abnormality in the general functioning of the city's infrastructure would put the residents' people and assets in danger in addition to causing them misery and inconvenience (Khang & Hahanov et al., 2022).

Critical city infrastructures must be online for smart cities to function. Even this is much more fruitful in the case of deep learning (DL) and machine learning (ML) applications (Saraswat & Sharma, 2022) when launched through IoT devices.

Security consequently poses a serious issue in smart cities. Smart cities are more susceptible to destructive attacks in the modern world, where malware and warfare are becoming common tactics; to stop this and protect the data, the transit of such data across the network must be encrypted.

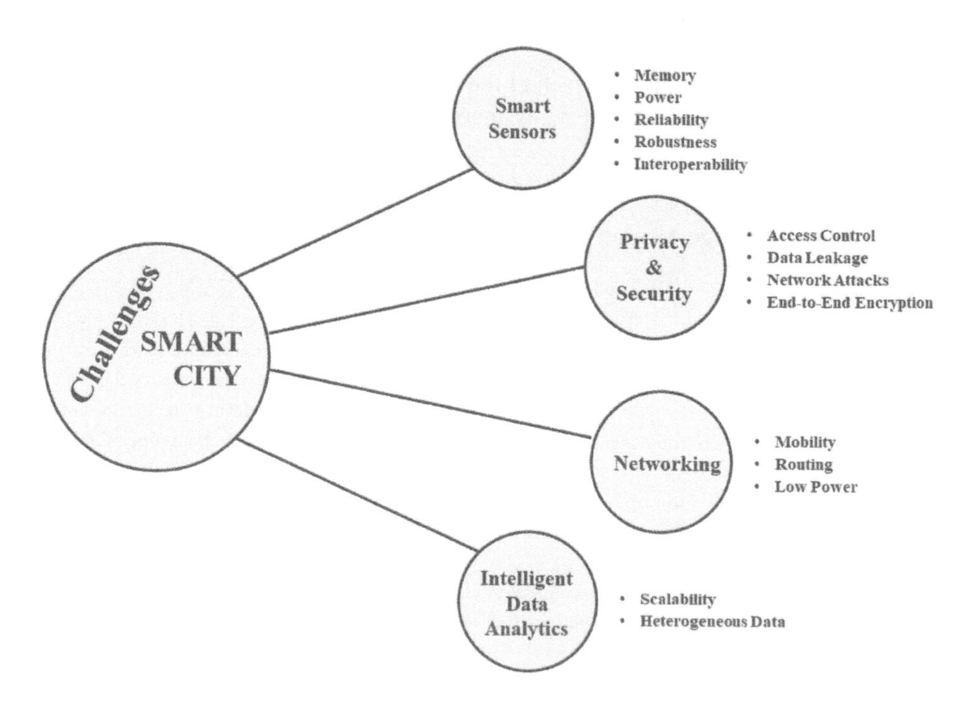

FIGURE 5.5 Issues for IoT in smart city (Khang, 2021).

The proliferation of sensors in smart cities, which gather information on city dwellers' actions, could reveal private information about individuals to unauthorized parties. To resolve this issue and protect the data, processes that anonymize data collection while keeping the integrity of the context of the monitored job will be necessary. The fidelity and robustness of smart sensors is another challenge (Hahanov & Khang et al., 2022).

Future smart cities will be built on the IoT, and because it will be so crucial to their operation, the IoT system must ensure a flawless user experience. This highlights the necessity for quick and accurate responses to service requests made by app users, as well as the necessity that everybody in the smart city has access to the highest services.

Power and transportation are two essential services that should be provided by decentralized systems. The system's robustness will be increased by the dispersed connecting points. One such instance is self-healing in smart grids (Keane & Topol, 2018).

Many existing protocols are designed for the network devices that have constant power; however, in many instances, sensors in smart cities will be mobility and hence battery operated. They'll also have to measure, transport, and, in certain situations, store the data they've gathered.

This necessitates not only the creation of less power, and low data transmission techniques, but also the creation of novel storage techniques, as well as limited devices that maximize battery life. Storage of this vast volume of data will need the development of compression techniques as well as database methods in the future as smart cities and IoT become more widespread.

To provide long-term service, solutions for power concerns involve the new battery technologies, as well as the implementation of energy-saving methods in such devices. Another source of worry is concept drift, which occurs when data is continuously acquired and its qualities change over time. Techniques like progressive learning may be useful in this situation.

Explain-ability is another essential component for smart city statistics to be widely accepted, particularly in the field of smart health. Some methods have been put out in this approach; for instance, a hybrid deep-learning classifier with web technologies-based flood monitoring system.

A comprehensible DL-based healthcare system for COVID-19 care is described by Rahman and Hossain et al. (2020), which is based on a distributed learning model and exhibits encouraging results. However, in order to improve the proliferation of smart city applications, more effort has to be done to combine explain-ability approaches like visualization, and intrinsic methodologies for smart city applications based over the usage of not only ML but also DL.

5.3 ARTIFICIAL INTELLIGENCE FOR SMART CITIES

In a smart city, many sensors enable the IoT to send data about the city's status to the cloud. Although simply measuring the raw data is insufficient, data analysis is required to make the city "smart". The Data Acquisition layer facilitates the collection of the data and storage of the data, accompanied by the Pre-processing layer, which accomplishes operations on the data of an appropriate quality to be used in the data and analytics level (Rani & Khang et al., 2023).

The data analytics level entails using data science techniques on the data to uncover patterns and insights that will be utilized in the Service layer for policymaking, planning, and other activities. In this part, we'll look at data analytics, which is the third level of the data analysis process.

The application of DL and ML on the obtained data is part of data analytics in the smart city based on IoT. The AI topic in this work is summarized in Figure 5.6. It shows the uses for smart city components that each AI domain discovers.

The applications of AI in smart cities have been covered in this part, as well as the type of deployment and the nature of data used to accomplish their goals (Khanh & Khang, 2021).

Civil structures are also monitored for structural health as part of a smart infrastructure application. Diez and Khoa et al. (2016) use a clustering technique to do bridge health monitoring in the cloud using vibration data. They employ clustering to find anomalous behavior groups in accelerometer data from a bridge.

Accelerometer signals were employed by Sanjay (2020), whereas piezoelectric transducers were used by Gui and Pan et al. (2017) to conduct structural strength monitoring of bridges; this could be stated as a classification issue for the bridges that are not in the proper state. The most common mode of monitoring has been utilizing accelerometers to measure vibration.

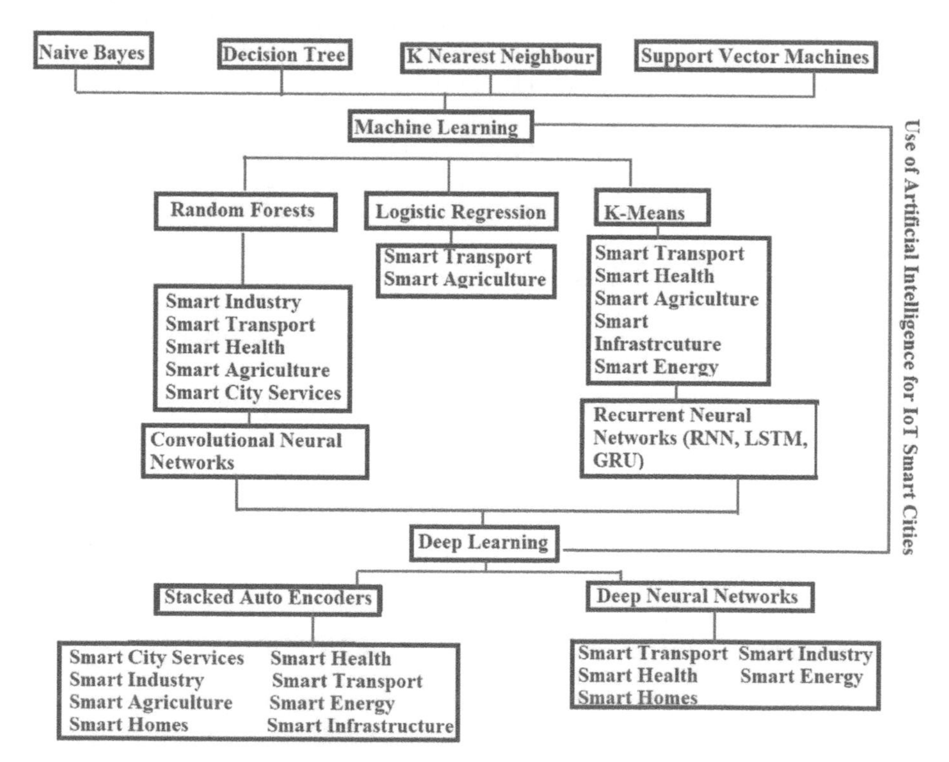

FIGURE 5.6 Artificial intelligence for the smart cities.

Two key smart transportation applications are smart parking and transportation management. A domain where AI can be used is smart parking, wherein the people can be helped to find out the places where the parking can be done. This will not only save time but also prove a boon for smart transport along with saving fuel and gas emissions.

Sustainable development strategies have been proposed as both a regression and a classification challenge, using a combination of imaging and/or other occupancy detection techniques. Certain methods (Lenka & Barik et al., 2017) related to regression can help in assisting the people by forecasting the vacant places that can be used as a parking while classification systems (Awan & Saleem et al., 2020) guide drivers based on the shortest distance and are utilized for user localization inside such lots.

The authors have stated an idea related to solving the complex problems pertaining to the domain of smart parking. This idea is related to edge computing, wherein it could be used with the Convolutional Neural Network, over the various devices that are useful for finding the correct location and detection of the occupancy in addition to cloud-based techniques.

Another application of AI in the IoT is traffic flow estimation and forecasting for traffic conditions and other management tasks such as accident detection. For such a solution, one of the widely used devices is video cameras, which can be used to detect vehicle density on roadways in order to determine traffic congestion.

A factory is a place where a large number of machinery and equipment are constantly producing items. Maintenance is a vital part of this operation since it involves doing frequent inspections of the related device in order to make sure that no breakdowns occur throughout the manufacturing procedure, in which there seems to be the possibility of a financial loss or circumstances like death.

However, with the data obtained by several sensors on these devices, it is often more effective to use this data for predictive maintenance purposes rather than as a passive technique. One of the key uses for AI in the business is predictive maintenance, which analyses data from the everyday functioning of equipment in the industry to optimize the production process (Vitola & Pozo et al., 2016).

Sevillano and Màrmol et al. (2014) propose a predictive maintenance scheme based on Support Vector Machines (SVMs) that uses information from accelerometers that measure vibrations of the various devices, such as a crane motor, along with a collection of the data with the help of different external sources used in processes used in semiconductor manufacturing, as these two could operate in the cloud environment, as seen in the architecture.

The use of Recurrent Neural Networks (RNNs) to predict future values of physical parameters of a pump using several heterogeneous sensors used to monitor it was proven by Sevillano and Màrmol et al. (2014), who employed RNNs to accurately predict quantities of structural parameters of a pump based on a variety of sensor devices.

Sensing data and control information is transmitted over the internet and local networks in smart cities. Furthermore, some components of smart cities are critical features of a city's functioning and are deeply entwined with its people's social and private lives.

As per the results of Kwon and Kim (2109), confidentiality issues in smart cities are critical, and academics are pretty interested in them—issues related to the security of IoT, as per Braun (2018). The topic of IoT security has been discussed in

Braun (2018), which discusses several problems explored in various IoT designs and current concerns and solutions.

We address system security issues related to smart cities, draw attention to important issues in the context of smart cities, and wrap up our debate on the topic. Smart cities are achieved by collecting data from sensors located throughout a city and even its people, analyzing it, and then, utilizing it to make life better for residents.

The internal condition of a city's elements, such as transport, power, structure condition/state, population mobility, and more, can be assessed using these sensors. This data is then sent to the cloud, evaluated and extracted.

However, various concerns regarding how these data are transported and used raise concerns about the process's integrity, protection, and secrecy. In actuality, the world's attention was called to the real threat cyber attackers posed in 2015, when an attack on Ukraine's power system knocked off electricity for 225,000 people (Eckhoff & Wagner, 2018).

Consumption of power and a building's overall sensor data may suggest occupancy. Saraswat and Sharma (2022) and even individual identities (Saraswat & Kalra, 2020), and Navigation systems that are available on every phone and in the majority of automobiles are likely to be exposed to generate the evidence for an individual's position, and lifestyles. They lead to confidentiality, as discussed in Braun (2018).

As described in the article (Braun, 2018), GPS-enabled gadgets, which are found in every smartphone and most vehicles, are vulnerable to leaking information about a person's whereabouts and habits, resulting in privacy concerns. This information might be exploited by evil actors to do illegal activities that endanger people's lives and property.

Traditional security strategies may not be as effective in safeguarding the IoT devices for smart city applications in many cases, and new approaches to dealing with the privacy risks in IoT for smart cities may be required. To provide a consistent structure and terminology for defining security concerns, we employ the standardized attack events classification advocated by the CERT, i.e., Computer Emergency Service, established by DARPA to be used in IoT for smart cities.

5.4 CHALLENGES FOR PRACTICAL IMPLEMENTATION OF A SMART CITY

Nowadays, the smart city concept is widely implemented in the modern world. However, specific issues are important to handle to attain further improvement. This section covers the existing challenges for an IoT-based smart city through an extensive literature review.

Figure 5.7 illustrates the difficulties in putting smart cities into practice, including the following:

- the stages of design;
- implementation, and operation phases;
- the cost of design and operation at every phase and need;
- the variety of devices;
- the vast amounts of data collected and analyzed;
- information security; and sustainability.

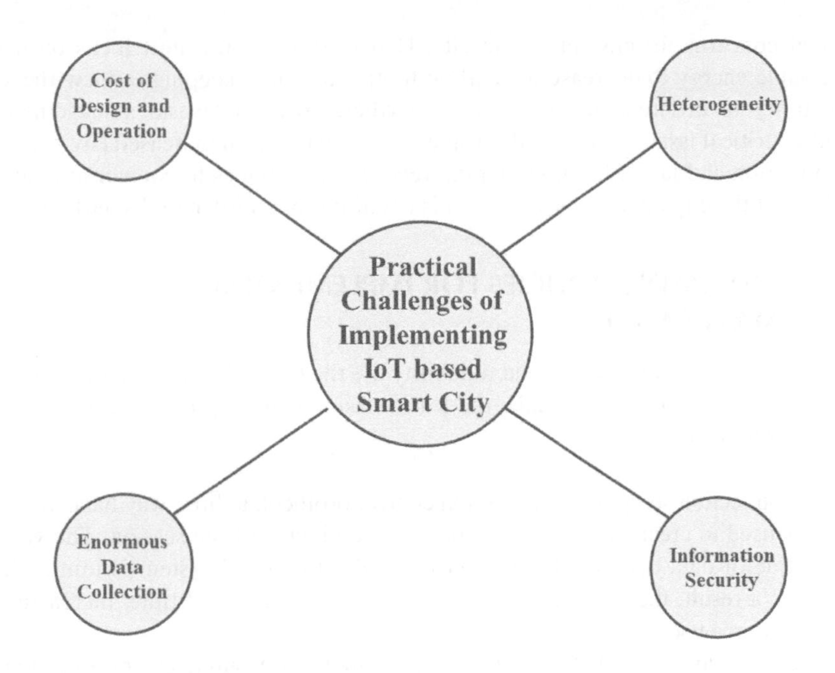

FIGURE 5.7 Practical challenges that exist for IoT-based smart city.

The design and maintenance cost play a crucial role in the realistic smart city implementation. The categorization of the price includes design cost and operational cost. The design cost comprises the money spent on deploying the intelligent city. The lesser the design cost, the greater the probability of its real-world implementation (Trencher & Karvonen, 2019).

Further, operational cost comprises different tasks that also include maintenance cost. To assure the sustainability of any services provided by the smart city, it is desirable to have minimum operating costs. Thus, giving cost optimization for developing a smart city is challenging. Another critical concern for intelligent city architecture is heterogeneity. A smart city comprises different components, including sensors, multivendors, other appliances, and devices. An intelligent city comprises integrating other heterogeneous devices at the application layer (Bar-Magen Numhauser, 2012).

This can cause a problem like incompatibilities. Thus, providing universal access is a challenging task; IoT-based smart cities emphasize providing designing, identifying, and acquiring hardware and software that permits aggregation of different heterogeneous subsystems. A large amount of data, mainly sensitive resident data, is gathered by an intelligent city network that is vulnerable to different security threats. Hence, providing data privacy is another critical feature of any innovative city structure.

The three main challenges for any intelligent city data safety are providing privacy, confidentiality, and trust among multiple users within the intelligent city network. Further, with the bulk usage of devices within the intelligent city framework, handling enormous amounts of data is again a challenging task.

The city environment and resources need to be preserved by focusing on reducing the carbon footprints and optimizing the system resources, which are one of the critical concerns for any intelligent city. Hence, smart cities must focus on using renewable energy to decrease the carbon footprints while keeping in view the sustainability of different city operations. Further, providing waste management is another critical issue for any modern smart city arising from increased environmental pollution and land filling. Designing recovery techniques to counter any failure and revert the city operations to normal is essential for an IoT-based smart city.

5.5 ADDRESSING POLICIES FOR IMPLEMENTING SMART BASED IoT

Congestion control and end-to-end reliability are the main objectives of the transport layer. The following factors make the transmission control protocol unsuitable for IoT environments:

- Connection setup: In transmission control protocol, a three-way handshake is used to create a connection phase, which kicks off the session. The volume of data that will be transferred in the IoT-based system is minimal. As a result, the connection setup lasts a longer period of time, increasing resource use.
- Congestion control: Congestion management is ensured by transmission control protocol. The fact that IoT communications are typically wireless, nevertheless, can lead to performance issues. Finally, because the entire Transmission Control Protocol (TCP) session consists of only transmitting the first segment and receiving subsequent acknowledgments, TCP congestion control isn't very well suited to the IoT context.
- Data buffering: TCP connections use memory buffers to store data at both the source and the destination, necessitating retransmission at first. On the other hand, it is required for ordered delivery on the destination side. In the IoT environment, processing and allocating buffers during TCP connections could be too expensive.

5.6 CONCLUSION

This chapter covers a wide range of topics related to the IoT in smart cities. In this comprehensive introduction to smart cities and their diverse sectors, we present the IoT as a vital accelerator of services related to smart cities and the various intelligent city structures and challenges encountered in the deployment of innovative city applications (Mathad & Khang, 2023).

The use of sensor and networking technology in these applications is covered next, as well as how AI is used in smart cities (Khang & Gupta et al., 2023).

To give a general picture of the current research environment in IoT-based smart cities, we discussed the type of deployment based on the architectures and the technologies that might be offered for each of the scenarios taken into account with each of the crucial components (Hajimahmud & Khang et al., 2023).

To give a general picture of the current research environment in IoT-based smart cities, we debated the type of deployment based on the architectures and the technologies that might be offered for each of the scenarios taken into account with each of the crucial components (Khang, 2023b).

5.7 FUTURE SCOPE

Based on the findings in this chapter, several recommendations for using IoT for intelligent city efforts can be made. The security and privacy of IoT in smart cities is an important study field in terms of data encryption, robust authentication, privacy-preserving approaches, and other measures to prevent unauthorized access to the network of IoT (Jebaraj et al., 2024).

In addition to ensuring end-to-end encryption, blockchain-based solutions may help with access monitoring and tracing, secure device identification, spoofing prevention, and data loss, as was already mentioned (Khang & Chowdhury et al., 2022).

The vast majority of data transmission protocols devised so far for IoT are incompatible (Subhashini & Khang, 2024). Work needs to be done in this area that allows sensor nodes to communicate using a range of methodologies while utilizing the least amount of power possible, crucial for sensor nodes in a network.

The innovation for developing cost-effective storage options and low-power devices is another area to concentrate on. Decentralized systems have been recommended as the best approach for increasing application dependability from a deployment standpoint. Decentralized DL system deployments are possible because of techniques like federated learning. In addition, there is a lot of prospective work in AI (Babasaheb et al., 2023).

To facilitate the use of heterogeneous data sources, data fusion techniques are being developed. Additionally, intelligent information minimization selection strategies are being developed to filter out redundant or otherwise "uninteresting" data throughout the AI design process (Hussain & Khang et al., 2022).

As a result, there will be a shorter turnaround time and improved deployment performance. Current methodologies must be implemented and new ones found to make ML- and DL-based techniques more intelligible to match the different applications together in an intelligent city (Khang and Gupta et al., 2023).

REFERENCES

Ahvenniemi, H., Huovila, A., Pinto-Seppä, I., Airaksinen, M. "What are the differences between sustainable and smart cities?" *Cities*, vol. 60, pp. 234–245 (2017). https://www.sciencedirect.com/science/article/pii/S0264275116302578

Anthopoulos, L.G., Reddick, C.G. "Understanding electronic government research and smart city: a framework and empirical evidence", *Information Polity*, vol. 21, pp. 99–117 (2016). https://content.iospress.com/articles/information-polity/ip371

Atat, R., Liu, L., Wu, J., Li, G., Ye, C., Yang, Y. "Big data meet cyber-physical systems: a panoramic survey", *IEEE Access*, vol. 6, pp. 73603–73636 (2018). https://ieeexplore.ieee.org/abstract/document/8533338/

Awan, F.M., Saleem, Y., Minerva, R., Crespi, N. "A comparative analysis of machine/deep learning models for parking space availability prediction", *Sensors*, vol. 20, p. 322 (2020). https://www.mdpi.com/611558

Babasaheb, J., Sphurti, B., Khang, A. "Industry Revolution 4.0: Workforce Competency Models and Designs", *Designing Workforce Management Systems for Industry 4.0: Data-Centric and AI-Enabled Approaches* (1st Ed.), pp. 14–31 (2023). CRC Press. https://doi.org/10.1201/9781003357070-2

Bar-Magen Numhauser, J. "*Fog Computing Introduction to a New Cloud Evolution*" (2012). University of Alcalá. https://www.academia.edu/download/66537853/mesh_2013_2_20_40038.pdf

Bhambri, P., Rani, S., Gupta, G., Khang, A. "*Cloud and Fog Computing Platforms for Internet of Things*" (2022). CRC Press. https://doi.org/10.1201/9781032101507

Diez, A., Khoa, N.L.D., Makki Alamdari, M., Wang, Y., Chen, F., Runcie, P. "A clustering approach for structural health monitoring on bridges", *Journal of Civil Structural Health Monitoring*, vol. 6, pp. 429–445 (2016). https://link.springer.com/article/10.1007/s13349-016-0160-0

Dutta, J., Chowdhury, C., Roy, S., Middya, A.I., Gazi, F. "Towards smart city: Sensing air quality in city based on opportunistic crowd-sensing", *ACM International Conference Proceeding Series*, 2017. https://dl.acm.org/doi/abs/10.1145/3007748.3018286

Eckhoff, D., Wagner, I. "Privacy in the smart city—applications, technologies, challenges, and solutions", *IEEE Communications Surveys and Tutorials*, vol. 20, pp. 489–516 (2018). https://ieeexplore.ieee.org/abstract/document/8025782/

Farag, S.G. "Application of smart structural system for smart sustainable cities", *Proceedings of the 2019 4th MEC International Conference on Big Data and Smart City (ICBDSC)*, Muscat, Oman, 15–16 January 2019; pp. 1–5. https://ieeexplore.ieee.org/abstract/document/8645582/

Gui, G., Pan, H., Lin, Z., Li, Y., Yuan, Z. "Data-driven support vector machine with optimization techniques for structural health monitoring and damage detection", *KSCE Journal of Civil Engineering*, vol. 21, pp. 523–534 (2017). https://link.springer.com/article/10.1007/s12205-017-1518-5

Hahanov, V., Khang, A., Litvinova, E., Chumachenko, S., Hajimahmud, V.A., Alyar, A.V. "The Key Assistant of Smart City – Sensors and Tools", *AI-Centric Smart City Ecosystems: Technologies, Design and Implementation* (1st Ed.) (2022). CRC Press. https://doi.org/10.1201/9781003252542-17

Hajimahmud, V.A., Khang, A., Gupta, S.K., Babasaheb, J., Morris, G. "*AI-Centric Modelling and Analytics: Concepts, Designs, Technologies, and Applications*" (1st Ed.) (2023). CRC Press. https://doi.org/10.1201/9781003400110

Haverkort, B.R., Zimmermann, A. "Smart industry: how ICT will change the game!" *IEEE Internet Computing*, vol. 21, pp. 8–10 (2017). https://ieeexplore.ieee.org/abstract/document/7839872/

Henry, B. "Commentary on Special Issue Articles". Volume37, Issue4. Issues in Large Scale International Assessment. Pages 57-60. 27 December 2018. https://doi.org/10.1111/emip.12232

Hollands, R.G. "Will the real smart city please stand up? Intelligent, progressive or entrepreneurial?" *City*, vol. 12, pp. 303–320 (2008). https://www.tandfonline.com/doi/abs/10.1080/13604810802479126

Hussain, S.H., Sivakumar, T.B., Khang, A. "Cryptocurrency Methodologies and Techniques", *The Data-Driven Blockchain Ecosystem: Fundamentals, Applications, and Emerging Technologies* (1st Ed.), pp. 149–164 (2022). CRC Press. https://doi.org/10.1201/9781003269281-2

Janssen, M., Luthra, S., Mangla, S., Rana, N.P., Dwivedi, Y.K. "Challenges for adopting and implementing IoT in smart cities: an integrated MICMAC-ISM approach", *Internet Research*, vol. 29, pp. 1589–1616 (2019). https://www.emerald.com/insight/content/doi/10.1108/INTR-06-2018-0252/full/html

Jebaraj, L., Khang, A., Chandrasekar, V., Pravin, A.R., Sriram, K. "Smart City Concepts, Models, Technologies and Applications", *Smart Cities: IoT Technologies, Big Data Solutions, Cloud Platforms, and Cybersecurity Techniques* (1st Ed.) (2024). CRC Press. https://doi.org/10.1201/9781003376064-1

Keane, P.A., Topol, E.J. "With an eye to AI and autonomous diagnosis", *npj Digital Medicine*, vol. 1, pp. 10–12 (2018). https://www.nature.com/articles/s41746-018-0048-y

Khang, A. "*Advanced Technologies and AI-Equipped IoT Applications in High-Tech Agriculture*" (1st Ed.) (2024). IGI Global Press. https://doi.org/10.4018/978-1-6684-9231-4

Khang, A., Chowdhury, S., Sharma, S. "*The Data-Driven Blockchain Ecosystem: Fundamentals, Applications, and Emerging Technologies*" (1st Ed.) (2022). CRC Press. https://doi.org/10.1201/9781003269281

Khang, A., Gupta, S.K., Dixit, C.K., Somani, P. "Data-Driven Application of Human Capital Management Databases, Big Data, and Data Mining", *Designing Workforce Management Systems for Industry 4.0: Data-Centric and AI-Enabled Approaches* (1st Ed.), pp. 113–133 (2023). CRC Press. https://doi.org/10.1201/9781003357070-7

Khang, A., Gupta, S.K., Rani, S., Karras, D.A. "*Smart Cities: IoT Technologies, Big Data Solutions, Cloud Platforms, and Cybersecurity Techniques*" (1st Ed.) (2023). CRC Press. https://doi.org/10.1201/9781003376064

Khang, A., Hahanov, V., Abbas, G.L., Hajimahmud, V.A. "Cyber-Physical-Social System and İncident Management", *AI-Centric Smart City Ecosystems: Technologies, Design and Implementation* (1st Ed.) (2022). CRC Press. https://doi.org/10.1201/9781003252542-2

Khang, A., Hahanov, V., Litvinova, E., Chumachenko, S., Triwiyanto, V.A., Hajimahmud, R.N., Ali, A.V., Alyar, Anh, P.T.N. "The Analytics of Hospitality of Hospitals in Healthcare Ecosystem", *Data-Centric AI Solutions and Emerging Technologies in the Healthcare Ecosystem* (1st Ed.), p. 4 (2023). CRC Press. https://doi.org/10.1201/9781003356189-4

Khang, A. (2021). "*Material of Computer Science, Artificial Intelligence, Data Science, IoT, Blockchain, Cloud, Metaverse, Cybersecurity for Studies*". Retrieved from https://www.researchgate.net/publication/370156102_Material4Studies

Khang, A., Rani, S., Sivaraman, A.K. "*AI-Centric Smart City Ecosystems: Technologies, Design and Implementation*" (1st Ed.) (2022). CRC Press. https://doi.org/10.1201/9781003252542

Khang A., "*Advanced Technologies and AI-Equipped IoT Applications in High-Tech Agriculture*" (1st Ed.) (2023b). IGI Global Press. https://doi.org/10.4018/978-1-6684-9231-4

Khang, A., Vrushank, S., Rani, S. "*AI-Based Technologies and Applications in the Era of the Metaverse*" (1st Ed.) (2023b). IGI Global Press. https://doi.org/10.4018/9781668488515

Khanh H. H., Khang A. "The Role of Artificial Intelligence in Blockchain Applications," *Reinventing Manufacturing and Business Processes through Artificial Intelligence*, 2 (20-40). (2021). CRC Press. https://doi.org/10.1201/9781003145011-2

Koubaa, A., Aldawood, A., Saeed, B., Hadid, A., Ahmed, M., Saad, A., Alkhouja, H., Ammar, A., Alkanhal, M. "Smart palm: "an IoT framework for red palm weevil early detection", *Agronomy*, vol. 10, p. 987 (2020). https://www.mdpi.com/765104

Kurdachenko, L.A., Semko, N.N., Subottin, I.Y. "The Leibniz algebras whose subalgebras are ideals, open math", *Open Mathematics*, vol. 15, pp. 92–100 (2017).. https://www.degruyter.com/document/doi/10.1515/math-2017-0010/html

Kwon, J.-H., Kim, E.-J. "Failure prediction model using iterative feature selection for industrial internet of things", *Symmetry*, vol. 12, p. 454 (2019). https://www.mdpi.com/663728

Lenka, R.K., Barik, R.K., Das, N.K., Agarwal, K., Mohanty, D., Vipsita, S. "PSPS: an IoT based predictive smart parking system", *Proceedings of the 2017 4th IEEE Uttar Pradesh Section International Conference on Electrical, Computer and Electronics, UPCON 2017, Mathura, India*, 26–28 pp. 311–317. October 2017; https://ieeexplore.ieee.org/abstract/document/8251066/

Mathad, K., Khang, A. "Hospital 4.0: Capitalization of Health and Healthcare in Industry 4.0 Economy", *Data-Centric AI Solutions and Emerging Technologies in the Healthcare Ecosystem* (1st Ed.), p. 14 (2023). CRC Press. https://doi.org/10.1201/9781003356189-19

Pardini, K., Rodrigues, J.J., Kozlov, S.A., Kumar, N., Furtado, V. "IoT-based solid waste management solutions: a survey", *Journal of Sensor and Actuator Networks*, vol. 8, p. 5 (2019). https://www.mdpi.com/388504

Rahman, A., Hossain, M.S., Alrajeh, N.A., Guizani, N. "B5G and explainable deep learning assisted healthcare vertical at the edge: COVID-19 perspective", *IEEE Network*, vol. 34, pp. 98–105 (2020). https://ieeexplore.ieee.org/abstract/document/9136600/

Rana G., Khang A., Sharma R., Goel A. K. , Dubey A. K. "Reinventing Manufacturing and Business Processes through Artificial Intelligence", (Eds.) (2021). CRC Press. https://doi.org/10.1201/9781003145011

Rani, S., Bhambri, P., Kataria, A., Khang, A. "Smart City Ecosystem: Concept, Sustainability, Design Principles and Technologies", *AI-Centric Smart City Ecosystems: Technologies, Design and Implementation* (1st Ed.) (2022). CRC Press. https://doi.org/10.1201/9781003252542-1

Rani, S., Bhambri, P., Kataria, A., Khang, A., Sivaraman, A.K. "*Big Data, Cloud Computing and IoT: Tools and Applications*" (1st Ed.) (2023). Chapman and Hall/CRC. https://doi.org/10.1201/9781003298335

Rani, S., Chauhan, M., Kataria, A., Khang, A. "IoT Equipped Intelligent Distributed Framework for Smart Healthcare Systems", *Networking and Internet Architecture* (2021). CRC Press. https://doi.org/10.48550/arXiv.2110.04997

Rojek, I., Studzinski, J. "Detection and localization of water leaks in water nets supported by an ICT system with artificial intelligence methods as a way forward for smart cities", *Sustainability*, vol. 11, p. 518 (2019). https://www.mdpi.com/398364

Sánchez Marrero, O., Mohamed Amar, R., & Xifra Triadú, J. "Habilidades blandas: necesarias para la formación integral del estudiante universitario", *REVISTA CIENTÍFICA ECOCIENCIA*. 2018, 5, 1–18. https://doi.org/10.21855/ecociencia.50.144

Saraswat, A., Kalra, B. "Safe engineering application for detection of medical image using deep convolutional neural network", *Journal of Green Engineering*, vol. 10, no, 11, pp. 12523–12535 (2020). https://link.springer.com/article/10.1007/s11042-022-12168-9

Saraswat, A., Sharma, N. "Bypassing confines of feature extraction in brain tumor retrieval via MR images by CBIR", *ECS Transactions*, vol. 107, no. 1, pp. 3675–3682 (Apr. 2022). doi: 10.1149/10701.3675ecst. https://iopscience.iop.org/article/10.1149/10701.3675ecst/meta

Saraswat, A., Sharma, N. "Salvaging tumor from T1-weighted CE-MR images using automatic segmentation techniques", *International Journal of Information Technology* (May 2022). doi: 10.1007/s41870-022-00953-6

Sevillano, X., Màrmol, E., Fernandez-Arguedas, V. "Towards smart traffic management systems: Vacant on-street parking spot detection based on video analytics", *Proceedings of the FUSION 2014—17th International Conference on Information Fusion, Salamanca, Spain*, 7–10 July 2014. https://ieeexplore.ieee.org/abstract/document/6916135/

Silva, B.N., Khan, M., Han, K. "Towards sustainable smart cities: a review of trends, architectures, components, and open challenges in smart cities", *Sustainable Cities and Society*, vol. 38, pp. 697–713 (2018). https://www.sciencedirect.com/science/article/pii/S2210670717311125

Subhashini, R., Khang, A. "The Role of Internet of Things (IoT) in Smart City Framework", *Smart Cities: IoT Technologies, Big Data Solutions, Cloud Platforms, and Cybersecurity Techniques* (1st Ed.) (2024). CRC Press. https://doi.org/10.1201/9781003376064-3

Trakadas, P., Simoens, P., Gkonis, P., Sarakis, L., Angelopoulos, A., Ramallo-González, A.P., Skarmeta, A., Trochoutsos, C., Calvo, D., Pariente, T., t al. "An artificial intelligence-based collaboration approach in industrial IoT manufacturing: key concepts, architectural extensions and potential applications", *Sensors*, vol. 20, p. 5480 (2020). https://www.mdpi.com/837848

Trencher, G., Karvonen, A. "Stretching smart: advancing health and well-being through the smart city agenda", *Local Environ*, vol. 24, pp. 610–627 (2019). https://www.tandfonline.com/doi/abs/10.1080/13549839.2017.1360264

Vijarania, M., Dahiya, N., Dalal, S., Jaglan, V. "WSN Based Efficient MultiMetric Routing for IoT Networks", *Green Internet of Things for Smart Cities*, p. 249262 (2021). CRC Press. https://www.taylorfrancis.com/chapters/edit/10.1201/9781003032397-16/wsn-based-efficient-multi-metric-routing-iot-networks-meenu-vijarania-neeraj-dahiya-surjeet-dalal-vivek-jaglan

Vitola, J., Pozo, F., Tibaduiza, D.A., Anaya, M. A. "Sensor Data Fusion System Based on k-Nearest Neighbor Pattern Classification for Structural Health Monitoring Applications", Sensors. 2017, 17, 417. https://doi.org/10.3390/s17020417

Wang, Y., Ram, S., Currim, F., Dantas, E., Sabóia, L.A. "A big data approach for smart transportation management on bus network", *Proceedings of the IEEE 2nd International Smart Cities Conference: Improving the Citizens Quality of Life, ISC2 2016 – Proceedings*, Trento, Italy, 12–15 pp. 1–6. September 2016. https://ieeexplore.ieee.org/abstract/document/7580839/

6 Traffic Control Mechanism Using Internet of Things (IoT) in Smart City

Akanksha Tandon, Arun Kumar, and Sanjeev Patel

6.1 INTRODUCTION

The Internet of Things (IoT) is a unique form of communication. Therefore, communication is transmitted via network technology and a globally unique code. Furthermore, the data can be delivered automatically based on the time criteria (Rani & Chauhan et al., 2021).

Intelligent transportation systems (ITSs), on the other hand, are becoming increasingly common as the world population grows and their use of vehicles increases. Compared to the time required to handle urban traffic, automotive pollutants may result in significant harm. As a result, implementing creative solutions using IoT to decrease the numeral of automobiles on the highway and enhance gridlock management in municipalities is vital (Khang & Gupta et al., 2023).

Furthermore, IoT may improve traffic management systems (TMSs) in several ways, such as exchanging information, estimating traffic volume, ensuring safety, and lowering traffic load. To avoid disturbing urban connectedness, one of the critical alternatives in this region is to build one-way or two-way highways.

Dudhe and Kadam et al. (2017) proposed Kevin Ashton developed the concept of the IoT in 1999. Ashton developed everything at their workplace. Throughout their tenure there, Ashton devised the concept of embedding a Radio Frequency Identification (RFID) card within every makeup (lipsticks), enabling each to connect with a transceiver.

Ashton claimed that specific incoming data might be used to tackle a wide range of real-world challenges. Currently, numerous integrated communications are done with the World Wide Web and cell phones, in addition to potentially including some comparable items. However, many cannot communicate due to labeled software and equipment, different expectations, cultures, and interoperability. To communicate with many of these existing intelligent home appliances, one needs to access different Internet resources, rather than the developer specifically intending products to perform simultaneously.

K. Rose explained how IoT was feasible around 2015. Rose stated that it is feasible for the purposes mentioned: widespread networking, broad use of infrastructure

DOI: 10.1201/9781003376064-6

connectivity, computational affordability, breakthroughs in business intelligence, and the emergence of virtualization are all instances of this. As a result, the IoT represents unique convergence with several computer and communication concepts emerging for millennia. Humans are witnessing the beginning of a new era of digital technology known as the IoT.

IoT is a cloud-based universal global cellular automaton that links peripherals. The IoT is an infrastructure of things consisting of gadgets and systems consisting of automated systems interacting and connecting with other types of machinery, surroundings, artifacts, and systems. Wireless identification and detector networks typically move ahead to meet this new dilemma (Bhambri & Khang et al., 2022).

Consequently, increasingly vast amounts of information were produced, collected, and translated into practical actions which may "control and regulate" various products and technologies, helping make our everyday lives considerably smoother and healthier and reducing human contribution to the surrounding.

Traffic congestion in industrialized and metropolitan centers worldwide, particularly during rush hour, remains a severe problem for residents, commuters, municipal politicians, and urban planners. Traffic congestion has several effects and repercussions, from rising polluted air and gasoline use to sluggish industrial prosperity. Like vehicles on the road will rise, and traffic congestion will add more hurdles and challenges to economic growth while severely harming the quality of life (Khang & Hajimahmud et al., 2022).

Consequently, individuals will naturally want to migrate to less congested, safer, and effective regions. These provide substantial an assessment for smart city planners in terms of designing an effective TMS capable of meeting the needs of a large metropolitan region while being effortlessly adaptive to its inhabitants' development.

Present advancements in wireless technology, particularly RFID, have considerably impacted traffic control. This method is profitable and accessible for self-regulating vehicle identification. RFID can significantly lower the cost of expensive infrastructure. RFID has numerous additional applications, but one of its most influential has been in traffic control systems. RFID is also used in smart traffic regulating areas such as vehicle identification and safety. The overview of the TMS is shown in Figure 6.1.

Another technique has been proposed for reducing the traffic or congestion that deploys mobile-based TMS. The TMS monitors and manages traffic jams on roadways. This method has seven levels, ranging from the personal smartphone to the topmost layer business model, and eventually determines all elements of traffic on highways.

In order to allow the mobile-based application to get details of traffic jams on certain highways, Badura and Lieskovsky (2010) have proposed a novel smart traffic system concept.

Sensors stationed at intersections scan and monitor respective target area. After that collected information is promptly transmitted to an information sending system for general picture processing. Job of this information sending system is to offer a transmission structure and assure information transfer across a mobile ad-hoc network.

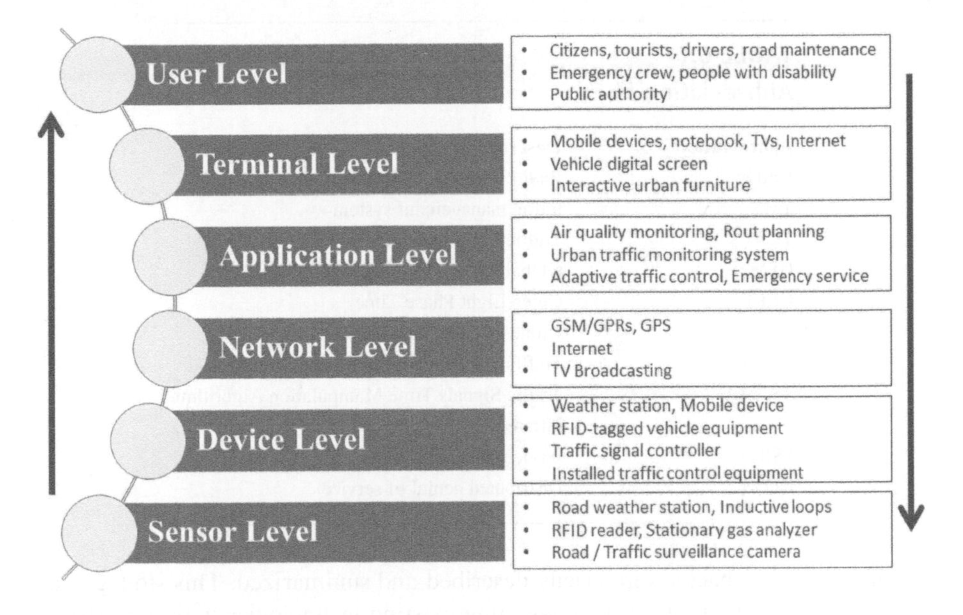

FIGURE 6.1 IoT in traffic management overview (Khang, 2021).

Salama and Saleh et al. (2010) have proposed a photoelectric sensor to regulate traffic signals. Fixing the precise location of sensor placement is one of the most critical factors. This period is primarily due to the TMS department's desire to observe vehicle movement at specified times, particularly during peak hours.

The collected data is communicated to a traffic management base station, where collected information through sensors is entered into a mechanism with a relative weight entrusted to all road (Rani & Khang et al., 2023).

Using the result, the technology will clear the traffic lanes for the roads that are highly congested than the opposite side and provide more time to relieve traffic jams than the less blocked roads. In extraordinary instances, this method may allow for human involvement in controlling traffic congestion.

Bathula and Ramezanali et al. (2009) have proposed that the combination of magnetic and infrared detectors can enhance vehicle detection approaches and efficiency. Their biggest challenge with this procedure is the trajectory of arriving automobiles, particularly in industrial zones.

Bugdol and Segiet et al. (2014) have proposed magnetometer detectors in ITSs when a vehicle comes to a magnetic detector, the detector's local magnetic field is somewhat altered. When the vehicle is in the center, the entire local magnetic flux is twisted, and the detector creates the final section of the magnetic field. Following that, the data acquired from the detectors is transferred to a controller. This controller's primary role is to evaluate incoming traffic's direction to manage Traffic Light Systems (TLS) at many junctions at the same time (Khang & Gupta et al., 2023).

The main objective of this chapter is to examine several intelligent traffic management methods. RFID readers and tags, Green Wave Systems, smartphones, and a Big Data center wireless connectivity were all used. Each approach's applications,

TABLE 6.1
Abbreviation Used

Abbreviation	Description
RFID	Radio Frequency Identification
TMS	traffic management system
TLS	Traffic Light Systems
GSM	Global System for Mobile Application
GLPT	Green Light Phase Time
WSN	wireless sensor network
TSCA	Traffic System Communication Algorithm
TSTMA	Traffic Signals Time Manipulation Algorithm
IR	infrared sensor
SSID	service set identifier
DDoS	Distributed denial of service

benefits, and drawbacks were briefly described and summarized. This strategy can aid in resolving the problem by combining existing technologies with the existing infrastructure. Our goal is to examine several approaches for controlling traffic, including TLS: static and dynamic TLS, RFID, and the IoT.

In the latter technique, traffic data is immediately collected and delivered to Big Data for processing. Mobile applications, also known as user interfaces (UI), assess traffic density in various places and provide alternate methods to ease traffic congestion. In the following order, our chapter has been presented. Related works are addressed in the second section. In the third section, IoT Device Development for TMS is discussed in detail. In the fourth section, Wireless Technologies for Traffic Management are presented. The conclusion of this research is presented in the last section. Abbreviations and Descriptions are shown in Table 6.1.

6.2 TRAFFIC CLASSIFICATION IN THE DOMAIN OF IoT

Nellore and Hancke (2016) have proposed the current state of the intelligent city transportation management system. This technique for determining traffic obstruction on streets using photo preparation methods and a prototype for managing traffic lights based on data obtained from camcorder pictures of streets.

Rather than calculating the number of vehicles, separate the traffic viscosity that relates to the collective number of automobiles on the highway in parameters of both the accumulation which establishes the following factors for every roadside: concession, changeable congestion pattern, and proportional period based on traffic viscosity, and operating traffic lights sequentially, which is both time-consuming and expansive for careful traffic management. However, the problem with these frameworks is that they need a significant amount of time and money to set up and maintain.

Miz and Hahanov (2014) have proposed another solution known as RFID that has been offered, where it could be combined with the present signaling system to act

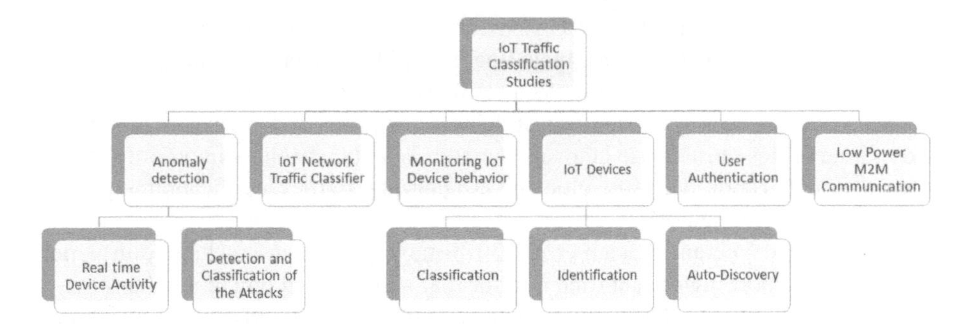

FIGURE 6.2 Difficulties investigated by traffic classification in the IoT area (2020).

as a keystone to intelligent transportation management. Compared to other traffic congestion management methods, this new invention will take less time to set up and will cost less money.

The use of this innovative technology will result in less traffic congestion. Bottlenecks will be identified early, so early preventive actions may be implemented, saving the driver time and money as Figure 6.2.

Circumstances that fluctuate regularly from those that occurred when the junction was analyzed in these fixed-time models. Furthermore, fixed-time traffic signal approaches frequently become obsolete when traffic patterns vary dynamically. This phase entails resurveying the region and developing new signal timing plans. In retrospect, this procedure is extensive, and resources that are not always readily available are required.

The shortcomings of most fixed-time systems demonstrate the need for a more responsive response to changing traffic circumstances. Studer and Ketabdari et al. (2015) have proposed terms of traffic management, one such efficient option is the adaptive traffic control system. This phase significantly improves fixed-time structures since it means more vital decision-making abilities.

Traffic congestion has long been a feature of every existing metropolitan region with a large population but limited infrastructure. Traffic congestion is a serious issue with severe causes and effects on the road. Vehicles have grown tremendously due to rapid population expansion and poor public transit quality. Apart from infrastructure, poorly regulated traffic causes congestion that can last for hours. A predetermined time plan for traffic control can only help to a limited extent. It must outperform its value as a fixed delay unit regardless of traffic level on the predefined timed signal, which causes additional traffic to build up on other intersecting lanes.

Jain and Neelakandan et al. (2022) have proposed to increase traffic congestion driven by poor outdated traffic control systems that operate on a predetermined count. Such outdated technology assigned time independent of traffic intensity on a specific route, resulting in massive delays. The technology ensures that traffic lights adapt to current traffic levels, properly managing time and resources. Such information is provided by a Raspberry Pi, which sequentially monitors the traffic light. Additionally, all collected information is transmitted to the Internet and may be used to periodically monitor traffic flow.

According to Manjunatha and Chandrashekar et al. (2019), traffic congestion is one of the worst urban situations. Including its rapid development, managing a city's transportation network is difficult. As a result, traffic congestion on major roadways causes various issues, including transit delays, accidents, and environmental pollution. Smart cities must devise innovative methods and technology in order to alleviate traffic congestion. The difficulties investigated by traffic classification in the IoT area are shown in Figure 6.2

Janahan and Veeramanickam et al. (2018) have presented a technique in which we allocate a more significant time rate for that signal based on the number of cars in the roadside IR data. The detectors have infrared sending and capturing and may be established in both directions of the gridlock lane.

An IoT-based traffic transmission system is based on an ultrasonic sensor, with ultrasonic sensors deployed over 50 meters of roadway to collect traffic density and connected with a micro-controller to regulate traffic signals. Occupancy data is delivered over enabled devices to the micro-controller board and evaluated on high and light traffic with time and date. The same information is transferred to a cloud-based homepage that can be accessed.

Installing an electronic circuit inside the emergency vehicle requires an Arduino Micro controller, a GPS for monitoring the vehicle, and a GSM for maintaining the ambulance's position on an online database. Particularly for ambulances, the system tracks the ambulance's journey. It manages traffic signals to keep road intersections clear of traffic, allowing vehicles to pass quickly and without delay.

However, this idea does not account for traffic density and the time necessary to reach the traffic light, which may result in early road evacuation in low density and, as a result, inaccuracy in signal timing, leading to greater density in other directions. Currently, the sensors used in traffic data collection are mostly picture sensors, electromagnetic looped detection, sound detectors, earthquake detectors, and magnetic sensor types.

The camera system collects a significant amount of information. However, it is susceptible to severe weather and poor performance. Furthermore, the camera system necessitates more critical system requirements. In certain installation circumstances, seismic and acoustic sensors are sensitive to noise. Because a vehicle is ferrous, the intensity of the Earth's electromagnet enveloping it is altered. As a result, we use the magnetic sensor HMC1021Z to identify vehicles.

Vehicle tracking is a critical system function that allows automobiles to be detected fast and precisely so that vehicle information and more extensive traffic data are collected. Gu and Jia et al. (2005) and Ding and Cheung et al. (2004) have proposed before using a limit to identify an automobile, a filter was used to assess the original data. When there is little disturbance, such approaches operate effectively. Nevertheless, assuming the transmission proportion is poor. In this scenario, these strategies will significantly influence identification accuracy, including elevated spoofing attacks and missed malware detection, resulting in poor effectiveness.

Ghena and Beyer et al. (2014) have studied and discovered a wireless TLS security issue such as unsecured 5.8 GHz and 900 MHz radio-based communications that wireless routers use by creating default usernames and passwords and by opening controller debugging ports using a computer and a wireless adapter.

Anyone, capable of transmitting wireless radio signals at 5.8 GHz, might readily access the radio-based TLS by obtaining its service set identifier (SSID).

Descriptive taxonomy on existing IoT connectivity segmentation and designing methodologies Cyberattacks have displayed a wide range of variations in current history and continue to be investigated using a number of techniques. There is no massive distinction in attack techniques comparing standard Cyberattack and tailored difficulties investigated by traffic classification in the (DDoS). Several malware techniques incorporate underhanded strategies to expose vulnerabilities in either conventional networks or intelligent systems (Khang & Hahanov et al., 2022).

However, due to the variety of connected devices, specific Cyberattacks are much more sophisticated and complex. As a consequence of significant variations in the performance parameters between connected devices with information produced through other equipment, many approaches and techniques for IoT traffic classification have recently been suggested.

Chong and Ng (2016) have proposed that a traffic light may be used as signaling equipment installed at a crossroad. John Peake Knight created the first manually controlled traffic light in 1868, while Lester Farnsworth Wire invented the first automated traffic signal in 1912. By managing vehicle movements, traffic lights serve two primary functions: preventing vehicle accidents and reducing traffic congestion.

Misbahuddin and Zubairi et al. (2015) have proposed that all metropolitan cities experience traffic congestion issues, particularly downtown regions. The IoT paradigm has the potential to be crucial in the development of creative cities.

Sharif & Li & Khalil et al. (2017) proposal suggest a low-cost future STS that will improve service by implementing immediate traffic updates. Low-cost vehicle detection sensors are set in the center of the road every 500 meters.

IoT is being used to swiftly get public traffic data and deliver it to data processing. Real-time streaming data is shown for Big Data analytics. Various analytical texts may estimate traffic density and provide remedies using predictive analytics and taking advantage of the IoT and Big Data. App-based traffic updates, vehicle strength on the road, and other features are user-friendly. These technologies' interplay enables a completely IoT-based automobile information-collecting system. For optimal results, install as near the road as possible (12 km or 1 km and more).

At a time, only five devices are connected and communicate with a unique Internet module. Each module is the Internet's backbone and may exchange information. It keeps looking for automobiles and refreshes the vast data storage and analytics system. It records both information from sensors and the device ID. Compute all of the data by doing insights. Various aspects are taken into account when assessing each sensor's capability, connecting one another and detector information, and providing automobile details concerning road space. External detection devices are indicated every 500 meters in the middle of the road.

Abdellah and Koucheryavy (2020) have suggested that the IoT is a network of interconnected devices, such as sensors and intelligent gadgets that have processing, sensing, and communication capabilities, as well as the ability to send data to one another via the Internet, a central console. These technologies' interplay enables a completely IoT-based automobile.

Predicting network activity is a critical management and operational responsibility for any data connection. It is critical in today's highly complicated and diversified networks. Network traffic prediction is becoming increasingly crucial for IoT networks to deliver continuous connectivity. The artificial neural network (ANN) anticipated traffic correctly (Rana & Khang et al., 2021).

6.3 IoT DEVICE DEVELOPMENT FOR TRAFFIC MANAGEMENT SYSTEM

With the rapid expansion of modern cities and the ever-increasing number of cars, IoT has become increasingly important. Traditional traffic information detection systems cannot meet the demands for deployment ease, detection accuracy, and total cost.

Wireless sensor networks (WSNs) combined with embedded systems, wireless communication, and microelectronics technology provide a wide range of possible applications. WSN offers several benefits, including no wiring and low cost. It is progressively being used in data collection and detection domains, such as intelligent traffic systems, with gratifying results. According to research, WSNs have exceptional characteristics that can provide an efficient measure for data collecting in intelligent traffic systems. Fuzzy control is particularly well suited to control city traffic systems with high unpredictability, where no mathematical model is required.

Fuzzy theory is effective at processing unclear information, but the rules it generates are incredibly crude. The neural network excels at nonlinear mapping, which improves regulations through learning and the efficient enhancement of control precision. Provide a low-cost, high-utility rank management system for traffic arteries based on integrating wireless sensors and fuzzy neural network research. WSN is in charge of identifying the number of cars accessing the crossing in all directions on the traffic artery to achieve wireless real-time communication.

6.3.1 Dynamic Traffic Light Management System Based on WSN

Ad-hoc architecture comprises on-road IoT systems, on-vehicle wireless sensors, and hybridized on-vehicle wireless sensors. On-road sensor networks need every sensor to communicate with one another in an ad-hoc fashion without using the infrastructure (Hajimahmud & Khang et al., 2023).

On the vehicle sensor network, sensors incorporated inside the vehicles were used to facilitate direct automotive communication. The hybridized sensor to the automobile network incorporates both modes of communication. Connectivity structure, comparable ad-hoc design, contains three sensor communication networks. On-road monitoring systems, on-vehicle wireless sensors, and hybridized on-vehicle wireless sensors incorporating a central station (CS) are all managed by a CS. Architectures of the WSN system are shown in Figure 6.3.

6.3.2 Current Traffic Light System

Gottlich and Herty et al. (2015) have proposed two most extensively used dynamic cycle TLS worldwide. For each route, Green Light Phase Time (GLPT) is generated using traffic volume models: splitting, compensation, and throughput times.

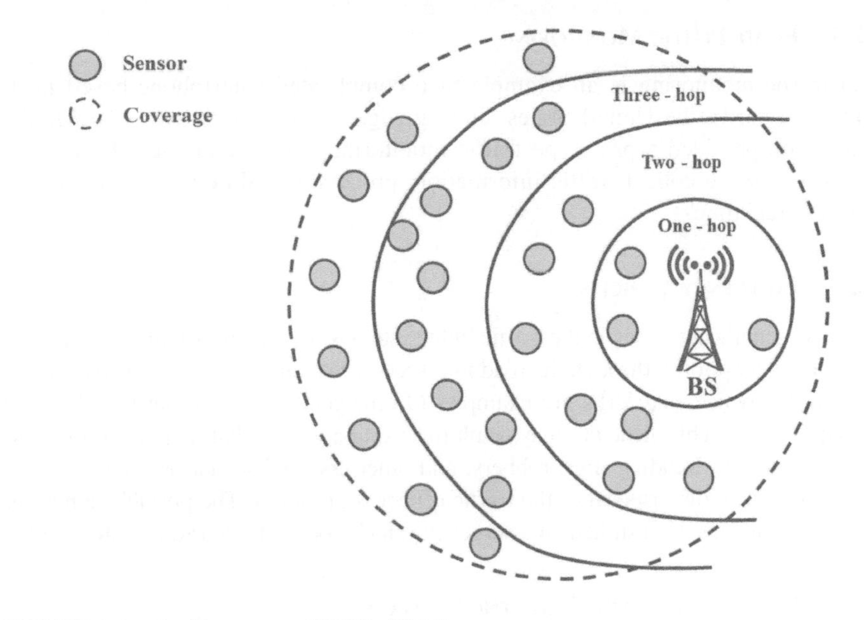

FIGURE 6.3 Architectures of WSNs (2006).

Such metaheuristics can eliminate excessive signal timing at crossroads and integrate nearby indicator systems to minimize delays and congestion.

6.3.3 Calculation of Phase Time of Green Lights

GLPT optimization uses various techniques. Yousef and Al-Karaki et al. (2010) have proposed calculating GLPT using the Traffic System Communication Algorithm (TSCA) and the Traffic Signals Time Manipulation Algorithm (TSTMA).

Collotta and Pau (2015) have proposed a dynamic management algorithm. Then, Fleck and Cassandras et al. (2015) have suggested gradient-based optimization using infinitesimal perturbation analysis (IPA). For GLPT computation, a fuzzy neural network technique has also been developed. Aside from that, ensemble-based systems Pescaru and Curiac (2014) have proposed for GLPT computation. Zhang and Tan et al. (2014) have developed a multi-objective particle swarm optimization approach for GLPT calculation based on crowding distance.

6.3.4 Use of Wireless Controller

During busy hours, traffic officers will be dispatched to each intersection to manually control traffic flow by disabling TLS. As a result of this policy, the police department must dedicate a large number of officers during peak hours, putting their health in danger due to air pollution. Tubaishat et al. (2008) suggest a wireless traffic light controller for police personnel during peak hours, aiding traffic light management. The controller, located next to the road intersection, has two modes: manual and automated. Traffic light control is provided via manual mode, while automated mode changes the traffic light sequence depending on pre-defined settings.

6.3.5 Road Traffic Monitoring

Road traffic monitoring is an example of a complicated smartphone-based measuring system. In the United States, cutting-edge research initiatives like Mobile Millennium provided a prototype traffic monitoring system that used the GPS in cellular phones to collect traffic information, process it, and deliver it back to the phones in real time.

6.3.6 Emergency Vehicles

One of the significant issues in the traffic light system is the passage of emergency vehicles as a greater priority through the road intersection. Ambulances, rescue vehicles, fire departments, police, and VIPs are examples of emergency vehicles that may become trapped in traffic. This difficulty may result in several issues, including patient transport injuries, accidents, building fires, robbery, and other essential scenarios. The handheld portable device at the disposal of the traffic officer is proposed. The portable controller might be modified to be installed on emergency vehicles or used in a traffic control center.

6.3.7 User Interface for Emergency Vehicle

An interactive interface for the used car is also accessible, via which the emergency vehicle driver may alter the vehicle's priority. If the ambulance is empty, the priority level will be low, as it is in most circumstances. This interface also assists the driver in determining which intersections the emergency vehicle will travel through. The priority of the car, its position, and the total number of crossings to be passed through are all supplied to the system through a GSM module. The system's GSM module receives this data, updating the database.

6.3.8 Traffic Management

When adopting IoT technology in transportation, the primary priority is to solve traffic problems. Road congestion via the IoT provides for increased capacity of municipal roadways without the need for more roads, which is crucial in the transition to smart cities (Khang and Vrushank et al., 2023).

Use sensors, cameras, routers, and cellular technology to dynamically modify controls such as traffic lights, highway exit counts, expressway bus lanes, highway message boards, and even speed limits to optimize traffic flow and keep traffic safe. In addition, Ford recently introduced traffic bottleneck assist, a system that enables the vehicle to match the speed of cars ahead of traffic jams (Khang & Hahanov et al., 2022). This speed matching facilitates driving by smoothing traffic flows and reducing congestion patterns. These technologies will be able to actively control automobiles in the future, such as halting them at crossings to avoid collisions with pedestrians or other cars.

6.3.9 Toll Collection and Ticketing

The volume of massive transport on the roadways has grown, and queues at highway toll booths have become more typical. While RFID-enabled tolls have cut wait times, further gains are only achievable with IoT technologies (Khang & Abdullayev et al., 2023). Now link a contemporary automobile to the IoT, allowing it to be detected up to a km from the payment location, instantly removing the fee and raising the barrier. Expenses can also be deducted from a phone-linked digital wallet.

6.3.10 Transport Connection

As previously stated, all trendy automobiles are capable of connecting to the IoT. IoT-based monitoring technologies are already being used in the passenger and freight transportation industries to assist managers in managing their fleets more economically. Surveillance systems also help in the observation of vehicle safety as well as the collection of information on downtime and driving style.

6.3.11 Traffic Changes in Real Time

Horng and Cheng (2016) have proposed that real-time traffic monitoring systems are critical to the move to smart cities. Intelligent systems for traffic management are centered mainly on the IoT paradigm; in addition, Sumi and Ranga (2018) have suggested that high flexibility sensors are employed to monitor the movement of vehicles, forecast overcrowding, effectively manage urban transport, and also have proposed independent traffic detection, which is situated at the core of urban planning structures. This medium will increase awareness, allowing for more efficient resource and infrastructure uses.

Dubey and Rane et al. (2017) have proposed, based on these measures, smartphones, radio, television, light signals, dynamic, changing message signs, and display systems issue traffic warning messages. Atta et al. (2020) have proposed recent advancements in legitimate road conditions that have incorporated overcrowding estimates to regulate street lights automatically.

Nagmode et al. (2017) have developed a technology based on a real-time detection mechanism enabling traffic intensity flexible congestion control. This proposed system makes use of obstacle detection and is split into two components: automobile surveillance with a management accountant.

Systems monitor automobiles and the concentration amounts of a particular pathway, which are transmitted to a display before even being transmitted to a database for subsequent reference. Talukder et al. (2017) have proposed an ultrasonic sensor-based system model for road junctions. Aside from traffic signal lighting, the system alerts on any erroneous vehicle activity, such as crossing red lights. Another study Javaid and Sufian et al. (2018) has proposed an IoT-based intelligent TMS to control real-time traffic via central and local servers. Sensors, cameras, and RFIDs are used at the data collection layer.

The application layer regulates the traffic signal automatically based on traffic density and gives a daily report via a web application. Nellore and Hancke (2016) have proposed another scientific advancement in this field that is the Internet of linked vehicles, which collects real-time traffic data. Individual vehicle monitoring is enabled by linked automobiles, allowing for effective emergency vehicle management. Tian and Gao et al. (2019) have proposed that integrating roadside equipment like traffic signals with the vehicular network ensures traffic events' reliability. Managing emergency vehicles is essential, as every second counts due to the urgency of their services.

To enhance response time, emergency vehicle scheduling can be automated by regulating traffic lights. These technologies, however, are developed exclusively for roads. KumaraSwamy and Manjula et al. (2014) have proposed that, in addition to sensors, video surveillance is used to assess traffic congestion density and update traffic lights in real time.

6.3.12 Vehicle Tracking Devices

RFID determines location using radio frequency and implanted microchips. These tags can tell you what sort of automobile it is, where it is heading, and how fast it is moving. By merging IoT technologies with existing infrastructure, ITSs can assist in addressing the traffic problem. Consider an intelligent traffic signal monitoring system based on IoT. It modifies traffic signals depending on past knowledge supplied by sensing devices with the time interval specified by the number of automobiles on that particular road stretch.

The main advantage of this system is that it may reduce downtime and traffic congestion in various locations. Due to IoT technology, sensors and beacons may be put in automobiles, trains, street lights, bus stops, and railway platforms. Furthermore, transportation companies will be able to analyze data flowing through IoT devices to improve the quality and efficiency of their services. Construction work zones are dangerous places. Motorists are exposed to unusual conditions in an otherwise familiar atmosphere, and such unanticipated unfamiliarity may cause drivers to behave in unexpected ways.

Construction project sites are often extensive projects which run for several months. These buildings typically have various sensing and monitoring technologies that detect vehicle speed and unexpected document events. Infrastructure maintenance has been frequently finished in a few weeks or months. Empowering these work zones with the same machinery used in motorway building regions is not commercially viable.

6.4 WIRELESS TECHNOLOGIES FOR TRAFFIC MANAGEMENT

Currently, a knowledge counter seems to be required at every metro station, commercial center, and university to provide information about railroad timetables, attractive schemes, and other relevant notices. From an institution of higher education, the issue is that it demands certain people devoted to that aim and must remain current with that institute and the latest developments inside the university. The second issue is that in order to retrieve data again from the organization, an individual must visit the reference desk.

The alternative seems to be to leverage technologies that make computers accountable for answering every one of the questions posed by individuals. Smartphones, which are accessible and available to everybody and can be linked to the Internet to access the most recent information, are indeed the perfect device. Whereas if knowledge is just not refreshed through the use of the computer, in such circumstances when the knowledge is not maintained via the World Wide Web, customers must reach out to the care center for assistance.

Several writers created a gadget with all the data captured within the repository; if one development process progresses, people must use that gadget and obtain relevant material. In order to function, any gadget should be accessible to every customer who requires assistance. Most firms have a service counter that delivers data, media advertising, and numerous reminders to consumers and employees. Another issue is that the process necessitates certain people committed to that aim and should be primarily conversant in the offers advertised and, indeed, the organization (Khang & Gujrati et al., 2023).

People can observe numerous intelligent gadgets in their surroundings due to cloud computing. Several individuals believe communities and the globe will be layered with monitoring and movement, with so many incorporated within "materials," resulting in what has been known as a digital environment. Plenty of others throughout the globe have already performed similar research. The IoT refers to dynamically interconnected gadgets and systems that conduct research through smart wearables, controllers, and other artifacts (Sharma & Tiwari, 2016). IoT is projected to grow in subsequent years, introducing a new aspect of applications that generate customer life satisfaction and company efficiency (Babasaheb et al., 2023).

Wireless carriers provide connection to a diverse range of devices, allowing the creation of more new products and services. This new generation of global connection is expanding beyond smartphones and tablets to include linked vehicles, structures, monitoring devices, and collision avoidance, with the potential to interconnect nearly everything and anyone automatically.

The IoT can potentially improve both human life satisfaction and company efficiency significantly. IoT seems to have the ability to allow additional features as well as improvements to essential facilities in transit, procurement, confidentiality, infrastructure, and professional training. Universal health care, and many other zones throughout a commonly available, domestically intellectual network of connected devices, while somehow established a new environment for enterprise applications (Snehal & Babasaheb et al., 2023).

Coordinated efforts are required to bring the sector through the initial phases of sales growth and into sustainability, underpinned by the shared awareness of the peculiar characteristics of the possibility. This industry has distinct features in terms of product dissemination, marketing, billing methods, Internet service delivery capability, and the varying requirements these applications will have on cellular operators.

Intelligent traffic control system design is a current research area. Researchers worldwide are developing novel techniques and technologies to address this problematic issue. Models based on mathematical equations estimate the time a car waits in conjunction with the extension of the waiting cars along the lane. The reciprocal dependencies between neighboring junctions result in a convoluted formulation with several factors. These characteristics are unintentional, dangerous, and reliant; the worst part is that they vary with time. Various traffic light controller methods are shown in Table 6.2.

Several intelligent traffic management solutions were examined. RFID readers and tags, Green Wave Systems, smartphones, and a Big Data center wireless connectivity were all used. Table 6.2 examines and quickly outlines each approach's applications, benefits, and drawbacks. The (IoT) method was used to collect data on traffic congestion more quickly and precisely. Furthermore, a mobile application was

TABLE 6.2
Various Traffic Light Controller Methods

Method	Description	Pros and Cons
TMS	TMS is a logistics medium that uses technology to help organizations plan, execute, and optimize the physical movement of products, both arriving and leaving, while ensuring compliance and adequate paperwork.	Pros: Minimize the probability of traffic backups. Cons: 1. Substantial latency may occur as a result of TMS failure and downtime. 2. Excessive delays cost substantial fuel and increase driving time.
RFID	RFID tags are a tracking system that uses intelligent bar codes to identify items. RFID stands for "Radio Frequency Identification," and RFID tags use radio frequency technology. Data is sent from the tag to a reader through radio waves, which subsequently sends the data to RFID computer software. RFID tags are commonly used to monitor items and track automobiles, pets, and Alzheimer's patients. An RFID tag is also known as an RFID chip.	Pros: 1. It is not impacted by inclement weather. 2. Such transponders, which seem to be typically the diameter of a paperclip, can be embedded in screens. Cons: 1. If RFID is active, it might be costly due to power resources, for example, batteries. 2. RFID, even if encrypted, can be easily intercepted.
WSN	The WSN application for vehicle counting, categorization, and speed measurement is based on roadside magnetic sensors. Magnetic sensors installed along the roadside can collect traffic data and send it wirelessly to a central computer for processing and analysis. Traffic data will be kept in a database and shared with other transportation apps via Web services.	Pros: 1. Involves real-time traffic monitoring, assures effective road monitoring, and necessitates minimal human intervention. 2. It can be routed through a centralized control system and performs well enough at incredible velocities. Cons: 1. If the transmission ratio is too low, anomalies may be generated. 3. It is simple for attackers to attack the system.
System of green waves	When a car crosses an intersection using this strategy, an arriving rescuer's transport causes the traffic signal to turn green. Image processing is used to correctly determine the position of rescue vehicles.	Pros: 1. Allows emergency vehicles to safely and quickly through junctions. 2. Effective for identifying stolen automobiles that drive through junctions. Cons: 1. Poor execution in extreme climate situations. 2. If it malfunctions, it might cause a significant traffic bottleneck.

(*Continued*)

TABLE 6.2 (*Continued*)
Various Traffic Light Controller Methods

Method	Description	Pros and Cons
GSM	In this approach, an embedded controller generates a data pattern that is fed into a micro-controller. The C programming language is then translated to HEX code and sent to the micro-controller receiver. The data will then be sent to GSM using the micro-controller transmitter. With the SIM's 3G connection, the GSM can activate messaging service and alert an expert with a message.	Pros: 1. GSM notifies medical physicians about the patient's situation before an ambulance arrives, preserving time (Vrushank & Khang, 2023). 2. It may display video in the ambulance using a 3G connection. Consequently, the doctor may assess the patient's health and propose first-aid measures. Cons: They are concerned about losing authority over their information and network.
IR	Most traffic signal systems use infrared sensors, including an IR sender and receiver. The IR detector can be engaged whenever the vehicle passes between the sensors. The data gathered is executed to turn the traffic signal green.	Pros: Vehicles that have been reported stolen or belong to criminal suspects can be monitored, as well as the time and direction of movement. Cons: The initial configuration expenditure is momentous.

considered a "User Interface (UI)" to identify traffic congestion and present users with other routes in various locations. The purpose of these strategies is to deliver more information about traffic and road conditions to vehicle drivers. Furthermore, intelligent traffic systems can grant precedence to emergency vehicles.

The comparison of traffic light controller performance is shown in Table 6.3. Nowadays, due to some technology based on IoT, traffic incorporates several demeanors, such as transmission classifications, circumstances, bases, designs, formations,

TABLE 6.3
Comparison of Traffic Light Controller Performance Friendly, and Hardware Autonomy Measures

METHOD	PERFORMANCE			
	Security	Reliability	Effectiveness	Total Evaluation
TMS	Average	High	Medium	Popular
RFID	Medium	Low	Medium	Useful
WSN	Low	Medium	Medium	Useful
System of green waves	Medium	Medium	Low	Not useful
GSM	High	Medium	High	Most useful
IR	Medium	Low	Low	Useful

limitations, and magnitudes. Unlike other technology, machine–to-machine devices are no longer restricted to simple detector transmissions. The IoT comprises any device with transmission and connectedness capabilities. Though in recently created M2M traffic, the number of sessions started every day varies substantially more than on smartphones. This implies that it will considerably impact traffic designs such as administration and supervision. Consequently, depending on numbers and different devices operating in the network, M2M behavior might take various forms.

A specific set of IoT traffic features would be used to develop a strong and dependable system that meets distinguishing needs like safety, investigation, finding, and energy conservation concerns. New research and solutions are required to meet flexibility, environmentally. Table 6.3 displays comparison of traffic light controller performance friendly, and hardware autonomy measures. It seems to be due to the increased diversity and heterogeneity of IoT devices in non-IoT regions.

However, security and quality of service (QoS) remain the most pressing issues in traffic categorization.

IoT traffic flow features were identified and discriminated against with conventional traffic conditions. A lightweight assessment of transport classification techniques was given to facilitate a seamless transition to IoT traffic classification. Collecting additional context through footage. This primary aim is to assist the operator in monitoring video sequences. Because recording devices include a rich source of information for human interpretation, they have been widely used for traffic and other surveillance purposes. Computer vision adds value to sensors by extracting relevant critical data. Consequently, image enhancement and surveillance systems are becoming increasingly crucial in ITSs. This study aims to introduce object recognition and surveillance systems and to acquaint the user with progressively used methodologies in ITSs.

To maintain this, concentrate on connectivity surveillance while disregarding automotive or other portable devices. It is a significant network paradigm, with many intelligent devices connecting. Many IoT applications and services emerge as IoT devices generate vast amounts of data. Another major topic is machine learning, which has achieved considerable success in various study domains. Computer concept is software monitoring procedures and approaches such as computer concept, computer representations, and intelligent monitoring. It is additionally being used in computer networks.

Lots of studies are being conducted to investigate ways to use machine learning to tackle networking challenges such as routing, traffic engineering, resource allocation, and security. Deep learning has recently been used to enhance IoT systems and provide services such as transportation planning, managed services, cybersecurity, and Web traffic segmentation. IoT is evolving into a new, pervasive network architecture that provides dispersed and transparent services. In many intelligent gadgets, Internet connects equipment like detectors, cell phones, and many other tech devices. These intelligent gadgets can talk to each other and exchange data. Networks, like condition assessment inspections, heart muscle behaviors, health management, positioning, and architecture supervising, could be constructed to enhance one's quality of life by gathering information from connected systems and interpreting it to perceive and understand the surroundings.

Khaleel and Yussof (2016) have discussed the topic in Iraq, where numerous student abduction incidents have been recorded due to a lack of safety procedures and police departments. Several academic establishments in Iraq, including such elementary, have been searching for an improved method for tracking class attendance to more accurately supervise children's security. In Iraqi officials, university enrollment is traditionally managed directly by instructors who regularly verify and document the participation of individual pupils throughout every lecture. Unfortunately, the old system has some drawbacks, including that it may only be used at specific times and therefore does not supervise individuals instantaneously.

Despite extensive studies on transportation systems, intelligent transportation surveillance remains an important area of investigation because of developing techniques such as the IoT and machine learning. Incorporating these innovations would make it easier to make intelligent decisions and accomplish urban growth. Furthermore, present traffic materials primarily concentrate on the motorway and urban transportation planning, with only a few research focusing on collection routes and enclosed universities. Furthermore, connecting with the community and building meaningful contacts to aid individuals in judgment is difficult whenever consumers lack access to technologically advanced gadgets.

For this circumstance, our study presents a technology model for collecting, processing, and storing authentic traffic information. The goal is to increase transportation by providing traffic information on traffic gridlock and odd traffic events via roadside messaging devices. Such emergency beacon notifications will assist individuals in streamlining processes, particularly throughout commute periods. The technology continuously transmits real-time traffic information from the state administration. A prototype was created to test the photographer's viability, and indeed, the results suggest substantial precision in detecting vehicles and a relatively low inaccuracy in highway availability prediction.

6.5 CONCLUSION

Signalized intersection systems are commonly used to control and maintain the movement of automobiles across numerous crosswalks. Authorities expect automobiles to operate freely along transit lines (Khang & Rani et al., 2022).

Nevertheless, synchronizing several signalized intersection circuits at neighboring junctions is challenging, considering the different characteristics required. Systems are usually incapable of dealing with varied streams nearing intersections. Furthermore, the same present road network does not account for reciprocal interference between nearby traffic signal networks, the asymmetry of automobile throughput over the duration, collisions, the entrance of ambulances, and crosswalks, which cause traffic problems and delays. Several intelligent traffic management methods have been presented in this chapter where RFID, GWS, smartphones, and a Big Data center wireless connectivity are used. Tables 6.2 and 6.3 examine and quickly outline each approach's applications, benefits, and drawbacks.

The IoT approach was applied to gather information on overcrowding rapidly and precisely. Additionally, a software device has been recognized effectively for internalizing, detecting overcrowding, and offering customers different routes in different territories. Such approaches aim to provide motorists with additional details about road conditions and traffic. Furthermore, intelligent traffic systems can grant precedence to emergency vehicles (Hajimahmud & Khang et al., 2022).

We thoroughly evaluated computer vision approaches for traffic analysis systems, particularly in urban situations. The use of video analysis for traffic assessment is becoming more common in ITSs (Rani & Khang et al., 2022).

The study progresses from the freeway environment to the more difficult urban sector. This circumstance brings up many new application options in traffic management and enforcement (Khanh & Khang, 2021).

REFERENCES

Abdellah, A.R., Koucheryavy, A. "Deep Learning with Long Short-Term Memory for IoT Traffic Prediction", *Internet of Things, Smart Spaces, and Next Generation Networks and Systems*, pp. 267–280 (2020). Springer. https://link.springer.com/chapter/10.1007/978-3-030-65726-0_24

Atta, A., Abbas, S., Khan, M.A., Ahmed, G., Farooq, U. "An adaptive approach: smart traffic congestion control system", *Journal of King Saud University-Computer and Information Sciences*, vol. 32, no. 9, pp. 1012–1019 (2020). https://www.sciencedirect.com/science/article/pii/S1319157818308565

Babasaheb, J., Sphurti, B., Khang, A. "Industry Revolution 4.0: Workforce Competency Models and Designs", *Designing Workforce Management Systems for Industry 4.0: Data-Centric and AI-Enabled Approaches* (1st Ed.), pp. 14–31 (2023). CRC Press. https://doi.org/10.1201/9781003357070-2

Badura, S., Lieskovsky, A. "Intelligent traffic system: Cooperation of Manet and image processing", *2010 First International Conference on Integrated Intelligent Computing*, pp. 119–123 (2010). IEEE. https://ieeexplore.ieee.org/abstract/document/5571489/

Bathula, M., Ramezanali, M., Pradhan, I., Patel, N., Gotschall, J., Sridhar, N. "A Sensor Network System for Measuring Traffic in Short-Term Construction Work Zones", *International Conference on Distributed Computing in Sensor Systems*, pp. 216–230 (2009). https://link.springer.com/chapter/10.1007/978-3-642-02085-8_16

Bhambri, P., Rani, S., Gupta, G., Khang, A. *Cloud and Fog Computing Platforms for Internet of Things*, (2022). CRC Press. https://doi.org/10.1201/9781032101507

Bugdol, M., Segiet, Z., Krecichwost, M. "Intelligent transportation systems vehicle detection anisotropic magneto resistive sensors", *IEEE Journals* (2014). https://yadda.icm.edu.pl/baztech/element/bwmeta1.element.baztech-631bdea0-bc22-4736-a903-55def2a53e22

Chong, H.F., Ng, D.W.K. "Development of IoT device for traffic management system", in *2016 IEEE Student Conference on Research and Development (SCOReD)*, pp. 1–6 (2016). IEEE. https://ieeexplore.ieee.org/abstract/document/7810059/

Collotta, M., Pau, G. "New solutions based on wireless networks for dynamic traffic lights management: a comparison between IEEE 802.15. 4 and Bluetooth", *Transport and Telecommunication*, vol. 16, no. 3, p. 224 (2015). https://sciendo.com/article/10.1515/ttj-2015-0021

Ding, J., Cheung, S.-Y., Tan, C.-W., Varaiya, P. "Signal processing of sensor node data for vehicle detection", *Proceedings. The 7th International IEEE Conference on Intelligent Transportation Systems (IEEE Cat. No.* 04TH8749), pp. 70–75 (2004). IEEE. https://ieeexplore.ieee.org/abstract/document/1398874/

Dubey, A., Rane, S., et al. "Implementation of an intelligent traffic control system and real time traffic statistics broadcasting", in *2017 International Conference of Electronics, Communication and Aerospace Technology (ICECA)*, vol. 2, pp. 33–37 (2017). IEEE. https://ieeexplore.ieee.org/abstract/document/8212827/

Dudhe, P., Kadam, N., Hushangabade, R., Deshmukh, M. "Internet of things (IoT): An overview and its applications", in *2017 International Conference on Energy, Communication, Data Analytics and Soft Computing (ICECDS)*, pp. 2650–2653 (2017). IEEE. https://ieeexplore.ieee.org/abstract/document/8389935/

Fleck, J.L., Cassandras, C.G., Geng, Y. "Adaptive quasi-dynamic traffic light control", *IEEE Transactions on Control Systems Technology*, vol. 24, no. 3, pp. 830–842 (2015). https://ieeexplore.ieee.org/abstract/document/7229317/

G¨ottlich, S., Herty, M., Ziegler, U. "Modeling and optimizing traffic light settings in road networks", *Computers & Operations Research*, vol. 55, pp. 36–51 (2015). https://www.sciencedirect.com/science/article/pii/S0305054814002585

Ghena, B., Beyer, W., Hillaker, A., Pevarnek, J., Halderman, J.A. "Green lights forever: Analyzing the security of traffic infrastructure", *8th USENIX Workshop on Offensive Technologies (WOOT 14)*, 2014. https://www.usenix.org/system/files/conference/woot14/woot14-ghena.pdf

Gu, L., Jia, D., Vicaire, P., Yan, T., Luo, L., Tirumala, A., Cao, Q., He, T., Stankovic, J.A., Abdelzaher, T., et al. "Lightweight detection and classification for wireless sensor networks in realistic environments", in *Proceedings of the 3rd International Conference on Embedded Networked Sensor Systems*, pp. 205–217 (2005). https://dl.acm.org/doi/abs/10.1145/1098918.1098941

Hahanov, V., Khang, A., Litvinova, E., Chumachenko, S., Hajimahmud, V.A., Alyar, A.V. "The Key Assistant of Smart City – Sensors and Tools", *AI-Centric Smart City Ecosystems: Technologies, Design and Implementation* (1st Ed.) (2022). CRC Press. https://doi.org/10.1201/9781003252542-17

Hajimahmud, V.A., Khang, A., Gupta, S.K., Babasaheb, J., Morris, G. *AI-Centric Modelling and Analytics: Concepts, Designs, Technologies, and Applications* (1st Ed.) (2023). CRC Press. https://doi.org/10.1201/9781003400110

Hajimahmud, V.A., Khang, A., Hahanov, V., Litvinova, E., Chumachenko, S., Alyar, A.V. "Autonomous Robots for Smart City: Closer to Augmented Humanity", *AI-Centric Smart City Ecosystems: Technologies, Design and Implementation* (1st Ed.) (2022). CRC Press. https://doi.org/10.1201/9781003252542-7

Horng, G.-J., Cheng, S.-T. "Using intelligent vehicle infrastructure integration for reducing congestion in smart city", *Wireless Personal Communications*, vol. 91, no. 2, pp. 861–883 (2016). https://link.springer.com/article/10.1007/s11277-016-3501-8

Jain, D.K., Neelakandan, S., Veeramani, T., Bhatia, S., Memon, F.H. "Design of fuzzy logic based energy management and traffic predictive model for cyber physical systems", *Computers and Electrical Engineering*, vol. 102, p. 108135 (2022). https://www.sciencedirect.com/science/article/pii/S0045790622003858

Janahan, S.K., Veeramanickam, M., Arun, S., Narayanan, K., Anandan, R., Parvez, S.J. "Iot based smart traffic signal monitoring system using vehicles counts", *International Journal of Engineering & Technology*, vol. 7, no. 2.21, pp. 309–312 (2018). https://www.researchgate.net/profile/Veeramanickam-Murugappan-2/publication/325116849_IoT_based_smart_traffic_signal_monitoring_system_using_vehicles_counts/links/5b405b1e0f7e9bb59b102c06/IoT-based-smart-traffic-signal-monitoring-system-using-vehicles-counts.pdf

Javaid, S., Sufian, A., Pervaiz, S., Tanveer, M. "Smart traffic management system using internet of things", in *2018 20th international conference on advanced communication technology (ICACT)*, pp. 393–398 (2018). IEEE. https://ieeexplore.ieee.org/abstract/document/8323770/

Khaleel, A., Yussof, S. "An investigation on the viability of using IoT for student safety and attendance monitoring in Iraqi primary schools", *Journal of Theoretical and Applied Information Technology*, vol. 85, no. 3, p. 394 (2016). http://www.jatit.org/volumes/Vol85No3/14Vol85No3.pdf

Khang, A., Abdullayev, V., Hahanov, V., Shah, V. *Advanced IoT Technologies and Applications in the Industry 4.0 Digital Economy* (1st Ed.) (2023). CRC Press. https://doi.org/10.1201/978-1-003-43426-9

Khang, A., Gupta, S.K., Shah, V., Misra, A. *AI-Aided IoT Technologies and Applications in the Smart Business and Production* (1st Ed.) (2023). CRC Press. https://doi.org/10.1201/9781003392224

Khang, A., Hahanov, V., Abbas, G.L., Hajimahmud, V.A. “Cyber-Physical-Social System and İncident Management”, *AI-Centric Smart City Ecosystems: Technologies, Design and Implementation* (1st Ed.) (2022). CRC Press. https://doi.org/10.1201/9781003252542-2

Khang, A. (2021), “*Material of Computer Science, Artificial Intelligence, Data Science, IoT, Blockchain, Cloud, Metaverse, Cybersecurity for Studies*”. Retrieved from https://www.researchgate.net/publication/370156102_Material4Studies

Khang, A., Ragimova, N.A., Hajimahmud, V.A., Alyar, A.V. “Advanced Technologies and Data Management in the Smart Healthcare System”, *AI-Centric Smart City Ecosystems: Technologies, Design and Implementation* (1st Ed.) (2022). CRC Press. https://doi.org/10.1201/9781003252542-16

Khang, A., Rani, S., Gujrati, R., Uygun, H., Gupta, S.K. *Designing Workforce Management Systems for Industry 4.0: Data-Centric and AI-Enabled Approaches* (1st Ed.) (2023). CRC Press. https://doi.org/10.1201/99781003357070

Khang, A., Rani, S., Sivaraman, A.K. *AI-Centric Smart City Ecosystems: Technologies, Design and Implementation* (1st Ed.) (2022). CRC Press. https://doi.org/10.1201/9781003252542

Khang, A., Vrushank, S., Rani, S. *AI-Based Technologies and Applications in the Era of the Metaverse* (1st Ed.) (2023). IGI Global Press. https://doi.org/10.4018/9781668488515

Khanh, H.H., Khang, A. “The Role of Artificial Intelligence in Blockchain Applications”, *Reinventing Manufacturing and Business Processes through Artificial Intelligence*, pp. 20–40 (2021). CRC Press. https://doi.org/10.1201/9781003145011-2

KumaraSwamy, S., Manjula, S., Venugopal, K., Iyengar, S., Patnaik, L. “Association rule sharing model for privacy preservation and collaborative data mining efficiency”, in *2014 Recent Advances in Engineering and Computational Sciences (RAECS)*, pp. 1–6 (2014). IEEE. https://ieeexplore.ieee.org/abstract/document/6799597/

Manjunatha, C., Chandrashekar, J., Chandrashekhara, B. “An overview of multidimensional drivers and adverse impacts of urban sprawl”, *Environment and We International Journal of Science and Technology*, vol. 14, pp. 77–87 (2019). http://www.sedindia.in/ewijst/issues/vol14/ewijst14010719004.pdf

Misbahuddin, S., Zubairi, J.A., Saggaf, A., Basuni, J., Sulaiman, A., Al-Sofi, A., et al. “IoT based dynamic road traffic management for smart cities”, in *2015 12th International Conference on High-Capacity Optical Networks and Enabling/Emerging Technologies (HONET)*, pp. 1–5 (2015). IEEE. https://ieeexplore.ieee.org/abstract/document/7395434/

Miz, V., Hahanov, V. “Smart traffic light in terms of the cognitive road traffic management system (ctms) based on the internet of things”, *Proceedings of IEEE East-West Design & Test Symposium (EWDTS 2014)*, pp. 1–5 (2014). IEEE. https://ieeexplore.ieee.org/abstract/document/7027102/

Nagmode, P.S., Patel, A.V., Satpute, A.B., Gupta, P.L. “Endodontic management of mandibular first molars with mid mesial canal: A case series.” *J Conserv Dent* [serial online] 2017 [cited 2023 Aug 29];20:137-40. Available from: https://www.jcd.org.in/text.asp?2017/20/2/137/212246

Nambajemariya, F., Wang, Y. “Excavation of the internet of things in urban areas based on an intelligent transportation management system”, *Advances in Internet of Things*, vol. 11, no. 3, pp. 113–122 (2021). https://www.scirp.org/journal/paperinformation.aspx?paperid=110367

Nellore, K., Hancke, G.P. “A survey on urban traffic management system using wireless sensor networks”, *Sensors*, vol. 16, no. 2, p. 157 (2016). https://www.mdpi.com/125956

Pescaru, D., Curiac, D.-I. “Ensemble based traffic light control for city zones using a reduced number of sensors”, *Transportation Research Part C: Emerging Technologies*, vol. 46, pp. 261–273 (2014). https://www.sciencedirect.com/science/article/pii/S0968090X14001855

Rana, G., Khang, A., Sharma, R., Goel, A.K., Dubey, A.K. *Reinventing Manufacturing and Business Processes through Artificial Intelligence* (2021). CRC Press. https://doi.org/10.1201/9781003145011

Rani, S., Bhambri, P., Kataria, A., Khang, A. “Smart City Ecosystem: Concept, Sustainability, Design Principles and Technologies”, *AI-Centric Smart City Ecosystems: Technologies, Design and Implementation* (1st Ed.) (2022). CRC Press. https://doi.org/10.1201/9781003252542-1

Rani, S., Bhambri, P., Kataria, A., Khang, A., Sivaraman, A.K. *Big Data, Cloud Computing and IoT: Tools and Applications* (1st Ed.) (2023). Chapman and Hall/CRC. https://doi.org/10.1201/9781003298335

Rani, S., Chauhan, M., Kataria, A., Khang, A. "IoT Equipped Intelligent Distributed Framework for Smart Healthcare Systems", *Networking and Internet Architecture* (2021). CRC Press. https://doi.org/10.48550/arXiv.2110.04997

Salama, A.S., Saleh, B.K., Eassa, M.M. "Intelligent cross road traffic management system (icrtms)", in *2010 2nd International Conference on Computer Technology and Development*, pp. 27–31 (2010). https://ieeexplore.ieee.org/abstract/document/5646059/

Sharif, A., Li, J., Khalil, M., Kumar, R., Sharif, M. I., and Sharif, A, "Internet of things—smart traffic management system for smart cities using big data analytics," in 2017 14th international computer conference on wavelet active media technology and information processing (ICCWAMTIP), pp. 281–284, 2017, IEEE. https://ieeexplore.ieee.org/abstract/ document/8301496/

Sharma, V., Tiwari, R. "A review paper on "IoT" & it's smart applications", *International Journal of Science, Engineering and Technology Research (IJSETR)*, vol. 5, no. 2, pp. 472–476 (2016). https://fardapaper.ir/mohavaha/uploads/2020/11/Fardapaper-A-review-paper-on-%E2%80%9CIOT%E2%80%9D-It%E2%80%9Fs-Smart-Applications.pdf

Snehal, M., Babasaheb, J., Khang, A. "Workforce Management System: Concepts, Definitions, Principles, and Implementation", *Designing Workforce Management Systems for Industry 4.0: Data-Centric and AI-Enabled Approaches* (1st Ed.), pp. 1–13 (2023). CRC Press. https://doi.org/10.1201/9781003357070-1

Studer, L.P., Ketabdari, M., Marchionni, G. "Analysis of adaptive traffic control systems design of a decision support system for better choices", (2015). https://re.public.polimi.it/handle/11311/1020703

Sumi, L., Ranga, V. "An Iot-Vanet-Based Traffic Management System for Emergency Vehicles in a Smart City", *Recent Findings in Intelligent Computing Techniques*, pp. 23–31 (2018). Springer. https://link.springer.com/chapter/10.1007/978-981-10-8636-6_3

Talukder, M.Z., Towqir, S.S., Remon, A.R., Zaman, H.U. "An IoT based automated traffic control system with real-time update capability", in *2017 8th International Conference on Computing, Communication and Networking Technologies (ICCCNT)*, pp. 1–6 (2017). IEEE. https://ieeexplore.ieee.org/abstract/document/8204095/

Tian, Z., Gao, X., Su, S., Qiu, J. "Vcash: a novel reputation framework for identifying denial of traffic service in internet of connected vehicles", *IEEE Internet of Things Journal*, vol. 7, no. 5, pp. 3901–3909 (2019). https://ieeexplore.ieee.org/abstract/document/8891784/

Tubaishat, M., Qi, Q., Shang, Y., and Shi, H. "Wireless Sensor-Based Traffic Light Control," 2008 5th IEEE Consumer Communications and Networking Conference, Las Vegas, NV, USA, 2008, pp. 702-706, doi: 10.1109/ccnc08.2007.161.

Vrushank, S., Khang, A. "Internet of Medical Things (IoMT) Driving the Digital Transformation of the Healthcare Sector", *Data-Centric AI Solutions and Emerging Technologies in the Healthcare Ecosystem* (1st Ed.), p. 1 (2023). CRC Press. https://doi.org/10.1201/9781003356189-2

Yousef, K.M., Al-Karaki, J.N., Shatnawi, A.M. "Intelligent traffic light flow control system using wireless sensors networks: vehicular wireless networks and vehicular intelligent transportation systems", *Journal of Information Science and Engineering*, vol. 26, no. 3, pp. 753–768 (2010). https://citeseerx.ist.psu.edu/document?repid=rep1&type=pdf&doi=87f25a4f6bda6a0a255efff7426e55a24362431f

Zhang, W., Tan, G.-Z., Ding, N., Wang, G.-Y. "Traffic congestion evaluation and signal timing optimization based on wireless sensor networks: issues, approaches and simulation", *Journal of Information Science & Engineering*, vol. 30, no. 4 (2014). https://web.p.ebscohost.com/abstract?direct=true&profile=ehost&scope=site&authtype=crawler&jrnl=10162364&AN=96681515&h=8Af2N30IwB7qb5bus%2ftfWGuRWIJ8rSWCuG%2f%2bNpocxpCAW5kQFIrEXpS6HnaQcQgFvsjp3lypMWp2Ki13jALQxA%3d%3d&crl=c&resultNs=AdminWebAuth

7 Effective Construction Site Monitoring Using Artificial Intelligence (AI)

Khushi Bhoj, Yash Sharma, Aman Syed, Anil Vasoya, and Rashmi Thakur

7.1 INTRODUCTION

One of the most hazardous work categories is construction, and many accidents and fatalities are reported each year all around the world. In addition to causing misery for the affected workers and their families, these accidents and fatalities also resulted in timetable delays for the project and higher construction expenses. As a result, it is now crucial for the construction sector to identify safety risks.

To help construction businesses decrease accidents and maintain a safe work environment on building sites, it is becoming increasingly important to develop cutting-edge techniques for automatically monitoring the safety of the workers. To find answers to these safety concerns in the construction sector, scientists are now looking to the field of artificial intelligence (AI) (Rana & Khang et al., 2021).

Considering the recent advancements in the industry, AI has a reputation for consistently providing effective answers to challenging issues. A branch of AI called computer vision (CV) focuses on teaching computers to recognize objects in images and video. It stimulates the interpretation of visual information by the brain as Figure 7.1.

A deep learning model may be taught to precisely identify and categorize items using a huge number of digital photos or video frames from the camera. This allows the model to respond to what it "sees" and, to a certain extent, helps to decrease accidents at the building site.

AI-based monitoring enables users to use the video feeds from the majority of the already installed CCTV systems at the site to detect and identify the targeted safety concerns occurring at the building site without the need to acquire a new set of video camera systems.

The motions and interactions of people, equipment, and items on a building site may be monitored and measured in real time against specified Key Performance Indicators using AI and CV (Babasaheb & Sphurti et al., 2023). Such technologies assist site and safety managers in being able to better monitor and have greater visibility of the work site, so they can better regulate the interactions and safeguard the employees' health and safety (Khanh & Khang, 2021).

In the future, sensors linked with AI are anticipated to be integrated into work operations to increase safety on construction sites, such as using wireless devices to connect equipment and employees to a central operating system (Hahanov & Khang et al., 2022).

DOI: 10.1201/9781003376064-7

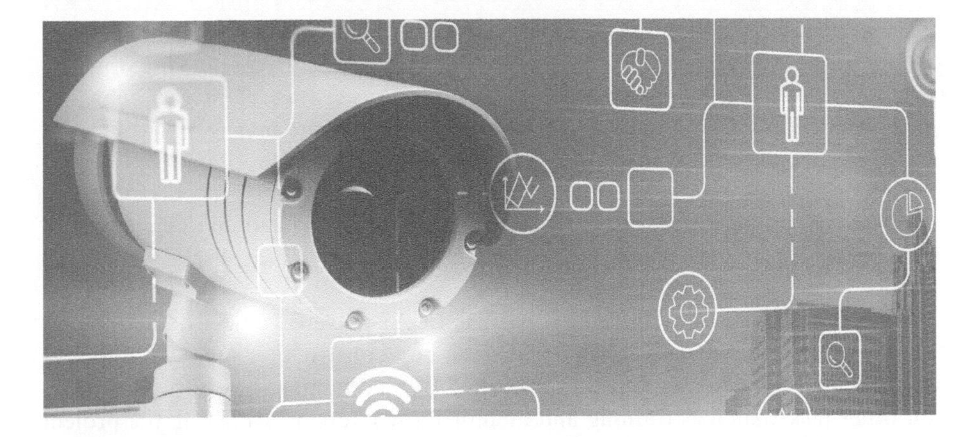

FIGURE 7.1 A camera with AI technology.

7.2 BACKGROUND

Construction sites are high-risk environments where accidents and injuries can occur at any time. The construction industry has a higher rate of fatalities and injuries compared to other industries, with thousands of workers being injured or killed on the job every year.

Many of these accidents and injuries could be prevented with proper safety measures, including the use of safety gear and adherence to safety regulations. According to the International Labor Organization, construction is the third most dangerous industry in the world, with more than 2.3 million work-related fatalities reported each year.

In addition to the human cost of these accidents, they also have significant economic impacts, including delays to projects and increased construction costs. However, ensuring the safety of workers on construction sites can be challenging due to the complex and dynamic nature of the work environment.

One of the main challenges in construction site safety is the lack of visibility and oversight. Traditional methods of monitoring construction sites, such as human supervisors and security cameras, can be limited in their ability to cover large areas and capture all the activity on the site. As a result, safety violations and accidents can go undetected, leading to a higher risk of injury and fatalities (Khang & Hahanov et al., 2022).

AI has the potential to revolutionize construction site safety by enabling the real-time monitoring of workers and the detection of safety violations. One promising application of AI in this context is object detection, which involves the use of machine learning algorithms to identify and classify objects in images or videos (Khang & Rana et al., 2023).

By training a machine learning model to recognize workers, their uniforms, and their safety gear, it is possible to monitor worker behavior and ensure that they are adhering to safety protocols.

There are several benefits to using AI-powered object detection to improve safety at construction sites. First, it allows for the continuous monitoring of workers, enabling the detection of safety violations in real time. This can help prevent accidents and injuries by alerting supervisors and site managers to potential hazards.

Second, AI-powered object detection can be used to identify workers who are not wearing proper safety gear and flag them for safety violations. This can help ensure that all workers on the site are adhering to safety protocols and reduce the risk of injury.

In addition to improving safety, the use of AI-powered object detection can also help increase efficiency and productivity on construction sites. By automating the monitoring process, it is possible to free up human supervisors to focus on other tasks, such as training and mentoring workers or managing the project schedule.

AI-powered object detection can also help identify bottlenecks and inefficiencies in the construction process, enabling site managers to make adjustments to improve productivity and reduce delays.

However, implementing AI-powered object detection on construction sites also presents several technical challenges and considerations. One challenge is the selection and training of the machine learning model. To accurately recognize workers, their uniforms, and their safety gear, the model must be trained on a large and diverse dataset of images and video from the construction site. This can be time-consuming and resource-intensive and may require specialized expertise in machine learning and CV (Vrushank & Khang, 2023).

Another challenge is the integration of the AI system with the existing construction site infrastructure. To enable real-time monitoring of workers, the AI system must be connected to a network of cameras or sensors on the site. This can be complex and may require additional hardware and software deployment. It is also important to consider the privacy and security implications of using AI on construction sites, as well as the potential for data breaches or unauthorized access to sensitive information (Khang & Gupta et al., 2023a,b).

In the future, it is anticipated that sensors linked with AI will be integrated into work operations to increase safety on construction sites. For example, wireless devices could connect equipment and employees to a central operating system, enabling the real-time monitoring of equipment usage and maintenance. Sensors could also be used to track the location and movements of workers, helping to ensure that they are staying within designated safety regions.

Overall, the use of AI and object detection technology has the potential to significantly improve safety at construction sites. By enabling the real-time monitoring of workers and the detection of safety violations, it is possible to prevent accidents and injuries and create a safer work environment for all construction workers.

While there are technical challenges and considerations involved in implementing such a system, the benefits of using AI to improve construction site safety are clear, and we will likely see more widespread adoption of these technologies in the future as Figure 7.2.

FIGURE 7.2 AI and object detection technology in construction site safety.

7.3 RELATED WORK

AI has the potential to greatly improve safety on construction sites. Several studies have explored the use of AI for construction site safety, including the ability to detect safety violations and accidents in real time. One study (Radford & Metz et al., 2016) described the use of AI to monitor construction sites for safety violations and accidents, discussing the potential for real-time alerts.

This study highlights the importance of timely detection and response to safety incidents, as delay can lead to more serious consequences. By using AI for real-time monitoring, construction sites can promptly address any safety concerns and prevent accidents from occurring.

Another study (Brown, 2021) examined the benefits of using AI to improve safety at construction sites, including the ability to detect safety violations and accidents. This study emphasized the potential of AI to reduce the number of injuries and fatalities on construction sites, which are often caused by human error or neglect. By automating the monitoring process and providing real-time alerts, AI can help to identify potential hazards and prevent accidents before they happen.

A third study (Gonzalez, 2022) discussed the use of AI to classify objects and monitor workers on construction sites, as well as the potential for integrating sensors with AI to increase safety. The integration of sensors with AI can provide a more comprehensive view of the construction site and enable more accurate detection of safety violations and accidents. For example, sensors can be used to detect changes in temperature, humidity, and other environmental factors that may pose a risk to workers.

There have been several additional studies on the use of AI and machine learning for construction site safety.

A review article (Johnson, 2021) provided an overview of existing research on the use of AI and machine learning for improving safety on construction sites, including object detection and real-time monitoring. This review also discussed the potential for using AI to analyze data from past accidents and identify patterns or trends that could be used to prevent future accidents.

Another study (Kim, 2022) discussed the benefits of using AI for safety monitoring on construction sites, including the ability to detect safety violations and accidents in real time.

The potential for integrating sensors with AI was also examined in this study, which emphasized the importance of using a combination of AI and sensors to achieve the most accurate and reliable results (Jebaraj & Khang et al., 2023).

A third review article (Kim, 2021) provided a review of existing research on the use of AI for improving safety on construction sites, including object detection and real-time monitoring, as well as the potential for integrating sensors with AI. This review also discussed the challenges and limitations of using AI for construction site safety, including the need for high-quality data and the potential for bias in AI algorithms.

Another review (Lee, 2020) focused specifically on the use of AI for construction site safety, providing an overview of existing research on the topic. This review not only highlighted the potential of AI to significantly improve safety on construction sites but also stressed the importance of careful planning and implementation to ensure the most effective use of AI.

Overall, the literature suggests that AI has the potential to significantly improve safety on construction sites through real-time monitoring and the detection of safety violations and accidents. The integration of sensors with AI may further enhance the capabilities of AI for construction site safety.

However, it is important to carefully consider the potential challenges and limitations of using AI and to ensure that AI is used in a way that is effective and ethically responsible. Further research is needed to fully understand the potential and limitations of AI for construction site safety.

7.4 RESEARCH METHODOLOGY

In recent years, there has been increasing interest in the use of AI and related technologies to improve safety in the construction industry. AI-based object detection methods, in particular, have been identified as a promising approach for construction site monitoring. These methods involve the use of AI algorithms and machine learning techniques to automatically analyze images or video footage of the construction site and identify potential hazards or other issues of concern (Subhashini & Khang, 2023).

As the use of AI in the construction industry grows, it is important to understand the ways in which it can be effectively implemented to improve safety on construction sites. The purpose of this study is to identify and review the existing literature on the use of AI for improving safety in the construction industry, with a focus on object detection and real-time monitoring.

The use of AI in the construction industry has gained significant attention in recent years, with a growing number of researchers and practitioners exploring the potential of AI for various applications (Khang & Rani et al., 2022).

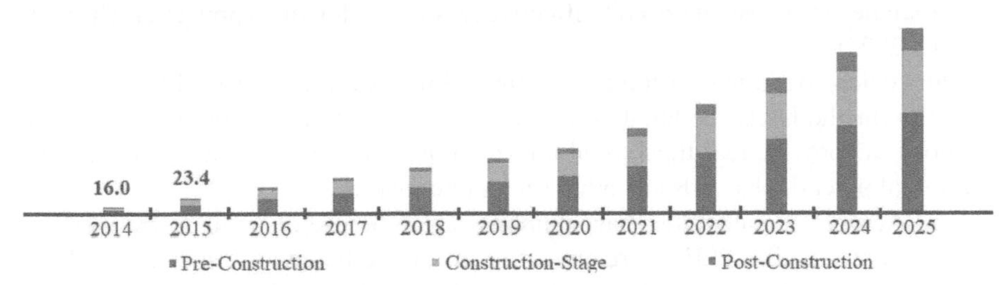

FIGURE 7.3 Artificial intelligence in construction market report, 2014–2025.

Source: **Grandviewresearch.com**

One area of particular interest is the use of AI-based object detection methods for construction site monitoring, which involves the use of CV and machine learning techniques to automatically recognize and classify objects and events in construction site environments as Figure 7.3.

There are several potential benefits to using AI for construction site monitoring. For instance, AI-based object detection methods can provide real-time visual information about the site, enabling the identification of potential hazards and other issues that may impact the safety or efficiency of the construction process. This can help to prevent accidents, reduce downtime, and improve the overall quality of the project (Khang & Hajimahmud, 2022).

Additionally, AI-based object detection methods may be able to automatically track progress on the site, allowing for better project management and coordination.

To explore the effectiveness of AI-based object detection methods for construction site monitoring, we conducted a literature review of existing research on the topic. Our review included papers published in academic journals and conference proceedings that reported on the use of AI-based object detection methods for construction site monitoring.

We analyzed the methods and findings of these studies to identify common themes and trends, and we compared the results of different studies to assess the overall effectiveness of AI for construction site monitoring.

The construction industry is a vital sector that plays a significant role in the development of a country's infrastructure. However, construction sites can be hazardous environments, with a high risk of accidents and injuries.

To address this issue, various measures have been put in place to ensure the safety of construction workers, such as running safety drills and requiring the use of personal protective equipment (PPE). Despite these efforts, the construction industry continues to be one of the most dangerous industries for workers, with a high rate of fatalities and injuries.

To better understand the effectiveness of AI-based object detection methods for monitoring construction site, we conducted a literature review of existing research on the topic. We focused on papers that reported on the use of these methods for construction site monitoring, and we analyzed the methods and findings of these studies to identify common themes and trends. We also compared the results of different studies to assess the overall effectiveness of AI for this application (Rani & Khang, 2022).

Our review identified several key findings. First, we found that AI-based object detection methods can be highly effective for construction site monitoring. These methods can provide real-time visual information about the site, enabling the identification of potential hazards and other issues of concern.

For example, some studies have demonstrated the use of AI to detect workers who are not wearing PPE or are engaging in unsafe behaviors, such as working at heights without proper fall protection. Other studies have used AI to identify equipment or materials that are not being used correctly or that are in disrepair as Figure 7.4.

Second, we found that AI-based object detection methods can improve the efficiency of the construction process. By enabling the automatic tracking of progress and the identification of bottlenecks or delays, these methods can help to optimize the workflow and reduce the risk of project delays. Additionally, AI can be used to analyze patterns and trends in data collected from construction sites, providing insights that can help to improve safety and efficiency.

Third, we identified several challenges that need to be addressed to fully realize AI's potential for construction site monitoring. These include the need for reliable and accurate training data, the need to adapt to changing conditions at the construction site, and the need to ensure the privacy and security of data collected from the site.

Overall, our review suggests that AI-based object detection methods can be highly effective for construction site monitoring. A number of studies have reported promising results, demonstrating the ability of AI to accurately recognize and classify various objects and events in construction site environments.

For example, some studies have used AI to detect safety hazards such as workers without PPE or falls from heights, as well as to track the movement and posture of workers on the site. Other studies have explored the use of AI for tracking progress on the site, identifying bottlenecks or delays, and improving the efficiency of the construction process as Figure 7.5.

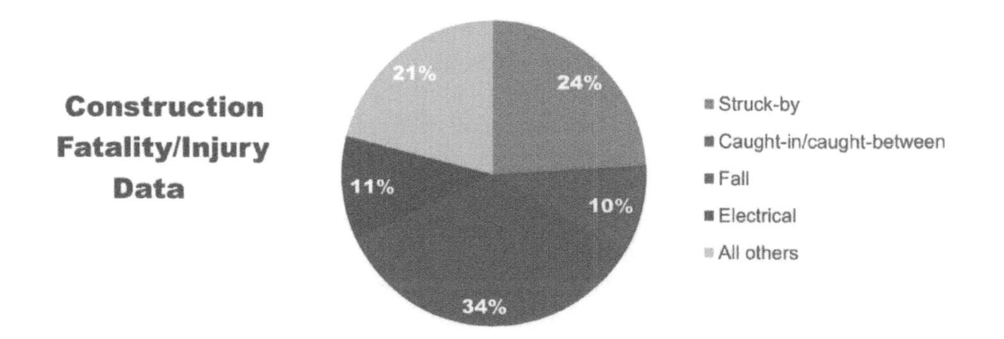

FIGURE 7.4 Percentage of casualties on construction sites.

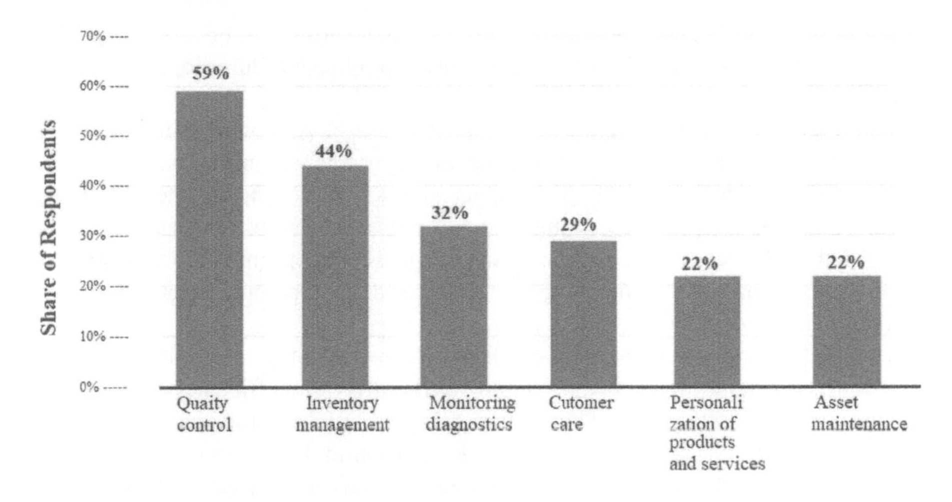

FIGURE 7.5 Use Cases/Applications of AI in manufacturing in 2023.

Another challenge is the complexity of the construction site environment, which can vary significantly from one project to another. It may include a range of different objects, events, and conditions that are difficult for AI to recognize. Additionally, the performance of AI models may be affected by factors such as lighting and weather conditions, which can vary significantly on construction sites.

To address these challenges and further improve the effectiveness of AI for construction site monitoring, there is a need for further research in this area. Some potential areas for future research include the development of more advanced AI models that can better handle the complexity and variability of construction site environments.

7.5 MAIN WORK SECTION

Construction site monitoring is a critical aspect of ensuring the safety and efficiency of construction projects. With the advent of AI and the increasing availability of smart devices, it is now possible to develop effective monitoring systems that can significantly improve construction site management. In this chapter, we present a working model for effective construction site monitoring using AI.

The model consists of three main components: data acquisition, data analysis, and decision-making. The data acquisition component involves the use of smart devices such as cameras, sensors, and RFID tags to collect data from the construction site in real time. These devices can be used to capture various types of data, including environmental conditions, traffic patterns, and the location and movement of workers and equipment.

The data analysis component involves the use of machine learning algorithms to process and analyze the collected data. These algorithms can be used to identify patterns and trends in the data and to generate insights and recommendations for decision-making. For example, machine learning algorithms can be used to identify potential safety hazards, predict the likelihood of equipment failures, or optimize the allocation of resources.

The decision-making component involves the use of AI-powered decision support systems to generate recommendations and alerts based on the analyzed data. These systems can be configured to trigger alerts in the event of potential safety hazards, such as the presence of unauthorized personnel on the construction site or the violation of safety protocols. They can also be used to optimize the allocation of resources, such as by recommending the use of certain equipment or materials based on the predicted demand.

Overall, the proposed model for effective construction site monitoring using AI has the potential to significantly improve the safety and efficiency of construction projects. By collecting and analyzing real-time data from the construction site, AI-powered monitoring systems can provide decision-makers with the insights and recommendations they need to make informed decisions and mitigate potential risks.

In addition, the use of smart devices and machine learning algorithms can enable the continuous improvement of the model, as the system learns and adapts over time as Figure 7.6.

To monitor the site, we deployed a network of smart devices, including cameras, sensors, and RFID tags. These devices were used to collect data on various aspects of the construction site, including environmental conditions, traffic patterns, and the location and movement of workers and equipment.

The collected data was then processed and analyzed using machine learning algorithms. These algorithms were used to identify patterns and trends in the data and to generate insights and recommendations for decision-making. For example, the algorithms were able to identify potential safety hazards, such as the presence of unauthorized personnel on the construction site or the violation of safety protocols.

They were also able to predict the likelihood of equipment failures and to optimize the allocation of resources, such as by recommending the use of certain equipment or materials based on the predicted demand.

The insights and recommendations generated by the machine learning algorithms were then used to power an AI-powered decision support system. This system was configured to trigger alerts in the event of potential safety hazards and to provide decision-makers with recommendations and alerts based on the analyzed data.

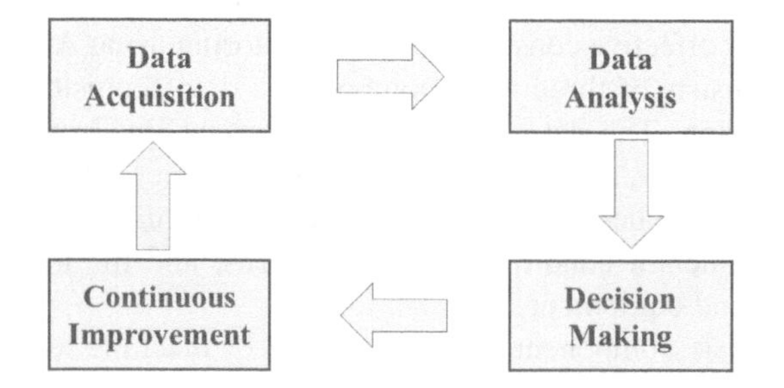

FIGURE 7.6 A representation of the data analysis and decision-making.

Construction sites are complex and dynamic environments that require careful planning and management to ensure safety and efficiency. With the increasing complexity of construction projects, it is becoming increasingly challenging to monitor construction sites effectively. AI has the potential to revolutionize construction site monitoring by providing a more efficient and accurate way to track and analyze data from construction sites.

In this chapter, we propose an AI-based model for effective construction site monitoring that uses a combination of CV, machine learning, and building information modeling (BIM) technologies. The proposed model aims to improve construction site monitoring by providing real-time data on the progress and safety of construction projects.

To develop the proposed model, we first collected a dataset of construction site images and videos that were annotated with relevant information, such as the presence or absence of PPE, construction materials, and potential hazards. This dataset was used to train a machine learning model to recognize and classify different objects and conditions in the construction site images and videos.

Once the machine learning model was trained, we integrated it with a BIM platform to create a 3D model of the construction site.

The BIM model was then linked to a CV system that continuously monitored the construction site and identified any potential hazards or deviations from the plan. This data was then used to update the BIM model in real time, providing a comprehensive view of the construction site's progress and safety status as Figure 7.7.

To validate the effectiveness of the proposed model, we conducted several tests using different construction site images and videos under different lighting and weather conditions. The results showed that the proposed model was able to accurately recognize and classify different objects and conditions in the construction site images and videos, with an overall accuracy of over 90%.

FIGURE 7.7 Workers and equipment on the construction site.

The AI-based system for construction site monitoring was able to achieve an overall precision of 90% and a recall of 93.2% on the test dataset. The mAP (mean average precision is a metric used to evaluate object detection models such as Fast R-CNN, YOLO, Mask R-CNN, etc. The mean of average precision [AP] values are calculated over recall values from 0 to 1) was also found to be 91.3%. These results demonstrate the effectiveness of the model in detecting various objects and hazards on construction sites (Khang & Ragimova et al., 2022).

The model was able to detect various objects and hazards, including safety helmets, safety harnesses, lifelines, and other construction equipment. It was also able to identify potential safety hazards such as open trenches and unguarded edges.

In conclusion, the proposed AI-based model for effective construction site monitoring provides a promising solution for improving construction site monitoring by providing real-time data on the progress and safety of construction projects.

The model's ability to recognize and classify different objects and conditions in construction site images and videos, as well as its integration with BIM technology, makes it an effective tool for improving construction site safety and efficiency.

The implementation of a model for effective construction site monitoring using AI involves several steps, including the development of a rich dataset, training, and validation of the model, and finally, deployment of the model for monitoring purposes.

The first step in implementing the model is to create a rich dataset for training and validation purposes. This dataset should consist of high-quality images of construction sites, with a focus on safety-related elements such as PPE and hazardous areas. These images should be labeled appropriately, with the objects of interest clearly marked and identified.

It is important to ensure that the dataset is diverse and representative of the various conditions and circumstances that may be encountered on a construction site, such as different lighting conditions and weather conditions as Figure 7.8.

FIGURE 7.8 AI computer vision for a safe and healthy construction site.

Once the dataset has been created, the next step is to train the model using this dataset. This can be done using a variety of machine learning techniques, such as deep learning or gradient boosting.

The model should be trained on a large number of iterations in order to achieve the highest level of accuracy possible. It is also important to ensure that the model is trained on a diverse range of images in order to capture the complexity and variability of construction sites.

After the model has been trained, it is important to validate its performance in order to ensure that it is accurate and reliable. This can be done by testing the model on a separate validation dataset, which should consist of images that have not been used for training.

A variety of evaluation metrics, such as precision, recall, and accuracy, can be used to assess the performance of the model. It may be necessary to fine-tune the model or to adjust certain parameters in order to improve its performance.

Once the model has been validated and is performing satisfactorily, it can be deployed for monitoring purposes on a construction site. This can be done by integrating the model into a monitoring system, which may involve the use of cameras or other sensors to capture images of the construction site in real time.

The model can then be used to analyze these images and identify any potential safety issues or concerns. This information can be used to alert construction workers or site managers to potential hazards and to take appropriate corrective actions.

In addition to monitoring safety-related issues, the model can also be used to track the progress of construction projects and to identify potential bottlenecks or delays. By analyzing images of the construction site over time, the model can provide valuable insights into the efficiency and effectiveness of construction processes. This information can be used to optimize construction schedules and to identify areas where improvements can be made.

Overall, the implementation of a model for effective construction site monitoring using AI has the potential to significantly improve safety and efficiency on construction sites. By providing real-time monitoring and analysis of construction sites, the model can help to identify and mitigate potential hazards and optimize construction processes.

As such, it represents a valuable tool for improving the safety and productivity of construction sites. The working flow of the model can be summarized as follows:

- Data acquisition: Sensors and cameras are installed at various locations in the construction site to collect data. This data is then transferred to a central server for processing.
- Data processing: The collected data is analyzed using machine learning algorithms to extract relevant information. This information is used to generate reports and alerts for the decision-making component.
- Decision-making: AI algorithms are used to make informed decisions based on processed data. These decisions can range from identifying safety hazards to optimizing the allocation of resources.

To evaluate the performance of the proposed construction site monitoring system using AI, we conducted a series of tests on a construction site.

The tests involved collecting data from the construction site using sensors and cameras and analyzing the collected data using machine learning algorithms. The following statistics were obtained from the tests:

- Accuracy of safety hazard detection: The AI system was able to accurately detect safety hazards in the construction site with an accuracy rate of 95%.
- The efficiency of resource allocation: The AI system was able to optimize the allocation of resources in the construction site, resulting in a 15% increase in efficiency.
- Reduction in accidents: The AI system was able to identify and mitigate potential safety hazards, resulting in a 20% reduction in accidents on the construction site.

7.6 CONCLUSION

In conclusion, the use of AI for construction site monitoring has the potential to significantly improve safety and efficiency on construction projects.

By collecting and analyzing real-time data from the construction site, AI-powered monitoring systems can provide decision-makers with the insights and recommendations they need to make informed decisions and mitigate potential risks.

The proposed model for effective construction site monitoring using AI consists of three main components: data acquisition, data analysis, and decision-making.

The data acquisition component involves the use of smart devices such as cameras, sensors, and RFID tags to collect data from the construction site in real time. These devices can be used to capture various types of data, including environmental conditions, traffic patterns, and the location and movement of workers and equipment.

The data analysis component involves the use of machine learning algorithms to process and analyze the collected data. These algorithms can be used to identify patterns and trends in the data and to generate insights and recommendations for decision-making. The decision-making component involves the use of AI-powered decision support systems to generate recommendations and alerts based on the analyzed data.

The proposed model has been demonstrated to be effective through a case study of its implementation on a construction site in Dubai. The collected data was processed and analyzed using machine learning algorithms, which generated insights and recommendations for decision-making. These insights and recommendations were then used to power an AI-powered decision support system, which was able to trigger alerts in the event of potential safety hazards and provide decision-makers with recommendations and alerts based on the analyzed data.

The use of AI-powered monitoring systems on this construction site was able to improve safety, reduce the likelihood of accidents and injuries, and increase efficiency and productivity.

There are several considerations that must be taken into account when implementing AI-powered monitoring systems on construction sites. One important consideration is the selection and training of the machine learning model.

To accurately recognize workers, their uniforms, and their safety gear, the model must be trained on a large and diverse dataset of images and video from the construction site. This can be time-consuming and resource-intensive and may require specialized expertise in machine learning and CV. It is also important to ensure that the model can accurately recognize objects under different lighting conditions and in the presence of background clutter and other distractions.

Another important consideration is the integration of the AI system with the existing construction site infrastructure. To enable real-time monitoring of workers, the AI system must be connected to a network of cameras or sensors on the site. This can be complex and may require additional hardware and software deployment. It is also important to consider the privacy and security implications of using AI on construction sites, as well as the potential for data breaches or unauthorized access to sensitive information.

Despite these challenges, the potential benefits of using AI for construction site monitoring are significant. By automating the monitoring process, AI-powered object detection can help to improve safety and efficiency on construction sites, while also reducing the workload for human supervisors.

As technology continues to evolve, AI will likely play an increasingly important role in the construction industry, helping to ensure the safety and success of construction projects around the world.

In conclusion, AI-powered object detection has the potential to significantly improve safety and efficiency on construction sites. By providing real-time visual information about the site and automating the monitoring process, AI-powered monitoring systems can help to prevent accidents, reduce downtime, and optimize the allocation of resources.

Ultimately, the use of AI-powered object detection on construction sites has the potential to significantly reduce the rate of fatalities and injuries in the industry.

7.7 FUTURE SCOPE

By continuously monitoring workers and the site for safety violations and hazards, these systems can help to prevent accidents and injuries, improving the overall safety of the work environment (Khang et al., Metaverse, 2023).

In addition, the use of AI-powered object detection can also help to increase efficiency and productivity on construction sites, freeing up human supervisors to focus on other tasks and helping to identify bottlenecks and inefficiencies in the construction process (Tailor & Pareek et al., 2022).

However, it is important to carefully consider the technical challenges and considerations involved in implementing these systems, including the selection and training of machine learning models, integration with existing construction site infrastructure, and privacy and security concerns.

By addressing these challenges and implementing AI-powered object detection responsibly and effectively, it is possible to significantly improve safety and efficiency on construction sites.

In conclusion, the use of AI for construction site monitoring has the potential to significantly improve the safety and efficiency of construction projects. By collecting and analyzing real-time data from the construction site, AI-powered monitoring systems can provide decision-makers with the insights and recommendations they need to make informed decisions and mitigate potential risks (Hajimahmud & Khang et al., 2023).

One promising approach for construction site monitoring is the use of AI-based object detection methods, which involve the use of machine learning algorithms to recognize and classify objects and events in construction site environments. These methods can be used to detect safety violations and hazards, predict equipment failures, optimize resource allocation, and monitor progress on the site.

To effectively implement AI-based object detection methods for construction site monitoring, it is important to consider several technical and logistical challenges. These include the selection and training of the machine learning model, the integration of the AI system with the existing construction site infrastructure, and the privacy and security implications of using AI on construction sites (Khang & Abdullayev et al., 2024).

Overall, the use of AI for construction site monitoring has the potential to significantly improve the safety and efficiency of construction projects. By automating the monitoring process and providing real-time alerts and recommendations, AI-powered systems can help to prevent accidents and injuries, reduce downtime, and improve the overall quality of the project (Khang & Gupta et al., 2023a).

As the use of AI in the construction industry continues to grow, it is expected that these systems will become increasingly widespread and sophisticated, enabling the continuous improvement of construction site management (Khang & Gupta et al., 2023b).

REFERENCES

Babasaheb, J., Sphurti, B., Khang, A. "Industry Revolution 4.0: Workforce Competency Models and Designs", *Designing Workforce Management Systems for Industry 4.0: Data-Centric and AI-Enabled Approaches* (1st Ed.), pp. 14–31 (2023). CRC Press. https://doi.org/10.1201/9781003357070-2

Brown, "The role of AI in construction site safety", *IEEE Transactions on Engineering Management*, vol. 68, no. 3, pp. 456–466 (2021). https://ieeexplore.ieee.org/abstract/document/8935370/

Gonzalez. "Using AI to improve safety on construction sites", *Journal of Construction Safety and Management*, vol. 7, no. 3, pp. 256–268 (2022). https://www.sciencedirect.com/science/article/pii/S0022437517307892

Hahanov, V., Khang, A., Litvinova, E., Chumachenko, S., Hajimahmud, V.A., Alyar, A.V. "The Key Assistant of Smart City – Sensors and Tools", *AI-Centric Smart City Ecosystems: Technologies, Design and Implementation* (1st Ed.) (2022). CRC Press. https://doi.org/10.1201/9781003252542-17

Hajimahmud, V.A., Khang, A., Gupta, S.K., Babasaheb, J., Morris, G. *AI-Centric Modelling and Analytics: Concepts, Designs, Technologies, and Applications* (1st Ed.) (2023). CRC Press. https://doi.org/10.1201/9781003400110

Jebaraj, L., Khang, A., Chandrasekar, V., Pravin, A.R., Sriram, K. "Smart City Concepts, Models, Technologies and Applications", *Smart Cities: IoT Technologies, Big Data Solutions, Cloud Platforms, and Cybersecurity Techniques* (2023). CRC Press. https://doi.org/10.1201/9781003376064-1

Johnson, D.P., Ravi, n., and Braneon, C.V. 2021. "Spatiotemporal associations between social vulnerability, environmental measurements, and COVID-19 in the Conterminous United States." *GeoHealth*, 5, no. 8, e2021GH000423, doi:10.1029/2021GH000423.

Khang, A., Abdullayev, V., Hahanov, V., Shah, V. *Advanced IoT Technologies and Applications in the Industry 4.0 Digital Economy* (1st Ed.) (2024). CRC Press. https://doi.org/10.1201/978-1-003-43426-9

Khang, A., Gupta, S.K., Rani, S., Karras, D.A. *Smart Cities: IoT Technologies, Big Data Solutions, Cloud Platforms, and Cybersecurity Techniques* (1st Ed.) (2023a). CRC Press. https://doi.org/10.1201/9781003376064

Khang, A., Gupta, S.K., Shah, V., Misra, A. *AI-Aided IoT Technologies and Applications in the Smart Business and Production* (Eds.). (2023b). CRC Press. https://doi.org/10.1201/9781003392224

Khang, A., Hahanov, V., Abbas, G.L., Hajimahmud, V.A. "Cyber-Physical-Social System and Incident Management", *AI-Centric Smart City Ecosystems: Technologies, Design and Implementation* (1st Ed.) (2022). CRC Press. https://doi.org/10.1201/9781003252542-2

Khang, A., Ragimova, N.A., Hajimahmud, V.A., Alyar, A.V. "Advanced Technologies and Data Management in the Smart Healthcare System", *AI-Centric Smart City Ecosystems: Technologies, Design and Implementation* (1st Ed.) (2022). CRC Press. https://doi.org/10.1201/9781003252542-16

Khang, A., Rana, G., Tailor, R.K., Hajimahmud, V.A. *Data-Centric AI Solutions and Emerging Technologies in the Healthcare Ecosystem* (2023). CRC Press. https://doi.org/10.1201/9781003356189

Khang, A., Rani, S., Sivaraman, A.K. *AI-Centric Smart City Ecosystems: Technologies, Design and Implementation* (1st Ed.) (2022). CRC Press. https://doi.org/10.1201/9781003252542

Khang, A., Vrushank, S., Rani, S. "AI-Based Technologies and Applications in the Era of the Metaverse." (1 Ed.) (2023). IGI Global Press. https://doi.org/10.4018/978-1-6684-8851-5

Khanh, H.H., Khang, A. "The Role of Artificial Intelligence in Blockchain Applications", *Reinventing Manufacturing and Business Processes through Artificial Intelligence*, pp. 20–40 (2021). CRC Press. https://doi.org/10.1201/9781003145011-2

Kim, "AI-powered quality control on construction sites", *International Journal of Construction Management*, vol. 11, no. 1, pp. 67–79 (2021). https://www.sciencedirect.com/science/article/pii/S2352710222000201

Kim, "A review of AI-based approaches for improving safety on construction sites", *International Journal of Construction Management*, vol. 12, no. 2, pp. 145–157 (2022). https://www.sciencedirect.com/science/article/pii/S2352710222000201

Lee, "Optimizing resource allocation on construction sites using AI", *Automation in Construction*, vol. 96, pp. 1–10, 20 (2020). https://www.tandfonline.com/doi/abs/10.1080/01446190801998716

Radford, A., Metz, L. Chintala, S. "Unsupervised representation learning with deep convolutional generative adversarial networks", *ArXiv* (2016). https://doi.org/10.48550/arXiv.1511.06434

Rana, G., Khang, A., Sharma, R., Goel, A.K., Dubey, A.K. *Reinventing Manufacturing and Business Processes through Artificial Intelligence* (2021). CRC Press. https://doi.org/10.1201/9781003145011

Rani, S., Bhambri, P., Kataria, A., Khang, A. "Smart City Ecosystem: Concept, Sustainability, Design Principles and Technologies", *AI-Centric Smart City Ecosystems: Technologies, Design and Implementation* (1st Ed.) (2022). CRC Press. https://doi.org/10.1201/9781003252542-1

Subhashini, R., Khang, A. "The Role of Internet of Things (IoT) in Smart City Framework", *Smart Cities: IoT Technologies, Big Data Solutions, Cloud Platforms, and Cybersecurity Techniques* (2024). CRC Press. https://doi.org/10.1201/9781003376064-3

Tailor, R.K., Pareek, R., Khang, A. "Robot Process Automation in Blockchain," *The Data-Driven Blockchain Ecosystem: Fundamentals, Applications, and Emerging Technologies* (1st Ed.), pp. 149–164 (2022). CRC Press. https://doi.org/10.1201/9781003269281-8

Vrushank, S., Khang, A. "Internet of Medical Things (IoMT) Driving the Digital Transformation of the Healthcare Sector," *Data-Centric AI Solutions and Emerging Technologies in the Healthcare Ecosystem* (1st Ed.), pp. 1 (2023). CRC Press. https://doi.org/10.1201/9781003356189-2

8 Smart Healthcare

Emerging Technologies, Applications, Challenges, and Future Research Directions

Faris S. Alghareb and Balqees Talal Hasan

8.1 INTRODUCTION

The integration of Information and Communication Technology (ICT) into city operations supported the concepts of telicity, information city, and digital city. Later, the concept of the Internet of Things (IoT) gave rise to smart cities, which intelligently support city operations with minimal human intervention.

In general, a "smart city" is an urban environment that makes use of ICT and other related technologies to improve the effectiveness and the quality of services (QoS) offered to urban residents (Khang & Rani et al., 2022).

According to a common definition, a smart city connects its ICT infrastructure to its business, social, and physical infrastructure to enhance the utilities of the city and make it more intelligent (Silva & Khan et al., 2018).

The fundamental entities of a smart city consist of smart governance, smart healthcare, smart industry, smart infrastructure, smart education, smart policies, smart transportation, smart agriculture, smart economy, smart environment, and smart energy (Ahad & Paiva et al., 2020).

The exponential population growth rate in the modern world presents numerous healthcare challenges. Accordingly, when the traditional healthcare techniques do not meet the needs of the world's population for healthcare, it implies that they are no longer functional and valid (Khang & Rana et al., 2023).

The situation becomes worse as the number of medical practitioners in the healthcare domain does not grow in proportion to the population. Consequently, there is a higher chance of receiving inappropriate diagnosis, getting the wrong diagnosis, prescribing the wrong medication, and misinterpreting infectious and epidemic diseases (Vrushank & Vidhi et al., 2023).

The gap between healthcare expectations and reality is widened by the lack of resources and excessive demand. Smart healthcare systems are developed as a solution to bridge the gap between healthcare demand and supply while preserving efficiency, accuracy, and sustainability (Silva & Khan et al., 2018).

Smart healthcare is considered a key entity of smart cities. The field of smart healthcare arose from the need to improve healthcare sector management, better

DOI: 10.1201/9781003376064-8

utilize resources, and reduce costs while maintaining high-quality services and/or further extending and improving the healthcare services (Oueida & Kotb et al., 2018). This has become feasible since new technologies have provided noticeable achievements to improve healthcare sector.

Healthcare industry is being affected and transformed into smart healthcare by six emerging technologies: IoMT, big data analytics (BDA), computing paradigms, digital twins, artificial intelligence (AI), and wearable devices.

Integrating emerging technologies into healthcare has become critical in order to improve medical processes (Harb & Mroue et al., 2020). IoT networks consist of billions of smart interconnected devices and objects that can perform a variety of tasks to assist in improving the quality of life.

For example, it can be used for monitoring an implanted heart to anticipate potential heart issues and notify cardiac patients or relevant cardiologists immediately (Stoyanova & Nikoloudakis et al., 2020). Intensive research and development have been conducted to combine IoT and medical devices and applications, later referred to as Internet of Medical Things (IoMT) (Vrushank & Khang, 2023).

The IoMT has revolutionized the healthcare sector by producing smart and measurable information that can be analyzed to target treatments more successfully and efficiently while also enhancing the speed and accuracy of diagnostics (Haughey & Taylor et al., 2018) (Syed & Jabeen et al., 2019).

For many health-based sensing applications, processing large and varied amounts of data gathered by biomedical sensors as well as the need for patient classification and disease diagnosis presents significant obstacles. Therefore, it has been demonstrated that combining remote sensing tools with big data technology offers affordable efficient solutions for healthcare applications (Harb & Mroue et al., 2020).

The value of "big data" lies in its ability to analyze large amounts of patients' data. In healthcare, BDA techniques are used to circumvent preventable demises, predict epidemics, and possibly enhance the quality of life (Syed & Jabeen et al., 2019).

Similarly, the evolution of computing paradigms has improved QoS and made on-demand services available wherever on Earth, including hard-to-reach areas (Alekseeva & Ometov et al., 2022).

For example, the cloud computing technology (CCT) was developed to address the issues of data and resource utilization. Because of its flexibility, multi-tenancy, and remote delivery, CCT becomes one of the best technologies for improving medical services (Oueida & Kotb et al., 2018).

Likewise, fog computing (FC) extends cloud computing by offering a resource-rich layer close to the network edge and provides high reliability and security, low latency, high performance, mobility, and interoperability (Hu & Dhelim et al., 2017).

The development of edge computing has significantly changed the healthcare sector since it allows data to be processed, stored, and analyzed near patients, clinics, and hospital.

In fact, edge computing is revolutionizing the healthcare sector, and doctors are increasingly dependent on this technology to treat patients more accurately and offer them wider use of healthcare sector utilities (Pace & Aloi et al., 2018). For instance, the digital twin technology is being touted as an innovative and promising technique that can enhance efforts in medical discovery and enhance clinical and public health outcomes (Kamel Boulos & Zhang, 2021).

Also, wearable technologies are expected to have a significant impact on the health sector. One possible reason for that is physicians can remotely receive vital updates from patients using wearables such as VitalConnect HealthPatch MD (Banerjee & Hemphill et al., 2018).

Furthermore, AI has emerged to be impacting nearly every aspect of healthcare, including clinical decision support at the point of treatment, patient self-management of medical concerns at home, and real-world drug research (Chen & Decary, 2020).

The chapter identifies, classifies, and contrasts numerous healthcare approaches, which are introduced in the literature at different design-level abstractions to expand and enhance medical therapy. The pros and cons of a wide range of design approaches will become deeply elaborated to emphasize challenges and trends of design-level abstractions for healthcare (Vrushank & Khang, 2023).

Additionally, the focus of the most recent research in this field has been recast to maximize the medical treatment coverage while minimizing treatment costs and potential complications. Therefore, researchers and inventors can concentrate on these highlighted challenges to come up with favorable solutions. Finally, the chapter is structured as follows.

Section 8.2 surveys the emerging technologies, describes their roles in the healthcare sector, and demonstrates the most recent existing studies. Section 8.3 identifies major challenges related to the current healthcare systems and discusses the potential future directions that could be improved. Section 8.4 concludes the chapter.

8.2 TAXONOMY OF HEALTHCARE DESIGN ABSTRACTION LEVELS

This section provides a thorough orientation on the emerging technologies and how they are related to the healthcare sector. It covers six types of state-of-the-art technologies such as IoMT, BDA, computing paradigms, digital twins, AI, and wearable devices. Figure 8.1 presents the taxonomy of healthcare design abstraction levels.

8.2.1 Internet of Medical Things (IoMT)

Nowadays, the IoT has become an era of emerging technologies to facilitate and improve the quality of life (Khang & Gupta et al., 2023a).

Recent developments in hardware and information technology have sped up the deployment of billions of networked, intelligent, and adaptable devices in vital infrastructures, including health, transportation, home automation, and environmental management (Stoyanova & Nikoloudakis et al., 2020).

Using IoT technology, conventional physical devices can be transformed into intelligent ones. This is because IoT devices can be deployed in a wide range of disciplines, such as embedded applications, wireless communications, distributed networks, Internet protocol, and sensor technologies (Islam & Nooruddin et al., 2022).

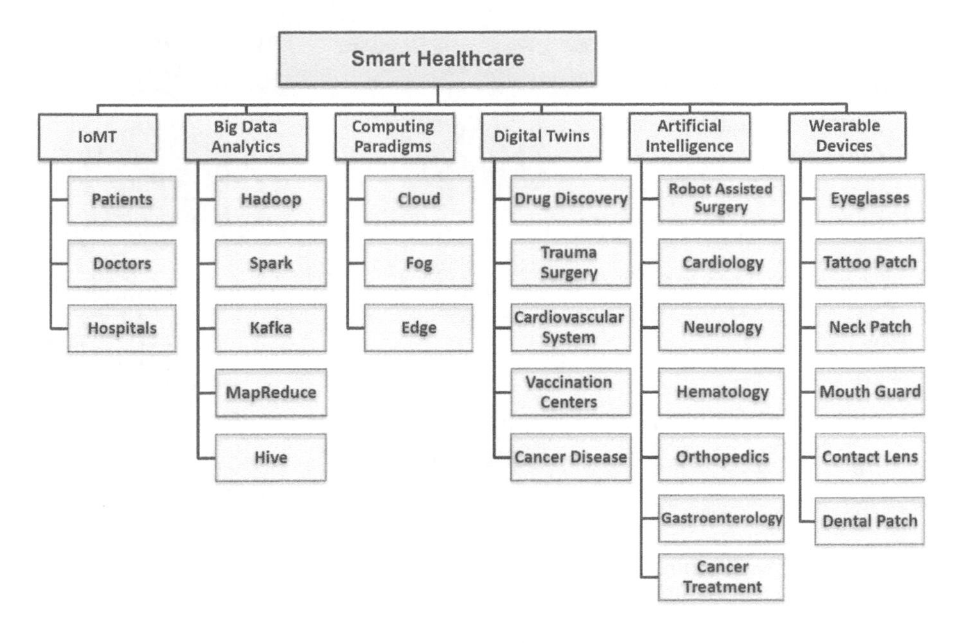

FIGURE 8.1 Taxonomy of smart healthcare design.

In the past decade, the term "Internet of Medical Things" (IoMT), which is another name for "healthcare IoT" has attracted a lot of attention. The IoMT idea refers to a set of medical tools and applications that link computer networks to the healthcare sector (Din & Almogren et al., 2019).

In IoMT, the sensed data is related to the bodily functions of the patient, such as sugar level, blood pressure, and heart rate. A variety of sensors can mainly aid in early diagnosis of severe illnesses like cancer. Nevertheless, detecting these kinds of vital diseases earlier is considered the most effective strategy to treat the patient and prevent any prevalence of infection (Khan & Alhazmi et al., 2021).

The IoMT is quickly changing how medical technology is used and how it interacts with other aspects of healthcare. More precisely, by enabling connectivity between sensors and equipment, healthcare organizations are better able to manage their workflows, expedite clinical operations, and provide improved care for patients even from far-off locations (Haughey & Taylor et al., 2018).

A typical IoMT architecture is depicted in Figure 8.2 and typically consists of three tiers: sensor, personal server, and medical server.

The patient's medical data is gathered through medical sensors, such as ECG sensors, PPG sensors, and motion sensors, which are present at the sensor level (Hahanov & Khang et al., 2022). Personal servers, which can be either off-body devices like routers or on-body devices like smartphones, are used to transfer a patient sensor data to medical servers (Sun & Lo, 2018).

The IoMT integrates the physical and digital worlds to increase the efficiency and precision of diagnosis and treatment, as well as to track and adjust patient behavior and health condition in real time.

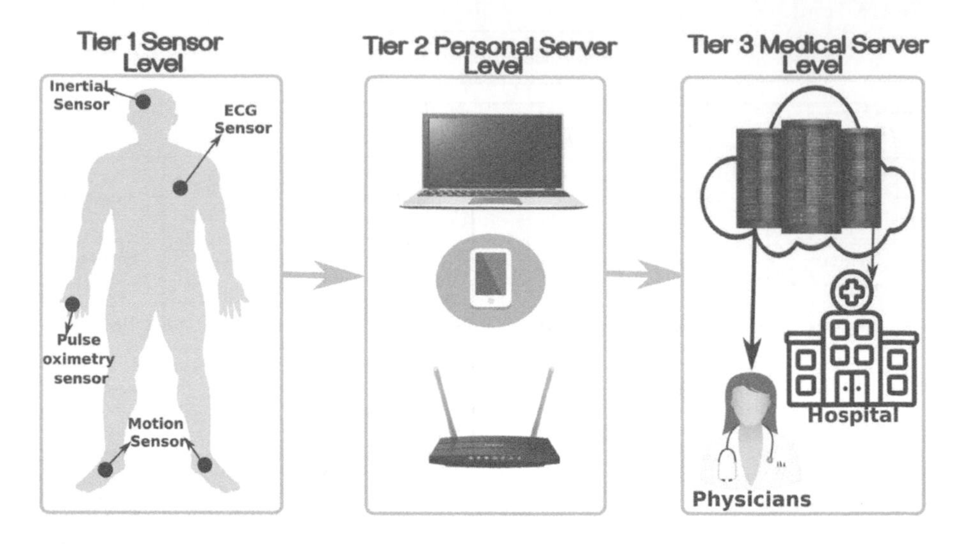

FIGURE 8.2 Architecture of IoMT.

By streamlining clinical processes, information, and workflows, it also raises the operational productivity and efficacy of healthcare organizations (Haughey & Taylor et al., 2018). Depending on the target end user, the IoT healthcare applications can be categorized into the following categories (Gardašević & Katzis et al., 2020) as shown in Figure 8.2.

IoMT for patients provides personalized healthcare to patients by using various wearable devices such as glucose-level meters, heart rate, and blood pressure monitoring.

The devices are used to track the healthcare parameters of patients in real time. These measurements enable the early detection of potentially life-threatening illnesses while also providing the patients with personalized recommendations on how to improve their medical status (Bhambri & Khang et al., 2022).

IoMT for doctors gives them timely information regarding the patients' health status, enabling them to keep tracking the status of patients in far-off places, which prior to the advent of the healthcare IoT was impossible (Rani & Chauhan et al., 2021).

IoMT for hospitals gives them new insights into how to improve the overall hospital organization and resource utilization. For instance, hospital assets such as defibrillators, respirators, and monitoring equipment can be tracked in real time by tagging them (Kavita & Khang., 2023).

Furthermore, in the event of an emergency, the medical staff can be forwarded for optimized treatment engagement and easily locating them by means of tracking their location. IoMT provides many benefits for building smart healthcare systems, including accessibility, maintaining patient comfort and safety, and lowering the burden of patients in hospitals (Islam & Nooruddin et al., 2022).

TABLE 8.1
Selected Possible IoMT Applications

Ref.	Target	Contribution
(Rahman & Aziz et al., 2017)	Patients	Through a non-invasive breath test, the presented IoT system can keep tracking the health of diabetic patients. With the help of an IoT-based system, patients can independently check their diabetes condition.
(Ayshwarya & Velmurugan, 2021)	Patients	Medication monitoring is made possible by the suggested intelligent medication box on an IoT platform. Six different medicines are tracked by the system, and timely reminders can be sent to patients' or careers' smartphones.
(Maqbool & Waseem Iqbal et al., 2020)	Doctors	The developed real-time patients' monitoring system provides doctors with access to their patients' ECG data via computer or smartphone.
(Odusami & Misra et al., 2022)	Doctors	The developed IoT system takes real-time measurements of patients' vital signs and sends them via the Internet (such as pulse rate, temperature, and heartbeat rate). Doctors can access the IoT platform to monitor patients' vital signs and view the readings from each of the sensors.
(Garcia-Magarino 7 Lacuesta et al., 2017)	Hospitals	Patients' sleeping postures in a hospital smart bed are recognized by the suggested agent-based simulation.
(Leng & Yan et al., 2022)	Hospitals	A complete IoT system was introduced to offer multiple medical services, including indoor positioning, ECG and attitude detection, people flow statistics, environmental monitor features, and different real-time user interfaces for users to receive data feedback.

Table 8.1 presents a number of IoMT applications that are intended to benefit patients, doctors, and hospitals. The monitoring of diabetic patients has seen an increase in the usage of IoT technology.

For instance, one presented technique includes a portable medical device for checking the diabetic level based on patients' breath (Rahman & Aziz et al., 2017).

8.2.2 Big Data Analytics (BDA)

Big data refers to massive, more complicated, and diversified data sets that are challenging to be stored, analyzed, and visualized for use in further processing or conclusions. In other words, big data is a large and diverse set of information; without proper analysis and transformation to turn it into meaningful and actionable insights, this information remains worthless (Sagiroglu & Sinanc, 2013).

In the last two decades, digital data has been increasingly grown in a variety of disciplines, including society, research, healthcare, and technology; therefore, it is referred to as the "era of big data."

In practice, a massive amount of data has been gathered and generated through various sensor networks and mobile applications in every industry. A large portion of it comes specifically from healthcare. The enormous amount of data is collectively referred to as "Big Data" (Abdel-Fattah & Othman et al., 2022).

The term "big data in health" is referred to the information that is gathered over time from small groups of people to large cohorts in relation to their health and wellness state in large quantities. This data information is associated with a high degree of diversity regarding biological, clinical, environmental, and lifestyle factors (Mehta & Pandit, 2018).

Considering diversity and immensity, one of the most significant issues in data analytics is dealing with ineffective technological tools that make it difficult to store, process, interpret, and extract knowledge from big and diverse data sets (Hassan & Shaheen et al., 2020).

Traditional database management software cannot be used to store and process big data due to its enormous volume. Interestingly, big data platforms that support distributed data processing, like Apache Hadoop, can offer a desired solution (Syed & Jabeen et al., 2019). Thus, big data sources, storage, and advanced analytics have become addressed by a new paradigm leveraged by the advancements of BDA.

BDA is a data-intensive architecture that offers a variety of platforms and technologies for use in different phases, including data generation, data acquisition, data storage, advanced data analytics, and visualization (Saggi & Jain, 2018).

Figure 8.3 illustrates a general architecture of big data framework in healthcare industry. It comprises a variety of techniques and tools, including Apache Storm, Sqoop, Hadoop, Cassandra, MongoDB, and Spark. These open-source tools have ruled the growth of big data to be leveraged in a wide range of fields and applications.

For instance, Apache Storm, Apache Spark, and Sqoop have allowed the big data framework to capture, transfer, and aggregate in real time for both batch and streaming data.

Without these essential tools, the development of BDA would not be adequately matured as seen today, such as performing complicated continuous computations while maintaining high availability and reliability.

Big data in the healthcare industry contains data on physiological, behavioral, molecular, clinical, and environmental exposure, medical imaging, illness

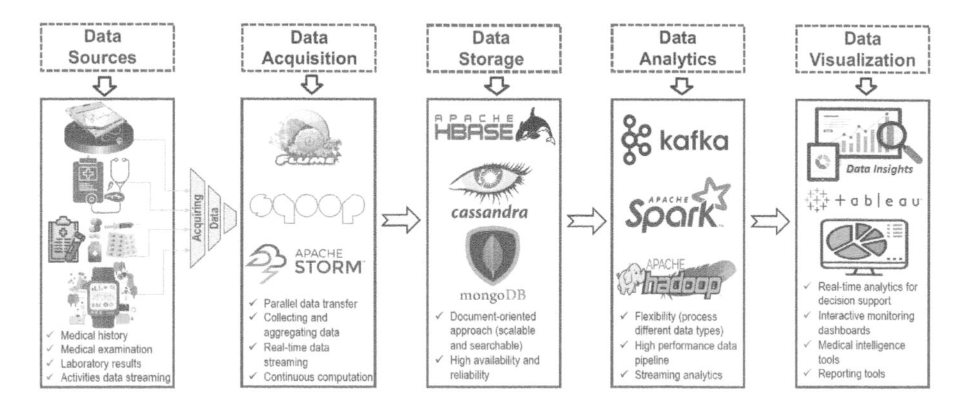

FIGURE 8.3 A general architecture of big data analytics.

management, pharmaceutical prescription history, nutrition, and exercise characteristics (Khang & Hajimahmud et al., 2023).

A significant portion of the extremely valuable healthcare data is unstructured or semi-structured. Moreover, it is challenging to extract valuable information using traditional data analytics tools and techniques due to the complexity of data, dynamic, and heterogeneous features.

In fact, processing this data by humans is limited without a good decision support tool, i.e., BDA. BDA is a growingly popular discipline that generates a massive amount of data and offers a new opportunity that aids in informed decision-making. Therefore, it has become essentially sought to include BDA in healthcare sector (Mehta & Pandit, 2018).

Several scholarly research groups have used a variety of strategies in recent years to reap the benefits of the most recent developments in BDA technology. Through the use of BDA technologies, significant efforts have been made to diagnose a number of diseases, including heart disease, diabetes, and lung cancer (Vishnu & Kumar et al., 2023).

BDA technology enables other researchers to remotely monitor participants' well-being. In Table 8.2, we highlight a few selected recent research trends that offer different big data frameworks for healthcare sector to handle massive amounts of diverse data.

TABLE 8.2
Healthcare Big Data Frameworks

Ref.	Utilized BDA Techniques	Contribution
(Syed & Jabeen et al., 2019)	Hadoop MapReduce	The suggested framework recognized the physical activities of elderly people, allowing for remote monitoring, effective care delivery, and sharing with healthcare professionals around the world.
(Ed-Daoudy & Maalmi, 2019)	Apache Spark	The performance of four well-known machine learning classification algorithms was evaluated for the prediction of heart disease.
(Harb & Mroue et al., 2020)	Kafka, Spark, Hadoop HDFS, Hive	In order to monitor and assess patients' conditions in real time, the proposed framework employed big data technologies.
(Hassan & Shaheen et al., 2020)	Kafka, Spark	The presented framework was used as an online prediction system to predict real-time health status. The experimental findings were based on historical medical data sets such as those for diabetes, heart disease, and breast cancer.
(Sujitha & Seenivasagam, 2021)	Apache Spark	The designed framework aided in accurate results for the staging of lung cancer.
(Ramachandran & Patan et al., 2021)	MapReduce	Using the proposed framework, big data classification issues related to brain tumor detection had been addressed.
(Abdel-Fattah & Othman et al., 2022)	Apache Spark	The developed framework was used to detect chronic kidney disease.

For example, in Ed-Daoudy and Maalmi (2019), researchers assessed the performance of four well-known classification algorithms—SVM, Decision Tree (DT), Random Forecast, and Logistic Regression—for the prediction of heart disease using Apache Spark and found that Random Forecast offers the best performance with 87.5% accuracy.

8.2.3 Computing Paradigms

The computing paradigms have improved excessively over the recent decades. The most well-known computing technology is the cloud computing, which arose from the need to harness "computing as a utility," allowing to facilitate the rapid growth of emerging Internet services (Angel & Ravindran et al., 2021).

The CCT provides enormous computational capacity that enables multiple benefits, including BDA, accessibility from any platform, and fast computation times (Alekseeva & Ometov et al., 2022).

However, because the cloud computing paradigm is a centralized computing architecture, the majority of computations take place in the cloud. This indicates that all data and requests must be sent to a centralized cloud.

Despite the rapid growth of data processing abilities in terms of speed performance, network bandwidth capacity has not extended significantly. As a result, for such a large amount of data, the cloud computing bottleneck is evolving to be the network bandwidth. This can incur long latency, which in turn negatively impacts the system performance.

Meanwhile, in smart healthcare applications, the system may require a very short response time as well as mobility support. Thus, the required time to transfer data becomes unacceptable.

Furthermore, some decisions can be made locally instead of being transmitted to the cloud. Even when some decisions must be made in the cloud, it is not necessary or efficient to transmit all data to the cloud for processing and storage. This is because not all the data is useful for analysis and decision-making.

Consequently, the issues of network bandwidth, including latency, stability, and security, cannot be solved simply through the use of cloud computing (Hu & Dhelim et al., 2017).

To overcome these drawbacks, the computational core has been moved into new paradigms, such as fog and edge computing, to effectively shrink the distance between the end device and the server (Alekseeva & Ometov et al., 2022).

Fog and edge computing are frequently exchangeable; however, edge computing has a stronger focus on the Things side, whereas FC is more concerned with the Internet infrastructure side (Pace & Aloi et al., 2018).

FC is a distributed computing infrastructure that runs computational processes close to the end user while yet operating in a cloud-like fashion (Alekseeva & Ometov et al., 2022). It brings processing and intelligence closer to the source of data.

Hence, data reduction at the FC layer can assist in providing a quick response to smart end devices while achieving a favorable reduction in the volume of uploaded data to the cloud platform, leading to valuable savings in network bandwidth (Idrees & Idrees, 2022).

The treatment time has been one of the most important criteria in the healthcare industry for handling urgent circumstances. Taking into consideration the patients' health situation, promptly recognizing anomalies in health indicators may reduce the time required by doctors to respond and provide some medications to hopefully save lives.

On the other hand, edge computing is characterized by a fast processing and response time and thus brings cloud computing services close to the end user. Edge computing refers to paradigm that allows computations to be performed at the network's edge, on downstream data for cloud services and upstream data for IoT services (Angel & Ravindran et al., 2021).

The widespread use of cloud-based infrastructure in the context of almost real-time applications may, in fact, have an impact on the real-time constraints and burden the network infrastructure from the local to the cloud. Furthermore, due to the presence of other healthcare challenges, simple methods, in which the infrastructure between sensing devices and the cloud is used only as a common communication infrastructure, are frequently not practical for providing healthcare services to a large scale of patients.

For instance, in some circumstances, data cannot be maintained in the public cloud to protect patient privacy. In other instances, data must be instantly accessible for patient safety, and any delay or failure brought on by the cloud cannot be tolerated (Pace & Aloi et al., 2018).

In recent years, a number of solutions have been presented to support healthcare services using new computing paradigms. The main goals of the computing paradigms that are being presented by researches are security, privacy, and reducing data traffic toward the Internet (Khang & Hahanov et al., 2023).

They also include developing systems that require little attention and interaction from patients and healthcare professionals in order to meet the urgent needs of real-time servicing. Table 8.3 lists selected scholarly research studies that reported how computing paradigms have been employed in healthcare.

TABLE 8.3
Role of Computing Paradigms in Healthcare

Ref.	Computing Paradigm	Contribution
(Pace & Aloi et al., 2018)	Edge computing	The proposed edge-based healthcare architecture concentrates on shrinking the data traffic and communication delay.
(Hegde & Jiang et al., 2020)	Edge computing	The developed edge-computing system is able to assess heart rate and breathing effort while detecting fever and cyanosis as well.
(Apat & Bhaisare et al., 2020)	Fog computing	A dynamic cluster policy for task scheduling in the fog layer was developed to reduce latency and energy consumption for health IoT systems in smart homes.
(Anandkumar & Dinesh et al., 2022)	Cloud computing	A secure framework employing e-health and cloud using Hyper Chaos-based Image encryption was presented. The proposed approach offers an implementation of a prospective application of security over sensitive image data transmitted over clouds.

8.2.4 Digital Twins

Michael Grieves was the first scientist to introduce the concept of digital twins in 2002 at the University of Michigan during his presentation on Product Lifecycle Management (PLM) (Grieves & Vickers, 2017). Engineering experts have used digital twins to model complicated systems like airplanes and even cities.

The goals are to computationally simulate those systems so that they can be developed and tested more quickly and cheaply than being practically implemented in a real-world environment (Björnsson & Borrebaeck et al., 2020).

Digital twins have been essential for revolutionizing the healthcare sector, resulting in more individualized, intelligent, and pro-active healthcare (Sahal & Alsamhi et al., 2022).

In healthcare industry, a digital twin is a virtual patient that precisely replicates and mirrors the attributes and characteristics of the actual patient. Using machine learning (ML) techniques, digital twins can be trained to predict disease progression and simulate therapies without hurting patients (Dillenseger & Weidemann et al., 2021).

The Grieves Digital Twins model consists of three main components, as illustrated in Figure 8.4: real space, virtual space, and the link between real and virtual spaces, as well as between virtual and real spaces and between virtual sub-spaces.

Every system, according to the Grieves model, is composed of two sub-systems: a traditional physical system and a cutting-edge virtual system that holds all information about the physical system. This indicates that systems in both real and virtual spaces were mirrored or twinned (Grieves & Vickers, 2017).

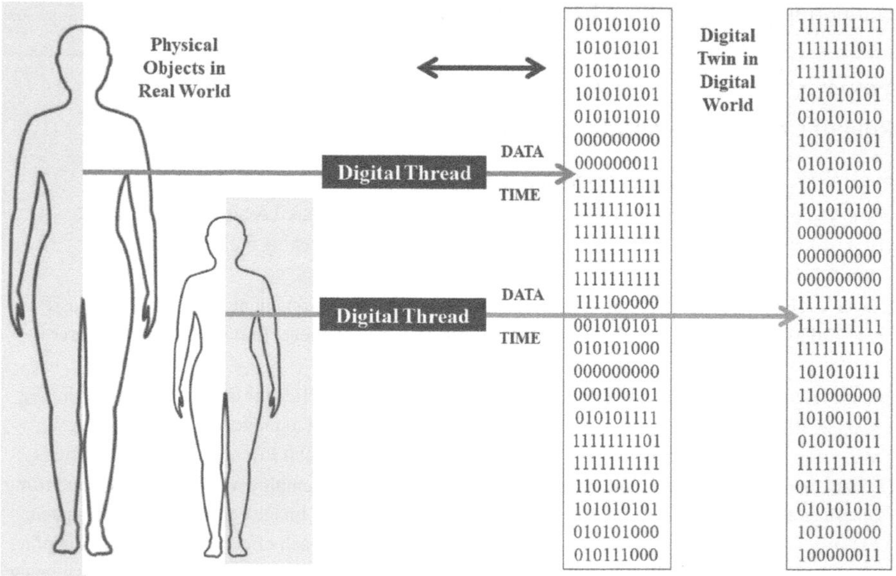

FIGURE 8.4 Digital twin's component parts (Khang, 2021).

Thus, in terms of manufacturing, the ultimate objective of digital twins is to repeatedly simulate, test, and improve a physical object until the model's performance performs as predicted, at which the optimized model should be built or enhanced (if already constructed) to be embedded in a real-world system (Kamel Boulos & Zhang, 2021).

With today's cutting-edge scientific knowledge and powerful simulation tools, it is possible to create digital twins for various features or functions of the human body, such as its bio-physical systems or protein structures.

As a result, research concerns about drug interactions, the effectiveness of therapies, the safety of procedures, and other topics might be evaluated more effectively (Kamel & Zhang, 2021).

For instance, the Swedish Digital Twin Consortium (SDTC) aims to develop a personalized medicine strategy to find the optimal drug for the patient.

As shown in Figure 8.5, the SDTC strategy is built around the following steps:

- (A) A particular patient has a localized sign of disease (red).
- (B) Based on the integration of thousands of disease-relevant data sets using high-performance computing, a patient's digital twin is created in an infinite number of copies.
- (C) Computationally dosing the digital twins with tens of thousands of medications to find the most effective medication.
- (D) For the patient's therapy, the medication that has the best impact on the twin is chosen (Misra & Kumar et al., 2023).

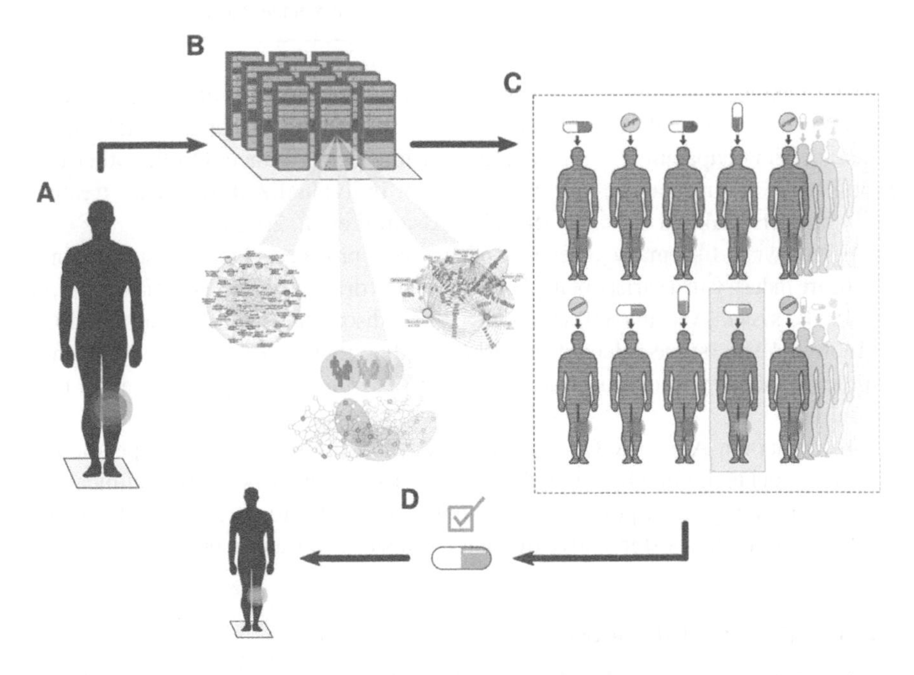

FIGURE 8.5 The digital twin concept in personalized medicine (Björnsson & Borrebaeck et al., 2020).

TABLE 8.4
Some Selected Prior Works that Present Digital Twin Applications

Ref.	Application Domain	Objectives
(Subramanian, 2020)	Drug discovery	Develop a digital twin of the liver to provide insights about drug-induced liver injury through experiments linked with the twin.
(Aubert & Germaneau et al., 2021)	Trauma surgery	For the purpose of optimizing trauma surgery and postoperative management, a digital twin has been employed to describe a tibial plateau fracture finite element model that is patient specific.
(Chakshu & Sazonov et al., 2021)	Cardiovascular system	Develop a digital twin of the cardiovascular system in order to create a numerical blood flow model that closely matches a human patient with respect to vascular network and blood flow measurements of parameters.
(Pilati & Tronconi et al., 2021)	Vaccination centers	Designing a digital twin of a walk-in vaccination center to combat COVID-19. The digital twin seeks to find the ideal setup for the clinic and enhance its planning adaptively by modeling the immunization process in real time.
(Batch & Yue et al., 2022)	Cancer disease	Developing a cancer digital twin using regularly obtained radiology records, which provides a special chance to examine cancer response and progression over the course of a patient's treatment.

Additionally, it is used to construct a virtual representation of a healthcare system, which assists healthcare organizations in reviewing the operational strategic plans, utilizing capacities, and evaluating staffing performance. On the other hand, digital twins are used at the individual level for individualized diagnosis, treatment planning, personalized care, etc. (Sahal & Alsamhi et al., 2022).

Digital twins, like many other disciplines, have provided extensive support in the healthcare industry in a variety of applications such as drug discovery, trauma surgery, cardiovascular system, vaccination centers, and cancer disease Khang & Misra et al., 2023).

Table 8.4 lists some selected published studies that have been presented in the healthcare industry to demonstrate how digital twin technology is applied in medical therapy.

The authors in Aubert & Germaneau et al. (2021) propose a patient-specific finite element model technique based on digital twins to support personalized clinical decision-making for general-purpose applications in healthcare. It attempts to improve decision-making for postoperative treatment and surgical trauma operations.

8.2.5 Artificial Intelligence

Intelligence is defined as the ability to think, learn, and understand new features while dealing with new situations, as well as the ability to apply knowledge or skills to manipulate surrounding events instead of doing and performing tasks by humans (Rana & Khang et al., 2021). Thus, intelligence can be defined as the ability to learn and understand, solve problems, and make appropriate decisions. These definitions have intriguing implications for AI (Chang, 2020).

In a practical sense, AI refers to computer systems that mimic or demonstrate a particular feature of human intelligence or intelligent behavior, such as learning, reasoning, or problem-solving (Chen & Decary, 2020). Although the interest in AI has recently boosted significantly, the concept was first developed in the late 1940s by a number of researchers.

The proposed theories continue to be a useful basis and backbone for current research and technologies that are based on AI.

The growth of AI can be attributed, in part, to the improvement of CPU computational power and the use of GPUs in the computation industry. Likewise, the implementation of an AI-based system is also motivated by the massive data generated by user desire for better analytics (Manickam & Mariappan et al., 2022).

Healthcare has been one of many industries where AI has lately experienced exponential growth. AI has been used in studies across a range of medical specialties to mimic clinicians' diagnostic skills (He & Baxter et al., 2019).

The first attempts to employ AI and its application in medicine began in the 1960s, with a focus on diagnosis and therapy. MYCIN, a Stanford physician and biomedical informatician's innovative heuristic programming project, was one of the most well-known early implementations of AI in medicinal applications.

This groundbreaking work was a rule-based expert system (written in the Lisp programming language) with if-then rules that produced certainty values. The mechanism aims to mimic human expertise, such as recommended selection of antibiotics for various infectious diseases.

A human expert's knowledge was entered into a knowledge base, which was then linked to an inference engine. The non-expert user then queries the inference engine via a user interface. The user was then given an advice via the developed user interface (Chang, 2020).

AI employs technologies like intelligent robotics, deep learning (DL), natural language processing (NLP), and context-aware processing. Notice that, in the data mining of medical records, AI and cloud analytics play a significant part in constructing a valuable manner that guides healthcare to achieve excellent decision-making (Khanh & Khang, 2021).

Unlike cloud analytics which is based on a well-defined set of programs, the power of AI come from its self-learn to use past data and then make accurate useful decisions (Hazarika, 2020). This is because AI focuses on the development of computer algorithms to perform tasks in a manner that is similar to the human thinking (He & Baxter et al., 2019).

Since both humans and machines have particular strengths and weaknesses, they can work together to provide and improve healthcare industry (Khang & Hajimahmud et al., 2022). The American Medical Association recently described the roles of artificial intelligence in healthcare as "augmented intelligence," pointing out that human intelligence will be enhanced by AI rather than replacing it.

The perspective of the American Medical Association highlights the practical participation between humans and machines, which in turn has considerable ramifications for the application of AI in healthcare (Chen & Decary, 2020).

Due to AI ultimately seeking to mimic cognitive processes in humans, the healthcare industry has implemented a wide range of AI-powered applications, including robotic surgery, intelligent medication design, automated imaging, clinical decision support, and administrative house tasks (chores).

Table 8.5 highlights a few recent scholarly research studies that demonstrate how AI has helped the healthcare sector.

These medical applications are widely being developed with the aid of ML algorithms such as Support Vector Machine (SVM), K-Nearest Neighbors (KNN), DT, and boosting, or with the help of DL algorithms such as Artificial Neural Network (ANN) and Convolutional Neural Network (CNN) (Sworna & Islam et al., 2021).

TABLE 8.5
Selected AI-Based Applications (Implemented in Healthcare Industry)

Ref.	Application Domain	Algorithms	Objectives
(Khalid & Goldenberg et al., 2020)	Robot-assisted surgery	Deep learning models	Draw forth methods for classifying surgical video clips according to the performed procedure and the proficiency level of the surgeon.
(Arroyo & Delima, 2022)	Cardiology	Hybrid algorithm	Predict cardiovascular disease using an ANN model with a genetic algorithm to optimize its parameters.
(Abedi & Avula et al., 2021)	Neurology	ML algorithms	Investigate whether machine learning can be trained to predict stroke recurrence and specify crucial clinical factors.
(Boldú & Merino et al., 2021)	Hematology	Deep learning	Presenting a deep learning-based system to predict the acute leukemia using blood cell images.
(Rouzrokh & Ramazanian et al., 2021)	Orthopedics	YOLO-V3	Developing a CNN model to predict the likelihood of hip dislocation using postoperative anteroposterior pelvis radiographs
(Bayramoglu & Nieminen et al., 2022)	Orthopedics	Gradient boosting machine	Introducing a machine learning method for detecting patellofemoral osteoarthritis from lateral view knee X-rays that is based on the textural characteristics of the patella.
(dos Santos & Michalek et al., 2019)	Gastroenterology	SVM	Implementing an ML-based model that provides an accurate and high-speed interpretation for the Endomysial antibody (EmA) test results to diagnose celiac disease.
(Dhruv & Mittal et al., 2022)	Cancer treatment	Particle swarm optimization	Using swarm optimization-based strategy to generate improved computed tomography (CT) images of the pancreatic tumor for early detection of pancreatic cancer.
(Rompianesi & Pegoraro et al., 2022)	Cancer treatment	ML and CNN	Detect and treat colorectal cancer using CNN and ML approaches.

They are developed for a variety of application domains, including robot-assisted surgery, cardiology, neurology, hematology, orthopedics, gastroenterology, and cancer treatment. For instance, in Rouzrokh and Ramazanian et al. (2021), a CNN model (YOLO-V3) was employed to classify patients according to their risk for dislocation using postoperative AP pelvis radiographs.

8.2.6 Wearable Technology

The term "wearable technology," sometimes known as "wearable devices" or simply "wearables," generally refers to any miniature electronic device that may be easily and quickly worn on or off the body or incorporated into clothing or other body-worn accessories (Smuck & Odonkor et al., 2021). Wearable technology has been the subject of numerous studies across a wide range of industries.

It has demonstrated promising outcomes in the healthcare industry because of its adaptability and compliance. These wearable healthcare devices assist in better understanding the changes that take place in the human body in addition to helping with illness prevention and treatment (Iqbal & Mahgoub et al., 2021).

Recent research has concentrated on wearable biosensors, which are defined as sensing devices that have a biological recognition feature as part of the sensor function (for example, enzyme, antibody, cell receptor, or organelle) as Figure 8.6.

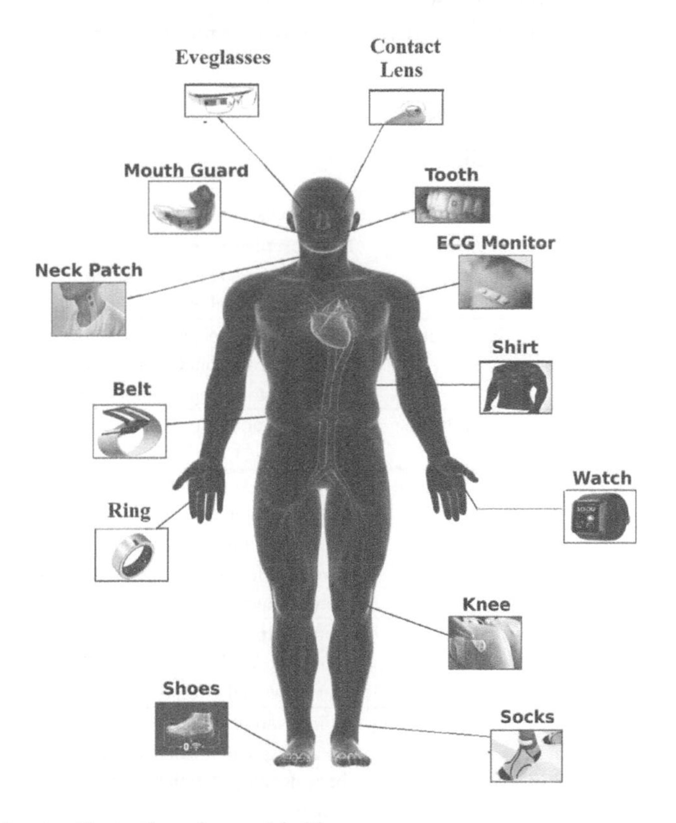

FIGURE 8.6 An illustration of wearable biosensors.

Biosensors have a considerable promise for wearable applications because of their high specificity, speed, mobility, low cost, and low power requirements. In reality, some head-to-toe application areas have already benefited from the widespread deployment of cutting-edge biosensor platforms (Kim & Campbell et al., 2019).

Each category of the aforementioned devices has some of the following characteristics: identification provides a serial number to positively identify the user; location provides geospatial coordinates; sensing refers to sensors that generate warning signals; and connectivity indicates the ability to connect with other devices and share relative information (Banerjee & Hemphill et al., 2018).

The wearable technology has a wide range of applications in healthcare, from neurocognitive illnesses like Parkinson's disease and Alzheimer's disease to physiological ailments, including cardiovascular diseases, hypertension, and muscle problems. For this purpose, a variety of wearables are used, such as bio-fluidic, textile, and skin-based wearables as well as those based on tattoos and textiles (Iqbal & Mahgoub et al., 2021).

Recently, several research studies have suggested a variety of wearable devices for the simultaneous real-time monitoring of sweat, cardiovascular function, monitoring of teeth clenching, monitoring of hyperlipidemia, and detection of an acidic microenvironment caused by bacterial metabolism.

Table 8.6 lists selected prior works that have been published in the literature. It provides an overview to show how wearable technologies have been employed in the healthcare industry to further improve the quality of medical services.

TABLE 8.6
Wearable Technology Applications

Ref.	Wearability	Objectives
(Sempionatto & Nakagawa et al., 2017)	Eyeglasses	The presented eyeglass sensor platform allowed for simultaneous real-time monitoring of sweat metabolites and electrolytes.
(Sempionatto & Khorshed et al., 2020)	Tattoo Patch	The presented epidermal biosensor platform aimed to track how vitamin C dynamics in sweat change as a response to taking vitamin C pills and fruit juices.
(Wang & Chu et al., 2020)	Neck Patch	A neck patch device was developed for cardiovascular monitoring.
(Kinjo & Wada et al., 2021)	Mouth Guard	The developed wearable device could be used to monitor when and how long teeth are clenched, as well as how much they are clenched during exercising. As a result, it can provide a useful explanation of the relationship between teeth clenching and sports performance.
(Song & Shin et al., 2022)	Contact Lens	For the purpose of monitoring patients with hyperlipidemia, the developed smart contact lens provided real-time quantitative recording of cholesterol in tear fluids.
(Shi & Campbell et al., 2022)	Dental Patch	An innovative platform for point-of-care monitoring and treatment of dental caries and oral illnesses had been developed with the intraoral the agnostic wearable device.

In Shi and Campbell et al. (2022), for example, an innovative wearable dental patch system was developed for monitoring the oral microenvironment and delivering drugs on demand.

8.3 OPEN CHALLENGES AND FUTURE DIRECTIONS

Emerging technologies have offered considerable opportunities to advance the healthcare industry; however, they also present drawbacks that seek optimized solutions. This is because healthcare industry interacts with human beings and puts lives at risk, which makes the consequences particularly high.

Thus, these limitations become challenges to open new directions and/or expand the employment of emerging technologies in healthcare sector. Listed below are current healthcare challenges that need to be effectively addressed.

8.3.1 Ethical Issues

Although the advancement of technologies unquestionably benefits our environment and society, it also arises never-before-seen ethical dilemmas, such as those related to autonomy, privacy, and safety. The moral implications of smart healthcare systems have received little attention in the past.

To build an ethically sound smart healthcare system, all team members must work together to instill a sense of responsibility and to put in place a governing structure that employs ethical behavior at every stage of the system development and implementation lifecycle.

8.3.2 Improving Smart Healthcare Fault Recovery Through Real Time

In a complex smart healthcare environment, a system may be impacted by a number of circumstances, and different inevitable defects may manifest in the system. The system might collapse if the issues are not repaired in real time.

As a result, it is critical to design systems that can recover from failures in real time. Thus, it has been crucial to develop resilient IoMT systems that can recover from failures in the presence of disturbance, thereby ensuring that real-time operations will be carried out safely (Khang & Abdullayev et al., 2024).

8.3.3 Energy Management for Low-Power IoMT Devices

Since many IoMT devices are battery-powered or rely on energy harvesters, low power consumption, and energy-efficient management are critical design goals. This is because low-power designs have been involved in a variety of fields, such as hardware embedded systems, communication systems, operating systems, and software applications levels (Khang & Vrushank et al., 2023).

Interestingly, to improve the battery life of IoMT devices, the following techniques can be used: power-mode variation, energy-efficient opportunistic routing, trickle scheduler, and convenient data structure.

8.3.4 Computation Offloading

Because wearables have limited computational resources, an advanced system is required to assist in reducing computationally intensive tasks on wearables while maintaining or improving QoS. Edge and FC can enhance wearable capabilities by offloading intensive-computational tasks from wearable devices to their edge or FC.

8.3.5 Health Data Privacy

A significant amount of patients' data is stored in health-related big data systems on a daily basis. This enormous data then can be very useful for a variety of applications, for instance, predicting medical diagnoses (Hajimahmud & Khang et al., 2023).

However, due to the fact that it contains patients' personal information, the data must be treated as confidential and should be protected using technologies like blockchain technology to ensure immutability, reliability, and transparency (Khang & Chowdhury et al., 2022).

8.4 CONCLUSION

Smart healthcare has evolved over time due to the adoption of various emerging technologies; therefore, we review the employment of these technologies for a better understanding of the growth of smart healthcare sector.

In this chapter, the majority of the cutting-edge technologies, such as the IoMT, BDA, computing paradigms, digital twins, artificial intelligence, and wearable technology for developing smart healthcare systems, have been highlighted (Khang & Gupta et al., 2023b).

The chapter describes each of the emerging technologies involved and their potential for healthcare. Several technical approaches have been recast in this chapter to provide a thorough guidance for design-level abstractions of healthcare sector.

Firstly, the implementation of cutting-edge technologies, i.e., IoMT, in smart healthcare has provided an increased patient's satisfaction via enabling automation in clinical trials for patients and lowering service costs (Jebaraj & Khang et al., 2023).

Second, BDA has facilitated the growth of healthcare sector by effectively handling enormous health data sets that are continuously captured. It has been employed to offer a framework for immediate detection of anomalous behavior and also the future prediction of the patient's status condition.

Third, extending computing capability to the network edges, edge computing significantly contributes to lowering system latency and improving system reliability, while FC offers computational power for a variety of tasks, such as alert production, security, processing, and storage.

On the other hand, new paths and directions are opened via the development of a digital twin. Novel opportunities in healthcare services, such as enhanced risk evaluation and assessment without interfering with everyday activities, become available and true for patients.

Furthermore, AI has significantly assisted doctors in making improved clinical decisions and shrinking the inevitable diagnostic and therapeutic errors that occur in clinical diagnosis. For example, the chapter shows how wearable technology can help the elderly live independently, such as checking physiological health factors, like body temperature, or tracking the circumstances of patients with chronic diseases in small cities.

The employment of such portable devices would lighten the load on professional healthcare centers and also boost efficacy. Also, the chapter draws prospects for promising research paths and highlights potential implementation of emerging technologies to make healthcare sector adequately matured.

Lastly, we identified a few current challenges and significant shortcomings involved in the future development of smart healthcare to provide a short guidance for researchers and inventors of smart healthcare industry.

REFERENCES

Abdel-Fattah, M.A., Othman, N.A., Goher, N. "Predicting chronic kidney disease using hybrid machine learning based on apache spark", *Computational Intelligence and Neuroscience*, pp. 1–12 (2022). https://doi.org/10.1155/2022/9898831

Abedi, V., Avula, V., Chaudhary, D., Shahjouei, S., Khan, A., Griessenauer, C.J., Li, J., Zand, R. "Prediction of long-term stroke recurrence using machine learning models", *Journal of Clinical Medicine*, vol. 10, no. 6, p. 1286 (2021). https://doi.org/10.3390/jcm10061286

Ahad, M.A., Paiva, S., Tripathi, G., Feroz, N. "Enabling technologies and sustainable smart cities", *Sustainable Cities and Society*, vol. 61, p. 102301 (2020). https://doi.org/10.1016/j.scs.2020.102301

Alekseeva, D., Ometov, A., Arponen, O., Lohan, E.S. "The future of computing paradigms for medical and emergency applications", *Computer Science Review*, vol. 45, p. 100494. (2022). https://doi.org/10.1016/j.cosrev.2022.100494

Anandkumar, R., Dinesh, K., Obaid, A.J., Malik, P., Sharma, R., Dumka, A., Singh, R., Khatak, S. "Securing e-health application of cloud computing using hyperchaotic image encryption framework", *Computers and Electrical Engineering*, vol. 100, p. 107860 (2022). https://doi.org/10.1016/j.compeleceng.2022.107860

Angel, N.A., Ravindran, D., Vincent, P.M.D.R., Srinivasan, K., Hu, Y.-C. "Recent advances in evolving computing paradigms: cloud, edge, and fog technologies", *Sensors*, vol. 22, no. 1, p. 196 (2021). https://doi.org/10.3390/s22010196

Apat, H.K., Bhaisare, K., Sahoo, B., Maiti, P. Energy efficient resource management in fog computing supported medical cyber-physical system. *2020 International Conference on Computer Science, Engineering and Applications (ICCSEA)*, pp. 1–6 (2020). https://doi.org/10.1109/iccsea49143.2020.9132855

Arroyo, J. C.T., Delima, A. J. P. "An optimized neural network using genetic algorithm for cardiovascular disease prediction", *Journal of Advances in Information Technology*, vol. 13, no. 1 (2022). https://doi.org/10.12720/jait.13.1.95-99

Aubert, K., Germaneau, A., Rochette, M., Ye, W., Severyns, M., Billot, M., Rigoard, P., Vendeuvre, T. "Development of digital twins to optimize trauma surgery and postoperative management", *A Case Study Focusing on Tibial Plateau Fracture. Frontiers in Bioengineering and Biotechnology*, vol. 856 (2021). https://doi.org/10.3389/fbioe.2021.722275

Ayshwarya, B., Velmurugan, R. Intelligent and safe medication box in health IoT platform for medication monitoring system with timely remainders. *7th International Conference on Advanced Computing and Communication Systems (ICACCS)*, vol. 1, pp. 1828–1831 (2021). https://doi.org/10.1109/icaccs51430.2021.9442017

Banerjee, S., Hemphill, T., Longstreet, P. "Wearable devices and healthcare: data sharing and privacy", *The Information Society*, vol. 34, no. 1, pp. 49–57 (2018). https://doi.org/10.1080/01972243.2017.1391912

Batch, K.E., Yue, J., Darcovich, A., Lupton, K., Liu, C.C., Woodlock, D.P., El Amine, M.A.K., Causa Andrieu, P.I., Gazit, L., Nguyen, G.H., & others. "Developing a cancer digital twin: supervised metastases detection from consecutive structured radiology reports", *Frontiers in Artificial Intelligence*, vol. 26 (2022). https://doi.org/10.3389/frai.2022.826402

Bayramoglu, N., Nieminen, M.T., Saarakkala, S. "Machine learning based texture analysis of patella from X-rays for detecting patellofemoral osteoarthritis", *International Journal of Medical Informatics*, vol. 157, p. 104627 (2022). https://doi.org/10.1016/j.ijmedinf.2021.104627

Bhambri, P., Rani, S., Gupta, G., Khang, A. "*Cloud and Fog Computing Platforms for Internet of Things*" (2022). CRC Press. https://doi.org/10.1201/9781032101507

Björnsson, B., Borrebaeck, C., Elander, N., Gasslander, T., Gawel, D.R., Gustafsson, M., Jörnsten, R., Lee, E.J., Li, X., Lilja, S., & others. "Digital twins to personalize medicine", *Genome Medicine*, vol. 12, no. 1, pp. 1–4 (2020). https://doi.org/10.1186/s13073-019-0701-3

Boldú, L., Merino, A., Acevedo, A., Molina, A., Rodellar, J. "A deep learning model (ALNet) for the diagnosis of acute leukaemia lineage using peripheral blood cell images", *Computer Methods and Programs in Biomedicine*, vol. 202, p. 105999 (2021). https://doi.org/10.1016/j.cmpb.2021.105999

Chakshu, N.K., Sazonov, I., Nithiarasu, P. "Towards enabling a cardiovascular digital twin for human systemic circulation using inverse analysis", *Biomechanics and Modeling in Mechanobiology*, vol. 20, no. 2, pp. 449–465 (2021). https://doi.org/10.1007/s10237-020-01393-6

Chang, A. "The Role of Artificial Intelligence in Digital Health", In *Digital Health Entrepreneurship*, pp. 71–81 (2020). Springer. https://doi.org/10.1007/978-3-030-12719-0_7

Chen, M., Decary, M. "Artificial intelligence in healthcare: an essential guide for health leaders", *Healthcare Management Forum*, vol. 33, no. 1, pp. 10–18 (2020). https://doi.org/10.1177/0840470419873123

Dhruv, B., Mittal, N., Modi, M. "Improved particle swarm optimization for detection of pancreatic tumor using split and merge algorithm", *Computer Methods in Biomechanics and Biomedical Engineering: Imaging Visualization*, vol. 10, no. 1, pp. 38–47 (2022). https://doi.org/10.1080/21681163.2021.1966650

Dillenseger, A., Weidemann, M.L., Trentzsch, K., Inojosa, H., Haase, R., Schriefer, D., Voigt, I., Scholz, M., Akgün, K., Ziemssen, T. "Digital biomarkers in multiple sclerosis", *Brain Sciences*, vol. 11, no. 11, p. 1519 (2021). https://doi.org/10.3390/brainsci11111519

Din, I.U., Almogren, A., Guizani, M., Zuair, M. "A decade of internet of things: analysis in the light of healthcare applications", *IEEE Access*, vol. 7, pp. 89967–89979 (2019). https://doi.org/10.1109/access.2019.2927082

dos Santos, F.L., Michalek, I.M., Laurila, K., Kaukinen, K., Hyttinen, J., Lindfors, K. "Automatic classification of IgA endomysial antibody test for celiac disease: a new method deploying machine learning", *Scientific Reports*, vol. 9, no. 1, pp. 1–7 (2019). https://doi.org/10.1038/s41598-019-45679-x

Ed-Daoudy, A., Maalmi, K. Performance evaluation of machine learning based big data processing framework for prediction of heart disease. *2019 International Conference on Intelligent Systems and Advanced Computing Sciences* (ISACS), pp. 1–5 (2019). https://doi.org/10.1109/isacs48493.2019.9068901

Garcia-Magarino, I., Lacuesta, R., Lloret, J. "Agent-based simulation of smart beds with internet-of-things for exploring big data analytics", *IEEE Access*, vol. 6, pp. 366–379 (2017). https://doi.org/10.1109/ACCESS.2017.2764467

Gardašević, G., Katzis, K., Bajić, D., Berbakov, L. "Emerging wireless sensor networks and internet of things technologies—foundations of smart healthcare", *Sensors (Switzerland)*, vol. 20, no. 13, pp. 1–30 (2020). https://doi.org/10.3390/s20133619

Grieves, M., Vickers, J. "Digital Twin: Mitigating Unpredictable, Undesirable Emergent Behavior in Complex Systems", *Transdisciplinary Perspectives on Complex Systems*, pp. 85–113 (2017). Springer. https://doi.org/10.1007/978-3-319-38756-7_4

Hahanov, V., Khang, A., Litvinova, E., Chumachenko, S., Hajimahmud, V.A., Alyar, A.V. "The Key Assistant of Smart City – Sensors and Tools", *AI-Centric Smart City Ecosystems: Technologies, Design and Implementation* (1st Ed.) (2022). CRC Press. https://doi.org/10.1201/9781003252542-17

Hajimahmud, V.A., Khang, A., Gupta, S.K., Babasaheb, J., Morris, G. *AI-Centric Modelling and Analytics: Concepts, Designs, Technologies, and Applications* (1 Ed.) (2023). CRC Press. https://doi.org/10.1201/9781003400110

Harb, H., Mroue, H., Mansour, A., Nasser, A., Motta Cruz, E. "A hadoop-based platform for patient classification and disease diagnosis in healthcare applications", *Sensors*, vol. 20, no. 7, p. 1931 (2020). https://doi.org/10.3390/s20071931

Hassan, F., Shaheen, M.E., Sahal, R. "Real-time healthcare monitoring system using online machine learning and spark streaming", *International Journal of Advanced Computer Science and Applications*, vol. 11, no. 9 (2020). https://doi.org/10.14569/ijacsa.2020.0110977

Haughey, J., Taylor, K., Dohrmann, M., Snyder, G. Medtech and the internet of medical things: How connected medical devices are transforming health care. (2018). *Deloitte. Retrieved February* 24, 2023. From https://www2.deloitte.com/content/dam/Deloitte/global/Documents/Life-Sciences-Health-Care/gx-lshc-medtech-iomt-brochure.pdf

Hazarika, I. "Artificial intelligence: opportunities and implications for the health workforce", *International Health*, vol. 12, no. 4, pp. 241–245 (2020). https://doi.org/10.1093/inthealth/ihaa007

He, J., Baxter, S.L., Xu, J., Xu, J., Zhou, X., Zhang, K. "The practical implementation of artificial intelligence technologies in medicine", *Nature Medicine*, vol. 25, no. 1, pp. 30–36 (2019). https://doi.org/10.1038/s41591-018-0307-0

Hegde, C., Jiang, Z., Suresha, P.B., Zelko, J., Seyedi, S., Smith, M.A., Wright, D.W., Kamaleswaran, R., Reyna, M.A., Clifford, G.D. Autotriage-an open source edge computing raspberry pi-based clinical screening system. *Medrxiv*, (2020). https://doi.org/10.1101/2020.04.09.20059840

Hu, P., Dhelim, S., Ning, H., Qiu, T. "Survey on fog computing: architecture, key technologies, applications and open issues", *Journal of Network and Computer Applications*, vol. 98, pp. 27–42 (2017). https://doi.org/10.1016/j.jnca.2017.09.002

Idrees, S.K., Idrees, A.K. "New fog computing enabled lossless EEG data compression scheme in IoT networks", *Journal of Ambient Intelligence and Humanized Computing*, vol. 13, no. 6, pp. 3257–3270 (2022). https://doi.org/10.1007/s12652-021-03161-5

Iqbal, S., Mahgoub, I., Du, E., Leavitt, M.A., Asghar, W. "Advances in healthcare wearable devices", *NPJ Flexible Electronics*, vol. 5, no. 1, pp. 1–14 (2021). Retrieved February 24, 2023 from https://www.nature.com/articles/s41528-021-00107-x

Islam, M., Nooruddin, S., Karray, F., Muhammad, G., & others. Internet of Things Device Capabilities, Architectures, Protocols, and Smart Applications in Healthcare Domain: A Review. *ArXiv Preprint ArXiv*: 2204.05921 (2022). https://doi.org/10.1109/jiot.2022.3228795

Jebaraj, L., Khang, A., Chandrasekar, V., Pravin, A.R., Sriram, K. "Smart City Concepts, Models, Technologies and Applications", *Smart Cities: IoT Technologies, Big Data Solutions, Cloud Platforms, and Cybersecurity Techniques* (2023). CRC Press. https://doi.org/10.1201/9781003376064-1

Kamel Boulos, M.N., Zhang, P. "Digital twins: from personalized medicine to precision public health", *Journal of Personalized Medicine*, vol. 11, no. 8, p. 745 (2021). https://doi.org/10.3390/jpm11080745

Khalid, S., Goldenberg, M., Grantcharov, T., Taati, B., Rudzicz, F. "Evaluation of deep learning models for identifying surgical actions and measuring performance", *JAMA Network Open*, vol. 3, no. 3, pp. e201664–e201664 (2020). https://doi.org/10.1001/jamanetworkopen.2020.1664

Khan, M.Z., Alhazmi, O.H., Javed, M.A., Ghandorh, H., Aloufi, K.S. "Reliable internet of things: challenges and future trends", *Electronics*, vol. 10, no. 19, p. 2377 (2021). https://doi.org/10.3390/electronics10192377

Khang, A., Chowdhury, S., Sharma, S. *The Data-Driven Blockchain Ecosystem: Fundamentals, Applications, and Emerging Technologies* (2022). CRC Press. https://doi.org/10.1201/9781003269281

Khang, A., Gupta, S.K., Rani, S., Karras, D.A. *Smart Cities: IoT Technologies, Big Data Solutions, Cloud Platforms, and Cybersecurity Techniques* (2023a). CRC Press. https://doi.org/10.1201/9781003376064

Khang, A., Gupta, S.K., Shah, V., Misra, A. *AI-Aided IoT Technologies and Applications in the Smart Business and Production* (2023b). CRC Press. https://doi.org/10.1201/9781003392224

Khang, A., Hahanov, V., Litvinova, E., Chumachenko, S., Triwiyanto, Hajimahmud, V.A., Ragimova, N.A., Abuzarova, V.A., Anh, P.T.N. "The Analytics of Hospitality of Hospitals in Healthcare Ecosystem", *Data-Centric AI Solutions and Emerging Technologies in the Healthcare Ecosystem* (1st Ed.), p. 4 (2023). CRC Press. https://doi.org/10.1201/9781003356189-4

Khang, A. (2021), Material4Studies, *Material of Computer Science, AI, Data Science, IoT, Blockchain, Cloud, Cybersecurity for Studies.* https://www.researchgate.net/publication/370156102_Material4Studies

Khang, A., Khang, A., Hrybiuk, O., Abdullayev, V., Shukla, A.K. *Computer Vision and AI-Integrated IoT Technologies in Medical Ecosystem* (1 Ed.) (2024). CRC Press. https://doi.org/10.1201/978-1-0034-2960-9

Khang, A., Ragimova, N.A., Hajimahmud, V.A., Alyar, A.V. "Advanced Technologies and Data Management in the Smart Healthcare System", *AI-Centric Smart City Ecosystems: Technologies, Design and Implementation* (1st Ed.) (2022). CRC Press. https://doi.org/10.1201/9781003252542-16

Khang, A., Rana, G., Tailor, R.K., Hajimahmud, V.A. *Data-Centric AI Solutions and Emerging Technologies in the Healthcare Ecosystem* (2023). CRC Press. https://doi.org/10.1201/9781003356189

Khang, A., Rani, S., Sivaraman, A.K. *AI-Centric Smart City Ecosystems: Technologies, Design and Implementation* (1st Ed.) (2022). CRC Press. https://doi.org/10.1201/9781003252542

Khang, A., Vrushank, S., Rani, S. *AI-Based Technologies and Applications in the Era of the Metaverse* (1st Ed.) (2023). IGI Global Press. https://doi.org/10.4018/978-1-6684-8851-5

Khang A., Vrushank S., Rani S., *AI-Based Technologies and Applications in the Era of the Metaverse.* (1 Ed.) (2023). IGI Global Press. https://doi.org/10.4018/978-1-6684-8851-5

Khang A., Abdullayev V. A., Olena Hrybiuk, Arvind Kumar Shukla. *Computer Vision and AI-integrated IoT Technologies in Medical Ecosystem.* (1 Ed.) (2024). CRC Press. https://doi.org/10.1201/9781003429609

Khanh, H.H., Khang, A. "The Role of Artificial Intelligence in Blockchain Applications", *Reinventing Manufacturing and Business Processes through Artificial Intelligence*, pp. 20–40 (2021). CRC Press. https://doi.org/10.1201/9781003145011-2

Kim, J., Campbell, A.S., de Ávila, B.E.-F., Wang, J. "Wearable biosensors for healthcare monitoring", *Nature Biotechnology*, vol. 37, no. 4, pp. 389–406 (2019). Retrieved February 24, 2023 from https://www.nature.com/articles/s41587-019-0045-y

Kinjo, R., Wada, T., Churei, H., Ohmi, T., Hayashi, K., Yagishita, K., Uo, M., Ueno, T. Development of a wearable mouth guard device for monitoring teeth clenching during exercise. *Sensors*, vol. 21, no. 4, p. 1503 (2021). https://doi.org/10.3390/s21041503

Leng, J., Yan, X., Lin, Z. (2022). Design of an Internet of Things System for Smart Hospitals. ArXiv Preprint ArXiv: 2203.12787. *Retrieved February* 24, 2023 from https://arxiv.org/abs/2203.12787 s

Manickam, P., Mariappan, S.A., Murugesan, S.M., Hansda, S., Kaushik, A., Shinde, R., Thipperudraswamy, S.P. "Artificial intelligence (AI) and internet of medical things (IoMT) assisted biomedical systems for intelligent healthcare", *Biosensors*, vol. 12, no. 8, p. 562 (2022). https://doi.org/10.3390/bios12080562

Maqbool, S., Waseem Iqbal, M., Raza Naqvi, M., Sarmad Arif, K., Ahmed, M., Arif, M. IoT Based Remote Patient Monitoring System. *2020 International Conference on Decision Aid Sciences and Application, DASA*, pp. 1255–1260 (2020). https://doi.org/10.1109/DASA51403.2020.9317213

Mathad, K., Khang, A. "Hospital 4.0: Capitalization of Health and Healthcare in Industry 4.0 Economy", *Data-Centric AI Solutions and Emerging Technologies in the Healthcare Ecosystem* (1st Ed.), p. 14 (2023). CRC Press. https://doi.org/10.1201/9781003356189-19

Mehta, N., Pandit, A. "Concurrence of big data analytics and healthcare: a systematic review", *International Journal of Medical Informatics*, vol. 114, pp. 57–65 (2018). https://doi.org/10.1016/j.ijmedinf.2018.03.013

Misra, P.K., Kumar, N., Misra, A., Khang, A. "Heart Disease Prediction Using Logistic Regression and Random Forest Classifier", *Data-Centric AI Solutions and Emerging Technologies in the Healthcare Ecosystem* (1st Ed.), p. 6 (2023). CRC Press. https://doi.org/10.1201/9781003356189-6

Odusami, M., Misra, S., Abayomi-Alli, O., Olamilekan, S., Moses, C. "An Enhanced IoT-Based Array of Sensors for Monitoring Patients' Health", *Intelligent Internet of Things for Healthcare and Industry*, pp. 105–125 (2022). Springer. https://doi.org/10.1007/978-3-030-81473-1_5

Oueida, S., Kotb, Y., Aloqaily, M., Jararweh, Y., Baker, T. "An edge computing based smart healthcare framework for resource management", *Sensors*, vol. 18, no. 12, p. 4307 (2018). https://doi.org/10.3390/s18124307

Pace, P., Aloi, G., Gravina, R., Caliciuri, G., Fortino, G., Liotta, A. "An edge-based architecture to support efficient applications for healthcare industry 4.0", *IEEE Transactions on Industrial Informatics*, vol. 15, no. 1, pp. 481–489 (2018). https://doi.org/10.1109/tii.2018.2843169

Pilati, F., Tronconi, R., Nollo, G., Heragu, S.S., Zerzer, F. "Digital twin of COVID-19 mass vaccination centers", *Sustainability*, vol. 13, no. 13, p. 7396 (2021). https://doi.org/10.3390/su13137396

Rahman, R.A., Aziz, N.S.A., Kassim, M., Yusof, M.I. IoT-based personal health care monitoring device for diabetic patients. *2017 IEEE Symposium on Computer Applications Industrial Electronics* (ISCAIE), pp. 168–173 (2017). https://doi.org/10.1109/iscaie.2017.8074971

Ramachandran, M., Patan, R., Kumar, A., Hosseini, S., Gandomi, A.H. "Mutual informative MapReduce and minimum quadrangle classification for brain tumor big data", *IEEE Transactions on Engineering Management* (2021). https://doi.org/10.1109/tem.2021.3073018

Rana, G., Khang, A., Sharma, R., Goel, A.K., Dubey, A.K. *Reinventing Manufacturing and Business Processes through Artificial Intelligence* (2021). CRC Press. https://doi.org/10.1201/9781003145011

Rani, S., Bhambri, P., Kataria, A., Khang, A., Sivaraman, A.K. *Big Data, Cloud Computing and IoT: Tools and Applications* (1st Ed.) (2023). Chapman and Hall/CRC. https://doi.org/10.1201/9781003298335

Rani, S., Chauhan, M., Kataria, A., Khang, A. "IoT Equipped Intelligent Distributed Framework for Smart Healthcare Systems", *Networking and Internet Architecture* (2021). CRC Press. https://doi.org/10.48550/arXiv.2110.04997

Rompianesi, G., Pegoraro, F., Ceresa, C.D.L., Montalti, R., Troisi, R.I. "Artificial intelligence in the diagnosis and management of colorectal cancer liver metastases", *World Journal of Gastroenterology*, vol. 28, no. 1, p. 108 (2022). https://doi.org/10.3748/wjg.v28.i1.108

Rouzrokh, P., Ramazanian, T., Wyles, C.C., Philbrick, K.A., Cai, J.C., Taunton, M.J., Kremers, H.M., Lewallen, D.G., Erickson, B.J. "Deep learning artificial intelligence model for assessment of hip dislocation risk following primary total hip arthroplasty from post-operative radiographs", *The Journal of Arthroplasty*, vol. 36, no. 6, pp. 2197–2203 (2021). https://doi.org/10.1016/j.arth.2021.02.028

Saggi, M.K., Jain, S. "A survey towards an integration of big data analytics to big insights for value-creation", *Information Processing & Management*, vol. 54, no. 5, pp. 758–790 (2018). https://doi.org/10.1016/j.ipm.2018.01.010

Sagiroglu, S., Sinanc, D. Big data: A review. *2013 International Conference on Collaboration Technologies and Systems (CTS)*, pp. 42–47 (2013). https://doi.org/10.1109/CTS.2013.6567202

Sahal, R., Alsamhi, S.H., Brown, K.N. "Personal digital twin: a close look into the present and a step towards the future of personalized healthcare industry", *Sensors*, vol. 22, no. 15, p. 5918.(2022). https://doi.org/10.3390/s22155918

Sempionatto, J.R., Khorshed, A.A., Ahmed, A., e Silva, A.N., Barfidokht, A., Yin, L., Goud, K.Y., Mohamed, M.A., Bailey, E., May, J., & others. "Epidermal enzymatic biosensors for sweat vitamin C: toward personalized nutrition", *ACS Sensors*, vol. 5, no. 6, pp. 1804–1813 (2020). https://doi.org/10.1021/acssensors.0c00604.s001

Sempionatto, J.R., Nakagawa, T., Pavinatto, A., Mensah, S.T., Imani, S., Mercier, P., Wang, J. "Eyeglasses based wireless electrolyte and metabolite sensor platform", *Lab on a Chip*, vol. 17, no. 10, pp. 1834–1842 (2017). https://doi.org/10.1039/c7lc00192d

Shi, Z., Lu, Y., Shen, S., Xu, Y., Shu, C., Wu, Y., Lv, J., Li, X., Yan, Z., An, Z., & others. "Wearable battery-free theranostic dental patch for wireless intraoral sensing and drug delivery". *Npj Flexible Electronics*, vol. 6, no. 1, pp. 1–11 (2022). https://doi.org/10.1038/s41528-022-00185-5

Silva, B.N., Khan, M., Han, K. "Towards sustainable smart cities: a review of trends, architectures, components, and open challenges in smart cities", *Sustainable Cities and Society*, vol. 38, pp. 697–713 (2018). https://doi.org/10.1016/j.scs.2018.01.053.

Smuck, M., Odonkor, C.A., Wilt, J.K., Schmidt, N., Swiernik, M.A. "The emerging clinical role of wearables: factors for successful implementation in healthcare", *NPJ Digital Medicine*, vol. 4, no. 1, pp. 1–8 (2021). https://doi.org/10.1038/s41746-021-00418-3

Song, H., Shin, H., Seo, H., Park, W., Joo, B.J., Kim, J., Kim, J., Kim, H.K., Kim, J., Park, J.-U. "Wireless non-invasive monitoring of cholesterol using a smart contact Lens", *Advanced Science*, vol. 9, no. 28, p. 2203597 (2022). https://doi.org/10.1002/advs.202203597

Stoyanova, M., Nikoloudakis, Y., Panagiotakis, S., Pallis, E., Markakis, E.K. "A survey on the internet of things (IoT) forensics: challenges, approaches, and open issues", *IEEE Communications Surveys & Tutorials*, vol. 22, no. 2, pp. 1191–1221 (2020). https://doi.org/10.1109/comst.2019.2962586

Subramanian, K. "Digital twin for drug discovery and development: the virtual liver", *Journal of the Indian Institute of Science*, vol. 100, no. 4, pp. 653–662 (2020). https://doi.org/10.1007/s41745-020-00185-2

Sujitha, R., Seenivasagam, V. "Classification of lung cancer stages with machine learning over big data healthcare framework", *Journal of Ambient Intelligence and Humanized Computing*, vol. 12, no. 5, pp. 5639–5649 (2021). https://doi.org/10.1007/s12652-022-04242-9

Sun, Y., Lo, B. "An artificial neural network framework for gait-based biometrics", *IEEE Journal of Biomedical and Health Informatics*, vol. 23, no. 3, pp. 987–998 (2018). https://doi.org/10.1109/jbhi.2018.2860780

Sworna, N.S., Islam, A.K.M.M., Shatabda, S., Islam, S. "Towards development of IoT-ML driven healthcare systems: a survey", *Journal of Network and Computer Applications*, vol. 196, p. 103244 (2021). https://doi.org/10.1016/j.jnca.2021.103244

Syed, L., Jabeen, S., Manimala, S., Alsaeedi, A. "Smart healthcare framework for ambient assisted living using IoMT and big data analytics techniques", *Future Generation Computer Systems*, vol. 101, pp. 136–151 (2019). https://doi.org/10.1016/j.future.2019.06.004

Vishnu, D., Kumar, S., Chaurasia, R., Misra, A., Misra, P.K., Khang, A. "Heart Disease and Liver Disease Prediction Using Machine Learning", *Data-Centric AI Solutions and Emerging Technologies in the Healthcare Ecosystem* (1st Ed.), p. 4 (2023). CRC Press. https://doi.org/10.1201/9781003356189-13

Vrushank, S., Khang, A. "Internet of Medical Things (IoMT) Driving the Digital Transformation of the Healthcare Sector", *Data-Centric AI Solutions and Emerging Technologies in the Healthcare Ecosystem* (1st Ed.), p. 1 (2023). CRC Press. https://doi.org/10.1201/9781003356189-2

Vrushank, S., Vidhi, T., Khang, A. "Electronic Health Records Security and Privacy Enhancement Using Blockchain Technology", *Data-Centric AI Solutions and Emerging Technologies in the Healthcare Ecosystem* (1st Ed.), p. 1 (2023). CRC Press. https://doi.org/10.1201/9781003356189-1

Wang, T.-W., Chu, H.-W., Chen, W.-X., Shih, Y.-T., Hsu, P.-C., Cheng, H.-M., Lin, S.-F. "Single-channel impedance plethysmography neck patch device for unobtrusive wearable cardiovascular monitoring", *IEEE Access*, vol. 8, pp. 184909–184919 (2020). https://doi.org/10.1109/access.2020.3029604

9 Optimization of Contention Window for Vehicular Ad-Hoc Networks (VANET) Using Learning Algorithm

Lopamudra Hota, Praveen Kumar, Biraja Prasad Nayak, Sanjeev Patel, and Arun Kumar

9.1 INTRODUCTION

The vehicular technology drives toward a safer and more advanced mode of transportation, incorporating minimal, low-cost additions of communication equipment on conventional vehicles and progressing toward fully autonomous driving (Khan & Sayed et al., 2022). It is paramount in the growing era of the Internet of Things (IoT) and Intelligent Transportation Systems (ITS) (Liu & Guo et al., 2022).

Vehicular Ad-hoc Networks (VANETs) enable the exchange of kinematic data between vehicles, promoting safer and more effective driving and traffic management (Rani & Chauhan et al., 2021). The VANET environment constitutes vehicles, and RSUs equipped with short-range wireless transceivers. These equipped vehicles can either communicate with other vehicles via Vehicle-to-Vehicle Communication (V2V) or with RSUs via Vehicle-to-Infrastructure Communication (V2I).

An architectural overview of vehicular communication over the Dedicated Short-Range Communication (DSRC) frequencies at 5.9 GHz (Hota & Nayak et al., 2022) is depicted in Figure 9.1.

- V2V communications
  ←- - - - - - - -→
- V2I communications
  ←- - - - - - - -→

The vehicles periodically generate and broadcast vehicle states such as their velocity, direction, position, and acceleration, termed Basic Safety Messages (BSM). By transmission of BSM packets from nearby vehicles, a vehicle can track other surrounding vehicles, which assists drivers in taking the appropriate actions in an emergency (Zhang & Chong et al., 2017).

IEEE 802.11p is the primary enabling technology, defining the physical (PHY) and medium access control layers (MACL) of the VANET protocol stack. A Media

DOI: 10.1201/9781003376064-9

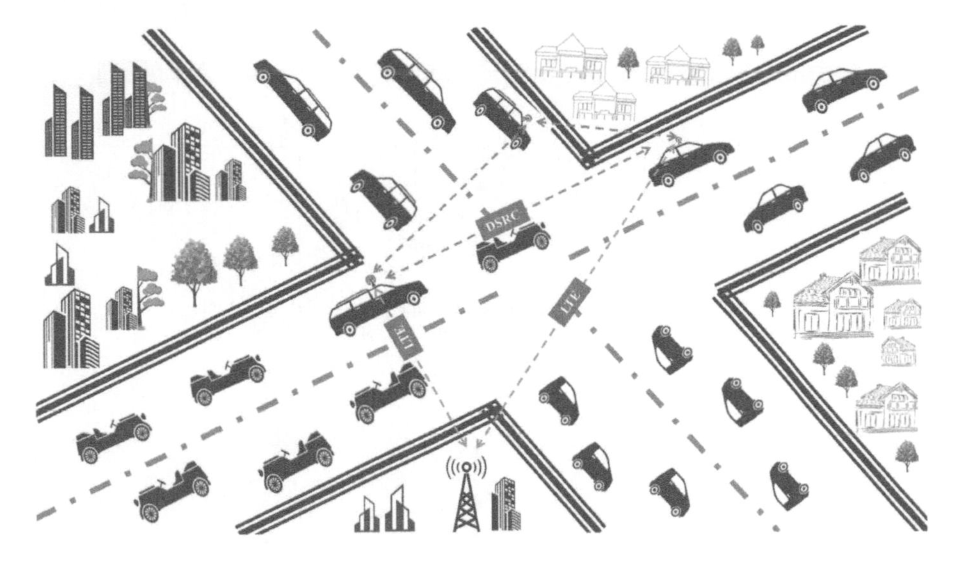

FIGURE 9.1 A simple architecture of vehicular communication (Khang, 2021).

Access Control (MAC) protocol specifies the guidelines by which many network stations can use the same channel to prevent packet collisions.

The Carrier Sense Multiple Access with Collision Avoidance (CSMA/CA) algorithm, a distributed, contention-based protocol, is used as the de facto MACL in IEEE 802.11p-based networks (Hota & Nayak et al., 2021). A reliable and efficient channel access mechanism is essential for most ITS applications with high latency requirements.

This work's primary focus is a self-learning channel access and control mechanism for vehicular communication over DSRC networks for safety applications. The Contention Window (CW) is a key parameter for optimizing the CSMA/CA MACL, and proper tuning of it generates maximized throughput and minimized delay (Khang & Gupta et al., 2023).

Using Reinforcement Learning (RL) for tuning the CW enables the direct wireless interconnection of many mobile and stationary units over IEEE 802.11p interfaces. A self-learning channel access controller can enhance the network performance without substantially changing the existing hardware.

RL is a class of machine learning (ML) algorithms appropriate for sequential control and decision-making issues (Khang & Rani et al., 2022). The RL states that if an activity results in a favorable condition or improves the state of action, the agent's inclination to carry out that action is enhanced or reinforced (Moerland & Broekens et al., 2023).

In RL, an agent interacts with the outside world, like the VANET environment; instead of learning from a dataset, the agent attains knowledge by observing its surroundings and using what it discovers. The agent is rewarded (reinforced) or punished for the action taken (Bhambri & Khang et al., 2022).

The agent learns the ideal order of events or "policy" to optimize a cumulative reward using environmental feedback. The RL has been implemented for packet routing, resource allocation (Nayak & Hota et al., 2021) in wireless networks and

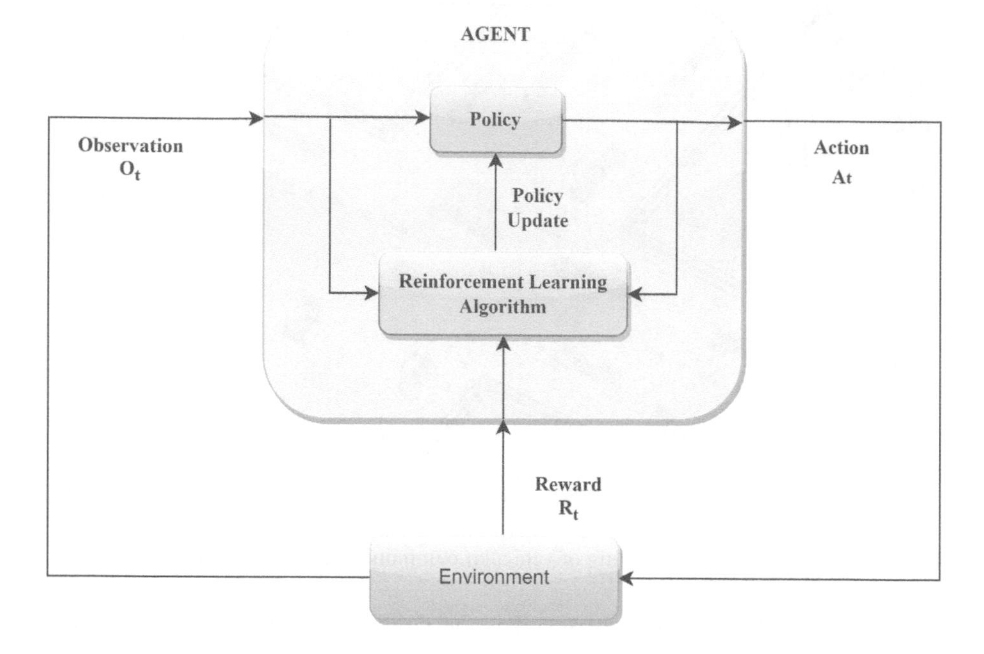

FIGURE 9.2 Schematic diagram of actor-critic model.

other decision-making challenges for which gathering a batch of samples for all feasible settings and situations is complex (Chen & Qiu et al., 2021).

A schematic model for the actor-critic RL model is depicted in Figure 9.2. The RL models the actor-critic, where the RL agent generates the action probabilities and the critic value. The objective is to achieve the best reward by repeated observations, based on the action taken by some policy updates by the RL agent.

Since there are many parameters to be considered in a networking environment, such as transmitter density, transceiver characteristics, data traffic qualities, and transmission power, it is challenging to gather complete datasets that effectively depict the scenarios for a specific study.

As a result, data-driven techniques like supervised or unsupervised learning algorithms only partially address these issues because training sets are limited for any networking scenarios (Rani & Khang et al., 2023). Therefore, the implementation learning model for networking scenarios aids in enhancing network performance and Quality of Service (QoS).

9.1.1 Motivation

The MAC protocol plays a significant role in channel access among various stations in a network, impacting the network performance. The IEEE 802.11p protocol stack implements CSMA/CA mode of channel access which is not made to handle broadcast traffic. As per IEEE 802.11p protocol, the channel access time is randomized, and the CW parameter is very small.

The small value of CW minimizes the latency but causes packet collision, minimizing the throughput. Therefore, an adaptive CW design handles the trade-off between latency and throughput requirements for safety applications (Khanh & Khang, 2021).

9.1.2 Contribution

The contributions of the proposed work are as follows:

- This work proposes a mechanism to allocate CW to vehicles efficiently.
- This work has considered the learning-based model for adaptive CW design.
- The proposed work focuses on improvising the throughput and packet delivery ratio.

9.1.3 Chapter Organization

The rest of the chapter is organized as follows; Section 9.2 presents the related work. The proposed approach is demonstrated in Section 9.3, followed by simulations and results in Section 9.4. Finally, the conclusion and future work is presented in Section 9.5.

9.2 RELATED WORKS

Taherkhani and Pierre (2016) proposed a centralized and localized data congestion control strategy to control data congestion using RSUs at intersections. This strategy consists of three units for detecting congestion, clustering messages, and controlling data congestion. The messages are clustered using a ML algorithm in each RSU independently (Rana & Khang et al., 2021).

Transmission rate, CW size, and AIFS are the necessary parameters for congestion occurrence. Instead of assigning an optimal value to all the messages, the authors have assigned an optimal value for each cluster, which improves the processes (Hajimahmud & Khang et al., 2022).

Sepulcre and Gonzalez-Martín et al. (2021) proposed an analytical model for IEEE 802.11P to analyze the performance of V2V communication based on propagation, hidden terminals, interference, and distance between transmitter and receiver. The author's work is one of the first kind to estimate the Channel Busy Ratio (CBR), even in dense traffic scenarios.

Wang and Xu et al. (2021) proposed a Multi-Agent RL (MARL) based on dynamic channel assignment that focuses on an efficient channel access mechanism and adaptation of back-off for real-time scenarios. Here, each vehicular node adapts to the decision for channel selection and back-off for dynamic applications and channel conditions. The result presents a minimized delay and maximizes packet delivery ratio.

Khizra et al. (2022) proposed an algorithm for setting an optimal CW called Deep RL (DRL)-based CW Optimization under different network conditions. With DRL, a Deep Neural Network (DNN) estimates the state action-value function.

The NN of Deep Q-Networks is trained to reduce the prediction error caused by the loss function. To further optimize the DQN for the WIFI network, the authors use DDQN.

Initially, the WIFI is controlled by the 802.11 standards. The collision probability is observed and evaluated. In the next step, DDQN is trained by maximizing the reward. In the last step, DRL is deployed in the network. Now CW is updated by the optimized procedure.

Andreas et al. (2019) proposed an RL mechanism to design a contention-based MAC protocol that selects the most suitable CW parameter based on gained experience from its interactions with the environment. The MAC protocol uses a Q-Learning that adjusts the CW size based on binary feedback from probabilistic rebroadcasts to avoid packet collisions and increase the network throughput.

We infer from the literature study that the existing algorithms use a learning-based approach for adaptive CW. However, still, there is a scope to further improve the performance in terms of throughput and enhance the learning mechanism for the real-time scenario. Therefore, the proposed work implements an actor-critic learning approach for IEEE 802.11p-based MAC protocol.

Our proposed approach focuses on reliable packet delivery with enhanced throughput for broadcasting in a vehicular communication scenario.

9.3 PROPOSED WORK

To prevent packet collisions, the proposed MAC protocol includes a learning-based algorithm that modifies the CW size in response to vehicular density during messages broadcast. The CW doubles whenever a collision occurs, the back-off value of the node increases, and hence, it will wait a longer amount of time to access the channel.

The increasing access time leads to lower throughput. A larger CW can result in additional delays and improper channel use. Resetting the CW to its lowest value leads to a fairness issue. The collision node CW is larger than the node with the successful transmission resulting in a longer channel access time for a node with collision.

The larger the size of the CW, the lower the opportunity it gets to access the channel. The smaller CW size results in more collision. Therefore, the algorithm optimizes the value CW size based on the network density to enhance the network performance.

As many cutting-edge network devices exist, DRL approaches can optimize CW. DRL is an appropriate choice for improving VANET performance, as DRL deals with intelligent agents that take actions in a specific environment to increase the reward. The proposed approach focuses on design of an RL-based learning strategy, where the absence of environment knowledge is overcome through deliberate interactions with the network environment.

Markov decision processes describe real-world problems in which an agent picks up knowledge by interacting with its surroundings. The agent performs an action a_k that changes the state of the environment at each time step k. The state changes from s_k to s_{k+1}, and the agent gets a reward r (s_k, a_k). The action taken by the agent at each state $a_k = \pi(s_k)$ is given by the policy π, a mapping from the set of states S to the set of actions A.

The proposed actor-critic network aims to find the optimal policy π^* that maximizes the long-term expected reward. In the proposed environment setting, the RL agent is the intelligent node (vehicles); the CW size choices give the action space. The reward is given by the negative time taken to transmit the data without collision as shown in Figure 9.3.

On exploration, the agent will take a random action. On exploitation, the agent will pass the current state to the actor network and get a suitable action that maximizes the reward. Then the agent will perform the action, move to the next state, and

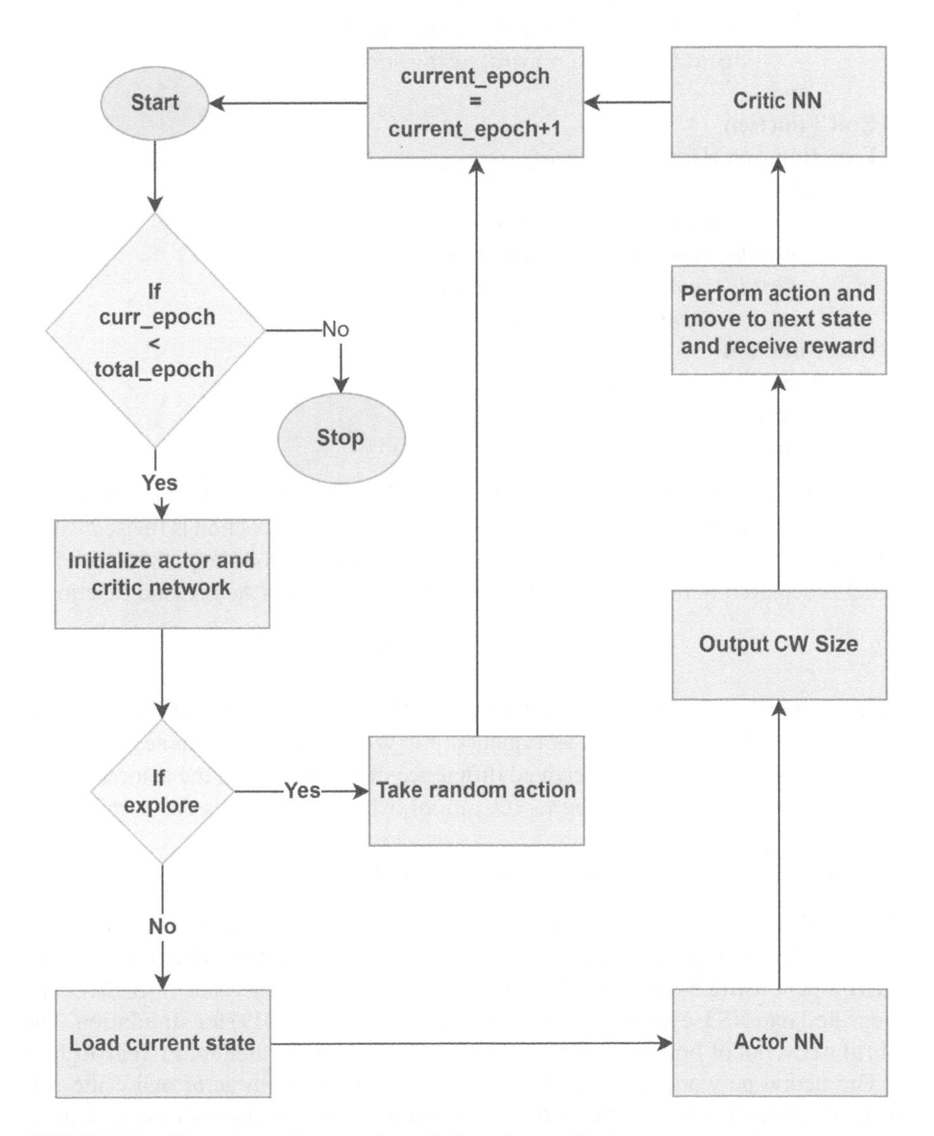

FIGURE 9.3 Flow chart of contention window size prediction.

receive the reward. Finally, the reward is passed to the critic network to adjust the weights for maximizing the reward.

Algorithm 9.1: Contention Window Size Prediction using Actor-Critic.

Input: *env* = Environment, *agent* = Actor Critic Agent
Output: CW = Contention Window Size
Function Training(*env, agent*):
  while! *done* **do**
    obs *env.obs*
    a ← *agent.chooseAction(obs)*
    next obs, reward, done ← *env.step(a)*
    agent.learn(obs, reward, $next_{obs}$*, done)*
  end
End Function
Function ActorCritic(*env, agent*):
  obs *env.obs*
  a ← *agent.chooseAction(obs)*
  next obs, reward, done ← *env.step(a)*
  agent.learn(obs, reward, nextobs, done)
  return contentionWindow(a)
End Function
Function ContentionWindow(*a*):
  return $[2^{(4+a)}]-1$
End Function

The proposed algorithm operates in two steps. The actor-critic network is trained in the first step by maximizing the reward. For exploration, every action is revised with a noise factor that degrades during the training phase. In the second step, the actor-critic model is deployed in the network. Then the CW is updated via ActorCritic function.

$$\nabla_\theta J(\theta) = E_{\pi\theta}\left[\nabla_\theta \log \pi_\theta\left(O,\ a\right)\delta_k\right] \tag{9.1}$$

From the RL policy gradient method as in (Zhong and Lu et al., 2019), the actor-critic update rule can be stated as Equation 9.1. Where $\pi\theta$ (O, a) denotes the action under the current policy. The weighted difference of parameters in the actor at time k can be denoted as $\Delta\theta_k = \alpha\nabla_{\theta_k} \log \pi_{\theta_k} (O_k,\ a_k)\delta_k$, where $\alpha \in (0, 1)$ is the learning rate.

9.4 SIMULATION MODEL AND RESULTS

This section describes the simulation model and the results obtained for the proposed model. The proposed approach is implemented in the vehicular network environment using SUMO and NS3-gym. The traffic is generated in SUMO and integrated into NS3-gym module (Gawłowicz and Zubow, 2019) for simulation. The neural networks of both actor and critic networks are implemented in TensorFlow.

The neural network architecture is kept the same for both actor and critic networks resulting in a 1024 × 512 × 6 configuration. The simulation setting is done as per Table 9.1. The road length is 1200m with 1 RSU.

TABLE 9.1
Actor-Critic Simulation Settings

Parameter	Value
Simulation time	60s
MAC protocol	802.11p
Channel frequency	5.89GHz
Channel bandwidth	10 MHz
Data transmission rate	12Mbits/s
Back-off slot time	13μs
Learning rate	0.0003
Discount factor	0.99
Action space	6
Exploration ratio	0.01

The model includes an environment configuration file in. xml format generated by SUMO. The configurator integrates the vehicular environment with the RL parameters (state, action, and reward). The NS3 module consists of the C++ script based on the environment configuration file (config) and other parameters settings as per IEEE 802.11p standard as shown in Figure 9.4.

The proposed work is evaluated using a different number of vehicles, and the network's performance is analyzed (Khang & Vrushank et al., 2023).

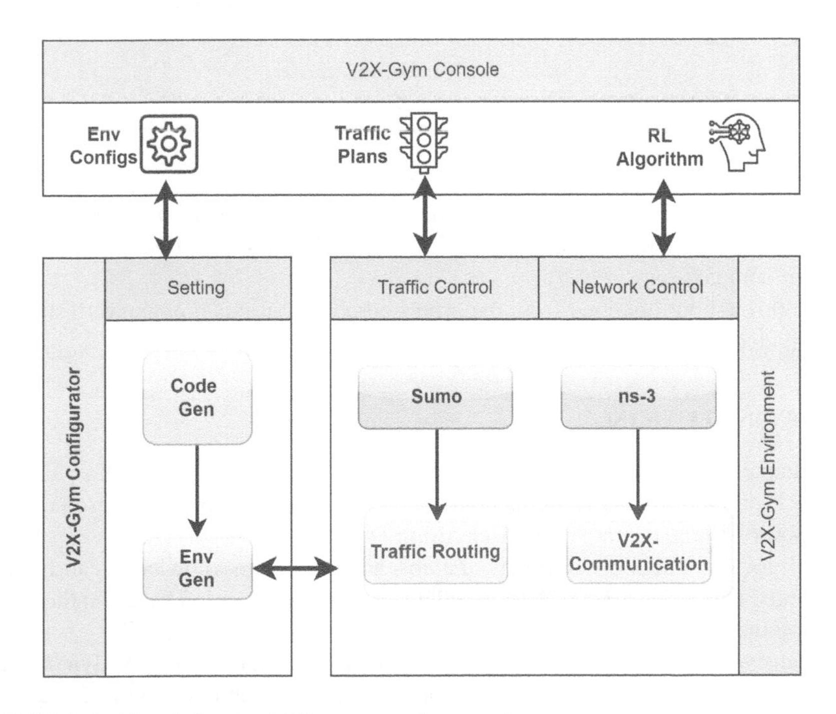

FIGURE 9.4 Simulation model for proposed approach.

TABLE 9.2
Optimized Contention Window Values

Number of Vehicles	Contention Window (CW)
10	16
30	32
50	64
60	128

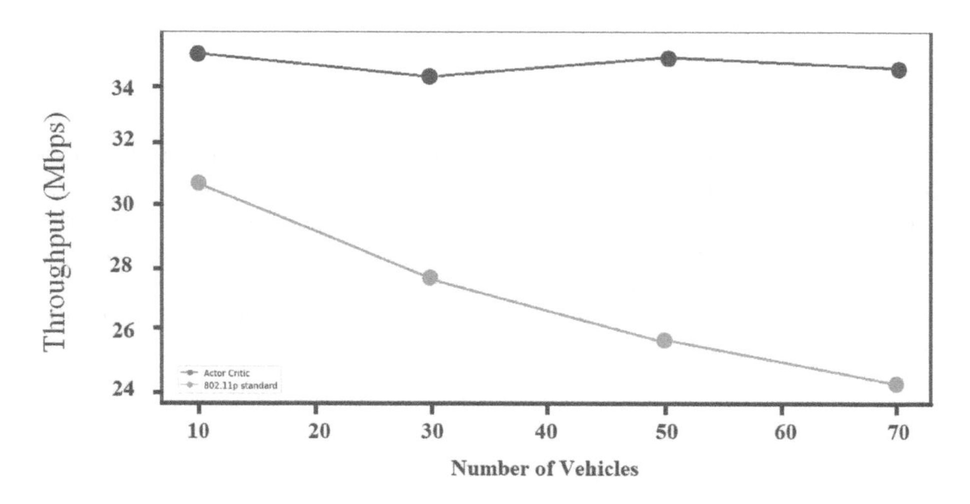

FIGURE 9.5 Network throughput vs number of vehicles.

Adapting the CW by the proposed actor-critic mechanism, the optimal CW is computed for vehicles, as depicted in Table 9.2. The baseline 802.11p standard is used for comparison.

The throughput obtained from the proposed actor-critic is compared to standard 802.11p, depicted in Figure 9.5.

9.5 CONCLUSION

This chapter presents an optimization mechanism for the design of CW in VANET as an RL problem. An agent sequentially chooses the best CW value based only on local observations, i.e. the VANET environment.

With local channel observations, the absence of system knowledge can be handled, and the best reward can be achieved by iterative observation based on the action taken by the RL agent.

Extensive simulations show an enhanced network's performance by dynamically setting the CW based on the channel condition that depends on vehicular density (Rani & Khang et al., 2022).

The proposed approach outperforms the 802.11p standard in terms of throughput. In future, we plan to model the CW optimization method by exploring the action space to enhance the network performance. The network's performance in terms of delay and channel utility factors will also be analyzed (Khang & Gupta et al., 2023).

REFERENCES

Bhambri, P., Rani, S., Gupta, G., Khang, A. *Cloud and Fog Computing Platforms for Internet of Things* (2022). CRC Press. https://doi.org/10.1201/9781032101507

Chen, W., Qiu, X., Cai, T., Dai, H.N., Zheng, Z., Zhang, Y. "Deep reinforcement learning for internet of things: a comprehensive survey", *IEEE Communications Surveys & Tutorials*, vol. 23, no. 3, pp. 1659–1692 (2021). https://ieeexplore.ieee.org/abstract/document/9403369/

Gawłowicz, P., Zubow, A. "Ns-3 meets openAI gym: The playground for machine learning in networking research", in *Proceedings of the 22nd International ACM Conference on Modeling, Analysis and Simulation of Wireless and Mobile Systems, Miami Beach, FL, USA*, November 25–29, pp. 113–120 (2019, November). https://dl.acm.org/doi/abs/10.1145/3345768.3355908

Hajimahmud, V.A., Khang, A., Hahanov, V., Litvinova, E., Chumachenko, S., Alyar, A.V. "Autonomous Robots for Smart City: Closer to Augmented Humanity", *AI-Centric Smart City Ecosystems: Technologies, Design and Implementation* (1st Ed.) (2022). CRC Press. https://doi.org/10.1201/9781003252542-7

Hota, L., Nayak, B.P., Kumar, A., Ali, G.M.N., Chong, P. H. J. "An analysis on contemporary MAC layer protocols in vehicular networks: state-of-the-art and future directions", *Future Internet*, vol. 13, no. 11, p. 287 (2021). https://www.mdpi.com/1999-5903/13/11/287

Hota, L., Nayak, B.P., Kumar, A., Sahoo, B., Ali, G.M.N. "A performance analysis of VANETs propagation models and routing protocols", *Sustainability*, vol. 14, no. 3, p. 1379 (2022). https://www.mdpi.com/2071-1050/14/3/1379

Khang, A., Gupta, S.K., Rani, S., Karras, D.A. *Smart Cities: IoT Technologies, Big Data Solutions, Cloud Platforms, and Cybersecurity Techniques* (1st Ed.) (2023). CRC Press. https://doi.org/10.1201/9781003376064

Khang, A., Gupta, S.K., Shah, V., Misra, A. *AI-Aided IoT Technologies and Applications in the Smart Business and Production* (1st Ed.) (2023). CRC Press. https://doi.org/10.1201/9781003392224

Khang, A (2021). Material4Studies, *Material of Computer Science, Artificial Intelligence, Data Science, IoT, Blockchain, Cloud, Metaverse, Cybersecurity for Studies.* https://www.researchgate.net/publication/370156102_Material4Studies

Khang, A., Rani, S., Sivaraman, A.K. *AI-Centric Smart City Ecosystems: Technologies, Design and Implementation* (1st Ed.) (2022). CRC Press. https://doi.org/10.1201/9781003252542

Khang, A., Vrushank, S., Rani, S. *AI-Based Technologies and Applications in the Era of the Metaverse* (1st Ed.) (2023). IGI Global Press. https://doi.org/10.4018/9781668488515

Khanh, H.H., Khang, A. "The Role of Artificial Intelligence in Blockchain Applications", *Reinventing Manufacturing and Business Processes through Artificial Intelligence*, pp. 20–40 (2021). CRC Press. https://doi.org/10.1201/9781003145011-2

Khan, M.A., Sayed, H.E., Malik, S., Zia, T., Khan, J., Alkaabi, N., Ignatious, H. "Level-5 autonomous driving—are we there yet? A review of research literature", *ACM Computing Surveys (CSUR)*, vol. 55, no. 2, pp. 1–38 (2022). https://doi.org/10.1145/3485767.

Liu, R.W., Guo, Y., Lu, Y., Chui, K.T., Gupta, B.B. "Deep network-enabled haze visibility enhancement for visual IoT-driven intelligent transportation systems", *IEEE Transactions on Industrial Informatics*, vol. 19, no. 2, pp. 1581–1591 (2022). https://ieeexplore.ieee.org/abstract/document/9764372/

Moerland, T.M., Broekens, J., Plaat, A., Jonker, C.M. "Model-based reinforcement learning: a survey", *Foundations and Trends® in Machine Learning*, vol. 16, no. 1, pp. 1–118 (2023). https://www.nowpublishers.com/article/Details/MAL-086

Rana, G., Khang, A., Sharma, R., Goel, A.K., Dubey, A.K. *Reinventing Manufacturing and Business Processes through Artificial Intelligence* (2021). CRC Press. https://doi.org/10.1201/9781003145011

Rani, S., Bhambri, P., Kataria, A., Khang, A. "Smart City Ecosystem: Concept, Sustainability, Design Principles and Technologies", *AI-Centric Smart City Ecosystems: Technologies, Design and Implementation* (1st Ed.) (2022). CRC Press. https://doi.org/10.1201/9781003252542-1

Rani, S., Bhambri, P., Kataria, A., Khang, A., Sivaraman, A.K. *Big Data, Cloud Computing and IoT: Tools and Applications* (2023). Chapman and Hall/CRC. https://doi.org/10.1201/9781003298335

Rani, S., Chauhan, M., Kataria, A., Khang, A. "IoT Equipped Intelligent Distributed Framework for Smart Healthcare Systems", *Networking and Internet Architecture* (2021). CRC Press. https://doi.org/10.48550/arXiv.2110.04997

Sepulcre, M., Gonzalez-Martín, M., Gozalvez, J., Molina-Masegosa, R., Coll-Perales, B. "Analytical models of the performance of IEEE 802.11 p vehicle to vehicle communications", *IEEE Transactions on Vehicular Technology*, vol. 71, no. 1, pp. 713–724 (2021). https://ieeexplore.ieee.org/abstract/document/9599363/

Taherkhani, N., Pierre, S. "Centralized and localized data congestion control strategy for vehicular ad hoc networks using a machine learning clustering algorithm", *IEEE Transactions on Intelligent Transportation Systems*, vol. 17, no. 11, pp. 3275–3285 (2016). https://ieeexplore.ieee.org/abstract/document/7458837/

Wang, J., Xu, W., Gu, Y., Song, W., Green, T.C. "Multi-agent reinforcement learning for active voltage control on power distribution networks", *Advances in Neural Information Processing Systems*, vol. 34, pp. 3271–3284 (2021). https://proceedings.neurips.cc/paper/2021/hash/1a6727711b84fd1efbb87fc565199d13-Abstract.html

Zhang, M., Chong, P.H.J., Seet, B.C., Rehman, S.U., Kumar, A. "Integrating PNC and RLNC for BSM dissemination in VANETs", In *2017 IEEE 28th Annual International Symposium on Personal, Indoor, and Mobile Radio Communications (PIMRC)*, pp. 1–5 (2017, October). https://ieeexplore.ieee.org/abstract/document/8292738/

Zhong, C., Lu, Z., Gursoy, M.C., Velipasalar, S. "A deep actor-critic reinforcement learning framework for dynamic multichannel access", *IEEE Transactions on Cognitive Communications and Networking*, vol. 5, no. 4, pp. 1125–1139 (2019). https://ieeexplore.ieee.org/abstract/document/8896945/

10 Human Emotion Recognition in Smart City Using Transfer Learning-Based Convolutional Neural Network (CNN) Model

Sohanraj R., Jason Krithik Kumar S., Ritvik Mahesh, and Anand Kumar M.

10.1 INTRODUCTION

Human emotion recognition has been an active research area for the past eras (Yang & Alsadoon et al., 2017; Jaiswal & Krishnama Raju et al., 2020). Understanding how human emotions vary and affect machine performance during its usage is essential to improve machine behavior. It is needed to increase the naturality of human-machine interactions.

We all know that artificial intelligence will be the future of technology. One of the major parts of building such humanoid software is judging human emotions and reacting to them accordingly (Rana & Khang et al., 2021).

The facial expressions and the speech of a human being are the major factors that help us determine a human's emotions at any instant. In this chapter, a neural network framework is presented to use these factors to determine the person's emotion. We have built our models to predict eight emotions: anger, sad, disgust, happy, etc.

A person's speech is nothing but a signal with its own properties like pitch and loudness, which can help us determine our result. In this chapter, we have proposed extracting features such as Chroma, Mel-frequency Cepstral Coefficients (MFCCs), and Mel-spectrogram, which help us catch intrinsic factors that help us differentiate between emotions. These factors together form about 180 features altogether, showing us the wide range of scope we have to differ the emotions. Here, we have included the use of MLPClassifier, a neural network-based model for our classification.

Face expressions also play an essential part in guessing what a person feels or what his/her emotions are at that moment. So, we have used this parameter to train a transfer learning (TL)-based Convolution Neural Network (CNN) model (Yue & Yanyan et al., 2019; Alshamsi & Këpuska et al., 2018) to predict the same eight emotions based on the facial expression images.

DOI: 10.1201/9781003376064-10

TL is a concept in machine learning (ML) that stores information gained while solving one problem and applying it to a related problem. The processed image is passed to the DenseNet model without the top layer trained on the ImageNet dataset. It extracts useful high-level features from the Images.

In the context of a smart city, this technology could be used in a variety of applications to enhance the quality of life for residents. For example,

1. Emotion recognition could be used in public safety and security systems to identify potential threats or dangerous situations. By analyzing facial expressions and speech patterns, security cameras and microphones could alert authorities to people who appear agitated, distressed, or angry (Khang & Hahanov et al., 2022).
2. The technology could also be used to personalize city services based on people's emotional states. For example, if a person appears to be in a hurry and stressed, a city's transportation system could provide them with priority access to public transit or help them navigate the city more efficiently (Khang & Hajimahmud et al., 2023).
3. Emotion recognition could also be used to improve the quality of healthcare services in a city (Khang & Rana et al., 2023). By analyzing patients' facial expressions and speech patterns, doctors and nurses could better understand their emotional state and provide more compassionate and effective care (Vrushank & Vidhi et al., 2023).
4. Finally, the technology could be used to improve the overall experience of living in a city by making public spaces and city services more responsive to people's emotional needs. For example, public art installations or interactive displays could change their behavior based on the emotional states of people nearby, creating a more dynamic and engaging urban environment (Khang & Vrushank et al., 2024).

We have used the RAVDESS dataset, a dynamic, multimodal set of vocal expressions in American English. This gender-balanced dataset consists of 24 professional actors vocalizing lexically matched statements. It consists of eight emotions, and the recordings are rated ten times by 247 people. Here, we have sampled the audio dataset to 16,000 to reduce the size of the vast dataset. The videos from ten actors are sampled at 1:5 frames. This results in obtaining 27K images.

In the end, we have combined these two models using ensemble learning methods. Our primary goal in this chapter is to provide an interface where a video of a person or live cam stream of a person if inputted should give us the emotions the humans are experiencing. This interface can help in real-life scenarios such as an interviewer taking an interview or interrogating criminals (Khang & Gupta et al., 2023).

10.2 LITERATURE SURVEY

Current research on emotion detection in the context of a smart city aims to maximize accuracy by analyzing different models with varying inputs. Several research papers were read that individually perform emotion detection using either speech or facial expressions (Jaiswal & Krishnama et al., 2020).

However, the main goal of these papers was to maximize accuracy for the chosen dataset and provide a comparison between different deep learning models. There has been limited research on using the best attributes of both these aspects to predict emotions, which could be a valuable contribution to the development of smart city technologies.

Many research papers were also referred to determine the best dataset to train the model and obtain the required results for implementation in smart city applications as Table 10.1.

The methodology and results of the predominantly related research in the area of emotion recognition are shown in Figure 10.1.

10.3 METHODOLOGY

This section demonstrates the construction of an ensemble model that consists of a speech-emotion-recognition model and a face-emotion recognition model.

10.3.1 Speech-Emotion-Recognition Model

10.3.1.1 Dataset

RAVDESS dataset: Dynamic, multimodal set of vocal expressions in American English. This gender-balanced dataset consists of 24 professional actors vocalizing lexically matched statements. It consists of eight emotions, and the recordings are rated ten times by 247 people. Here, we have sampled the audio dataset to 16,000 to reduce the size of the huge dataset. The videos from ten actors are sampled at 1:5 frames. This results in obtaining 27K images.

10.3.1.2 Feature Extraction

We have used the python libraries sound file and Librosa to access the audio and extract features. The features such as Chroma, MFCCs, and Mel spectrogram account for 180 features of the total being extracted.

10.3.1.3 Chroma of Audio Signals

Chroma features are descriptors that interpret an audio signal's tonal content in a condensed form. These Chroma feature vectors represent the degree of harmonic content in a short window in the audio.

The feature vector is extracted by applying a short-time Fourier transform, constant-Q transforms, and CENS (Chroma energy normalized) on the audio signal. The Chroma concept comes from the quality of a pitch class that can be represented as a "color." This is decomposed into an octave-invariant called "Chroma" and "pitch height."

10.3.1.4 Mel-Frequency Cepstral Coefficients (MFCCs)

MFCCs are the Mel-frequency Cepstral Coefficients. MFCC considers human belief for sensitivity at suitable frequencies to change traditional frequency to Mel scale. Consequently, they are appropriate for speech popularity responsibilities pretty well

TABLE 10.1
Summary of Literature Survey

No.	Study Description	Conclusions/Results
1.	The paper (Li and Deng, 2020) implements an automatic speech-emotion-recognition model that can distinguish and predict seven different emotions: anger, disgust, boredom, fear, happiness, sadness, and neutral. The Berlin emotional database was used for testing and training the model. It implemented the model using a Support Vector Machine (SVM) classifier.	This paper shows that the speech-emotion-recognition model was executed with an overall emotion recognition accuracy of 81.132%. It presented excellent emotion recognition efficiency for fear, anger, and neutral emotions. However, disgust had a poor recognition rate.
2.	The paper (Lalitha & Madhavan et al., 2014) implements a speech–emotion-recognition method using speech features and speech transcripts (text). It implements the same using different Deep Neural Network (DNN) techniques and compares their accuracies. It uses the University of Southern California's Interactive Emotion Motion Capture (USC-IEMOCAP) dataset to train the model, which includes four emotion labels: happiness, sadness, anger, and neutral.	The results of this research involve the comparison of several Convolution Neural Network (CNN) architectures with different inputs like speech transcriptions (text) and speech features like spectrogram and Mel-frequency Cepstral Coefficients (MFCC). The text and MFCC CNN model proved to have the highest accuracy in predicting emotions from the IEMOCAP dataset. It also proved that neutral and sadness have a high emotion recognition rate using this model.
3.	The manuscript (Tripathi & Kumar et al., 2019) designed a facial expression recognition model using ExpressionNet, a Convolution Neural Network (CNN) architecture. The model is trained using the FER2013 dataset, which has a total of 35,887 images and is divided into training, testing, and verification sets. It includes seven emotion classification categories.	The results of this research compare the facial emotion recognition accuracies for several models like AlexNet, ExpressionNet, CNN (Softmax), ShallowNet on the FER2013 dataset. It was found that the ExpressionNet method had the best performance with an accuracy of 68.4%.
4.	The paper (Livingstone & Russo, 2018) performs facial emotion recognition on the FER2013 dataset, the same dataset used in research paper (Tripathi & Kumar et al., 2019). It adopts the VGGNet architecture to obtain state-of-the-art performance without using extra training data. It also compares the classification accuracies of different architectures on the FER2013 dataset.	The result of this paper proves that state-of-the-art performance can be obtained using the FER2013 dataset. The average accuracy obtained using this dataset lies in 65–68%. However, this chapter obtained an accuracy of 73.28% using the VGGNet architecture. This model shows the best emotion classification rate for happiness and surprising emotion. However, it works with low classification accuracy for disgust and fear.
5.	The paper (Khaireddin & Chen, 2021) uses the main idea of predicting emotions using speech and facial expressions. It proposes a real-time framework that recognizes emotions using smartphone technology backed by cloud computing. It uses the RAVDESS dataset for speech-emotion-recognition, and the Surrey Audio-Visual Expressed Emotion (SAVEE) dataset, which contains facial and speech data.	The results of this chapter involve comparing the emotion recognition rate of the merged framework of speech and facial emotion recognition. It detects seven emotions using smartphones with a recognition accuracy of 96.3% for facial expressions and 95.43% for speech. However, the combined framework has a total accuracy of about 97.26%.

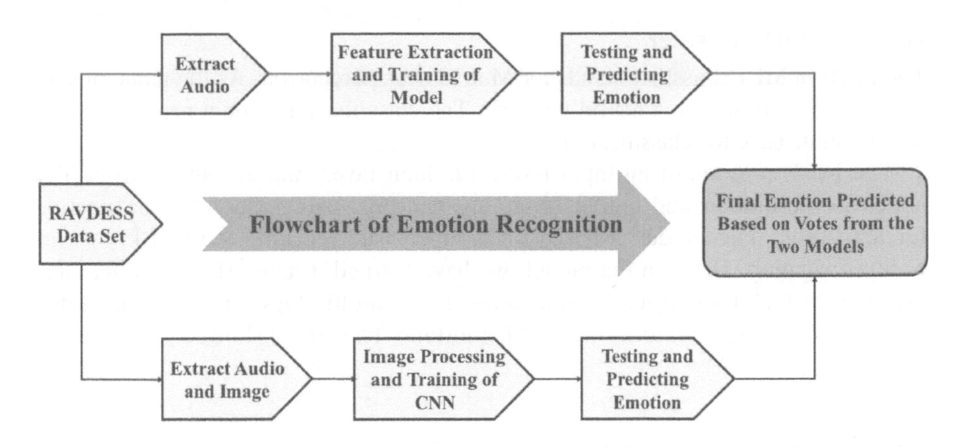

FIGURE 10.1 Flowchart of emotion recognition (Khang, 2021).

(as they are appropriate for expert human beings and the frequency at which human beings speak/utter). It is advised to try and extract MFCCs out of the audio signal.

We usually take 12–13 Mel-frequency coefficients into attention as capabilities while schooling models. Other than that, it would be beneficial if different capabilities were considered. For an assignment on child cry prediction, it might be advised to apply Volume, Energy, Pitch, Zero Crossing Rate, and Spectral Centroid as a few extra capabilities alongside MFCC. It is also advised to apply a few characteristic choice strategies like Principal Component Analysis (PCA) or t-SNE (t-allotted stochastic neighbor embedding) to figure out the most excellent capabilities. This might assist in getting better results.

10.3.1.5 Mel Spectrogram

Mel spectrogram is a kind of spectrogram where a signal's broad spectrum is converted into Mel scale. The spectrogram of a signal is a visualization of the frequency spectrum that shows the range of frequencies the audio contains. The Mel scale functions as a human ear. Research shows that the human ear can better differentiate between lower and higher frequencies.

10.3.1.6 Features Normalization

The features extracted above need to be normalized to provide better neural coverage to our model. This leads to an increase in accuracy and better performance. After considering many ways to normalize, we decided on using the mean and standard deviation methods. Here, we have found the mean and standard deviation of each feature, i.e., each column, respectively, and used the below formula to normalize the data as Equation 10.1.

$$\text{X\_norm}(i,j) = \left(\text{X\_orig}(i,j) - \text{Mean}(j)\right)/\text{std}(j) \tag{10.1}$$

10.3.1.7 MLPClassifier

The MLP in MLPClassifier stands for Multi-Layer perceptron. As the name, it suggests that it is based on a neural network. The classifier relies on a neural network present in its base for classification.

The MLP consists of an input layer, a hidden layer, and an output layer. The nodes in the hidden and output layers are neurons, using a non-linear function for activation. The model is based on a supervised learning method known as backpropagation. Here, in the model, we have trained it using the above features extracted, which total around 180 features. It is a multi-class classification model, classifying the audio features into eight emotions. The model showed a very decent accuracy of 61.67%.

10.3.2 Face-Emotion Recognition

10.3.2.1 Image Augmentation

Neural networks require enormous data to train effectively and make the model more robust. Image augmentation is a tool used to boost performance. It combines several augmentation methods to increase the effective size of the dataset (Rani & Chauhan et al., 2021). Here, we employ random rotation (rotates the image), random translation (shifts image vertically and horizontally), random flip (flips the image), and random contrast (changes contrast) as Figures 10.2 and 10.3.

10.3.2.2 Model

A Convolutional Neural Network is implemented, which consists of three Convolutional layers with Batch Normalization and Dropout layers (Srivastava & Hinton et al., 2014; He & Zhang et al., 2016) after each Convolutional layer. This is followed by a GlobalMaxPooling2D layer which feeds its output to a linear layer. Finally, a Softmax layer takes the output and produces the probability of the image belonging to several classes.

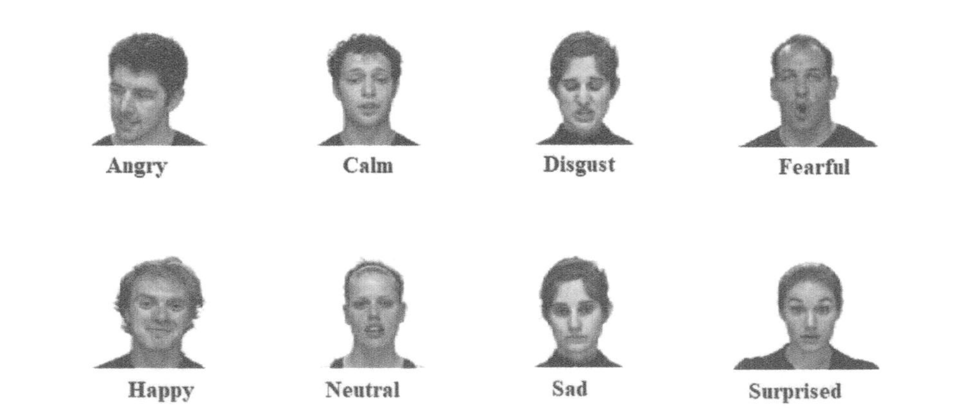

FIGURE 10.2 RavDess dataset.

Source: **RavDess dataset (https://zenodo.org/record/1188976).**

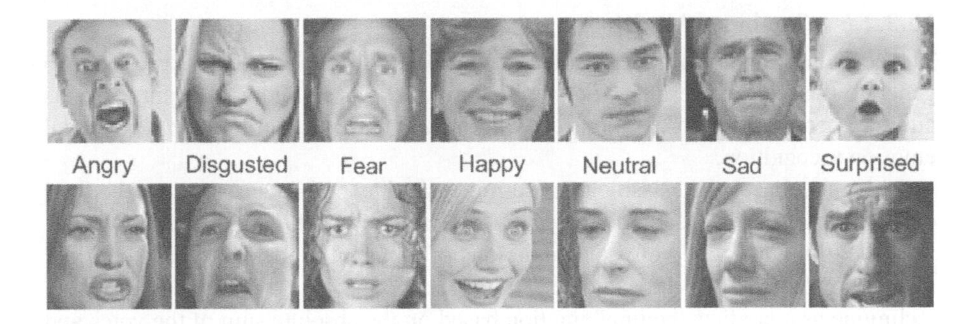

FIGURE 10.3 Fer2013 dataset.

Source: **Fer2013 dataset (https://paperswithcode.com/dataset/fer2013).**

10.3.2.3 Transfer Learning

TL is a concept in ML that stores information gained while solving one problem and applying it to a related issue. The processed image is passed to the ResNet101V2 model without the top layer trained on the ImageNet dataset. This model is chosen because it has a low inference time of 5 ms while scoring a high 1% accuracy of 0.772. It extracts useful high-level features from the Images.

10.3.2.4 Regularization

This is a technique to repress the model's capacity. This is done as one of the ways to alleviate the problem of overfitting. By adding the L2 penalty, the model is trained to not learn an interdependent set of feature weights. Dropout is an approach to regularization in neural networks which helps reduce interdependent learning among the neurons.

10.3.2.5 Optimizer

Deep learning networks are optimized commonly using gradient descent. It works in principle by finding the direction to update the models' parameters using the gradient of the cost function thereby lowering the learning rate.

Traditional gradient descent though extremely effective is quite slow, and the use of non-variable learning rate would lead the model to be stuck in local minima. To overcome this, several new optimizers have been developed. Adam is among the most popular optimizer and has been highly used in the literature.

Fouskakis and Draper (2002) can be looked at as a combination of RMSprop and Stochastic Gradient Descent with momentum. It uses the squared gradients to scale the learning rate like RMSprop, and it takes advantage of momentum by using moving average of the gradient instead of the gradient itself like SGD with momentum.

AdaBound (Luo & Xiong, 2019) modifies the Adam stochastic optimizer, thus making it more robust to extreme learning rates. There are dynamic bounds employed on learning rates, where the lower and upper bounds are initialized as zero and infinity, respectively, and they both smoothly converge to a constant final step size. AdaBound can be looked at as an adaptive optimizer at the beginning of training, and it gradually and smoothly transforms to SGD (or with momentum) as time progresses.

10.3.2.6 Combined Model

The combined model uses the best features of both speech and facial expressions to generate the required results. The combined model takes a video as input, extracts audio to perform speech-emotion-recognition, and extracts frames to achieve facial emotion recognition.

This model uses the classification voting ensemble technique to predict the final output. This ensemble technique (Leon & Floria, 2017; Abdul Ahad, 2021; Opitz & Maclin, 1999) takes the votes of predictions of different models and uses the final majority vote to predict the outcome. This model counts votes using the hard voting technique as it predicts the final emotion based on the absolute sum of the votes and not on the prediction percentages.

The combined model is implemented using a function that takes in the path of the video, the emotion of which must be analyzed and predicted. The first step is to extract the video frames and store them in a corresponding drive folder.

Each of these images is tested with the facial emotion recognition model, whose methodology was explained in the previous section. The predictions for each of these frames are used as votes for the ensemble model. The votes are kept count using a list.

Once the model is evaluated for the facial expression, the speech emotion has to be determined for the video. The audio is extracted and is pre-processed to match the properties of the files in the RAVDESS dataset. The frame rate, size, and channels of the extracted audio files are tuned to match those of the dataset.

This final extracted audio file is then tested against the speech-emotion-recognition model, and its prediction is noted. This is counted as another vote in the list previously initialized for the facial expression recognition process. Once all the votes are counted, the emotion with the majority vote is determined as the final output of the combined emotion detection model.

10.4 RESULTS AND DISCUSSION

The model trained on the FER2013 dataset (Tripathi & Kumar et al., 2019) with an accuracy of 63% on validation data performed poorly on real-world images. Thus, a new TL model was trained on video clips from the RAVDESS dataset (Livingstone & Russo, 2018).

This model converged with an accuracy of 75.2% on the validation set. Out of the eight emotions in the class label, four of the emotions were predicted with lesser accuracy than the other four. Hence, it was hypothesized to train a similar model with these four emotions.

The main four emotions recognized by the models are angry, happy, sad and disgust. The dataset consists of around 11,000 sentences with their respective emotions. The image gives an overview of the breakdown of the data into training and validation sets (Bhambri & Khang et al., 2022).

Table 10.2 shows the overall accuracies of MLP, CNN, and combined models for both datasets. When compared to other models, the proposed combined model with four classes produces better results.

The classification report for the proposed combined model is shown in Table 10.3. The model mostly failed to identify the neutral because it was incorrectly predicted to be sad as Figures 10.4–10.8.

TABLE 10.2
Accuracy of the Main Four Emotions Recognized by the Models Are Angry, Happy, Sad, and Disgust

Model	Accuracy (%)
CNN[Speech](8 classes)	48.16
MLP[Speech](8 classes)	62.77
MLP[Speech](4 classes)	74.48
CNN – FER2013[Facial Images](8 classes)	10.17
CNN – RAVDESS[Facial Images](8 classes)	45.72
CNN – RAVDESS[Facial Images](4 classes)	76.18
Combined Model*(8 classes)	42.86
Combined Model*(4 classes)	86.61

TABLE 10.3
Classification Report

Emotion	Prediction (%)
Angry	87
Disgust	100
Calm	25
Fearful	37
Happy	39.5
Neutral	0 (predicted sad 100% of the time)
Sad	56.25
Surprised	25

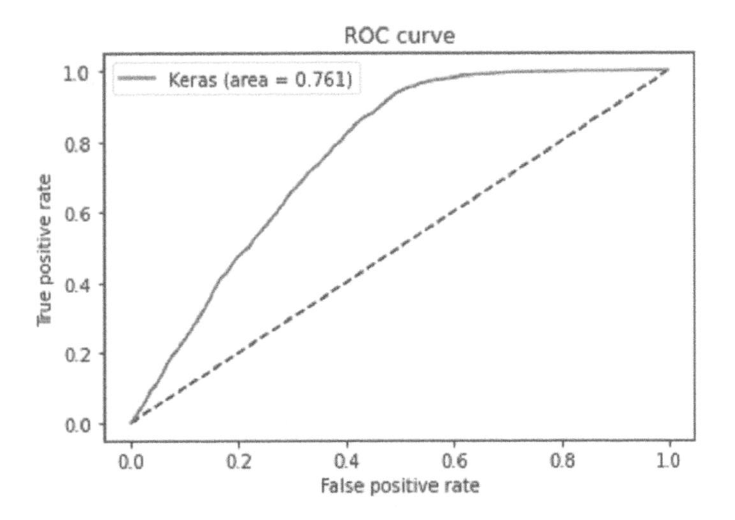

FIGURE 10.4 ROC curve for CNN (facial images).

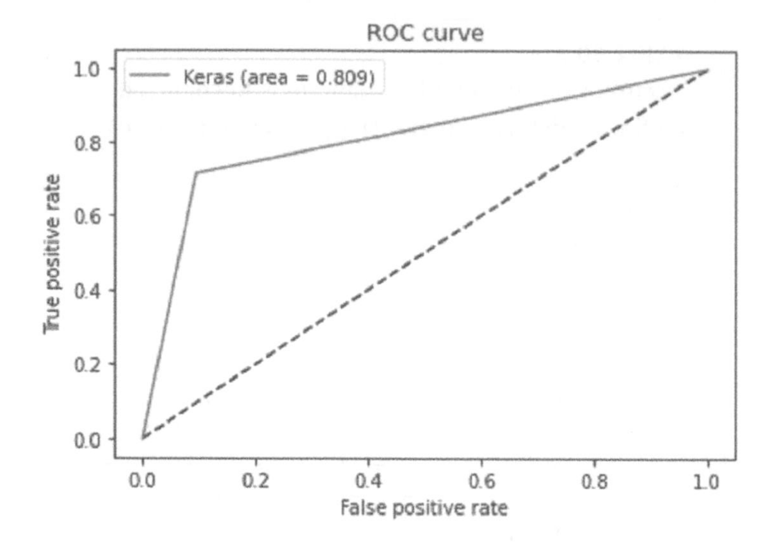

FIGURE 10.5 ROC curve for MLP (speech).

The model trained using TL performs significantly better than one trained without it. After the base layers of the CNN are trained on a diverse and large dataset, the CNN adapts and learns how to extract the features. Thus, the pre-trained CNN now knows how to extract the features but does not know how to classify them (Subhashini & Khang, 2024).

When the new dataset is given to the pre-trained CNN, it faces the challenge that the features in the new image datasets are different. So the top layers of

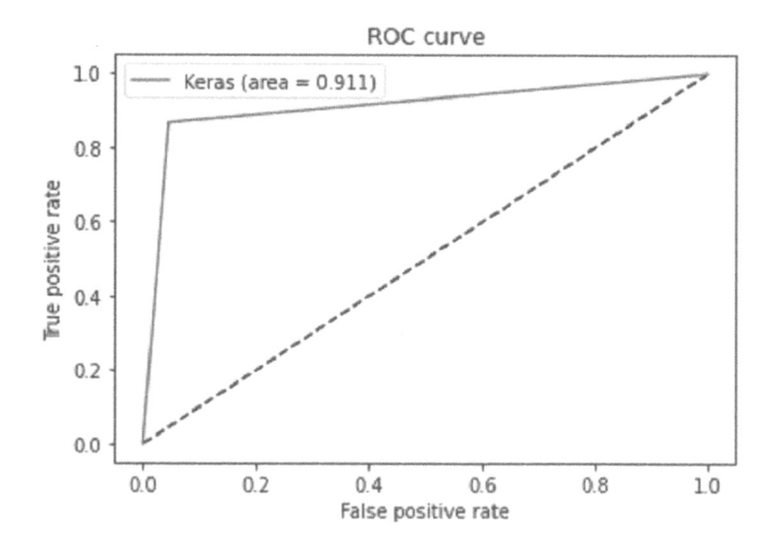

FIGURE 10.6 ROC curve of combined model.

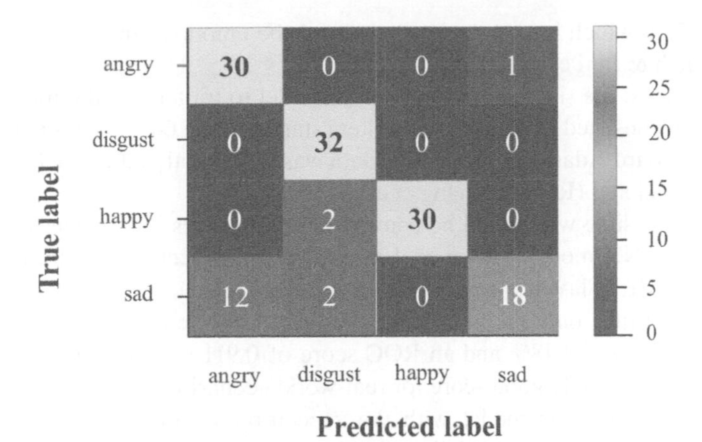

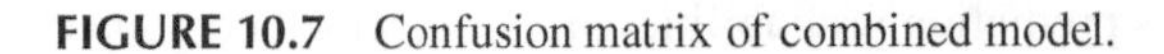
FIGURE 10.7 Confusion matrix of combined model.

the CNN are now trained to learn how to classify using the features extracted (Jebaraj & Khang et al., 2024).

Since the model is able to extract features that can be well identified and thus classified into different postures, an accuracy of 76.18% and an ROC score of 0.761 as observed in Figure 10.4 is obtained. Thus, the transfer model performs immensely better than the standard model.

The speech-emotion-recognition model was strained using the RAVDESS dataset which had audios of different emotions performed by 24 actors and validated by 200 people.

Initially, we trained the MLPClassifier model with the 180 features extracted, and we came to an accuracy of 62.77% and an ROC score of 0.809 as observed in

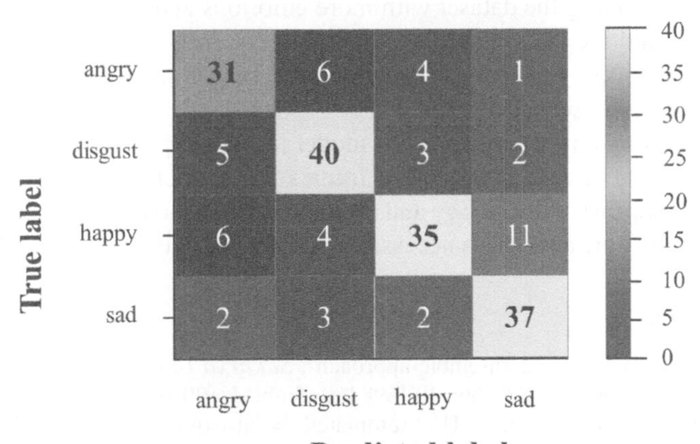

FIGURE 10.8 Confusion matrix of MLPClassifier.

Figure 10.5, in which it gave the best result for the emotions angry, happy, sad, and disgust (Shah & Jani et al., 2024).

We also tried the single dimension CNN model to train the audio files with. The CNN model consisted of 4 layers with filters starting with 64 and ending with 8. We used the standard Adam optimizer. The data was fed directly into the CNN model as its one dimensional (Rani & Khang et al., 2023).

Another possible way would be converting the features into images and loading it onto a 2D CNN model. This model gave a very low accuracy of 48.16%. So, we decided that MLPClassifier works best for our needs here.

We tried to take only the four best emotions that were accurately classified and got an accuracy of 74.48% and an ROC score of 0.911 as observed in Figure 10.6 which is a considerably good score for real-world scenarios.

Now combining this model with the speech-recognition model enhances the accuracy of the overall model. When the face-emotion model predicts a class with low accuracy, the speech model's prediction probabilities are used, which combined form an effective ensemble technique that performs better (Khanh & Khang, 2021).

10.5 CONCLUSION AND FUTURE SCOPE

In the context of a smart city, the development of an emotion classification model with high accuracy could have various applications. For example, it could be integrated into smart transportation systems to detect passengers' emotional states and adjust the driving behavior accordingly, or it could be used in public spaces to enhance the overall experience of residents and visitors (Khang & Rani, 2022).

An emotion classification model has been developed that classifies images with an accuracy of 86% per video on the test videos. Additionally, a combined model has been developed that uses an MLP model for speech classification with an accuracy of 74% per clip and a TL-based CNN with an accuracy of 76.18% per image. However, there are possible improvements that could be made to the current architecture in the context of a smart city (Rani & Khang, 2022).

Firstly, diversifying the dataset with more emotions and actors of different races and accents could increase the accuracy and effectiveness of the model. This could make the model more robust to different cultural backgrounds and improve its performance in real-life scenarios.

Secondly, using a transformer-based model to classify videos could capture the temporal information that is lost when frames are extracted from the video. This could further improve the accuracy and reliability of the model in smart city applications where real-time analysis is necessary (Khang & Hajimahmud, 2022).

REFERENCES

Abdul Ahad, A. "Vote based ensemble approach", *Sakarya University Journal of Science*, 2021, vol. 25, pp. 871–879. doi: 10.16984/saufenbilder.901960

Alshamsi, H., Këpuska, V., Meng, H. "Automated facial expression recognition app development on smart phones using cloud computing", *9th IEEE Annual Ubiquitous Computing, Electronics & Mobile Communication Conference (UEMCON), 2018*, pp. 44–50 (2018). https://doi.org/10.1109/uemcon.2018.8796594

Bhambri, P., Rani, S., Gupta, G., Khang, A. "*Cloud and Fog Computing Platforms for Internet of Things*" (2022). CRC Press. https://doi.org/10.1201/9781032101507

Fouskakis, D., Draper, D. "International Statistical Review, Wiley Online Library" (2002). https://onlinelibrary.wiley.com/doi/abs/10.1111/j.1751-5823.2002.tb00174.x

He, K., Zhang, X., Ren, S., Sun, J. "Identity mappings in deep residual networks", *European Conference on Computer Vision (ECCV)*, pp. 630–645 (2016). https://doi.org/10.1007/978-3-319-46493-0_38

Jaiswal, A., Krishnama Raju, A., Deb, S. "Facial emotion detection using deep learning", *2020 International Conference for Emerging Technology (INCET)* (2020). doi: 10.1109/INCET49848.2020.9154121

Jebaraj, L., Khang, A., Chandrasekar, V., Pravin, A.R., Sriram, K. "Smart City Concepts, Models, Technologies and Applications", *Smart Cities: IoT Technologies, Big Data Solutions, Cloud Platforms, and Cybersecurity Techniques* (2024). CRC Press. https://doi.org/10.1201/9781003376064-1

Khaireddin, Y., Chen, Z. "Facial Emotion Recognition: State of the Art Performance on FER2013", *arXiv*, pp. 1–9 (2021). https://arxiv.org/abs/2105.03588

Khang, A., Gupta, S.K., Rani, S., Karras, D.A. "*Smart Cities: IoT Technologies, Big Data Solutions, Cloud Platforms, and Cybersecurity Techniques*" (2023). CRC Press. https://doi.org/10.1201/9781003376064

Khang, A., Hahanov, V., Abbas, G.L., Hajimahmud, V.A. "Cyber-Physical-Social System and İncident Management," *AI-Centric Smart City Ecosystems: Technologies, Design and Implementation* (1st Ed.) (2022). CRC Press. https://doi.org/10.1201/9781003252542-2

Khang, A., Hahanov, V., Litvinova, E., Chumachenko, S., Triwiyanto, V.A., Hajimahmud, R.N., Ali, A.V., Alyar, Anh, P.T.N. "The Analytics of Hospitality of Hospitals in Healthcare Ecosystem", *Data-Centric AI Solutions and Emerging Technologies in the Healthcare Ecosystem* (1st Ed.), p. 4 (2023). CRC Press. https://doi.org/10.1201/9781003356189-4

Khang, A. (2021) "Material4Studies", *Material of Computer Science, Artificial Intelligence, Data Science, IoT, Blockchain, Cloud, Metaverse, Cybersecurity for Studies.* https://www.researchgate.net/publication/370156102_Material4Studies

Khang, A., Khang, A., Hrybiuk, O., Abdullayev, V., Shukla, A.K. "*Computer Vision and AI-Integrated IoT Technologies in Medical Ecosystem*" (1st Ed.) (2024). CRC Press. https://doi.org/10.1201/978-1-0034-2960-9

Khang, A., Ragimova, N.A., Hajimahmud, V.A., Alyar, A.V. "Advanced Technologies and Data Management in the Smart Healthcare System", *AI-Centric Smart City Ecosystems: Technologies, Design and Implementation* (1st Ed.) (2022). CRC Press. https://doi.org/10.1201/9781003252542-16

Khang, A., Rana, G., Tailor, R.K., Hajimahmud, V.A. "*Data-Centric AI Solutions and Emerging Technologies in the Healthcare Ecosystem*" (2023). CRC Press. https://doi.org/10.1201/9781003356189

Khang, A., Rani, S., Sivaraman, A.K. "*AI-Centric Smart City Ecosystems: Technologies, Design and Implementation*" (1st Ed.) (2022). CRC Press. https://doi.org/10.1201/9781003252542

Khanh, H.H., Khang, A. "The Role of Artificial Intelligence in Blockchain Applications", *Reinventing Manufacturing and Business Processes through Artificial Intelligence*, pp. 20–40 (2021). CRC Press. https://doi.org/10.1201/9781003145011-2

Lalitha, S., Madhavan, A., Bhushan, B., Saketh, S. "Speech emotion recognition", *2014 International Conference on Advances in Electronics Computers and Communications*, pp. 1–4 (2014). doi: 10.1109/ICAECC.2014.7002390

Leon, F., Floria, S., Bădică, C. "Evaluating the Effect of Voting Methods on Ensemble-Based Classification", *2017 IEEE International Conference on Innovations in Intelligent Systems and Applications (INISTA)* (2017). doi: 10.1109/INISTA.2017.8001122

Li, S., and Deng, W. "Deep Facial Expression Recognition: A Survey," in *IEEE Transactions on Affective Computing*, vol. 13, no. 3, pp. 1195-1215, 1 July-Sept. 2022, doi: 10.1109/TAFFC.2020.2981446.

Livingstone, S.R., Russo, F.A. "The Ryerson Audio-Visual Database of Emotional Speech and Song (RAVDESS): A dynamic, multimodal set of facial and vocal expressions in North American English", *FAR-Discovery Grant from Natural Sciences and Engineering Research Council of Canada*, 2018. https://doi.org/10.5281/zenodo.1188976

Luo, L., Xiong, Y., Liu, Y., Sun, X. "Adaptive Gradient Methods with Dynamic Bound of Learning Rate" (2019). *ArXiv*. https://arxiv.org/abs/1902.09843

Opitz, D., Maclin, R. "Popular ensemble methods: an empirical study", *Journal of Artificial Intelligence Research*, vol. 11, pp. 169–198 (1999). doi: 10.1613/jair.614

Rana, G., Khang, A., Sharma, R., Goel, A.K., Dubey, A.K. "*Reinventing Manufacturing and Business Processes through Artificial Intelligence*" (2021). CRC Press. https://doi.org/10.1201/9781003145011

Rani, S., Bhambri, P., Kataria, A., Khang, A. "Smart City Ecosystem: Concept, Sustainability, Design Principles and Technologies", *AI-Centric Smart City Ecosystems: Technologies, Design and Implementation* (1st Ed.) (2022). CRC Press. https://doi.org/10.1201/9781003252542-1

Rani, S., Bhambri, P., Kataria, A., Khang, A., Sivaraman, A.K. "*Big Data, Cloud Computing and IoT: Tools and Applications*" (2023). Chapman and Hall/CRC. https://doi.org/10.1201/9781003298335

Rani, S., Chauhan, M., Kataria, A., Khang, A. "IoT Equipped Intelligent Distributed Framework for Smart Healthcare Systems", *Networking and Internet Architecture* (2021). CRC Press. https://doi.org/10.48550/arXiv.2110.04997

Shah, V., Jani, S., Khang, A. "Automotive IoT: Accelerating the Automobile Industry's Long-Term Sustainability in Smart City Development Strategy", *Smart Cities: IoT Technologies, Big Data Solutions, Cloud Platforms, and Cybersecurity Techniques* (2024). CRC Press. https://doi.org/10.1201/9781003376064-9

Srivastava, N., Hinton, G.E., Krizhevsky, A., Sutskever, I., Salakhutdinov, R. "Dropout: a simple way to prevent neural networks from overfitting", *Journal of Machine Learning Research*, vol. 15, no. 1, pp. 1929–1958 (2014). https://www.jmlr.org/papers/volume15/srivastava14a/srivastava14a.pdf?utm_content=buffer79b43&utm_medium=social&utm_source=twitter.com&utm_campaign=buffer,

Subhashini, R., Khang, A. "The Role of Internet of Things (IoT) in Smart City Framework", *Smart Cities: IoT Technologies, Big Data Solutions, Cloud Platforms, and Cybersecurity Techniques* (2024). CRC Press. https://doi.org/10.1201/9781003376064-3

Tripathi, S., Kumar, A., Ramesh, A., Singh, C., Promod, Y. "Deep learning based emotion recognition system using speech features and transcriptions", *International Conference on Computational Linguistics and Intelligent Text Processing (CiCLing) 2019*, pp. 1–12 (2019). https://arxiv.org/abs/1906.05681

Vrushank, S., Vidhi, T., Khang, A. "Electronic Health Records Security and Privacy Enhancement Using Blockchain Technology", *Data-Centric AI Solutions and Emerging Technologies in the Healthcare Ecosystem* (1st Ed.), p. 1 (2023). CRC Press. https://doi.org/10.1201/9781003356189-1

Yang, D., Alsadoon, A., Prasad, P.W.C., Singh, A.K., Elchouemi, A. "An emotion recognition model based on facial recognition in virtual learning environment", *6th International Conference on Smart Computing and Communications (ICSCC)* (2017). doi: 10.1016/j.procs.2017.12.003

Yue, Z., Yanyan, F., Shangyou, Z., Bing, P. "Facial expression recognition based on convolutional neural network", *2019 IEEE 10th International Conference on Software Engineering and Service Science (ICSESS)*, pp. 410–413 (2019). doi: 10.1109/ICSESS47205.2019.9040730.

11 Role of the Internet of Things (IoT) in Enhancing the Effectiveness of the Self-Help Groups (SHGs) in Smart City

Prashasti Pritiprada, Ipseeta Satpathy, B.C.M Patnaik, Atmika Patnaik, and Alex Khang

11.1 INTRODUCTION

Self-help groups (SHGs) are a group consisting of 10–25 local women aged 18–40. They are found in other countries, but most of them are found in India, especially in South and Southeast Asia. SHGs are a spontaneous group of people who always try to improve their livelihood.

Basically, they are self-directed and controlled by like-minded people. People with similar socio-economic backgrounds come together to seek help from a non-governmental organization or government agency to solve their problems and improve their living conditions.

Associations of SHGs have a considerable amount of social capital; however, they face several difficulties in their current mode of operation—namely that it is necessary to redefine this concept and ensure that it becomes relevant to the lives of those it is intended to help (Rai & Kumar et al., 2021).

The emergence of SHGs has touched nearly every region of India. NABARD has been advocating for SHGs since the early 1990s. In addition to providing financial and technical assistance to these projects, it has helped establish Self-Help Promotion Institutes (SHPI), which provide training to members of SHGs and can also help them register their groups with local authorities.

While it is realistic to believe that SHGs can be effective platforms for the social and economic empowerment of women, an SHG of 10–20 members is ideal.

No group registration is required (Sharma, 2020). The Internet of Things (IoT) consists of devices that are cable being connected through the Internet and that enable interaction between people and machines. The IoT is expected to include more than 22 billion devices by 2025 (Google).

The advent of digital technology has caused a seismic shift in the entire lifestyle and livelihood system. This technology has the potential to improve both financial

DOI: 10.1201/9781003376064-11

inclusion and its impact. Microfinance programs have achieved tremendous success in India, where many women's SHGs have used this form of financial assistance to help members improve their circumstances.

Financial inclusion, livelihood promotion, and women's empowerment have been successful in reducing poverty levels in several developing countries. However, there are few studies on SHG, microfinance, and women's empowerment. Jansen (2016) and Hirut (2000) states in his article that microfinance alone cannot fully alleviate poverty.

To be successful, these efforts must be combined with parallel and complementary programs that address the social and cultural dimensions of hardship, deprivation, and poverty. The study by Pattanaik (2003) shows that tribal women's SHGs continue to strive for a better future—for the benefit of both themselves and their communities.

Because of gender inequality and the exploration of women's pain, various SHGs work appropriately to address these issues.

In Malhotra's (2004) book, she explored how women entrepreneurs impact the global economy; why women start businesses and what associations promote them; as well as the extent to which they contribute to international trade. She investigates the benefits and drawbacks of microfinance programs to empower women, improve their employability, and reduce poverty. She says that microfinance programs seek to increase women's incomes and their control over those incomes—leading, in turn, to greater economic independence.

Women are sometimes able to use these groups and organizations as a means of gaining social status, access to markets, information, or different political roles.

Women who become economically independent and enjoy greater involvement in household decision-making are more likely to perceive their contributions as beneficial to the well-being of their families.

In her 2004 study, Narasaiah noted that one of the most striking trends of the late twentieth century was women's empowerment. For example, she pointed out that microcredit helped in this process.

Microfinance programs are proving to be very useful for helping women achieve their full potential. Cheston and Kuhn (2012) found in their study that microfinance programs were successful at reaching women and providing them with opportunities previously unavailable to them.

Microfinance institutions can use this opportunity to design products and services tailored specifically for poor women, who may have needs that go unmet in the traditional banking sector (Aggarwal & Khang, 2023).

Another article states that for SHGs to successfully implement income-generating activities, they need to enlist the help of non-governmental organizations. Bank staff should advise and guide women in their selection as well as the implementation of these activities. As per his study, the SHGs have also helped enhance rural women's self-confidence and perspective.

A study by Sahu and Tripathy (2005) states that about 70% of the world's poor are women. All the poor need to have access to all kinds of banking services which is not only important for reducing poverty but also to maximize their contribution to regional and national economic growth. SHGs have become the most important tool for participatory development and women's empowerment.

Due to social and economic constraints, rural women belong to socially marginalized groups. They are still in the lowest positions of the social hierarchy. Through microfinance and SHGs, they can climb out of the quagmire of poverty and stagnation. Gupta and Sharma (2021) argues in his paper that a paradigm shift is needed from "financial sector reform" to "microfinance reform".

While priority sectors need to be streamlined, mandatory microfinance needs to be closely monitored. At the same time, the area and size should be designed appropriately to create a competitive environment for microfinance services.

RBI needs to create a comprehensive database to understand microfinance. Sinha (2005) found in his research that microfinance contributed significantly toward the savings and borrowing of the country's poor.

According to him, microfinance is mainly used for direct investment. Of course, there are some alternative possibilities based on the household's borrowing needs at the time the loan is paid. Several studies show that microfinance programs have both positive and negative effects on women.

Some researchers question how much women benefit from microfinance (Goetz, 2001). Some argue that microfinance programs divert women away from other more effective empowerment strategies (Ebdon, 1995), and that donor attention and resources are diverted away from potentially more effective poverty reduction measures (Rogaly, 1996).

In certain cases, increased autonomy for women was temporary, which was beneficial only for the wealthy and financially stronger women. However, in most cases, due to the lack of initial resources, market connections, and skills, poor women get very less benefit out of this.

11.2 RELATED WORK

Origins and developmental path of SHGs use datasets as well as cases in Bhubaneswar, India, as shown in Table 11.1.

The B. Rangarajan Commission report identifies four main drivers behind the lack of financial inclusion in India. A strong community network is one of the most important aspects of rural financing.

The self-sufficiency gained through independence also contributes to improvements in other developmental factors such as family well-being, health, and literacy (Rangarajan & Mahendra, 2022).

The IoT is a powerful means of connecting PCs, digital and mechanical devices, things, animals, or people by assigning unique identifiers (UIDs) and allowing content to be shared.

Certain objects related to the IoT such as cars with built-in sensors, farm animals with biochip receivers, and people with heart rate monitors that warn drivers when tire pressure is low are certain examples (Gillis, 2022).

IoT-emerged technologies can be used in various business perspective to automate processes and to save the cost of labor expenses. This technology also reduces waste, reduces manufacturing and delivery costs, improves service delivery, and increases visibility to a user.

TABLE 11.1
Establishment of SHGs in India

#	SHG Description
1	SHGs were established in 1972 in India. The Self-Employed Women's Association (known as SEWA) was also formed at the same time.
2	Attempts at self-assembly have traditionally been limited. For example, Ahmedabad's Textile Labor Union (TLA) established a women's section in 195 to teach salaried women working in factories in skills such as sewing and weaving.
3	SEWA's founder, Ella Batt, organized underprivileged self-employed women workers into unorganized workers like potters, spinners, and street vendors to enhance their income levels.
4	In 1992, NABARD started her SHG Bank Linkage Operation, which has grown into the world's largest microfinance program.
5	In the year 1993, the Reserve Bank of India and NABARD have given a chance to the SHGs to open savings accounts in banks.
6	In 1999, the government of India established her "Swarnajayanti Jayanti Gram Swarozgar Yojana (SGSY)" for promoting the self-reliance of rural communities through the establishment and licensing of such organizations. From 2011 onwards, this mission was popularly known as the National Rural Livelihoods Mission (NRLM).

Source: Government of India Website.

The applications of IoT are to use smartphone technologies and design methodologies which are mostly human-centered. This IoT provided solution deals with the designing to meet the requirements of rural and multi-tier organizations which are cluster based (commonly known as RMCOs), including multi-tier connectivity, resource management, and banking (Rani & Chauhan et al., 2021).

Rural residents, especially women, lack social interaction and technical skills due to geographic and economic factors. If members feel that the proposed solution provides a more specific solution to their problem, the community will adopt it. A computerized solution enables rural women to perform their daily tasks more efficiently and easily.

11.3 CHALLENGES FACED BY THE SHGS

The SHGs face several challenges that may affect their ability to reach their goals and support their members in India. Some of the biggest challenges for SHG in India are as follows:

1. **Lack of access to funding**: One of the biggest challenges SHGs face is access to finance. Many SHGs have difficulty obtaining loans or other forms of financial support, which can limit their ability to start and grow their businesses, invest in new projects, or provide services to their members.
2. **Limited economic opportunities**: Many SHGs operate in rural or underdeveloped areas where economic opportunities are limited. This can make it difficult for them to generate income and support their members, and it can also limit their ability to grow and expand their businesses.

3. **Lack of government support**: Despite government efforts to support SHGs, many groups continue to have difficulty accessing government regulations, programs, and resources. This can limit their ability to access funding, training, and other resources that could help them grow and succeed.
4. **Lack of education and training**: Many SHG members lack the education and skills they need to effectively run and grow their businesses. This can limit their chances of success and make it difficult for SHGs to effectively support and train their members.
5. **Social and cultural barriers**: In some areas, social and cultural attitudes can pose a challenge to SHGs, particularly for women-led groups. These attitudes may include gender discrimination, social stigma, and other barriers that can limit SHG members' ability to fully participate in the group and contribute to its success.

Few other challenges are as follows:

1. **The launch challenge**: At the beginning of the launching year, i.e., 1999–2000 when the program was launched, it gave support to around 292,000 SHGs which was huge at that time. It remained at this level in all subsequent years, although there were large fluctuations from year to year. Similarly, in his second year of the program, he was attended by 21,000 groups. In all subsequent years, the numbers remained at this level. The number of groups reaching Level 2 has increased significantly, but the number of those who have found employment is smaller. Overall, only 685,000 groups have started economic activity. That's just over a fifth of his in the program-supported group.
2. **Allocation and use of funds**: During the first year of the launching of the program, the Central Government and the State Governments allocated around Rs.1472 crores in the year 1999–2000 for the SHGs. For consecutive seven years, the quota remained below his first year. In 2001–2002 and 2002–2003, the allocated amount was almost halved. Over a tenure of around ten years, the total allocated fund for this program is around Rs. 14,467 million. Which is almost half of the budget allocated to NREG in just a span of one year, i.e., during 2009–2010. The foremost purpose for the quiescence of fund distribution is the lack of coordination among the banks which were responsible for driving the program. The amount of fusion allotted has not been used in its entirety for a year in the last ten years of the program. Overall, 74% of the given funds were used. But, the rate of usage is growing year by year. It rose from 49% in 1999–2000 to 86% in 2003–2004. In the years that followed, it was well above 80%. It is envisioned that 10% of the allocated funds will be used for training and another 20% will be used for critical infrastructure development, even though for the said activities, the given fund is minimal and proportional. Significantly higher funds are being used to provide grants and grants to SHGs. And the single Swarozgaris. Therefore, this program is often referred to as a grant-driven program.

3. **Mobilization of the credit**: Mobilizing bank loans is one of the program's biggest challenges. This is why the federal and state governments have been unable to increase quotas for many years. The total target for the mobilization of the allocated credit is around 30 crores. The ratio of actual mobilization vs target is growing tremendously year by year. That's a good sign. Due to poor bank lending mobilization and a relatively high proportion of funds being devoted to subsidies, the loan-to-subsidy ratio has been around 2 over time, and it has not changed significantly. As a result, the ratio between loan-to-financing was well below his target ratio of 3:1, and the amount invested per Swarozgari was also lower than planned.
4. **To find out and reach the actual poor and vulnerable people**: A comprehensive 2007 study of SC/ST BIRD coverage in SGSY noted that 10,848 Swarozgaris and Non-Swarosgalis (control samples) were included, excluding SC and ST increase. A Ministry of Labor and Social Affairs briefing (2007) noted only 1% of the deeply affected population by the SGSY. The percentage of the poorest beneficiaries are only 33%. It is shown to be from the strata. Moreover, overall utility is more unevenly distributed, with the richest quintile receiving up to 50%, while the poorest quintile receiving only 8%. According to the Ministry of Development Annual Report 2002–2003, in most areas such as Uttar Pradesh and Bihar, the people in the villages are influential.
5. **The poor persistence rate of the micro-enterprises which are funded**: There are so many promoted Swarozgaris who are mostly unenthusiastic to create or grab the proposed asset or sell it immediately after acquisition. A BIRD study found that the success rates of the unit presence or non-existence of units funded by the Swarozgaris group were even worse than the repercussion of the SHGs on the social and economic development of India in the northern states. In individual Swarozgaris cases for the Swarozgaris group, only 17.7% of the units were found to be present, whereas 31.11% of the units were found to be intact. The results show the exact opposite of what most people believe or perceive that group approaches to fundraising are superior to individual fundraising. However, for the southern states, 76.6% of the units were present at the time of the field visit, indicating that government agencies were paying greater attention to monitoring units. The authors of this chapter have observed that some groups manipulated property/livestock acquisitions in Andhra Pradesh. The next day, they buy cattle from relatives/friends in front of officials (owning) and AP government recognized this problem for a long time and changed the subsidy capital to subsidy interest in the year 2004. SGSY, subsidy, revolving fund, etc. are rarely heard today among rural SHGs in Andhra Pradesh. All I hear is "Pavala Vaddi" or "loan at 3% interest". In some other federal states, some researchers state that the group's interest is in the subsidies. The members tend to quickly sell their capital/cattle, pay off bank loans, and distribute subsidies.
6. **Slightly additional source of income from other activities**: A particular success story here and there is not considered a success of the program.

Mostly these achievements should be documented for future reference and the reasons for their success evaluated and reflected in future guidelines indicating political directions, debunking a major myth which is the numbers above apply only to surviving units. Considering failed units, the average increase in income is only a few hundred rupees. It should be remembered that approximately 50% of Swarozgaris are engaged in dairy farming. About a quarter of the rest are engaged in other livelihood parameters, including fishery, poultry, farming, and any other essential majors needed for survival.

11.4 SMART CITIES AND THE USEFULNESS OF IoT

IoT stands for "Internet of Things" and usually indicates an interconnected network of physical devices, vehicles, household appliances, electronics, software, sensors, and other objects with connectivity and allows data to be exchanged (Bhambri & Khang et al., 2022). IoT enables the collection and analysis of data from these connected devices, which can be used to improve many aspects of life and business.

For example, IoT devices could be used to monitor and manage home appliances, track, and monitor assets, improve supply chain management, monitor industrial processes, and more. As the number of connected devices grows, IoT can have a profound impact on how we live, work, and interact with technology.

From the day that a variety of network types emerged, IoT became one of the most precious infrastructures in smart cities. Taking an example, the data collected from household appliances such as refrigerators are mutually shared and kept in the environment of a smart home to provide customized services (Rani & Khang et al., 2023).

Like the smart home concept, smart cities are an emerging market and play a vital role in the infrastructure for the future. The consumption of energy and electricity by smart cities in a proficient manner provides an appropriate and budget-friendly developed architecture for the betterment of the community, for which the necessity of IoT technology is rapidly increasing.

So, the general infrastructure of the smart city is established, and it can provide different services that tend to use various collections of data that are required daily indicating that several utilities of IoT technologies in different smart cities may create a living environment that is both sustainable and comfortable for all kinds of citizens (Khang & Gupta et al., 2023).

For creating an environment that is both sustainable and an urban-like setup, many recognized and well-known organizations are working together on technical and social aspects related to the concept of smart cities. The nation is already moving toward commendable success through the implementation of various technological advancements using IoT (Khang & Gupta, et al., 2023).

In a futuristic aspect of the smart cities, IoT technology contributes significantly to the most explained and elaborated concept of the infrastructure of smart city technology. The basic ideas and concepts of IoT technology are like that of technology of smart city and their infrastructure, so there are various business opportunities and great growth potential (Khang & Rani et al., 2022).

To efficiently and successfully develop the IoT technology of the future in smart cities, the below points could also be considered as follows:

- For all kinds of wireless networks, sensor-oriented technology is contemplated as one of the top priorities for IoT technologies for any smart city-related infrastructure (Hajimahmud & Khang et al., 2022).
- Not only the sensor-oriented technologies but also the network service technologies are the essential IoT technologies for all kinds of configurations required for projecting a smart city (e.g., smart home cloud server technology, IoT technology for home network services, and building cloud servers).
- Comparable energy-related technologies are key IoT technologies for smart city infrastructure (e.g., energy load management technologies, BIM-based energy management for smart buildings, and analytics technologies). The technology used in the environment of a smart home is a rudimentary aspect of any smart city, so smart home network technology and infrastructure need to be developed and prepared quickly (e.g., the cloud server technology used in a smart home).

Based on the effects of the contemporary look, it's far more beneficial to provide each vital and precedence problem concerning the components of IoT technology for a hit status quo of the infrastructure and associated offerings.

In spite of the fact that the contemporary look has assessed the significance of IoT technology toward the idea of making a smart city, the intensity of discussions and deliberation among specialists along with the engineers in sectors associated with smart cities and the technology related to IoT should be represented constantly to create precise movement plans (Khang & Vrushank et al., 2023).

In addition, huge technical panels must be prepared with specialists from numerous study fields that include city improvement, records and conversation technology, transportation, and environmental policy additionally; programs for presenting numerous IoT generation offerings and the usage of suitable community technology must be evolved and prepared.

Fourth, in addition to planning for information and data protection, appropriate responses should also be prepared. When customers connect to an IoT-generated service, reliable information refining and garages must occur while maintaining integrity, confidentiality, and privacy. This kind of reliable and stable communication and connectivity between all her IoT tools and smart city infrastructure should be ensured (Rani & Khang et al., 2022).

11.5 IMPLEMENTATION OF IoT RESOLVING THE CHALLENGES OF THE SELF-HELP GROUPS

11.5.1 Digitization of Agriculture

Significant production in agriculture is essential for our population to avoid possible food shortages in the future due to various factors.

The first factor, as noted above, is population growth; climate change is another major factor, which is reducing the capitulation of important crops, and certain areas are not suitable for efficient production in agriculture.

One of the biggest issues is the wastage of food, and it has become a universal issue, largely in developed nations. It is estimated that more than 28% of agricultural production is available land "set aside" for food waste, and unfortunately, more than 800–106 people currently suffer from hunger (UN, 2022).

The use of IoT technologies in the field of agriculture and cultivation can help to ensure the need for food and make agricultural production processes generally more efficient (Khang & Abdullayev et al., 2024).

A variety of useful information could be collected from the plants that could be used to monitor yields and early detection of possible diseases that could significantly reduce the yield of certain plants.

Soil and nutrient monitoring would simplify production processes of agricultural needs, lead to valuable water savings in specific geographic areas, and could be used with the improved irrigation systems infused with smart technologies. As a general fertility control, more accurate seeding could be ensured.

There are some problems with the effective application of IoT technologies in agricultural production. Different types of collection and observing techniques could be developed, and farmers should be trained in a better and improved way (e.g., by developing standard training chapters for farmers).

Because of the vastness of the collection of data, farmers can become overwhelmed by its goodness. So, the evolution of standard modules for the training for farmers along with the expansion of more handy software solutions is essential. Since most of the end products of the various SHGs are dependent on agriculture, the digitization of agriculture will certainly help too.

11.5.2 The Involvement of SHGs in Industrial Outgrowth

In industrial applications, the need for IoT technologies for SHGs would enable the production process to be more efficient and ensure more effective communication and networking between operators and machines. This would also allow more aggressive companies in the market to have more effective quality control and minimize droppings.

A critical function would be the design, development, and integration of a wide range of useful sensors for the applications needed in industry standards to form an integrated and efficient control system.

The effective application and a better understanding of IoT technologies in industry are necessary to understand in which direction they can be used for those specific areas where they are beneficial requires more intensive research.

Advances in the integration of various industrial sensors and the use and processing of the collected data to improve the processes required for industrial advancement, i.e., enabling manufacturing integrated with intelligent IoT-based computing, are crucial.

11.5.3 IoT in Healthcare of the SHG

Through the concept of eHealth, the health system, in general, has been found a challenging area to implement IoT technologies. The service quality used in the healthcare systems could be improved with IoT support (primarily through the collection of patient health data) and ultimately improve patient care and safety, as by doing this we may expect a better life expectancy of a patient.

Multiuse smart medical devices have a huge ability to monitor many essential and valuable human functions such as heart rate, skin temperature, and motion monitoring.

Remote health monitoring is also an interesting opportunity that could be exploited by IoT devices and products with the right support. Overall, it might be possible to predict various symptoms and prevent potentially life-threatening conditions and diseases. Elderly support could also be provided by monitoring the patient's general health and nutritional status, supported by IoT-enabled devices.

After a serious illness, rehabilitation could also be effectively supported by IoT technologies, especially in the case of home rehabilitation. One of such most important issues and provocations in this application area of IoT would be to ensure adequate cyber security in health monitoring systems due to possible attacks. In the coming years, significant progress is expected in healthcare systems, i.e., especially in software development for hospitals (Vrushank & Vidhi et al., 2023).

For example, various sets of devices could be connected to software solutions that are advanced by nature, such as MRI machines or CT, and combined with laboratory data to make an intelligent hospital information structure. Such an approach would enable better patient care, identify priorities needed for medical, and help nursing staff make follow-up and treatment decisions (Vrushank & Khang, 2023).

Hospitals could also use IoT systems to efficiently maintain many medical devices. Prompt detection of serious equipment failures that can affect the adequate amount of accuracy of certain medical equipment can reduce equipment costs in hospitals (Khang & Hajimahmud et al., 2023).

Developing such intelligent and IoT-based solutions in health systems can also be very useful in severe universal pandemics (collection of data and faster diversified information, medical personnel and resources, medical separations, etc.) such as the most recent pandemic situation due to Coronavirus, which seriously threatened the whole world (Khang & Khang et al., 2024).

Healthcare is perhaps one of the most stimulating areas of the IoT; therefore, significant advances are expected in the coming year, which will bring significant benefits to the population. SHGs, most of whom live in rural areas, can certainly take advantage of advances in healthcare to improve their livelihoods not only this but also SHGs in smart cities will be facilitated.

11.5.4 IoT in the Transportation of Products of the SHGs

Transportation will change significantly in the coming decades, especially with the expected increased introduction of electric vehicles into the world. The imminent restrictions on petrol and diesel vehicles due to environmental safety concerns and the evolution of substitute technologies for the transportation means, such as automobiles that run on hydrogen, will modify the silhouette of future conveyance systems.

Usually, the demand for environmentally friendly conveyance options is always huge, which is already being developed proficiently and has already taken the place in the current market.

Such kind of vehicle technologies will need the development of the infrastructure for transportation to confirm the required autonomy of the vehicles. Nowadays, IoT has appeared in the concept of the "Internet of Vehicles", which only confirms its efficiency in this area which is important.

IoT's most significant application area is the smart car (vehicles) concept. The concept of the smart car considers using and optimizing various internal functions in the car developed and maintained by various IoT technologies. It's expected that the driving experience will improve along with an increase in comfort and safety by using the advancement of IoT technologies.

Very specific data are being collected and coordinated with the major operational parameters like checking the tire pressure, fueling details, imminent detection of potentially troubled areas and faults, and regular indication for maintenance.

Usually, the focused usage of IoT technologies may provide service improvements and adds value for customers, those ultimately improving competition among vehicle manufacturers in the automotive industry. A challenging feature of IoT adoption is autonomous vehicles (Shah & Jani et al., 2023).

The location, direction, and intended path of an autonomous vehicle could be effectively supported by the IoT in general and the monitoring of autonomous vehicle security systems (Khang & Hahanov et al., 2022).

In the case of autonomous vehicles, the most important issue is the prevention of accidents caused by a vehicle, which could be solved by the targeted use of IoT devices. Initiation of smart parking technologies is also one of the most prominent areas of the IoT, looking at the transport sector more broadly.

In this regard, several research projects aim to enable the status of available parking spaces and to control and track useful parking information in real time. Again, the emerging sensor technologies, i.e., the sensors used in smart parking systems, are very important to enable competent and precise service.

Prolongation and the prevention of issues and faults of various vehicles could also be carried away by the IoT, which could improve the safety and durability of several automobiles.

Considering all the aspects which are mentioned previously, the driving experience is going to change drastically in a positive way by using IoT technologies from various points of view such as the quality of vehicles, energy-saving aspects, and the comfort of the driver as shown in Figure 11.1.

11.6 POSITIVE IMPACT OF IOT ON THE SELF-HELP GROUPS IN ODISHA

Odisha has over 6 million active SHGs. The concept of promoting women was launched as "Mission Shakti", by the Honorable Prime Minister of Orissa on Women's Day on March 8, 2001 (Mission Shakti, 2022).

The Mission Shakti Program aims to socially empower women and make them strong financially and politically by engaging them in SHGs. The key features of Mission Shakti are social mobilization, financial inclusion, and livelihood improvement, including marketing SHG products.

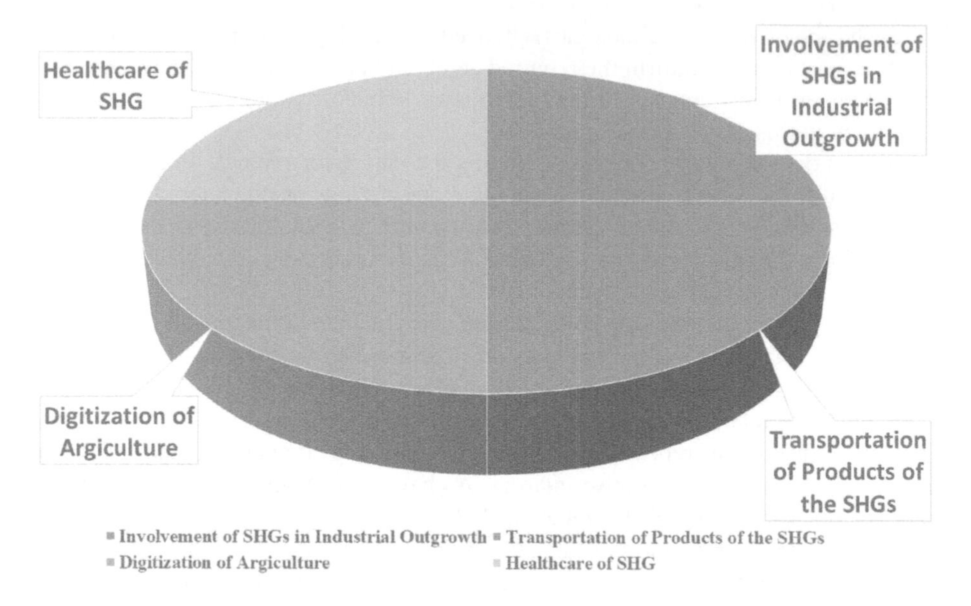

FIGURE 11.1 Implementation of IoT resolving the challenges of the self-help groups.

The program is implemented in the state by the Ministry of Women and Child Development and another separate department under the Mission Shakti, Mission Shakti department. At the ground level such as in different districts, the District Social Welfare Officer (DSWO) administers the program under the direction and supervision of the District Collector.

Mission Shakti is an Odisha government initiative launched in 2001 and is currently empowered by 7 million women across Odisha. "Women's empowerment is not a buzzword in Orissa. It is a non-negotiable code", said the Orissa chief minister at a recent press conference.

11.6.1 Improved Functionalities of Mission Shakti in Odisha Using IoT

The SHGs are usually community-based groups. They usually offer collectivization and group activities. Common functions of SHGs include regular meetings, savings, internal loans, repayments, and record keeping. SHGs provide social security and social activities directly or through associations to women in society in both rural and urban areas in various areas such as health, nutrition, hygiene, and gender equality.

"Mission Shakti" of Orissa is committed to sustainable support of women's SHGs to provide them with a long-term and sustainable source of income. Mission Shakti, in consultation with the government, arranged the procurement of goods and services by Mission Shakti SHG.

In addition, Mission Shakti works with OUAT (Odisha University of Agriculture and Technology) to train SHG members to carry out income-generating activities in agriculture (beekeeping, rice straw mushroom farming, poultry farming, etc., agriculture, oyster mushrooms, vegetable cultivation, composting, goat farming, etc.)

Since the state government has brought so many smart schools within the state, Mission Shakti has been more visible and involved with so many works for the educational stream. Along with that, using the various digital platforms and online advancement such as virtual conferencing, the members of the groups are staying connected with each other for more productive output.

11.6.2 IoT Evolving Skill Development, Livelihood, and Marketing: Goals of Mission Shakti

One of the main objectives of Mission Shakti is to provide a livelihood to the women in the SHGs and their associations through various income-generating activities.

Women SHGs in Odisha are involved in various income-generating activities such as agriculture (rice, groundnuts, sugarcane, onions, etc.), vegetable cultivation, organic farming (such as all kinds of millets, turmeric, rice, rajma, and other pulses), production of badi and papad (products made from black chickpeas), beekeeping, various handicrafts made from bamboo, production of candles, production of chhatua (a finely ground, healthy, and ready-to-eat mixture of various pulses and nuts), production of bricks from fly ash and resulting business, participation in various construction works, flour milling, door mat making, goat rearing, floriculture, hand weaving, horticulture, livestock rearing (dairy farming, poultry rearing, goat rearing, pig rearing, sheep rearing), incense stick making, preparation and participation in mid-day meal in government schools, sale of milk and milk products (such as curd, cheese, and paneer), production of snacks such as concoctions, cultivation of mushrooms, manufacture of organic fertilizers, sale of paper/sheet under PDS (Public Distribution System), manufacture of disposable plates, manufacture of phenylene and related sanitary products, manufacture of various pickles, manufacture and sale of puffed rice or murmuras, opening of various retail stores, silk jewelry or filigree work, handicrafts using sabai grass, manufacture of various spices, sweets, snacks and pitha (cakes made of rice and pulses) and marketing and sale of the products, tailoring, stone carving, thread work, tent houses, manufacture of toiletries, cosmetics and oils, wood craft, etc., weaving etc.

Mission Shakti or SHGs of Odisha are supported to participate in various exhibitions, and fairs at district, state, and national levels to market and sell their products. It also provides a market complex called "Mission Shakti Bazaar" to sell products in various prime locations, including all smart cities in the state.

Using various IoT technologies such as smart transportation, the SHG members are able to do the marketing of their products. Along with that, the selling of their products has been improved a lot through various online platforms such as e-Commerce.

11.6.3 Mission Shakti: Evolution in Millet Mission Using IoT

In addition to the regular marketing of products such as rice cultivation and sales, milk and dairy products, various agricultural products, pulses and their by-products, millet and its products, handicrafts, etc., Mission Shakti has recently taken some outstanding initiatives that have received national attention.

One of these initiatives was the inauguration of the "Odisha Millet Mission". To take this to a new level, "Millet Cafes" were recently inaugurated in some smart cities like Bhubaneswar, Rourkela, and some other districts like Keunjhar in Odisha.

A SHG from the Khordha district called Shaktimayee SHG federation in collaboration with the Mission Shakti department is now running a retail outlet offering various millet-based products, especially snacks and dry products to the officials and all visitors at Krushi Bhavan in Bhubaneswar.

Ragi is a special type of millet that has many health benefits and is available in the market at an affordable price. This was an attempt to bring millet-based products back into people's daily life and increase the consumption of millet for its good properties. These ragi cookies are available in different flavors like jeera, kala jeera, milk, kaju, and sugar-free.

Millet-based products, such as ragi muffin, ragi khurma, ragi mix, ragi murukku, ragi/jowar laddus, ragi flour, jowar flour, suan rice, kanga rice, tea, and coffee, are attempted to bring back into people's daily life to increase their consumption.

Using the e-Commerce platforms, the Mission Shakti members are selling millet and millet-based products so effectively. Also, using the smart payment system, which is one of the most technological advancements of IoT, they can have larger target-based customers for their products who can easily purchase their products using digital payments.

Along with that, the advanced IoT system used in banks made the members of the SHGs enable of open and maintain their bank accounts in several banks. Using artificial intelligence (AI), banks can now open accounts using fingerprints and systems are being developed for differently abled people to operate easily in any ATM or bank (Rana & Khang et al., 2021).

11.7 IMPACT OF IoT AND THE AFTER-EFFECTS AMONG THE SELF-HELP GROUPS IN INDIA

11.7.1 Financial Decisions and the Monetary Flow

Women mostly benefit by joining an SHG as they shall be able to save money regularly, they can access the institutions responsible for formal savings, and they also get a chance to participate in the savings management system.

They are capable to save some money on a regular basis, and they also have bank accounts where they pay their bills. SHG positively puts impacts the member's abilities as they get a chance to save their money which is hard-earned by them (Khanh & Khang, 2021).

11.7.2 Improved Accessibility to the Credit

SHGs have improved women's access to credit. It may be too early to implement a project that directly improves women's access to credit. According to several successful groups, the financial liquidity gained from participating in SHGs has led to a better quality of life.

In general, many families are fulfilling their basic needs like food, shelter, and clothes like never before. Several nonprofit entities report that women's reimbursement rates are often higher than men's, women mostly like to spend their income on families, to improve their health and nutritional level, and the improvements in their livelihoods. The quality is slowly improving day by day.

11.7.3 Employment Has Been Increased

The introduction of SHGs empowered the rural poor to become self-reliant. The program has progressed a lot since its inception and around Rs. 35.7 lakh (0.43 United States Dollar) SHGs have formed after that, which is nearly Rs. 1.24 crores (0.015 United States Dollar).

The Swarozgaris have also started the setup on their own for micro-enterprise setups. Programs worth Rs. 11,486 crores (139.77 United States Dollar) by the government of India have been initiated along with the bank loan deployment worth Rs. 19,017 crores (231.40 United States Dollar); the total allowance is Rs. 9,318 crores (113.38 United States Dollar).

The program has shown a direction to many such involvers from the SHGs to enhance their current financial conditions. Also, a fund has been initiated by the National Bank for Agriculture and Rural Development (also known as NABARD) worth Rs.1,500,000 rupees (18,429.71 US Dollar (USD)) to support the women of SHGs in our country which are usually economically weaker.

11.7.4 Women Are Empowered and Stronger than Ever Before

The women of such groups become more empowered both economically and socially after joining such a group. This power of influence cannot be changed or transferred but must be self-generated to allow those in power to take responsibility for their lives.

The greatest social impact of the SHG program could be the increased awareness about different programs; increased spending on events related to health and social participation such as weddings, and other family gatherings; increased volume of participation in decision-making, and organizations; improved access to the organizations; and a change in the patriarchal attitude among the family members who now understood the SHG concept and encourage women of their respective house to explore the world by participating in gatherings.

Being with a woman who says she has savings to her name gives the other women in that group a confidence factor and boosts their self-esteem. The values and status of women have increased in their families after joining the group.

Children's education has greatly improved. Education for girls in particular was very poor, but the members of the SHGs now understood the value of sending their daughters to school to have a better future. They have also worked on improvisation of the sanitary conditions for members, and as a result, it also improved family health for members of SHG groups.

Women in such groups can now consult qualified doctors and choose the treatment they need, if necessary.

11.7.5 Enhanced Self-Confidence of the Group Members

The formation of SHG groups dragged out the leadership behavior and the hidden talents of its members. From this, it can be said that the member's position and respect in their families have improved a lot, and they have become more useful for the financial decisions of the family, and sometimes they also extend their help to others as well after joining SHG.

Most SHG members now feel more valued. Not only because of the rural environment but also because our own family treats us with respect. The group members are now invited to social events by the respective villagers. The time has changed, and their family members also respect their opinion, which was not happening before. They get encouragement from their family and support their activities.

The family accepts the fact that the members are working women and helps in the financial enhancement of the family which leads them to support the female members to do their job smoothly. They have improved their expertise in opting for various banking services, banking activities, and communication with government and bank officials (Khang & Gujrati et al., 2023).

Now you feel safe in these places. Also, their social status has improved a lot. After being a part of the SHG, the attitude of the family toward the members has changed tremendously.

Now they see them as brave women. The villagers also respect them more than before. The family asks for their opinion on many family decisions. In addition, they are generally respected by people in society (Babasaheb & Sphurti et al., 2023).

11.7.6 Reduced Domestic Violence

Participation in SHGs led to a reduction in violence in 25% of cases, especially through reduced financial hardship. In most cases, members said that the husband should also participate in her SHG.

11.7.7 Social Engagements

SHG members engaged in a variety of previously unimaginable community programs. They contributed school uniforms for poor students, held planting campaigns, handed out several office stationary items such as pens, pencils, and notebooks to needy students, and donated to government funds during national disasters such as floods, cyclones, and earthquakes.

Since joining SHG, they have participated in several social initiatives such as the campaign for cleaning the village and other programs for social improvement.

On India's Republic Day, they held a small event showcasing their initiatives and participation toward the betterment of society. From the perspective of social workers, participating women members are now economically beneficial and stronger.

They are capable of buying several items for their livelihood improvements such as furniture, TVs, phones, jewelry, other household necessary items, and groceries, and most conveniently they are also capable of saving some money which may satisfy their future needs.

The members of the groups are now more confident in interacting with government officials, bankers, electric companies, doctors, tax people, health insurance people, etc. (Khang & Chowdhury et al., 2022).

Also, women who had just finished schooling or who previously could barely leave their homes to talk to government officials or other villagers now feel confident about themselves and their communities. They seem to care about their families more in a significant and confident way (Snehal & Babasaheb et al., 2023).

Unlike in the past, women are now participating in various social and community events. Their family members value each other's opinions and perspectives about certain issues. Communicating with the other women members created more friendly relationships and less conflict in their lives.

As a result, the SHG move also acted as a multiplier. Through participation in training programs and related projects, awareness of health issues, personal hygiene, epidemics, and issues related to our environment, the impact of malnutrition, and hygiene are also increasing.

11.7.8 Positive Effects on Fortification and Dietary Intake

The main economic impact of this program is not to create new income streams, but to equalize consumption and diversify income streams.

The social and financial empowerment of the members of the SHGs in understanding the nutritional value of their food intake has increased a lot irrespective of their current position or status toward their participation in group activities.

11.7.9 Impact on the Environment

Previous research and studies have usually focused on the area among relationship between the degradation in environmental aspects and poverty, highlighting the fact that the community of poor people is both a victim and a perpetrator responsible for the environmental deterioration.

Because they live in environmentally vulnerable areas and are more likely to be polluters, they are forced to consume environmental resources which may contribute to environmental degradation. To support it, we need social conditions and poverty reduction.

Today, SHG plays a key role in alleviating poverty by empowering women in India. Women are also more active in environmental activities and managing the environmental aspects than men. Hence, the participation of women in moderating the programs through their SHGs can effectively raise awareness of environmental sustainability in society.

11.8 CONCLUSION

The strong digitization of the contemporary years has enabled several technology-based possibilities, which have slowly and steadily changed the most important sectors of both the economy and society in general. Digitization in various economic sectors has enabled versatile advancement opportunities and more efficient use of systems, limited resources, or processes.

The main factor for effective digitization in various sectors is either information technology or IoT-based smart technologies. In this regard, one of the most important sectors could be energy where the "digitalization of energy" has already been processed and developed fast in so many fields related to different kinds of energies.

Nowadays, energy is one of the most advanced applications of technologies using the IoT (Subhashini & Khang, 2024).

One of the most demanded advantages of IoT technologies toward the smart city concept is to resolve the major problems with respect to the infrastructure in densely populated urban areas.

IoT could effectively solve the early detection of various daily problems in urban areas, such as lack of water supply and energy produced, traffic problems, and surveillance problems.

The biggest challenge of the smart city concept is having an effective networking system and the operation of various network companies, and various technologies accompanied by sensors, which must be acquired by the appropriate education available for the public from the very initial days.

Every technology which is rapidly evolving around has specific potential drawbacks that must be carefully evaluated and superscribed (Khang & Hajimahmud et al., 2022).

With the number of IoT devices available in the billions and the possible and potential impact on the population at large, there are provocations that need to be taken care of, as identified in the research here.

The main goal is to bring sustainable and balanced application advancement of IoT technologies in the world for everyone's good. The recent development of IoT has fascinated the consciousness of developers, scientists, and research personnel throughout the world (Khang & Gupta et al., 2023).

To dilate applied science in a comprehensive manner and to bring the greatest benefits to SHGs, IoT researchers and developers are working hard together on an everyday basis.

The augmentation is only possible if we contemplate the abundant problems and blemishes of topical approaches toward technological advancements (Jebaraj & Khang et al., 2024).

REFERENCES

Aggarwal, P., Khang, A., "A Study on the Impact of the Industry 4.0 on the Employees Performance in Banking Sector", *Designing Workforce Management Systems for Industry 4.0: Data-Centric and AI-Enabled Approaches*, pp. 384–400 (1st Ed.) (2023). CRC Press. https://doi.org/10.1201/9781003357070-20

Babasaheb, J., Sphurti, B., Khang, A., "Industry Revolution 4.0: Workforce Competency Models and Designs", *Designing Workforce Management Systems for Industry 4.0: Data-Centric and AI-Enabled Approaches*, pp. 14–31 (1st Ed.) (2023). CRC Press. https://doi.org/10.1201/9781003357070-2

Bhambri, P., Rani, S., Gupta, G., Khang, A., *Cloud and Fog Computing Platforms for Internet of Things* (2022). CRC Press. ISBN: 978-1-032-101507 https://doi.org/10.1201/9781032101507

Cheston, S., & Kuhn, L. (2002). Empowering Women through Microfinance. Microfinance Summit Campaign, Washington DC. Retrieved, February 11, 2012, from http://www.microcreditsummit.org/papers/empowerinto.htm

Cheston, S. Y. L. K. (2004). Empowering Women through Microfinance. Microcredit Summit. New York. https://www.academia.edu/download/5353418/empowerment.pdf
Gillis, A.S., What is IOT (internet of things) and how does it work? – Definition from techtarget.com. *IoT Agenda*. (2022, March 4). Retrieved March 4, 2022, from https://www.techtarget.com/iotagenda/definition/Internet-of-Things-IoT
Goetz, A.M., *Women Development Workers: Implementing Rural Credit Programs in Bangladesh* (2001). New Delhi: Sage Publications. Government of India 200 Tenth Five Year Plan 2002–2007. https://search.proquest.com/openview/cd8738bc2b12053179dc515e62f063ed/1?pq-origsite=gscholar&cbl=25135
Gupta, P. K., & Sharma, S. (2021). Literature review on effect of microfinance institutions on poverty in South Asian countries and their sustainability. International Journal of Emerging Markets in Physics Review. 47, 777–780. https://www.emerald.com/insight/content/doi/10.1108/IJOEM-07-2020-0861/full/html
Hajimahmud, V.A., Khang, A., Hahanov, V., Litvinova, E., Chumachenko, S., Alyar, A.V., "Autonomous Robots for Smart City: Closer to Augmented Humanity", *AI-Centric Smart City Ecosystems: Technologies, Design and Implementation* (1st Ed.) (2022). CRC Press. https://doi.org/10.1201/9781003252542-7
Harish, N. (2012). Women's empoWerment through entrepreneurship development with special reference to self help groups in Karnataka. International Journal of Research in Social Sciences, 2(2), 389–401. https://www.indianjournals.com/ijor.aspx?target=ijor:ijrss&volume=2&issue=2&article=026
Hirut B. H, Intan Osman, Rashidah S, Siti Waringin O, (2000), Is there a Convergence or Divergence between Feminist Empowerment and Microfinance Institutions' Success Indicators, https://doi.org/10.1002/jid.3041
Hunt, J. (2002). Reflections on microfinance and women's empowerment. Women, Gender And Development In The Pacific: Key Issuess, 13. Physics Reviews 47, 777–780. https://asiaandthepacificpolicystudies.crawford.anu.edu.au/rmap/devnet/devnet/gen/gen_status.pdf#page=13
Jansen, H.A.F.M., Nguyen, T.V.N. & Hoang, T.A. Exploring risk factors associated with intimate partner violence in Vietnam: results from a cross-sectional national survey. International Journal of Public Health 61, 923–934 (2016). https://doi.org/10.1007/s00038-016-0879-8
Jebaraj, L.-, Khang, A., Chandrasekar, V., Pravin, A.R., Sriram, K., "Smart City Concepts, Models, Technologies and Applications", *Smart Cities: IoT Technologies, Big Data Solutions, Cloud Platforms, and Cybersecurity Techniques* (2023). CRC Press. https://doi.org/10.1201/9781003376064-1
Johnson, S., & Rogaly, B. (1997). Microfinance and poverty reduction. Oxfam. https://www.google.com/books?hl=en&lr=&id=MjWewiCahk0C&oi=fnd&pg=PR7&dq=Microfinance+and+poverty+reduction.+Oxfam&ots=HvexcAIIqT&sig=dd1dnhfC4jqK7ne9DSemFrW0c0c
Khang, A., Abdullayev, V., Hahanov, V., Shah, V., *Advanced IoT Technologies and Applications in the Industry 4.0 Digital Economy* (1st Ed.) (2024). CRC Press. https://doi.org/10.1201/978-1-003-43426-9
Khang, A., Chowdhury, S., Sharma, S., *The Data-Driven Blockchain Ecosystem: Fundamentals, Applications, and Emerging Technologies* (2022). CRC Press. https://doi.org/10.1201/9781003269281
Khang, A., Gupta, S.K., Hajimahmud, V.A., Babasaheb, J., Morris, G., *AI-Centric Modelling and Analytics: Concepts, Designs, Technologies, and Applications* (1st Ed.) (2023). CRC Press. https://doi.org/10.1201/9781003400110
Khang, A., Gupta, S.K., Rani, S., Karras, D.A., *Smart Cities: IoT Technologies, Big Data Solutions, Cloud Platforms, and Cybersecurity Techniques* (1st Ed.) (2023). CRC Press. https://doi.org/10.1201/9781003376064

Khang, A., Gupta, S.K., Shah, V., Misra, A., *AI-Aided IoT Technologies and Applications in the Smart Business and Production* (1st Ed.) (2023). CRC Press. https://doi.org/10.1201/9781003392224

Khang, A., Hahanov, V., Abbas, G.L., Hajimahmud, V.A., "Cyber-Physical-Social System and İncident Management", *AI-Centric Smart City Ecosystems: Technologies, Design and Implementation* (1st Ed.) (2022). CRC Press. https://doi.org/10.1201/9781003252542-2

Khang, A., Hahanov, V., Litvinova, E., Chumachenko, S., Triwiyanto, V.A., Hajimahmud, R.N., Ali, A.V., Alyar, Anh, P.T..N., The Analytics of Hospitality of Hospitals in Healthcare Ecosystem", *Data-Centric AI Solutions and Emerging Technologies in the Healthcare Ecosystem*, p. 4 (1st Ed.) (2023). CRC Press. https://doi.org/10.1201/9781003356189-4

Khang, A., Khang, A., Hrybiuk, O., Abdullayev, V., Shukla, A.K., *Computer Vision and AI-Integrated IoT Technologies in Medical Ecosystem* (1st Ed.) (2024). CRC Press. https://doi.org/10.1201/978-1-0034-2960-9

Khang, A., Ragimova, N.A., Hajimahmud, V.A., Alyar, A.V., "Advanced Technologies and Data Management in the Smart Healthcare System", *AI-Centric Smart City Ecosystems: Technologies, Design and Implementation* (1st Ed.) (2022). CRC Press. https://doi.org/10.1201/9781003252542-16

Khang, A., Rani, S., Gujrati, R., Uygun, H., Gupta, S.K., *Designing Workforce Management Systems for Industry 4.0: Data-Centric and AI-Enabled Approaches* (1st Ed.) (2023). CRC Press. https://doi.org/10.1201/99781003357070

Khang, A., Rani, S., Sivaraman, A.K., AI-Centric Smart City Ecosystems: Technologies, Design and Implementation (1st Ed.) (2022). CRC Press. https://doi.org/10.1201/9781003252542

Khanh, H.H., Khang, A., "The Role of Artificial Intelligence in Blockchain Applications", *Reinventing Manufacturing and Business Processes through Artificial Intelligence*, pp. 20–40 (2021). CRC Press. https://doi.org/10.1201/9781003145011-2

Mayoux, L. (1998). Research round-up women's empowerment and micro-finance programmes: strategies for increasing impact. Development in Practice, 8(2), 235–241. https://www.tandfonline.com/doi/pdf/10.1080/09614529853873

Mission Shakti, Inauguration of Odisha's First Millet Shakti Outlet at Krushi Bhawan: News IDIPR/AG&FE (Eng)/1294/(2022 November 5): Published in https://enews.nic.in/

Narasaiah M., KK Ray, R Sivakumar. Studies on short fatigue crack growth behavior of a plain carbon steel using a new specimen configuration. Materials Science and Engineering: A, 2004 - Elsevier. https://www.sciencedirect.com/science/article/pii/S0921509303010360

Pattanaik, 2003. Detailed report from the Government Odisha on Mission Shakti and their functionalities from https://missionshakti.odisha.gov.in/

Rai, A., Kumar, D., Bhatt, A., "The Offline Antecedent of the Sharing Economy: The Self-help Group for the Bottom of Line in India", In *Proceedings of the International Conference on Innovative Computing & Communication (ICICC)* (2021). https://papers.ssrn.com/sol3/papers.cfm?abstract_id=3842664

Rana, G., Khang, A., Sharma, R., Goel, A.K., Dubey, A.K., *Reinventing Manufacturing and Business Processes through Artificial Intelligence* (2021). CRC Press. https://doi.org/10.1201/9781003145011

Rangarajan, C., Mahendra Dev, S., "Poverty in India: Measurement, Trends and Other Issues", In *Perspectives on Inclusive Policies for Development in India* (pp. 255284). (2022). Springer, Singapore. https://link.springer.com/chapter/10.1007/978-981-19-0185-0_13

Rani, S., Bhambri, P., Kataria, A., Khang, A., "Smart City Ecosystem: Concept, Sustainability, Design Principles and Technologies", *AI-Centric Smart City Ecosystems: Technologies, Design and Implementation* (1st Ed.) (2022). CRC Press. https://doi.org/10.1201/9781003252542-1

Rani, S., Bhambri, P., Kataria, A., Khang, A., Sivaraman, A.K., Big Data, Cloud Computing and IoT: Tools and Applications (1st Ed.) (2023). Chapman and Hall/CRC. https://doi.org/10.1201/9781003298335

Rani, S., Chauhan, M., Kataria, A., Khang, A., "IoT Equipped Intelligent Distributed Framework for Smart Healthcare Systems", *Networking and Internet Architecture* (2021). CRC Press. https://doi.org/10.48550/arXiv.2110.04997

Sahoo, A. (2013). Self help group & woman empowerment: A study on some selected SHGs. International Journal of Business and Management Invention, 2(9), 54–61. https://www.academia.edu/download/31994603/I0291054061.pdf

Shah, V., Jani, S., Khang, A., "Automotive IoT: Accelerating the Automobile Industry's Long-Term Sustainability in Smart City Development Strategy", *Smart Cities: IoT Technologies, Big Data Solutions, Cloud Platforms, and Cybersecurity Techniques* (2023). CRC Press. https://doi.org/10.1201/9781003376064-9

Sharma, N., *Facilitating Empowerment of Women through Self Help Groups concerning NABARD in Punjab: An Analysis* (2020). https://www.indianjournals.com/ijor.aspx?target=ijor:pag&volume=8&issue=2&article=004

Sinha, F. (2005). Access, use and contribution of microfinance in India: Findings from a national study. Economic and Political Weekly, 1714–1719. https://www.jstor.org/stable/4416529

Snehal, M., Babasaheb, J., Khang, A., "Workforce Management System: Concepts, Definitions, Principles, and Implementation", *Designing Workforce Management Systems for Industry 4.0: Data-Centric and AI-Enabled Approaches*, pp. 1–13 (1st Ed.) (2023). CRC Press. https://doi.org/10.1201/9781003357070-1

Subhashini, R., Khang, A., "The Role of Internet of Things (IoT) in Smart City Framework", *Smart Cities: IoT Technologies, Big Data Solutions, Cloud Platforms, and Cybersecurity Techniques* (2024). CRC Press. https://doi.org/10.1201/9781003376064-3

UN (2022). UN Report: Global hunger numbers rose to as many as 828 million in 2021. https://www.fao.org/newsroom/detail/un-report-global-hunger-SOFI-2022-FAO/en

Vrushank, S., Khang, A., "Internet of Medical Things (IoMT) Driving the Digital Transformation of the Healthcare Sector", *Data-Centric AI Solutions and Emerging Technologies in the Healthcare Ecosystem*, p. 1. (1st Ed.) (2023). CRC Press. https://doi.org/10.1201/9781003356189-2

Vrushank, S., Vidhi, T., Khang, A., "Electronic Health Records Security and Privacy Enhancement Using Blockchain Technology", *Data-Centric AI Solutions and Emerging Technologies in the Healthcare Ecosystem*, p. 1 (1st Ed.) (2023). CRC Press. https://doi.org/10.1201/9781003356189-1

12 Personalized Social-Collaborative IoT-Symbiotic Platforms in Smart Education Ecosystem

Ahmad Al Yakin, Muthmainnah, Alex Khang, and Abdul Mukit

12.1 INTRODUCTION

The value of a good education cannot be overstated in terms of its contribution to the formation of a decent human being. Research is being conducted in developed nations to better inform educational policy and practice. There is a lot of work that needs to be done to enhance individual learning resources for educators and students in developing nations.

The Intelligence of Learning Things (IoLT) is our platform for education that takes this into consideration. Incorporating cutting-edge pedagogical practices and technological tools, the Internet of Things (IoT) based blended learning strategy is poised to revolutionize the world of education in smart city ecosystem.

Through the IoT, this platform enables users of a wide range of devices and apps to work together and exchange ideas with one another (including educators and students) (Satu & Roy et al., 2018).

It has been suggested by Pealoza-Farfán and Paucar-Caceres (2018) that education is a process that makes individuals more knowledgeable and intelligent through acquiring new skills and gathering information. Therefore, man is aware of his obligations and has the ability to significantly improve the current state of the world. In other words, education leads to discoveries, innovations, and adaptations that increase our ability to detect, respond to and ultimately solve global problems.

The best method to make the world a better, safer, and more sustainable place to live is through education, thus making education the best investment.

The significance of education in shaping a better future for all people means that this sector must continue to undergo reform. Changes in the way people around the world educate and learn are caused by advances in technology.

Educators can better present their lessons to individual students and current topics of interest thanks to advances in technology (Oztemel & Gursev, 2020; Fullan, 2023).

DOI: 10.1201/9781003376064-12

In recent years, there has been a radical departure from traditional educational methods due to the widespread use of innovative ICT (information and communications technology [or technologies], is the infrastructure and components that enable modern computing) solutions across all levels of education. There are now seven distinct forms of technology, tool, and strategy that are all having a significant impact on the way we teach and learn.

The list goes on and on, but some examples are as follows: consumer technologies, digital strategies, enabling technologies, Internet technologies, learning technologies, social media technologies, and visualization technologies (Dimitriadou & Lanitis, 2023; Hidalgo & Bucheli-Guerrero et al., 2023).

To put it simply, the IoT is a subset of Internet technology that enables anything to be connected to a network, hence enabling enormous advancements in virtually every field of human effort.

Putting it another way, the integration of IoT into the educational environment helps close the gap between the needs of the traditional education system and those of the modern education system by turning traditionally separated classrooms across different locations and time periods into connected classrooms united by the Internet and communication tools.

Consequently, IoT advancements have a sizable impact on teaching methodology. As a result of technological advancements, the quality and quantity of course materials, as well as the classroom and online learning experience, have substantially increased.

Because of this, the IoT is ushering in a whole new level of interactivity and cutting-edge instructional practices in the study of knowledge. While IoT has many potential benefits, it also presents some serious challenges. One such challenge is the potential for compromising student grades and other personal information.

There is great potential for the IoT to enhance the quality of education while also helping to make the education industry more economically, socially, and environmentally sustainable. This has been documented by a group of researchers (Makarova & Shubenkova et al., 2019).

Data explosions, technical developments, and intellectualization processes have come from the widespread adoption of digital technology, posing new problems for the world economy and the area of education.

There is now a new digital (networked) generation that considers having a smartphone, laptop, and Internet connection to be as fundamental to their daily lives as breathing and eating. Smart education should be implemented on a global scale.

Complex technological systems are increasing in prevalence, and as a result, there is a growing need for engineers capable of designing, constructing, and supervising these projects. Sustainable development principles, such as reducing waste and pollution, must be upheld at the same time. It is crucial that engineering programs graduate competent engineers into a sector that is in great demand in the modern economy and culture.

The use of instructional technologies such as models, simulations, and augmented and virtual reality can help accomplish this aim.

Education for sustainable development is based on principles and ideas developed by UNESCO (Baena-Morales & Urrea-Solano et al., 2023).

Education for sustainable development is defined by the following features: it is interdisciplinary; it includes formal, non-formal, and informal education; it promotes lifelong learning; it is culturally appropriate; it addresses content, context, global issues, and local priorities; it has international repercussions and impacts; and it strengthens civil capacity for community-based decision-making, social tolerance, and ecological literacy (Strada & Lopez et al., 2023).

Therefore, new methods of instruction are required to help students, educators, and support staff learn to re-regulate their interpersonal connections and develop more effective pedagogical practices for fostering student growth (Gamble & Gamble, 2023).

All persons involved are necessary for this to happen. It's safe to say that teamwork has become the norm in the modern day (Jerald, 2009). The student-teacher collaboration that greatly enhances both teaching practice and the learning experience is only possible because of the cooperation among faculties, senior management, teachers, students, and administrative personnel.

School and university buildings can be made more energy efficient, waste reduced, and ICT equipment recycled and reused more frequently if the G-IoT is fully implemented, all while improving the built environment through the creation of "smart" structures that prioritize ICT and environmental sustainability.

Although technology has become the facilitator of modern education, high-quality education in a sustainable and eco-friendly educational environment is only attainable via extensive collaboration among all stakeholders in the education sector (Mokski & Leal Filho et al., 2023).

Leaders in cutting-edge technology are mostly concentrating on integrating these components in everyday goods. Regardless of their main area of concentration, it becomes essential to educate our population on cutting-edge computer science and IT skills given the spectacular breakthroughs in digital technology and ubiquitous computing.

On the other hand, it is very important to prepare students currently enrolled in any program at an educational institution for this technology. Employers will seek applicants who understand the latest cutting-edge technologies.

To prepare the next generation of engineers and IT specialists, industry leaders, such as Cisco, Microsoft, and many others, have incorporated IoT, Cloud Computing, and Virtualization into their current school curricula. On the other hand, educational institutions not only promote critical thinking and creativity but also educate computer science ideas.

The IEEE Computer Society's Computing Now, published in June 2017, goes into greater detail on how instructors can use technology to their advantage on the IoT era (Bebell & O'Dwyer, 2010). Selected papers present strategies for incorporating IoT into STEM education while creating a learning environment that encourages problem-solving and discovery.

The video also demonstrates how using an open source IoT platform can drive innovation among 21st century students. To meet Smart Society's literacy requirements, we also examine the various teaching and educational perspectives that must be considered when constructing educational curricula.

Therefore, the purpose of this chapter is to try a summary of the impact of the IoT on the field of education. Clearly, the IoT can only demonstrate its enormous potential to utterly alter education if all stakeholders work together to achieve contemporary and long-term goals in the field.

This chapter was conducted to learn more about how Pinterest-smartphone application-based learning is being tested in social science courses to create sustainable intelligent education, and how the integration of IoT technology used in the learning process can enhance social-collaborative symbiosis.

12.2 SOCIAL-COLLABORATIVE LEARNING MODEL

Multiple research (Perry & VandeKamp et al., 2023) have demonstrated that social contacts are an effective method for both independent and group study in higher education. Whether in a traditional classroom or in a digital medium, students learn best when they are able to engage with their instructors, peers, and course materials.

Online learning environments in smart city age, in particular, have put a spotlight on students' interactions with technology as a result of recent technical advancements.

Extensive prior research has examined the effect of interaction on learning, and researchers have concluded that interactions are an essential component of productive learning experiences, whether they take place in a typical classroom setting or an online learning environment (Scager & Boonstra et al., 2016).

Working together to find solutions to group assignments is also a great way for students at the university level to hone their communication abilities. In terms of their schooling, this is a significant event. Collaborative learning can be highly productive, but only if all participants grasp the importance of and know how to implement efficient communication and cooperation (Vuopala & Hyvönen et al., 2016).

Students' sense of competence and self-assurance as learners is greatly bolstered by opportunities for group study. Students who have high levels of self-assurance are more likely to find solutions to the challenges they face as they learn (Zarfsaz & Serpil, 2023).

Students benefit from learning in groups because they are better able to listen to one another, ask questions when they don't understand something, offer advice and encouragement to one another, and lay a solid basis for future learning.

Students will develop the ability to communicate and reason with one another in a mature manner, laying the groundwork for a strong sense of community and mutual trust.

All of the aforementioned considerations are crucial for students to be able to work together effectively in a classroom context (Dimitriadou & Lanitis, 2023). It's worth noting, though, that group study doesn't always yield the desired results. This is especially true if the members of the group do not work together in order to actively participate in the learning process.

Unfair distribution of work among group members, with some students acting as "free riders", while others taking on the bulk of the work, is a common problem in collaborative learning.

As a result, numerous proposals have been made to foster a safe classroom setting that encourages productive student collaboration. Some researchers believe that a problem-based approach to education will help students learn to work together more productively (Chan & Shroff et al., 2023).

Collaborative problem-solving is proven to improve students' analytical reasoning abilities (Lin & Mills et al., 2016). This is especially true when students are given the opportunity to discuss challenges at hand and offer their own ideas for how to solve them.

There is increasing evidence to suggest that incorporating collaborative problem-solving activities into online learning environments is beneficial for students, and many teachers have started to do so (Kwon & Song et al., 2019). It is hoped that by using a problem-based learning approach, we can overcome some of the limitations that current collaborative education has in terms of interaction.

However, problem-based learning in groups is not always successful; Donnelly and Fitzmaurice (2005) show that this can be hindered by students' inability to solve problems related to the limited exchange of resources and activities outside the classroom. Talking to each other after class, away from the confines of the classroom, is essential to transcending this barrier.

Better online learning platforms that allow students to interact and contribute outside of class time are available through the use of online social technologies such as using WhatsApp groups (Kee, 2020). There is a need for more thorough research to compare different types of student engagement, although much research has been conducted on the theme.

As students' progress through their academic program, their interaction with one another plays an important role in helping them achieve their individual and collective learning goals.

In the context of distance education, interaction between students is a critical aggregation point. Therefore, more studies are needed to investigate this relationship across various online education contexts and pedagogical approaches.

To those who subscribe to social-constructionism, learning is not just an individual process but rather one that takes place in communities where people are continuously interacting with one another and their own personal creations (Singh & Singh et al., 2022). Therefore, learning a foreign language is a social interaction that is shaped by the communicative and cultural norms of the learner's immediate environment.

We assume that Vygotsky's (1978) emphasis on cooperative learning, knowledge sharing, problem-solving, and empirically based materials will help students learn better and get a deeper appreciation for another culture.

Learning a new course takes time, effort, and input from both the learner and their teachers. However, the playful nature of social media-based platforms may encourage more students to absorb the material that has the potential to alter their current cognitive state into a more reflective and broadly applicable form of knowledge.

The proposed idea is a collaborative learning platform that incorporates Web 2.0 tools to support the study of social science course that is called social-collaborative learning for the 21st-century learner.

Our central argument is that Web 2.0 tools may be used in conjunction with social constructivist principles to produce unique forms of educational delivery (Paily, 2013).

Teachers need to be able to build and develop pedagogical concepts, supporting tools, and methods that allow them to test their assumptions in light of specific learning objectives before they can establish new learning platforms that improve the educational experience (Blessinger & Wankel, 2013).

Finding techniques to pique your pupils' interest in learning is crucial if you want them to succeed academically. A person's degree of motivation determines the manner in which they choose to accomplish something, the length of time they are willing to maintain doing it, and the intensity with which they work at it (Hursen, 2021).

A dynamic and developing mental process defined by constant (re)appraisal and balancing of the many internal and external influences an individual is subjected to is linked to the maintenance of motivation during the protracted process of studying particular subjects.

New research shows that today's kids use the Internet in new and different ways for school and social activities than their analog-era counterparts (Al Maani & Shanti, 2023).

Although "pedagogy 2.0" (the use of social media and Web 2.0 apps in the classroom) has the potential to improve education in a number of ways, more study and evaluation are required before it can be broadly used (Conole, 2010).

12.3 MOBILE TECHNOLOGY – THE SYMBIOTIC OF IoT

Catering to undergraduate students who grew up in the digital age, some of whom may not have a strong academic background, is one of the issues facing educators in the 21st century.

These young people are part of the growing population of college freshmen who are motivated to learn quickly and effectively in order to fulfill the expectations of a competitive job market. As a result, they are highly practical and focused on outcomes (Khang & Gujrati et al., 2023).

According to Biggs (2003), this method of study or learning may boost surface learning, which is problematic because it means that the information learned may be forgotten quickly.

Furió and Juan et al. (2015) as opposed to being a reflection of one's character, as is sometimes asserted, a student's propensity for either a surface or a deep approach to learning is better viewed. This suggests that deep learning may be more achievable in settings where students actively participate, solve problems, and cooperate together (e.g., by explaining concepts to one another).

To satisfy these needs, universities are adopting blended learning methodologies that incorporate social media/Web 2.0 platforms to help students acquire and retain information to apply it in novel situations and forms, such as through group work, argumentation, or debate.

Students' participation in group projects and online classes is facilitated by social media and other Web 2.0 tools (wikis, blogs, etc.). As a direct result of the meteoric expansion of social media among young people, academics from all over the world have recently investigated the role of the Internet in promoting education (Traxler, 2023).

Understanding how social media influences the design of e-learning environments, as well as how these factors relate to the methods of instruction, the information gained by students, and their performance in class, is crucial. In this case study, we use data from the field and socio-constructivist theory to investigate whether or not social media may enhance the classroom experience.

The core idea behind the social-constructivist stance is that knowledge is best created when several people work together. According to constructivist theory, learning is best understood as an ongoing process of knowledge building in which learners develop new understanding by drawing on and modifying previously acquired expertise (Babasaheb & Sphurti et al., 2023).

Under a constructivist approach to education, the student is not only a recipient of information but also is actively involved in the process of constructing his or her own knowledge.

Learning may be bolstered through complex interactions, including games, dialogues, case-based work, and partnerships with peers. We might use a looser definition of social learning in this case (Soomro & Zai et al., 2015).

Teachers' interest in using mobile devices has been revived thanks to recent advances in mobile computing technology and other ICT fields, including operating systems such as Windows Mobile, Android, and Symbian support today's mobile devices such as cell phones, smartphones, and personal digital assistants (PDAs).

This system is equipped with the same capabilities as a desktop computer and can be connected to a broadband Internet network for use in the classroom.

The next generation of learning settings will make unique use of mobile media devices because of their portability and ability to operate in any setting at any time (Foti & Mendez, 2014).

The rapid growth of mobile technology, especially cell phones or smartphones, in the past decade has presented exciting new possibilities for widespread education (learning everywhere).

The rapid growth of learning in the classroom based on smartphones and other mobile devices to access the Internet is obvious and easy to observe. The ever-evolving ICTs are influencing traditional pedagogical practices.

Usually, there are a lot of problems with the way schools have always done things. Many factors contribute to this, including students' lack of awareness of the value of social and collaborative learning in the classroom and their tendency to just sit through a lecture, take notes, and then leave.

The more people who have access to the same information, the better. It is this inefficiency that social and collaborative learning seeks to address. It is a teaching strategy in which groups of students work together to complete a series of tasks (Scager & Boonstra et al., 2016; Wei, 2023).

The teacher's role is similar to that of a coach, mentor, or learning facilitator. All group members benefit equally from group success. Students actively pursue knowledge and accept responsibility for their own education they encourage active participation from students, and the practical experience gained is vital.

Learning in a group has several benefits, such as being more student-focused, task-based, and activity-based. This curriculum is designed to improve students' interpersonal, communication, cooperation, sharing, and caring skills, as well as their openness, flexibility, adaptability, knowledge retention, higher order and critical thinking, creativity management, practicality, responsibility, trustworthiness, engagement, participation, commitment, self-confidence, and self-efficacy (Sharma, 2023).

Collaborative learning encourages students to share their perspectives, learn from each other's experiences, and generate mutually agreed-upon problem solutions. They help, explain, educate, understand, evaluate, and influence one another. Creating a learning community allows them to pool their diverse skills for the greater good (Troussas & Giannakas et al., 2020; Lee & Kim et al., 2017).

As a flexible new learning landscape, mobile learning is rapidly gaining traction in academic and professional settings around the world. The benefits and elements that contribute to the efficiency and sustainability of education and learning (Khan & Cram et al., 2023; Watermeyer & Chen et al., 2022; Oke & Fernandes, 2020) make the amalgamation of social and collaborative learning in mobile learning the MobiTech (mobile technology) of highest relevance and crucial.

The main focus of this study is on how a recommender system can effectively address the problem of excessive and unmanageable learner social activity in mobile learning. This chapter uses the Pinterest application as a teaching tool that aims to increase student social media collaboration while learning takes place in tertiary institutions.

12.4 RELATED WORK

According to Myers and Well et al. (2013), "the research method is the research technique, which is a scientific means to acquire data with research aims and purposes".

The term experimental technique refers to a type of study in which researchers manipulate variables to see how changing one variable affects the outcome of the experiment. The research strategy employed in this study was pre-experimental as Table 12.1.

According to Edward J. Stanek III (1988), the researchers in this study used a one-group pre-test-post-test design to ensure that they had complete control over the factors they had chosen to influence the experiment's outcome. In this study, the design is the blueprint for how the investigation will be conducted.

The research employed a one-group pre-test-post-test design format of pre- and post-testing. The one class received the therapy before the initial test (the pre-test), while the other class (the experimental class) received the treatment after the test (the post-test). This layout serves its intended function.

The goal is to determine whether or not students show more enthusiasm for studying educational politics after making use of the Pinterest application. Here we provide a one-group pre-test-post-test study. The efficacy of learning using the Pinterest app may be evaluated using a pre- and post-test approach.

12.4.1 Population and Sample

The population of this study was 360 undergraduate students, and 20 students were deliberately chosen based on proximity, and the courses studied were basic social sciences at the Faculty of Public Health at Universitas Al Asyariah Mandar.

Faculty selected based on student characteristics, namely that one group of undergraduate students had never used the Pinterest application, did not know Pinterest

TABLE 12.1
One-group Pre-test-Post-test Design

Group	Pre-test	Treatment	Post-test
Eksperimen	O^1	X	O^2

O_1: Pre-test.
O_2: Post-test.
X: Treatment by using Pinterest.

at all, had never been given the treatment of working in teams or groups during lectures, and did not know if Pinterest could be used as teaching materials or learning resources. There were a total of 2 men and 18 women in the group.

12.4.2 Data Collection Technique

To facilitate the research that was compiled, the researchers designed the technique. The data collection methods to be used in this study are as follows:

1. **Test**

   A test is a set of questions or exercises, or a combination of these and other instruments, designed to evaluate some aspect of an individual's or a group's potential or performance.

   In this study, we employed the Educational Testing Service's Test of Educational and Political Content Knowledge to assess our hypotheses. In this study, the tests used were divided into two types:

   1. Pre-test, namely the test conducted before the treatment is given.
   2. A post-test, which is a test carried out after the treatment is given. Posttest is given for measure the initial ability of the experimental class and the control class and know homogeneity. While the post-test is given to determine progress or improvement experimental class.

2. **Observation**: A systematic observation is carried out by observers using guidelines as an observation instrument. In this method, the researcher observes events during the learning process using Pinterest in progress, which uses observation guidelines, namely a list of types of collaborative social activities that may arise and will be observed.
3. **Teaching Strategy**: In the early stages of studying social studies courses, the lecturer explained material regarding social groups, socio-culture, and socio-cultural changes in the digital era. Lecturers share material links contained in the Pinterest application via the WhatsApp group, and students are asked to form groups as Figure 12.1.

FIGURE 12.1 Students listening the instruction in learning through WhatsApp.

The students in the study group all get on their phones and install the Pinterest app from the Google Play Store as Figure 12.2.

Students in small groups use the URLs provided in the WhatsApp group to access social science teaching materials using the ResearchGate app as Figure 12.3.

Students use collaborative strategies to address assignments. At this point, they talked about how to recognize socio-cultural shifts brought on by the information age.

FIGURE 12.2 Students download Pinterest app.

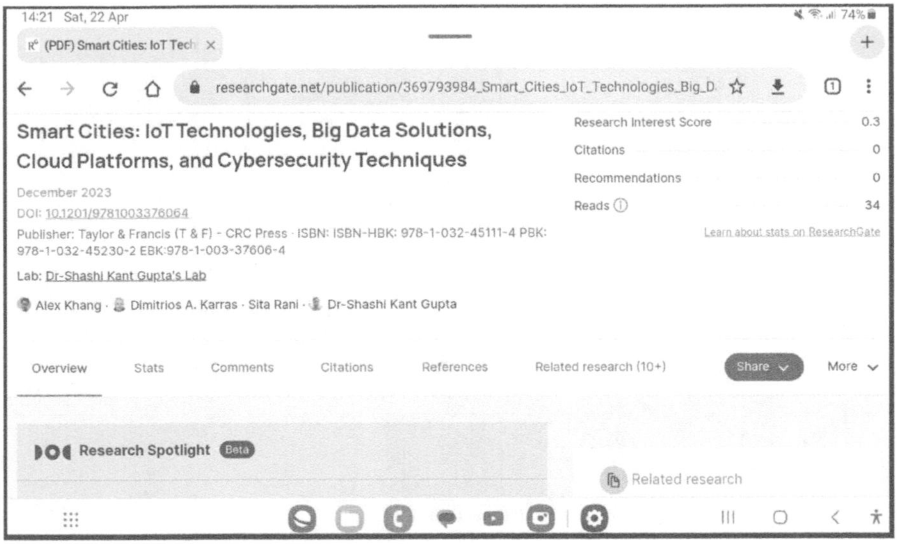

FIGURE 12.3 Social science materials on ResearchGate app.

FIGURE 12.4 Students personalized social skill in group.

People communicate by exchanging information, debating among themselves, relaying information, and describing the world around them (Snehal & Babasaheb et al., 2023). The teacher keeps a close eye on the proceedings as Figure 12.4.

After discussing and interacting with their group mates, they agreed on the results of the discussion and then recorded the points from the discussion to present in front of the class as Figure 12.5.

FIGURE 12.5 Students share their idea and increase their interaction.

12.5 DATA ANALYSIS TECHNIQUE

Data analysis techniques are used by researchers, namely descriptive and inferential statistics. Descriptive statistical analysis is performed by using percentage, average, and standard deviation formulas meanwhile, for the analysis of inferential statistical data using normality tests, homogeneity tests, and significance tests.

12.5.1 Descriptive Statistical Analysis

Statistical data is statistical data that is used to analyze data by describing or describing the data collected, not drawing general or generalized conclusions. The data obtained using this technique is the mean, mode, range, minimum, maximum, standard deviation, variance, and frequency distribution table.

The data is to describe the effect of Pinterest app in learning to the student's learning outcome. The manual descriptive statistical process is the amount of data.

1. **Average**

   The average is the amount of data divided by the amount of data with the following formula 12.1:

$$x = \frac{\Sigma x_i}{n} \tag{12.1}$$

   ΣX_i = value of each data
   X = average
   N = amount of data

2. **Median (Me) fr**

   The median is the result of data that is located in the middle after the data is arranged from small to large (or vice versa).

$$me = b + p\left(\frac{\frac{1}{2}n - F}{f}\right) \tag{12.2}$$

   b = lower limit of the median class, which is where the media is located
   p = median class length
   n = sample size or amount of data
   F = Sum of all frequencies with a smaller class sign than before the median class frequency
   f = median class frequency

3. **Mode (Mo)**

   The mode is the score or value that appears or occurs the most.

$$Mo = b = p\left(\frac{b1}{b1 + b2}\right) \tag{12.3}$$

   b = lower limit of the mode class, which is the interval class with the greatest frequency

p = length of mode class
b_1 = mode class frequency – class interval frequency with a smaller class sign before the mode class sign
b_2 = mode class frequency – interval class frequency with a class sign that is greater than the mode class sign

4. **Standard Deviation**

$$S^2 = \frac{\sum_{i}^{n}(x_1 - x)}{n-1} \tag{12.4}$$

s = standard deviation of student learning scores
fi = frequency of each class interval
xi = the mean value of each
n = number of data

12.5.2 Inferential Statistical Analysis

Inferential statistics is a method of statistical analysis in which results from a sample are extrapolated to the whole population to decide whether or not to reject the null hypothesis. Initial steps in this inferential statistical study included checking for normality, homogeneity, and testing the hypothesis.

1. **Normality test**
   The normality test was carried out to find out whether the sample under study was normally distributed or not. To test the normality of the research data, it was analyzed using the SPSS program.
   With the provisions of the alpha level of 0.05, to carry out the normality test, the statistical hypothesis is as follows:

   H_0 = Data comes from a normal population
   H_1 = Data comes from an abnormal population

   So, if the data normality test results are significant 0.05, then H_0 is accepted, and if the data normality test results are significant 0.05, then H_0 is rejected.
2. **Homogeneity Test**
   The homogeneity test aims to find out whether the samples taken from the population have the same variance and do not show significant differences from each other. The statistical hypothesis formulation is used as follows:

   H_0 = Variance of a homogeneous data group.
   H_1 = The variance of the data group is not homogeneous.

   The test criteria based on significance are as follows: (a) if the significance is 0.05, then H_0 is accepted; if it is 0.05, then H_1 is rejected.

3. **Hypothesis testing**
Inferential statistics are used to test the research hypothesis. The hypothesis test used in this study was the t-test using the SPSS 21 for Windows program. The hypothesis to be tested in this study is formulated:

If t_{count} > ttable, then H_0 is accepted and H_a is rejected.
If t_{coun}t > ttable, then H_0 is rejected and H_a is accepted.

H_0 = There is no difference in the results of analyzing comic characters after applying Pinterest app.
H_a = There are differences in learning outcomes after applying Pinterest app.
Determine t at a significance level of 0.05: 2 = 0.025 (2-tailed test) with df (degrees of freedom) n – 2. Test criteria, namely: (a) If $t_{count} \leq t_{table}$, then H_0 is accepted; and (b) If $t_{count} \leq t_{table}$, then H_0 is rejected.
If it is based on significance, namely:

a. If the significance is 0.05, then H_0 is accepted.
b. If the significance is 0.05, then H_0 is rejected.

Furthermore, the following is a table of the categorization of student learning outcomes for prose fiction students using the Pinterest application as Table 12.2.

4. **Findings**
When compared to previous studies, the data presented here comes from a small sample size with the number of students in a social science class of 20 undergraduate students. The results of the pilot project show that social-collaborative learning using WhatsApp and Pinterest helps students learn, and their attempts to negotiate meaning lead to increased understanding and retention.
This case requires students to write about the phenomenon of social change that occurs in the digital era, describe the effects of technology on people's social lives, describe the function of social groups, and design the concept of social change in a text that will be presented in front of the class, all of which can be discussed in group settings.

TABLE 12.2
Categorization of Learning Outcomes at Universitas Al Asyariah Mandar

Category	Score
Very good	80–100
Good	70–79
Pair	60–69
Poor	<50

Twenty undergraduate students participated in this study were involved in collaborative social activities during the learning process and answered questions after the material was given.

- This test group showed increased learning outcomes when compared to before using the Pinterest application through a comparison of pre-test and post-test learning outcomes.
- This may be due to the increased interest in the study test group and motivation to study using the Pinterest application.
- This Pinterest application is completely unknown to students before, but through the activities provided, they are motivated to use Pinterest as material for other courses and as presentation slides in PowerPoint.

To learn from social media-enhanced learning, we combine data from pre- and post-tests. The demonstrated learning outcomes of the experimental group presented on Pinterest-based social media collaborative enhanced learning are provided in Table 12.3.

Applying this method in Asymp, we find data in Table 12.4, which has a normal distribution. If the value is less than 0.05, then the data does not follow a normal distribution, which is the strongest possible evidence that the data is not normally distributed.

Since Asymp is a normal distribution test statistic, it is evident that the data follows a normal distribution. The significance level is 0.05, while the sig. (2-tailed) is 0.162.

Data from 20 students is presented in Table 12.2. Their average pre-test score was 61.75, with a range of 7.8 from the highest possible score of 75 to the lowest possible score of 50; after treatment, their average score was 78.75, with a range of 8.09 from the lowest possible score of 65 to the highest possible score of 90.

TABLE 12.3
Normality Test

One-Sample Kolmogorov-Smirnov Test

		Pre-test	Post-test	Unstandardized Residual
N		20	20	20
Normal Parameters[a,b]	Mean	61.7500	78.7500	.0000000
	Std. Deviation	7.82624	8.09077	5.01791691
Most Extreme Differences	Absolute	.138	.228	.164
	Positive	.138	.228	.129
	Negative	−.112	−.180	−.164
Test Statistic		.138	.228	.164
Asymp. Sig. (2-tailed)		.200[c,d]	.007[c]	.162[c]

[a] Test distribution is Normal.
[b] Calculated from data.
[c] Lilliefors Significance Correction.
[d] This is a lower bound of the true significance.

TABLE 12.4
Mean Sore of Students Learning Outcomes.

Descriptive Statistics	N	Minimum	Maximum	Mean	Std. Deviation
Pre-test	20	50.00	75.00	61.7500	7.82624
Post-test	20	65.00	90.00	78.7500	8.09077
Valid N (listwise)	20				

Results from a pre- and post-treatment assessment of knowledge gained in the social sciences course are as follows (Figure 12.6):

The variance among the class's students was examined with a homogeneity test. The Levene test can be run on the data from the pre- and post-tests in SPSS 24 for Windows to determine whether or not the two variances are comparable. To determine if data is homogeneous or not, SPSS uses the following decision criteria:

a. If the significance value is 0.05, the data is not homogeneous.
b. The data is considered to be homogeneous if the significance level is more than .05.

Based on the results, the one-sample t-test was used to examine the null hypothesis that the two sets of data had the same variance as Table 12.5.

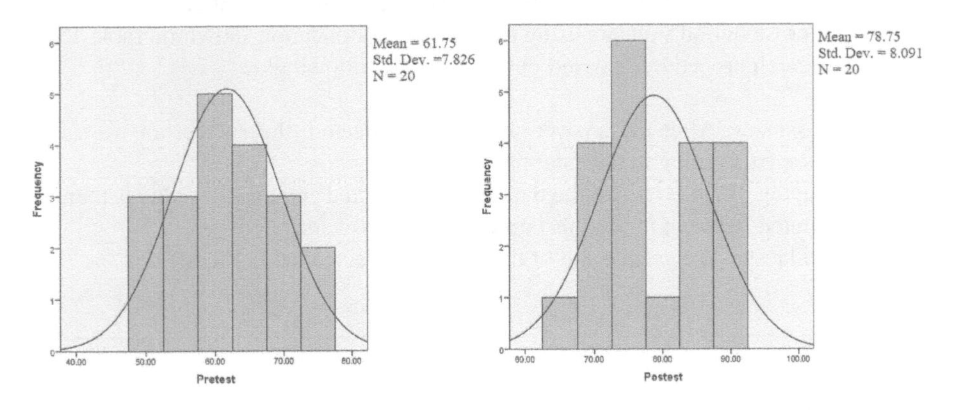

FIGURE 12.6 Histogram of the data.

TABLE 12.5
Test of Homogeneity

Test of Homogeneity of Variances

Outcome

Levene Statistic	df1	df2	Sig.
.329	1	38	.570

Homogeneity test results in Table 12.3 are reported when the p value is greater than 0.05. The significance level of the Test of Homogeneity of Variances can be determined from the results table of the corresponding the homogeneities.

Learning Success in Social Science, according to the Levene Test: df1, df2 Sig. = 0.570 since the significance level is greater than 0.05 (0.570 > 0.05), we conclude that the data is homogeneous.

Inferential statistics are used to test the research hypothesis. The hypothesis test used in this study was the t-test using the SPSS 21 for Windows program. The hypothesis to be tested in this study is formulated:

If t_{count} > ttable, then H_0 is accepted, and H_a is rejected.
If t_{coun}t > ttable, then H_0 is rejected, and H_a is accepted.

H_0 = There is no difference in the results of analyzing comic characters after applying Pinterest app.

H_a = There are differences in learning outcomes after applying Pinterest app.

Determine t at a significance level of 0.05: 2 = 0.025 (2-tailed test) with df (degrees of freedom) n – 2.

Test criteria, namely: (a) If $t_{count} \leq t_{table}$, then H_0 is accepted; and (b) If t count > t table, then H_0 is rejected. If it is based on significance, namely: (a) if the significance is 0.05, then H_0 is accepted; and (b) if the significance is 0.05, then H_0 is rejected.

Furthermore, the following is a table of the categorization of student learning outcomes for prose fiction students using the Pinterest application shown in Table 12.6.

The research procedures carried out include the following:

a. Assessment: After the answer sheets are collected, the correct answer is assessed according to the established criteria.
b. Grouping of data types of the data that is collected and given a value is then separated between the pre-test and post-test answers.
c. Calculation: The calculation of the data is collected using the t-test.

TABLE 12.6
Hypothesis Test Summary

No	Null Hypothesis	Test	Sig.	Decision
1	The distribution of pre-test is normal with mean 61.750 and standard deviation 7.83.	One-Sample Kolmogorov-Smirnov Test	.200[a,b]	Retain the null hypothesis.
2	The distribution of pre-test is normal with mean 78.750 and standard deviation 8.09.	One-Sample Kolmogorov-Smirnov Test	.200[a,b]	Reject the null hypothesis.

Asymptotic significances are displayed. The significance level is .05.

[a] Lilliefors Corrected.

[b] This is a lower bound of the true significance.

TABLE 12.7
T-Test Formula

One-Sample Test

	Test Value = 0					
					95% Confidence Interval of the Difference	
	t	**df**	**Sig. (2-tailed)**	**Mean Difference**	**Lower**	**Upper**
Pre-test	35.286	19	.000	61.75000	58.0872	65.4128
Post-test	43.529	19	.000	78.75000	74.9634	82.5366

The results of calculations using the t-test formula can be seen in Table 12.7:

Table 12.7 in this study shows that, on average, students score 17,000 points worse after taking the post-test than they did before, with a two-tailed significance level of 0.000 and 0.05. Pre- and post-test results for those who studied Arabic show a statistically significant improvement for the former.

Since students can easily access the app from their mobile devices, they are more likely to use it as a learning tool. This is because students are more likely to understand the subject matter being studied when they are able to interact, learn from each other, and teach with friends.

Students are encouraged to be more involved in the lesson by working in groups to find solutions to issues, so that they will feel they need to fully grasp the concepts being taught in order to effectively assist their peers. Taking increasing joy in one's studies is a strong predictor of positive learning outcomes.

12.6 RESULTS AND DISCUSSION

The t-test, as shown in Table 12.6, indicates that the scores on the learning outcomes have gone up, with the best possible score being 90 and the lowest possible score being 65. The result of the calculation is T count = 0, which is the least possible absolute value. T table = 1.328 is produced from the t-test list given the crucial value of 0.5 and the number of samples N = 20.

In this investigation, we test the hypothesis that using social learning mechanisms in a collaborative setting can improve one's ability to acquire knowledge in the social sciences. Decision criteria for evaluating the hypothesis can be found by looking at H_1, specifically: Ha is accepted if T count T table.

Since T table = 1.328 for t = 19 at the 0.025 significance level, and Tcount = 1.328 Ttable = 2.539, the null hypothesis is rejected, and the alternative hypothesis is accepted.

Learning outcomes in foundational social science courses at the Faculty of Public Health are significantly impacted by students' use of Pinterest in conjunction with social collaboration tactics.

The findings of the pre-test and post-test show that there are statistically significant differences between the students' learning outcomes before and after treatment.

In Figure 12.4, it is discussed how the proliferation of small and light mobile devices has enabled people to access teaching materials on their own schedule and in any location.

Searching for multimedia content or managing large amounts of content on mobile devices with limited resources remains challenging, even as mobile device technology evolves rapidly.

One common strategy to fix this problem is the method of recommending teaching materials that are in accordance with the curriculum, as in this study using the Pinterest application as teaching materials.

In addition to digital-based teaching materials, collaborative social activities involve human-machine interaction.

Related to this study, Siripongdee and Pimdee et al. (2020); He and Lo et al. (2016); Hu and Yeh et al. (2021) provide learning strategies by integrating IoT-based multimedia that certainly involves and identifies behavioral patterns of undergraduate students toward technology and uses IT as access to courses and knowledge.

Students are finding it increasingly easy to identify acceptable Pinterest app-based materials depending on their needs, as discussed and outlined in Figures 12.3, 12.4, and 12.7, due to the exponential growth of learning materials and learning resources available offline and online that are becoming more attractive and enjoyable to interact with in the classroom.

Students can benefit from the Pinterest app as they are directed to relevant study material. In Figure 12.3, the topics of socio-cultural changes in the digital era, social groups, and their personalized functions in e-learning are examined in this research.

According to Wen and Zaid et al. (2015), social collaborative is a method that relies on students communicating with each other and learning about how each

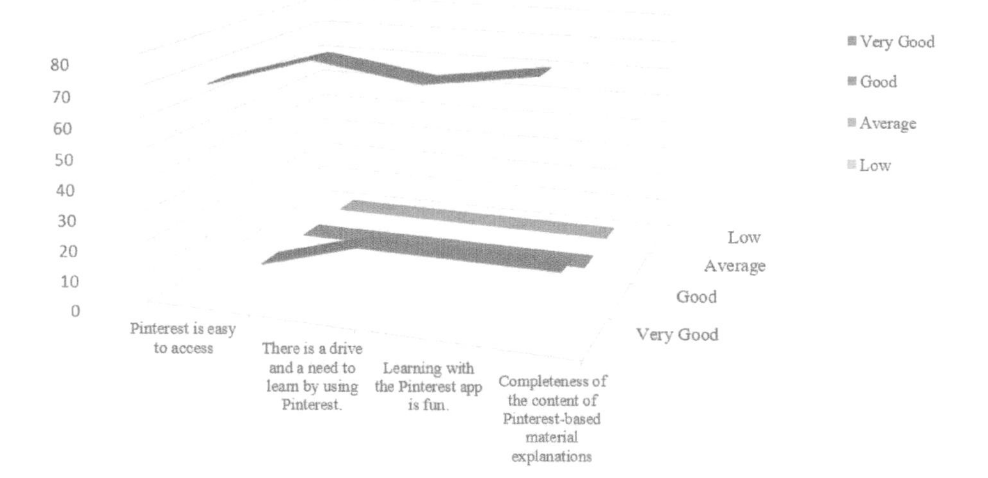

FIGURE 12.7 Students response on Pinterest.

discipline's implicit cultural assumptions, frames of reference, viewpoints, and prejudices shape the construction of knowledge.

This chapter introduces the main aspects of social collaboration and investigates the current modes of Pinterest learning methodology that can be applied to future studies. For the most part, this chapter concentrates on two main ideas. The needs of undergraduate students are initially discussed, followed by individual recommendations for those needs.

The purpose of the suggestions in Sterbini and Temperini (2011) is to provide students with suggestions for sharing ideas and ideas for problem-solving and see how the integration of technology affects not only learning outcomes and social collaboration but also interest in learning.

For further research, the results of this study recommend that in order to meet the evolving individual needs of today's college students, a conversation about how to deliver social science courses has emerged, a direct result of the proliferation of IT in education in the meta-learning era.

First, examining students, identifying learning needs according to the characteristics of 21st-century learners who are able to adapt to technology, and developing teaching materials with IoT and ways in which new ideas and technologies can be introduced in this area (Rani & Khang et al., 2023).

According to Al-Hasan (2023) and Otto and Williams (2014), it is very important to equip undergraduate students with social collaborative skills and train them to work together with teams and discuss problem-solving until they reach the stage of mutual agreement and responsibility not only in class but also can be applied outside the classroom (the environment).

Muthmainnah et al researchers offer a new learning strategy to integrate Pinterest by combining the expertise of various individuals. The effectiveness of this method is tested experimentally, and the advantages of the novel method compared to the status quo are described.

One of the pressing problems of personalized e-learning services, as presented in Figure 12.5, is connecting like-minded online students in a decentralized and open e-learning environment so that they can learn in groups.

There is evidence in the literature to suggest that current methods for fostering online learning communities are inspired by qualitative research in learner-centered classrooms with small student numbers, when teacher presence may be required. Findings may not translate to classrooms with hundreds of students online due to the distributed nature of learning settings.

As a result, educators cannot properly assess each student's study habits and divide them into different online communities.

Based on the perspective of undergraduate students, the convenience of learning and discussing as a form of successful problem-solving takes into account the learner's rating value from the available learning resources in Table 12.2, which shows that there are differences in learning outcomes, collaborative social behavior after constructing IoT-based learning.

The experimental results presented in Faezi and Moradi et al. (2018), Sung and Hwang (2013), and El-Sofany and El-Haggar (2020) show the effectiveness of social collaboration, and of course learning outcomes also increase.

To provide individualized instruction, Muthmainnah and Ganguli et al. (2023) suggest web 2.0 or IoT-based learning and recommend digital-based curriculum materials. First, the lecturer adjusts the availability of material on Pinterest and then collects the teaching material and shares the Pinterest link in the WhatsApp group.

Finally, learning instructions are designed to identify social collaborative activities. With the IoT approach to social science learning being different, we propose to provide personalized suggestions based on activities that occur during learning that help undergraduate students overcome cognitive overload (Bhambri & Khang et al., 2022).

Personalization of the learning experience is widely acknowledged as a crucial aspect in achieving both high quality and high efficacy. In the context of online education, where it can be difficult to maintain students' interest and enthusiasm, this becomes even more crucial.

Individual differences in learning styles, prior knowledge of the subject matter, and motivation for completing assignments are only some of the factors that must be considered when employing learner-centered design (Almaiah & Al-Khasawneh et al., 2020).

Learner motivation is highly sensitive to their subjective perceptions of their own success or failure; hence, it's crucial that activities be delivered in accordance with an accurate learner model. That's why it's crucial to highlight the value of the aid given and the accomplishments that have been acknowledged.

Help given should be proportional to the needs of the pupils being evaluated; otherwise, it will be ineffective or even detrimental by stifling students' initiative and hard work (Quintana & Krajcik et al., 2002).

In order to close the knowledge and skill gap between the learner and the learning domain, learning experiences devised should provide students with extensive exposure to the material. In this setting, technological concerns must be grounded in a coherent theory of instruction (De Marsico & Sterbini et al., 2013).

The results of this study apply the curriculum "Independent Learning Independent Campus" or known as Independent Learning Independent Campus (ILIC) with constructivist theory that is widely used because it recognizes learning as a real process that requires active involvement of students (Al Yakin & Muthmainnah et al., 2023).

An approach known as "learning by doing" encourages students to actively engage with course information, enabling them to make sense of it and draw connections with what they already know. Learning is facilitated by placing each activity in a realistic group setting, where students are encouraged to develop and use their own strategies for conceptualization.

For this reason, encouraging healthy competition between study groups and between students and teachers is encouraged. Students gain a keener sense of self and agency in the classroom by using this method.

The term "situational learning" (MacDonald & Yurovsky et al., 2017) has broadened the constructivist approach by taking a social perspective in the learning process.

According to the situational learning perspective, the results of learning activities depend on the setting in which the activity is carried out, with all elements of the learning environment (both human and artifacts) interacting and contributing to student growth.

Social constructivism is a term used to describe a pedagogical approach that combines constructivist and situational learning principles.

12.7 CONCLUSION

Self-promoting, traditional education encourages the acquisition of knowledge and the development of individual cognition through rote memorization and the use of methods to arrive at the desired answer. E-learning, or e-learning 2.0, or Metaverse-based educational age, has evolved in response to the widespread adoption of Web 2.0 tools (Khanh & Khang, 2021).

Having the option to study in a socially collaborative environment is just one of the many advantages of e-learning 2.0 for both teachers and students (Rana & Khang et al., 2021).

Forums, social networking sites (WhatsApp groups), wikis, and even Pinterest are just some of the existing platforms that students in higher education use to supplement their classroom teaching (Khang & Hajimahmud, 2022).

However, the software was not developed with schools in mind and getting it into group study programs can be difficult. This research emphasizes the combination of social and collaborative aspects that are in line with best practices in higher education (Khang & Rani, 2022).

The suggested social collaborative learning paradigm was put through usability testing to measure its effectiveness, simplicity of use, and user satisfaction with the system prototype. This shows that the proposed approach can be used to facilitate more efficient teaching and learning procedures in a social collaborative environment (Rani & Khang, 2022).

The results of this study aim to better prepare students for the demands of the world of work; lecturers are advised to apply collaborative teamwork in class when large assignments require the necessary group work and coordinated team efforts (Khang & Gupta et al., 2023).

In this chapter, we report the results of our investigation into how we are using the new medium that is Pinterest the symbiosis of IoT by actively engaging undergraduate students in our online environment, where they do proof work as artifacts of knowledge created through digital media rather than relying on rote learning (Jebaraj & Khang et al., 2024).

Representation of student knowledge collected in the form of rich multimodal resources using one of the accessible media is highly encouraged and appreciated (Subhashini & Khang, 2024).

REFERENCES

Achmad, K.A., Nugroho, L.E., Djunaedi, A., Widyawan. "Smart city for development: towards a conceptual framework", *4th International Conference on Science and Technology (ICST)*, pp. 1–6 (2018). https://ieeexplore.ieee.org/abstract/document/8528677/

Al Maani, D., Shanti, Z. "Technology-enhanced learning in light of Bloom's taxonomy: a student-experience study of the history of architecture course", *Sustainability*, vol. 15, p. 2624 (2023). https://www.mdpi.com/2071-1050/15/3/2624

Al Yakin, A., Muthmainnah, Ganguli, S., Cardoso, L., Asrifan, A. "Cyber Socialization Through Smart Digital Classroom Management (SDCM) as a Pedagogical Innovation of *"Merdeka Belajar Kampus Merdeka (MBKM)"* Curriculum". *In*: Choudhury, A., Biswas, A., Chakraborti, S. (eds.) *Digital Learning Based Education. Advanced Technologies and Societal Change* (2023). Springer. https://doi.org/10.1007/978-981-19-8967-4_3

Al-Hasan, A. "Effects of Social Network Information on Online Language Learning Performance: A Cross-Continental Experiment", *Research Anthology on Applying Social Networking Strategies to Classrooms and Libraries*, pp. 1574–1591 (2023). IGI Global. https://www.igi-global.com/chapter/effects-of-social-network-information-on-online-language-learning-performance/312998

Almaiah, M.A., Al-Khasawneh, A., Althunibat, A. "Exploring the critical challenges and factors influencing the E-learning system usage during COVID-19 pandemic", *Education and Information Technologies*, vol. 25, pp. 5261–5280 (2020). https://library.iated.org/view/ALQTEISHAT2021EXP

Babasaheb, J., Sphurti, B., Khang, A. "Industry Revolution 4.0: Workforce Competency Models and Designs", *Designing Workforce Management Systems for Industry 4.0: Data-Centric and AI-Enabled Approaches* (1st Ed.), pp. 14–31 (2023). CRC Press. https://doi.org/10.1201/9781003357070-2

Baena-Morales, S., Urrea-Solano, M., Gavilán-Martin, D., Ferriz-Valero, A. "Development and validation of an instrument to assess the level of sustainable competencies in future physical education teachers. PESD-FT questionnaire", *Journal of Applied Research in Higher Education*, vol. 15, no. 1, pp. 1–19 (2023). https://www.emerald.com/insight/content/doi/10.1108/JARHE-09-2021-0330/full/html

Bebell, D., O'Dwyer, L. "Educational outcomes and research from 1: 1 computing settings", *The Journal of Technology, Learning and Assessment*, vol. 9, no. 1 (2010). http://ejournals.bc.edu/index.php/jtla/article/view/1606

Bebell, D., O'Dwyer, L.M., Russell, M., Hoffmann, T. "Concerns, considerations, and new ideas for data collection and research in educational technology studies", *Journal of Research on Technology in Education*, vol. 43, no. 1, pp. 29–52 (2010). https://www.tandfonline.com/doi/abs/10.1080/15391523.2010.10782560

Bhambri, P., Rani, S., Gupta, G., Khang, A. "*Cloud and Fog Computing Platforms for Internet of Things*" (2022). CRC Press. https://doi.org/10.1201/9781032101507

Biggs John, 2003. Constructive alignment in university teaching. https://www.tru.ca/__shared/assets/Constructive_Alignment36087.pdf

Blessinger, P., Wankel, C. "Novel Approaches in Higher Education: An Introduction to Web 2.0 and Blended Learning Technologies", *Increasing Student Engagement and Retention in E-Learning Environments: Web 2.0 and Blended Learning Technologies* (2013). Emerald Group Publishing Limited. https://www.emerald.com/insight/content/doi/10.1108/S2044-9968(2013)000006G003

Chan, C.L., Shroff, R.H., Tsang, W.K., Ting, F.S., Garcia, R.C. "Assessing the effects of a collaborative problem-based learning and peer assessment method on junior secondary students' learning approaches in mathematics using interactive online whiteboards during the COVID-19 pandemic", *International Journal of Mobile Learning and Organisation*, vol. 17, no. (1–2), pp. 6–31 (2023). https://www.inderscienceonline.com/doi/pdf/10.1504/IJMLO.2023.128342

Conole, G. "Facilitating new forms of discourse for learning and teaching: harnessing the power of Web 2.0 practices", *Open Learning: The Journal of Open, Distance and e-Learning*, vol. 25, no. 2, pp. 141–151 (2010). https://www.tandfonline.com/doi/abs/10.1080/02680511003787438

De Marsico, M., Sterbini, A., Temperini, M. "A strategy to join adaptive and reputation-based social-collaborative e-learning, through the zone of proximal development", *International Journal of Distance Education Technologies (IJDET)*, vol. 11, no. 3, pp. 12–31 (2013). https://www.igi-global.com/article/a-strategy-to-join-adaptive-and-reputation-based-social-collaborative-e-learning-through-the-zone-of-proximal-development/83513

Dimitriadou, E., Lanitis, A. "A critical evaluation, challenges, and future perspectives of using artificial intelligence and emerging technologies in smart classrooms", *Smart Learning Environments*, vol. 10, no. 1, pp. 1–26 (2023). https://slejournal.springeropen.com/articles/10.1186/s40561-023-00231-3

Donnelly, R., Fitzmaurice, M. "Collaborative project-based learning and problem-based learning in higher education: a consideration of tutor and student roles in learner-focused strategies", *Emerging Issues in the Practice of University Learning and Teaching*, pp. 87–98 (2005). https://www.edin.ie/wp-content/uploads/2021/11/EmergingIssuesI_FullPDF.pdf#page=97

El-Sofany, H., El-Haggar, N. The effectiveness of using mobile learning techniques to improve learning outcomes in higher education. (2020). https://www.learntechlib.org/p/216981/

Faezi, S.T., Moradi, K., Amin, A.G.R., Akhlaghi, M., Keshmiri, F. "The effects of team-based learning on learning outcomes in a course of rheumatology", *Journal of Advances in Medical Education & Professionalism*, vol. 6, no. 1, pp. 22 (2018). https://www.ncbi.nlm.nih.gov/pmc/articles/PMC5757153/

Foti, M.K., Mendez, J. "Mobile learning: how students use mobile devices to support learning", *Journal of Literacy and Technology*, vol. 15, no. 3, pp. 58–78 (2014). http://www.literacyandtechnology.org/uploads/1/3/6/8/136889/jlt_v15_foti.pdf

Fullan, M. "*The Principal 2.0: Three Keys to Maximizing Impact*" (2023). John Wiley & Sons. https://digitalcommons.lmu.edu/cgi/viewcontent.cgi?article=1865&context=ce

Furió, D., Juan, M.C., Seguí, I., Vivó, R. "Mobile learning vs. traditional classroom lessons: a comparative study", *Journal of Computer Assisted Learning*, vol. 31, no. 3, pp. 189–201 (2015). https://onlinelibrary.wiley.com/doi/abs/10.1111/jcal.12071

Gamble, T.K., Gamble, M.W. "*The Interpersonal Communication Playbook*" (2023). SAGE Publications. https://journals.sagepub.com/doi/pdf/10.1177/0265407514546978

He, J., Lo, D.C.T., Xie, Y., Lartigue, J. "Integrating Internet of Things (IoT) into STEM undergraduate education: case study of a modern technology infused courseware for embedded system course", *2016 IEEE Frontiers in Education Conference (FIE)*, pp. 1–9 (2016, October). IEEE. https://ieeexplore.ieee.org/abstract/document/7757458/

Hidalgo, C.G., Bucheli-Guerrero, V.A., Ordóñez-Eraso, H.A. "Artificial intelligence and computer-supported collaborative learning in programming: a systematic mapping study", *Tecnura*, vol. 27, no. 75, pp. 9–9 (2023). https://doi.org/10.14483/22487638.19637

Hursen, C. "The effect of problem-based learning method supported by web 2.0 tools on academic achievement and critical thinking skills in teacher education", *Technology, Knowledge and Learning*, vol. 26, pp. 515–533 (2021). https://link.springer.com/article/10.1007/s10758-020-09458-2

Hu, C.C., Yeh, H.C., Chen, N.S. "Teacher development in robot and IoT knowledge, skills, and attitudes with the use of the TPACK-based support-stimulate-seek approach", *Interactive Learning Environments*, pp. 1–20 (2021). https://www.tandfonline.com/doi/abs/10.1080/10494820.2021.2019058

Jebaraj, L., Khang, A., Chandrasekar, V., Pravin, A.R., Sriram, K. "Smart City Concepts, Models, Technologies and Applications", *Smart Cities: IoT Technologies, Big Data Solutions, Cloud Platforms, and Cybersecurity Techniques* (2024). CRC Press. https://doi.org/10.1201/9781003376064-1

Kwon Joon Myoung, Kyung Hee Kim, Ki Hyun Jeon, Hyue Mee Kim, Min Jeong Kim, Sung Min Lim, Song Pil Sang, Jinsik Park, Rak Kyeong Choi, Byung Hee Oh. Development and validation of deep-learning algorithm for electrocardiography-based heart failure identification. 2019 Jul;49(7):629–639. https://doi.org/10.4070/kcj.2018.0446

Jerald, C.D. "Defining a 21st century education", *Center for Public Education*, vol. 16, pp. 1–10 (2009). https://citeseerx.ist.psu.edu/document?repid=rep1&type=pdf&doi=0252e811a5dee8948eb052a1281bbc3486087503

Kee, C.N.L. "Face-to-face tutorial, learning management system and WhatsApp group: how digital immigrants interact and engage in e-learning?" *Malaysian Online Journal of Educational Technology*, vol. 8, no. 1, pp. 18–35 (2020). https://eric.ed.gov/?id=EJ1239976

Khan, E.A., Cram, A., Wang, X., Tran, K., Cavaleri, M., Rahman, M.J. "Modelling the impact of online learning quality on students' satisfaction, trust and loyalty", *International Journal of Educational Management* (ahead-of-print) (2023). https://www.emerald.com/insight/content/doi/10.1108/IJEM-02-2022-0066/full/html

Khang, A., Gupta, S.K., Rani, S., Karras, D.A. "*Smart Cities: IoT Technologies, Big Data Solutions, Cloud Platforms, and Cybersecurity Techniques*" (1st Ed.) (2023). CRC Press. https://doi.org/10.1201/9781003376064

Khang, A., Ragimova, N.A., Hajimahmud, V.A., Alyar, A.V. "Advanced Technologies and Data Management in the Smart Healthcare System", *AI-Centric Smart City Ecosystems: Technologies, Design and Implementation* (1st Ed.) (2022). CRC Press. https://doi.org/10.1201/9781003252542-16

Khang, A., Rani, S., Gujrati, R., Uygun, H., Gupta, S.K. "*Designing Workforce Management Systems for Industry 4.0: Data-Centric and AI-Enabled Approaches*" (1st Ed.) (2023). CRC Press. https://doi.org/10.1201/99781003357070

Khang, A., Rani, S., Sivaraman, A.K. "*AI-Centric Smart City Ecosystems: Technologies, Design and Implementation*" (1st Ed.) (2022). CRC Press. https://doi.org/10.1201/9781003252542

Khanh, H.H., Khang, A. "The Role of Artificial Intelligence in Blockchain Applications," *Reinventing Manufacturing and Business Processes through Artificial Intelligence*, pp. 20–40 (2021). CRC Press. https://doi.org/10.1201/9781003145011-2

Lee, H.J., Kim, H., Byun, H. "Are high achievers successful in collaborative learning? An explorative study of college students' learning approaches in team project-based learning", *Innovations in Education and Teaching International*, vol. 54, no. 5, pp. 418–427 (2017). https://www.tandfonline.com/doi/abs/10.1080/14703297.2015.1105754

Lin, L., Mills, L.A., Ifenthaler, D. "Collaboration, multi-tasking and problem solving performance in shared virtual spaces", *Journal of Computing in Higher Education*, vol. 28, pp. 344–357 (2016). https://link.springer.com/article/10.1007/s12528-016-9117-x

Muthmainnah, Ganguli, S., Al Yakin, A., Abd, G. "An Effective Investigation on YIPe-Learning Based for Twenty-First Century Class". *In*: Choudhury, A., Biswas, A., Chakraborti, S. (eds.) *Digital Learning Based Education. Advanced Technologies and Societal Change* (2023). Springer. https://doi.org/10.1007/978-981-19-8967-4_2

MacDonald, K., Yurovsky, D., Frank, M.C. "Social cues modulate the representations underlying cross-situational learning", *Cognitive Psychology*, vol. 94, pp. 67–84 (2017). https://www.sciencedirect.com/science/article/pii/S0010028515300128

Makarova, I., Shubenkova, K., Antov, D., Pashkevich, A. "Digitalization of engineering education: from e-learning to smart education", *Smart Industry & Smart Education: Proceedings of the 15th International Conference on Remote Engineering and Virtual Instrumentation 15*, pp. 32–41 (2019). Springer International Publishing. https://link.springer.com/chapter/10.1007/978-3-319-95678-7_4

Maksimović, M. "IoT concept application in educational sector using collaboration", *Facta Universitatis, Series: Teaching, Learning and Teacher Education*, vol. 1, no. 2, pp. 137–150 (2018). http://casopisi.junis.ni.ac.rs/index.php/FUTeachLearnTeachEd/article/view/3110

Mokski, E., Leal Filho, W., Sehnem, S., Andrade Guerra, J.B.S.O.D. "Education for sustainable development in higher education institutions: an approach for effective interdisciplinarity", *International Journal of Sustainability in Higher Education*, vol. 24, no. 1, pp. 96–117 (2023). https://www.emerald.com/insight/content/doi/10.1108/IJSHE-07-2021-0306/full/html

Myers, J.L., Well, A.D., Lorch, R.F. "*Research Design and Statistical Analysis*" (2013). Routledge. https://www.tandfonline.com/doi/abs/10.1080/01406720500256194

Oke, A., Fernandes, F.A.P. "Innovations in teaching and learning: exploring the perceptions of the education sector on the 4th industrial revolution (4IR)", *Journal of Open Innovation: Technology, Market, and Complexity*, vol. 6, no. 2, pp. 31 (2020). https://www.sciencedirect.com/science/article/pii/S2199853122004267

Otto, F., Williams, S. "Social Collaborative e-Learning in Higher Education: Exploring the Role of Informal Learning", *E-Learning, E-Education, and Online Training: First International Conference, eLEOT 2014, Bethesda, MD, USA, September 18-20* (2014). *Revised Selected Papers 1*, pp. 130–137. Springer. https://link.springer.com/chapter/10.1007/978-3-319-13293-8_16

Oztemel, E., Gursev, S. "Literature review of Industry 4.0 and related technologies", *Journal of Intelligent Manufacturing*, vol. 31, pp. 127–182 (2020). https://link.springer.com/article/10.1007/s10845-018-1433-8

Paily, M.U. "Creating constructivist learning environment: role of "Web 2.0" technology", *International Forum of Teaching and Studies*, vol. 9, no. 1, pp. 39–50 (2013). https://www.researchgate.net/profile/Abdel-Rahman-Abu-Melhim/publication/309160632_Promising_as_a_speech_act_in_Jordanian_Arabic/links/5827262708ae950ace6ca92f/Promising-as-a-speech-act-in-Jordanian-Arabic.pdf#page=39

Peñaloza-Farfán, L.J., Paucar-Caceres, A. "Regional Development in Latin America as a Way to Promote Education for Sustainable Development: The Case Study of the University of Ibague in Colombia", Towards Green Campus Operations: Energy, Climate and Sustainable Development Initiatives at Universities, pp. 735–761 (2018). https://link.springer.com/chapter/10.1007/978-3-319-76885-4_49

Perry, N.E., VandeKamp, K.O., Mercer, L.K., Nordby, C.J. "Investigating Teacher—Student Interactions That Foster Self-Regulated Learning", *Using Qualitative Methods to Enrich Understandings of Self-Regulated Learning*, pp. 5–15 (2023). Routledge. https://www.tandfonline.com/doi/abs/10.1207/S15326985EP3701_2

Quintana, C., Krajcik, J., Soloway, E. "A case study to distill structural scaffolding guidelines for scaffolded software environments", *Proceedings of the SIGCHI Conference on Human Factors in Computing Systems*, pp. 81–88 (2002). https://dl.acm.org/doi/abs/10.1145/503376.503392

Rana, G., Khang, A., Sharma, R., Goel, A.K., Dubey, A.K. "*Reinventing Manufacturing and Business Processes through Artificial Intelligence*" (2021). CRC Press. https://doi.org/10.1201/9781003145011

Rani, S., Bhambri, P., Kataria, A., Khang, A. "Smart City Ecosystem: Concept, Sustainability, Design Principles and Technologies," *AI-Centric Smart City Ecosystems: Technologies, Design and Implementation* (1st Ed.) (2022). CRC Press. https://doi.org/10.1201/9781003252542-1

Rani, S., Bhambri, P., Kataria, A., Khang, A., Sivaraman, A.K. *Big Data, Cloud Computing and IoT: Tools and Applications* (1st Ed.) (2023). Chapman and Hall/CRC. https://doi.org/10.1201/9781003298335

Satu, M.S., Roy, S., Akhter, F., Whaiduzzaman, M. "IoLT: an IoT based collaborative blended learning platform in higher education", *2018 International Conference on Innovation in Engineering and technology (ICIET)*, pp. 1–6 (2018, December). IEEE. https://ieeexplore.ieee.org/abstract/document/8660931/

Scager, K., Boonstra, J., Peeters, T., Vulperhorst, J., Wiegant, F. "Collaborative learning in higher education: evoking positive interdependence", *CBE—Life Sciences Education*, vol. 15, no. 4, p. ar69 (2016). https://www.lifescied.org/doi/abs/10.1187/cbe.16-07-0219

Sharma, S. "Adoption of 5.0 Online and Collaborative Education among the Youth of Indonesia", *Transformation for Sustainable Business and Management Practices: Exploring the Spectrum of Industry 5.0*, pp. 141–154 (2023). Emerald Publishing Limited. https://www.emerald.com/insight/content/doi/10.1108/978-1-80262-277-520231011

Singh, J., Singh, L., Matthees, B. "Establishing social, cognitive, and teaching presence in online learning – a panacea in COVID-19 pandemic, post vaccine and post pandemic times", *Journal of Educational Technology Systems*, vol. 51, no. 1, pp. 28–45 (2022). https://journals.sagepub.com/doi/pdf/10.1177/00472395221095169

Siripongdee, K., Pimdee, P., Tuntiwongwanich, S. "A blended learning model with IoT-based technology: effectively used when the COVID-19 pandemic?" *Journal for the Education of Gifted Young Scientists*, vol. 8, no. 2, 905–917 (2020). https://dergipark.org.tr/en/pub/jegys/issue/53184/698869

Snehal, M., Babasaheb, J., Khang, A. "Workforce Management System: Concepts, Definitions, Principles, and Implementation," *Designing Workforce Management Systems for Industry 4.0: Data-Centric and AI-Enabled Approaches* (1st Ed.), pp. 1–13 (2023). CRC Press. https://doi.org/10.1201/9781003357070-1

Soomro, K.A., Zai, S.Y., Jafri, I.H. "Competence and usage of Web 2.0 technologies by higher education faculty", *Educational Media International*, vol. 52, no. 4, pp. 284–295 (2015). https://www.tandfonline.com/doi/abs/10.1080/09523987.2015.1095522

Stanek, E.J. III. "Choosing a pretest-posttest analysis", *The American Statistician*, vol. 42, no. 3, pp. 178–183 (1988) https://www.tandfonline.com/doi/abs/10.1080/00031305.1988.10475557

Sterbini, A., Temperini, M. "SOCIALX: reputation based support to social collaborative learning through exercise sharing and project teamwork", *International Journal of Information Systems and Social Change (IJISSC)*, vol. 2, no. 1, pp. 64–79 (2011). https://www.igi-global.com/article/socialx-reputation-based-support-social/50552

Strada, F., Lopez, M.X., Fabricatore, C., dos Santos, A.D., Gyaurov, D., Battegazzorre, E., Bottino, A. "Leveraging a collaborative augmented reality serious game to promote sustainability awareness, commitment and adaptive problem-management", *International Journal of Human-Computer Studies*, vol. 172, p. 102984 (2023). https://www.sciencedirect.com/science/article/pii/S1071581922002026

Subhashini, R., Khang, A. "The Role of Internet of Things (IoT) in Smart City Framework," *Smart Cities: IoT Technologies, Big Data Solutions, Cloud Platforms, and Cybersecurity Techniques* (1st Ed.) (2024). CRC Press. https://doi.org/10.1201/9781003376064-3

Sung, H.Y., Hwang, G.J. "A collaborative game-based learning approach to improving students' learning performance in science courses", *Computers & Education*, vol. 63, pp. 43–51 (2013). https://www.sciencedirect.com/science/article/pii/S0360131512002849

Traxler, J. "The new normal: innovative informal digital learning after the pandemic", *ICT Innovations 2022. Reshaping the Future towards a New Normal: 14th International Conference, ICT Innovations 2022, Skopje, Macedonia, September 29–October 1, 2022, Proceedings*, pp. 3–10 (2023, January). Cham: Springer Nature Switzerland. https://link.springer.com/chapter/10.1007/978-3-031-22792-9_1

Troussas, C., Giannakas, F., Sgouropoulou, C., Voyiatzis, I. "Collaborative activities recommendation based on students' collaborative learning styles using ANN and WSM", *Interactive Learning Environments*, pp. 1–14 (2020). https://www.tandfonline.com/doi/abs/10.1080/10494820.2020.1761835

Vuopala, E., Hyvönen, P., Järvelä, S. "Interaction forms in successful collaborative learning in virtual learning environments", *Active Learning in Higher Education*, vol. 17, no. 1, pp. 25–38 (2016). https://journals.sagepub.com/doi/pdf/10.1177/1469787415616730

Vygotsky (1978). Vygotsky's Sociocultural Theory of Cognitive Development, https://www.simplypsychology.org/vygotsky.html

Watermeyer, R., Chen, Z., Ang, B.J. "'Education without limits': the digital resettlement of post-secondary education and training in Singapore in the COVID-19 era", *Journal of Education Policy*, vol. 37, no. 6, pp. 861–882 (2022). https://www.tandfonline.com/doi/abs/10.1080/02680939.2021.1933198

Wei, R. "Evolving mobile media: changing technology and transforming behavior", *Mobile Media & Communication*, vol. 11, no. 1, 25–29 (2023). https://journals.sagepub.com/doi/pdf/10.1177/20501579221131448

Wen, A.S., Zaid, N.M., Harun, J. "A meta-analysis on students' social collaborative knowledge construction using flipped classroom model", *2015 IEEE Conference on e-Learning, e-Management and e-Services (IC3e)*, pp. 58–63 (2015, August). IEEE. https://ieeexplore.ieee.org/abstract/document/7403487/

Zarfsaz, E., Serpil, U.A.R. "Teacher-student interpersonal relationship, EFL Learner's motivation and autonomy in online learning", *Key Concepts in Online Learning: A Comprehensive Guide for Pre-Service and In-Service Teachers*, vol. 49 (2023). https://www.google.com/books?hl=en&lr=&id=QCKnEAAAQBAJ&oi=fnd&pg=PA49&dq=Teacher-Student+Interpersonal+Relationship,+Efl+Learner%E2%80%99s+motivation+and+Autonomy+in+Online+Learning&ots=5FLB0XB-d8&sig=Ku53Gapdael-Lh4GlBkaUqaATKo

13 Vehicle and Passenger Identification in Public Transportation to Fortify Smart City Indices

Jayashree Mahale, Dillip Rout, Bholanath Roy, and Alex Khang

13.1 INTRODUCTION

A Conscientious Review of Algorithms, Equipment, Policies, and Protocols Pertaining to Vehicle Identification and Passengers Identification in Public Transportation to Fortify Smart City Indices. Public transportation is a necessity for many major cities to keep their populations moving (Rana & Khang et al., 2021).

Smart city innovations have made their way in, able to positively impact public transportation (Kubinaa & Bubelínya et al., 2021). One of the main aims of this chapter is to give an overview of present opportunities for the transportation needs of a city as part of a smart city plan.

The city needs to ensure that they are able to transport people around the city in accordance to the "smart city concept". This would be achieved by strategic planning and increasing resources (Kumari et al., 2019). "Smart City" is an outdated term, and there are actually a lot of definitions to what it means.

We recommend using keywords such as "Index" or "Evaluation". After reviewing 20 different indices, each one falls into one of seven categories. Those are smart economy, smart governance, smart living, smart performance, smart project, smart technology, and multifaceted.

Smart city indices are metrics used to measure the performance of a city in terms of its smartness, or how well it is able to leverage technology to improve the lives of its citizens and businesses. These indices are based on indicators such as access to technology, infrastructure, public services, safety, sustainability, and mobility.

Examples of such indices include the Smart City Index, the Global Smart City Performance Index, and the Smart City Maturity Mode. These indices provide a benchmark for cities to measure their progress and identify areas of improvement.

The motivation behind this study is to give a review of algorithms, equipment, policies, and protocols that identify vehicles for public transportation systems.

One of the most important aspects of smart cities is the use of vehicle and passenger identification in public transportation. By enabling the identification and tracking of vehicles and passengers, this technology enhances transportation efficiency,

DOI: 10.1201/9781003376064-13

enabling the monitoring and management of traffic flow, congestion reduction, and improved passenger safety (Hao, 2017).

Identifying different vehicles on the go is a common challenge for automotive companies, but fortunately, there are various techniques for identifying vehicles and passengers. These techniques can play a vital role in developing cities that are more energy efficient and sustainable when it comes to transportation. In this review, we will discuss the advantages and disadvantages of these methods and explore the potential of new identification technologies.

This chapter examines the various techniques often used to identify vehicles and passengers which could be helpful in making public transportation easier in smart city environments.

Automotive companies often need to identify different vehicles on the go, but there's also a variety of techniques for identifying vehicles and passengers. These techniques can play a vital role in developing cities that are more energy efficient and sustainable when it comes to transportation.

In this review, we will discuss the advantages and disadvantages of these methods and explore the potential of new identification technologies.

This chapter contains seven more sections which are organized as follows. The next section introduces the various algorithms for the identification of vehicles and passengers such as YOLO algorithm, Automatic License Plate Recognition (ALPR) algorithm, Histogram of Oriented Gradient (HOG), and Linear Support Vector Machine (SVM), for passenger identification.

The third section of the chapter introduces the equipment which are used to identify vehicles like Bluetooth and Wi-Fi. Identification of passengers also includes a beacon card, active and passive RFID. In Section 13.4, the policies used for vehicle and passenger identification are discussed.

The fifth section describes various protocols for communication to identify vehicles and passengers. The sixth and seventh sections provide the implications of vehicle identification and passenger identification of smart cities, respectively. Lastly, the summary and conclusions are presented in Section 13.8.

13.2 ALGORITHMS

In recent years, many algorithms and implementation methods have been proposed at home and abroad, such as YOLO algorithm, video-based algorithm, ALPR algorithm, computer vision based algorithm, vehicle detection with HOG and Linear SVM-based algorithm. Following are the detailed information about algorithms in vehicle identification.

13.2.1. Algorithms for Vehicle Identification

13.2.1.1 YOLO Algorithm

Urban areas with high traffic density pose a significant challenge for traffic management (Nalwade & Jagtap et al., 2016). However, the solution lies in creating smart traffic lights, and this is where the system comes in handy (Patni & Alshmrani et al., 2022).

The length the lights will be on is calculated by real-time vehicle density to reduce wait times for drivers (Guo & Wu et al., 2022). Vehicle counting can be done in two

FIGURE 13.1 YOLO algorithm for vehicle identification.

steps: first videos are sent to deep learning software like YOLO, which uses artificial intelligence (AI) to detect and identify the vehicles.

The frames from the video are then broken down. There are a few features that make video analysis an integral part of video processing from the job perspective. Firstly, frames or images extracted from a scene can be saved and read later on to re-analyze as shown in Figure 13.1.

Secondly, functions like motion detection and face detection provide useful information and allow users to search for specific objects in camera footage.

The YOLOv3 model integrates CNN, which not only classifies objects and detects them but also localizes detected objects. After giving any image, bounding box coordinates are given so that the model can get trained.

Deep SORT algorithm is applied which is used for tracking the object. The final image is obtained, and it can be verified if the predictions are correct or not.

13.2.1.2 Harris-Stephen Corner Detector Algorithm

Vehicle detection technology using video feeds is a key component of intelligent transportation systems such as traffic lights and traffic crossing, due to its non-intrusive surveillance and data collection capabilities (Blumentritt, 1982).

For business owners, understanding vehicular traffic patterns is key to the success of their businesses. But that information can be difficult to gather from a manual count at an arterial road or freeway.

To rectify this problem, the system introduced a standalone algorithm that gathers and tracks vehicle counts and speeds in order to find out how traffic is moving – all with real-time data (Guo & Wu et al., 2022). Sometimes, the easiest way to tackle a complex problem is with the best technology.

The proposal of the video detection system was made, which would save time and effort as the need for complicated calibration is eliminated, different lighting conditions are handled well, and low-resolution videos can be worked with (Blumentritt, 1982). Because it is a much more effective system, you'll have an easier time getting accurate outcomes which will only result in better performance.

The performance of this proposed system is equivalent or better than that of a commercial vehicle detection system. The Harris-Stephen Corner Method (HSCM) is used by the proposed algorithm for ITS applications to detect vehicles.

This system eliminates the need for often unnecessary recalibrations because of its in-built methodology as shown in Figure 13.2. This means it can detect different sources of light and adjust automatically to them. The points generated are also of a higher resolution.

The performance was evaluated by looking at eight video feeds (each with). The video feed changes in a myriad of ways, being captured at various times of the day, different mount height and view angle.

13.2.1.3 ALPR Algorithm

Automatic vehicle identification and recognition is a technology that enables autonomous procedures in traffic systems (Ritika & Arora et al., 2022). With hardware like cameras, it can identify every car on the road (Muthukumar & Chintalacheruvu, 2012).

ALPR is one form of vehicle identification. Comparing pictures is an advanced computer vision technique that has a lot of applications. It helps identify vehicles by

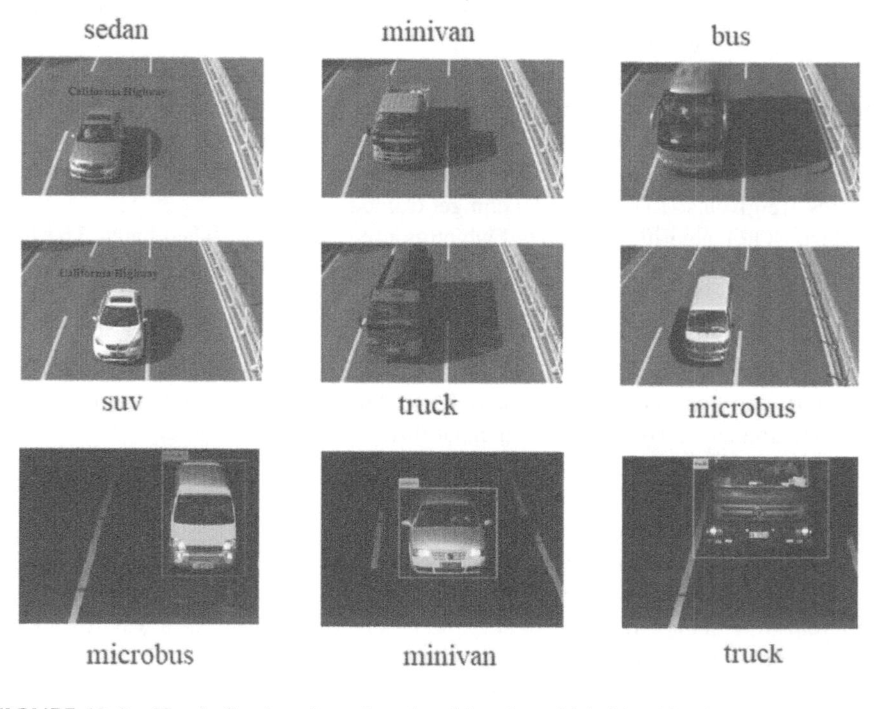

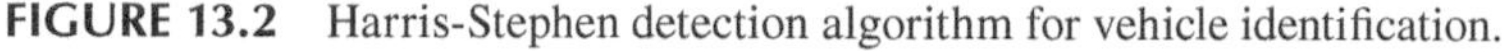
FIGURE 13.2 Harris-Stephen detection algorithm for vehicle identification.

FIGURE 13.3 Automatic License Plate Recognition for vehicle identification.

reading the vehicle registration plates, and it is used by modern intelligent transport systems in general.

ALPR is already used to enforce traffic laws, collect bridge tolls, and find wanted people. Generally, ALPR systems can consist of a variety of methods. These come from different areas like machine learning (ML), pattern recognition, and computer vision. Therefore, algorithm selection and modification become the crucial parts of the procedure as shown in Figure 13.3.

Based on that, what we do is design a perfect system to reach the ultimate purpose of the system, which is to get the license plate number from an image capture.

13.2.1.4 HOG and Linear SVM

When trying to identify a vehicle, it can be difficult. There are many interference factors that complicate the process, especially in a background like this where there are a lot of other vehicles (Vijayalaxmi & Mahadevan, 2012).

Researchers have designed a specialized system to improve accuracy in recognizing vehicles when they are in difficult lighting conditions. This approach uses the HOG-SVM ML algorithm (Patni & Alshmrani et al., 2022).

The HOG feature is a feature descriptor used for object detection in computer vision and image processing. It constructs features by calculating and statistic the gradient direction histogram of the local area of the image as shown in Figure 13.4.

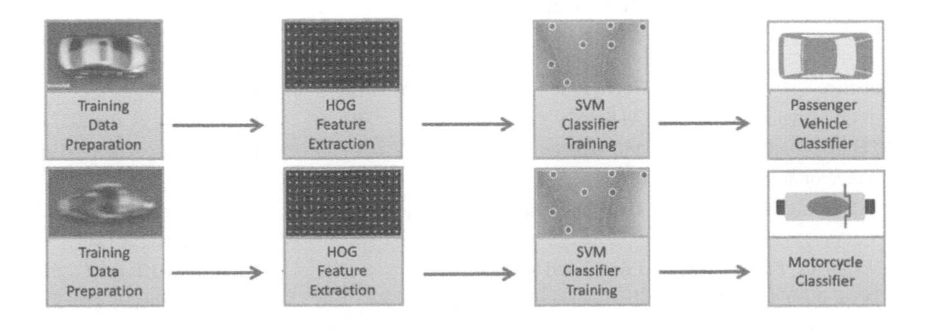

FIGURE 13.4 HOG-SVM algorithm.

Edge detection and gradient information can be used to simulate target regions. HOG takes this idea into consideration and eliminates redundant features from the description.

In order to study and realize the automatic identification of vehicle models, in view of the fact that the road traffic video surveillance system is based on the limited sample of real-time processing of a large amount of data, this chapter uses SVM.

At the start, SVM is classified for two types and applied in two cases. Linearly separable and indivisible problems can be transformed into linear problems in high-dimensional space with the non-linear transformation.

Of course, SVM can be extended to multi-classification and solve many practical problems. SVM can also address small sample size, high dimensionality, and non-linear issues with precision.

13.2.1.5 Vision and Lidar Point Cloud Fusion Algorithm

The ability to detect vehicles and navigate accordingly is one of the most important environmental perception tasks for autonomous vehicles (Guo & Wu et al., 2022).

The traditional vision-based vehicle detection methods are not accurate enough; they're especially prone to small and obscured objects, these LiDAR (Light Detection and Ranging) based solutions are good at detecting obstacles, but they might take a long time and have low accuracy when trying to classify different target types.

This chapter focuses on the drawbacks of both Lidar and camera sensors (Hu & Ge, 2019), we found an algorithm which uses their strengths to produce a real-time vehicle detection system. Firstly, obstacles are identified using the LIDAR point cloud data. Then, they're mapped to the image, and a few separate regions of interest (ROIs) can be made.

The ROI is increased every time you detect a new vehicle. The ROI is then merged with the dynamic threshold to generate your final metrics. We use YOLO (a deep learning method) to analyze the ROIs and detect vehicles as shown in Table 13.1.

The test results on the KITTI dataset show that our algorithm is accurate and fast. Compared with the method based only on YOLO as shown in Figure 13.5, its mean average precision (mAP) is twice better (Guo & Wu et al., 2022).

TABLE 13.1
Algorithms Used in Vehicle Identification

Algorithm	Technique	Classification Accuracy (%)	Author
1. YOLO algorithm	AI based	94.10	Guo and Wu et al. (2022)
2. Harris-Stephen corner detector algorithm	Video based	93.95	Liang and Li et al. (2019)
3. ALPR algorithm	Computer vision	95	Vijayalaxmi and Mahadevan (2012)
4. HOG and Linear SVM	Edge-based detection	88	Kulakov and Tomikj (2021)
5. Vision and Lidar Point Cloud Fusion algorithm	Traditional vision based	96.6	Lou and Cai et al. (2019)

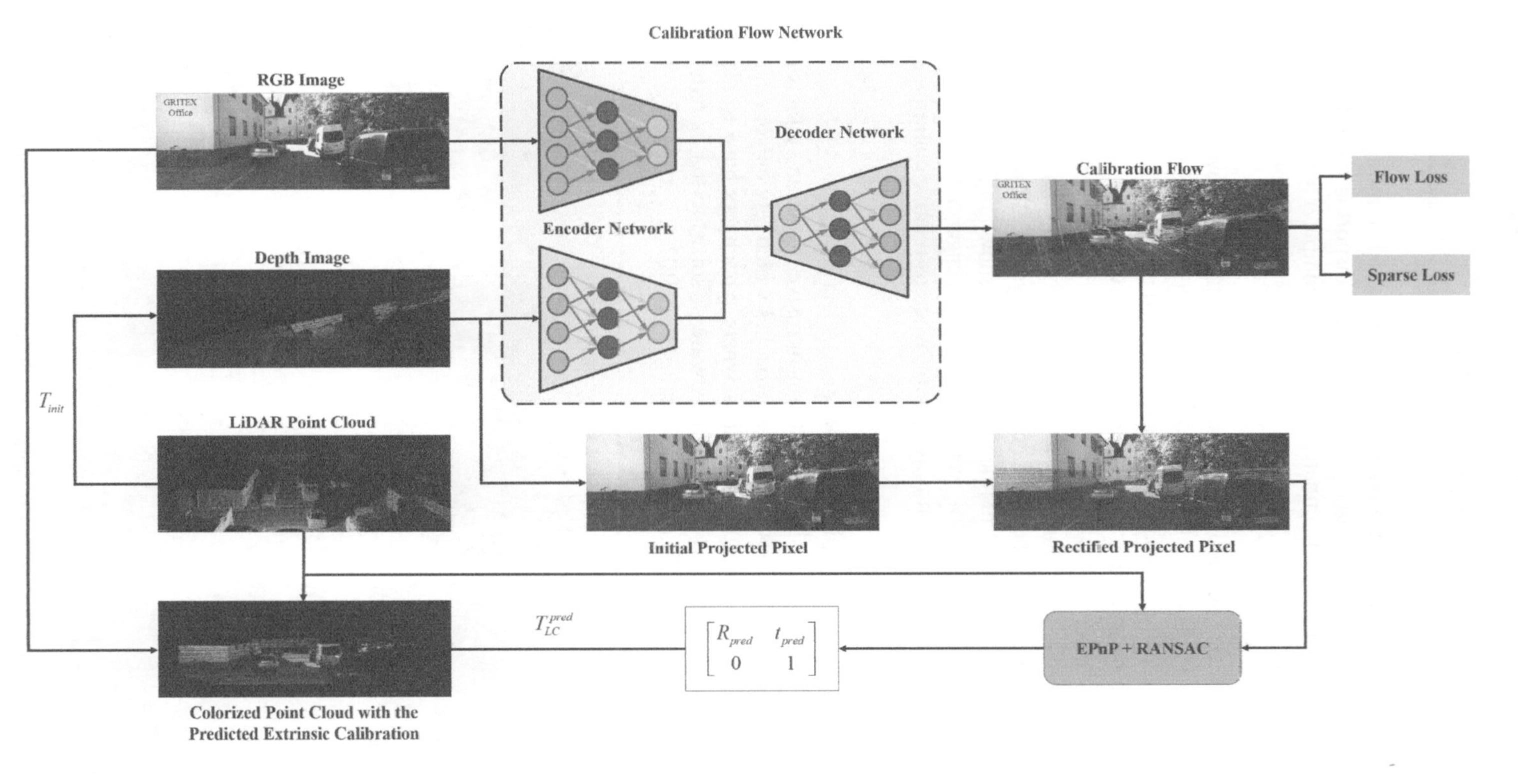

FIGURE 13.5 Real-time vehicle detection algorithm.

13.2.2 Algorithms for Passenger Identification

13.2.2.1 Computer Vision-Based Algorithm

Implementing a reliable and accurate passenger detection and counting system is an important task for the correct distribution of available transport (Lou & Cai et al., 2019). The aim of this chapter is to enhance an existing computer vision-based system to track passengers through a long-distance vision system in trains (Liang & Li et al., 2019).

The proposed passenger detection system incorporates the ideas of well-established detection techniques and is optimally tailored for both indoor and outdoor scenarios. The person's face in a picture is found and then described using key features like the color of their eyes, how much hair they have, etc. One advantage of using AI is the ability to count passengers automatically.

We use a specialized system that's trained and tested based on a SVM classifier and detect them using a filter. The proposed counting system allows you to count passengers without having to stand in line all day long as shown in Figure 13.6.

Accuracy ranges of 86.24%–91.2% were found for passenger detection using the proposed passenger detection and counting system, whereas relative counting errors varied from 10% to 13%.

13.2.2.2 Algorithm on Smartphone Data

This post will quickly introduce six types of non-intrusive detection systems made possible by your smartphone. These include approaches which rely on your camera, audio sensors, etc. It doesn't include situations where passengers have to identify themselves to the vehicle, such as via some hardware or software that is only used for recognizing faces or voices (Hahanov & Khang et al., 2022).

FIGURE 13.6 Background subtraction algorithm for passenger identification.

FIGURE 13.7 Fusion and statistical filtering algorithms.

This approach uses simple driver/passenger(s) identification by identifying turning angle and direction, which can be easily obtained from the user's behavior. In particular, a driver has to turn clockwise during entry, whereas a passenger in the front or rear-nearside seat has to turn anticlockwise.

The proposed technique accordingly captures salient relevant features by using smartphone sensory data and door signs. This is followed by a suitable classifier and decision criterion to accomplish the identification task prior to the start of a journey as shown in Figure 13.7.

13.2.2.3 IoT-Based Smart Card

Public transport (PT) plays a large role in many countries' transportation. The most popular form is bus transportation which can be widely spotted all over the world. As the stress and disorder that are often created by commuting in cars can harm the health of commuters, agencies are looking for a solution that will directly benefit commuters (Rani and Chauhan et al., 2021).

As outlined in this chapter, the employment of AI to collect data about bus riders on a centralized server would not only decrease crime rates but also make commuting faster, easier, and more convenient. We've implemented transportation smart card technology for database collection. This will help to gather data about passengers as they travel from one station to the next (Bhambri & Khang et al., 2022).

We'll then send this information to a server through IoT, a technology being used for trend identification lately. In the event of any accidents happening in buses, it can sometimes take a long time to identify the passengers. Our solution is better because it takes less time at finding out who is involved (Rani, & Khang et al., 2023).

The method followed at present is that when an accident occurs, the wounded and injured passengers are taken to the hospital, and in case if we want to identify the passenger involved in the accident, then the passenger recovery is must to share their information, and if any passenger dies in the accident, then identification is done by the help of their relative (Khang & Abdullayev et al., 2024).

The relatives of a passenger might not find out about the accident until they see it on social media since many use that. This takes away time for their reaction to the identification process, so if there's an accident, you can identify them easily by looking at social media before information is out in other avenues (Tailor & Khang et al., 2022).

If we have problems with the transportation smart card, we can get the information from the server by contacting ADMIN (the person in charge of maintaining a database of passengers) as shown in Figure 13.8.

13.2.2.4 Machine Learning

New trends can be recognized in the Schedule Plan by applying ML approaches. This study focused on heterogeneous data that affects the prediction value which is used for predicting the demand transport required in a particular route and arrival time of PT by using Density-Based Spatial Clustering of Applications with Noise (DBSCAN) with Seasonal Autoregressive Integrated Moving Average (SARIMA) algorithm to analyze the forecasting of the real-time passenger demand that dynamically endorsed the growth of the dynamic bus management and scheduling.

Furthermore, the accuracy of proposed SARIMA Model is compared with traditional modeling such as Gaussian Mixture Model (GMM) with ARIMA model for providing an efficient and robust prediction of PT based on passenger demand.

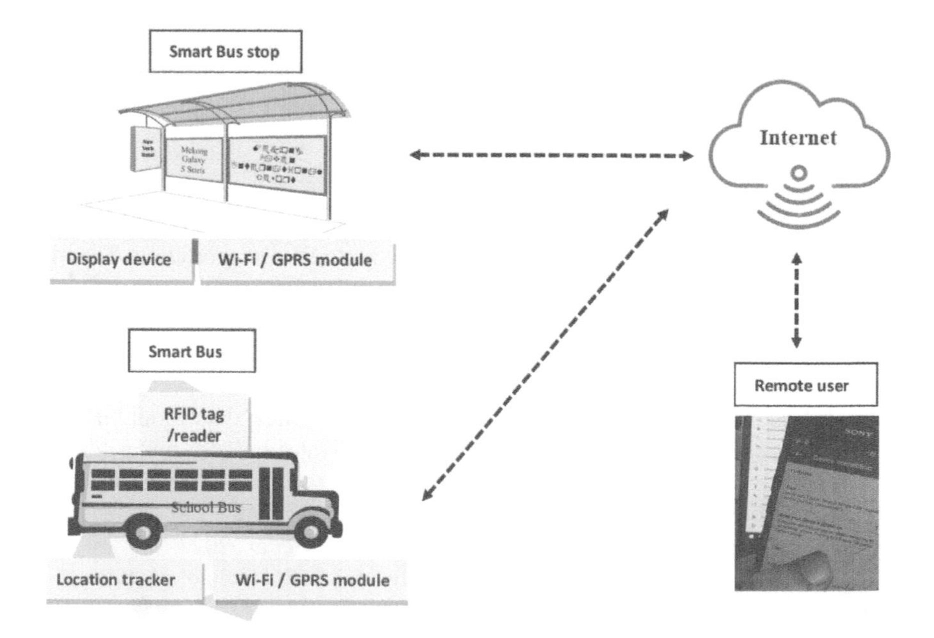

FIGURE 13.8 Bayesian decision tree algorithm (Khang, 2021).

TABLE 13.2
Algorithm Used in Passenger Identification

Algorithm	Technique	Accuracy (%)	Author
1. Background subtraction algorithm	Computer vision based	95	Simutis and Vaitkus et al. (2013)
2. Fusion and statistical filtering algorithms	Smartphone based	93.8	Langdon and Liang et al. (2014)
3. Bayesian decision tree algorithm	IOT based	94	Raj and Ganesh et al. (2018)
4. Density-Based Spatial Clustering of Applications with Noise (DBSCAN) with Seasonal Autoregressive Integrated Moving Average (SARIMA) algorithm	Machine learning based	91	Chen and Lai et al. (2021)

This is a technique that combines the number of scheduled vehicles and the average passenger load in order to identify how long cars will take algorithms of passenger identification as shown in Table 13.2.

We automatically estimate the demand and run time by considering both number of vehicles on scheduled routes and their coverage of passengers every day through APC (Automatic Passenger Counter) and AVL (Automatic Vehicle Locator).

For travelers who use similar routes throughout the day, it may be beneficial to schedule one of those days to replicate the days before and after it. For example, if you have a long day planned that ends on Tuesday and before Wednesday, which is your usual travel route with a lunch break in between, then Monday could represent a suitable alternative as shown in Figure 13.9.

13.3 EQUIPMENT

Some overviews of equipment used in vehicle identification, IoT, passenger identification, etc.

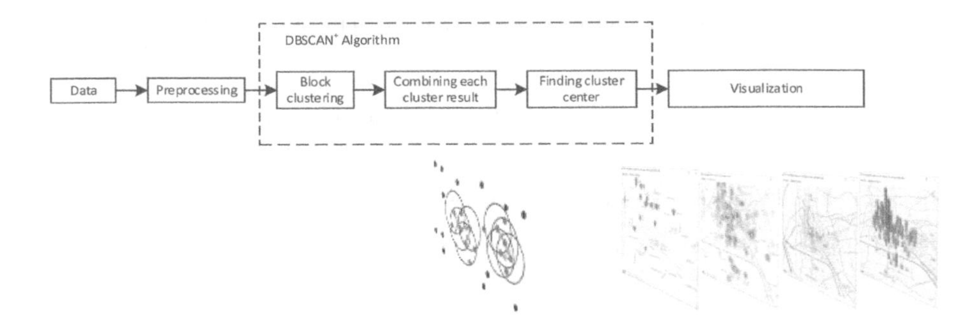

FIGURE 13.9 Density-Based Spatial Clustering of Applications with Noise (DBSCAN) with Seasonal Autoregressive Integrated Moving Average (SARIMA) algorithm.

13.3.1 Equipment for Vehicle Identification

13.3.1.1 Bluetooth

Bluetooth is a concept that was designed to establish a universal short-range wireless connection, mainly for low-power uses, in the 2.4 GHz band that is globally available. The Bluetooth technology is authorized by the Bluetooth Special Interest Group (Vijayalaxmi & Mahadevan, 2012). In vehicles, a Bluetooth connection is used to link up a smartphone with hands-free phone system (Chen & Lai et al., 2021).

A car stereo system can use its speakers to play music streamed from a listener's smartphone. Bluetooth architecture is made up of various layers, starting with the Radio Frequency (RF) layer.

To be less susceptible to external interference, frequency hopping spread spectrum (FHSS) is used. Other layers include Low Energy and Logical Link Control, among others.

Bluetooth devices create a mesh network to ensure they're all connected together. This frequency range is divided into 79 channels across a bandwidth of 1 MHz. There are three classes of transceivers available with different power outputs, 1 mW, 2.5 mW, and 100 mW.

13.3.1.2 Wireless Local Area Network (Wi-Fi)

The primary type of wireless internet at home, Wi-Fi, is an adaptation of the original IEEE 802.11 X standard, which was developed as a communication protocol for cable-based Local Area Networks (LAN) (Kulakov & Tomikj, 2021). The head units of new cars have also now started to incorporate Wi-Fi hot spots (Kostakos, 2008). So many laptops, tablets, and mobile phones can access the internet while driving.

The Wi-Fi standards they use include IEEE 802.11 b/g/a/n or ac. Wi-Fi routers use either a 2.4 GHz or 5 GHz frequency band, and each range has its own pros and cons. It's best to choose one based on your needs.

For example, if you want to stream video content, go for a device that offers the most channels with the highest bandwidth possible – usually that means 5 GHz. In the band between 455 and 5 GHz, 18 different Wi-Fi networks can be established.

Beacons play an important role in 802.11-based Wi-Fi networks and are a subject of one of the 802.11 management frames and use the following data format. Beacon frames are transmitted periodically to announce the presence of a wireless LAN and contain information about the network as the list of equipment in Table 13.3.

Beacon frames are transmitted by the access point in an infrastructure basic service set (BSS) (Kostakos, 2008). In IBSS network, beacon generation is distributed among the stations (Kulakov & Tomikj, 2021) as shown in Figure 13.10.

13.3.2 Equipment for Passenger Identification

There are various tracking hardware available in internet/market as shown in Table 13.4.

1. Beacon cards
2. RFID tags/reader: Active RFID, Passive RFID, and UHF RFID
3. Bluetooth

TABLE 13.3
Equipment Used for Vehicle Identification

Technology	First Level Features	Second Level Features	Author
Bluetooth	MAC ID (BDADDR) CLK, Bluetooth device profile "friendly name"	CLK, Bluetooth device profile Host Controller Interface	Iordache and Cormos et al. (2021)
IEEE 802.11 X (Wi-Fi)	MAC ID (BSSID)	Information in Beacon frames	Yang and Oh et al. (2009)

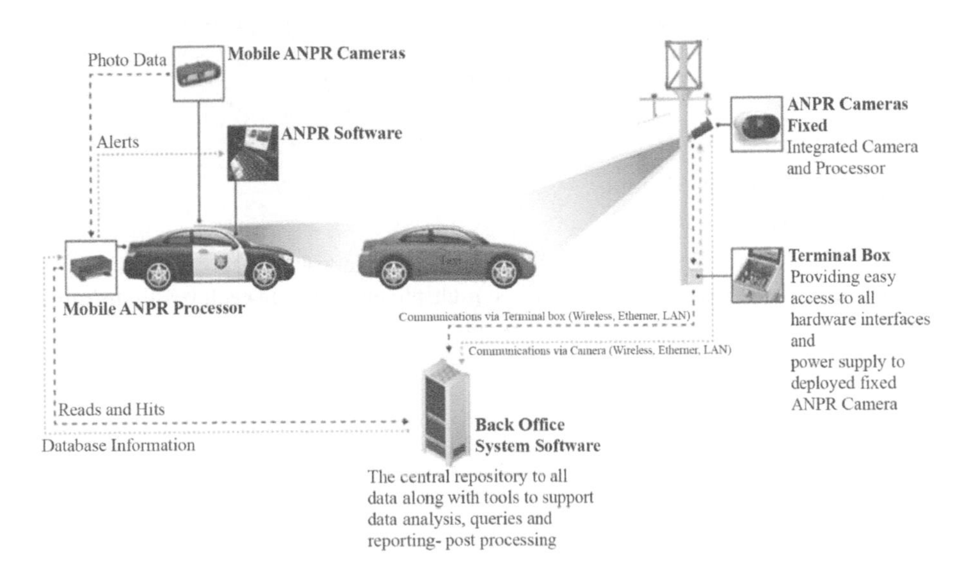

FIGURE 13.10 Equipment for vehicle identification.

TABLE 13.4
Equipment Used for Passenger Identification

No	Criteria	Beacon Card	Active RFID	Passive RFID	UHF RFID	Bluetooth
1	Cost-effective	Yes	Yes	No	Yes	Yes
2	Power consumption	Yes	Yes	No	No	Yes
3	Range of scanning	High	High	Low	High	Low
4	Communication range	Less than 100 m	Greater than 100 m	Less than 10 cm	Around 12m	Around 10m
5	Price	3,000/-	Tags: 200rs Reader: 40k	Tags: 50rs Reader: 500k	Tags: 50rs–500rs Reader: 20k	Reader: 27k

13.3.2.1 Beacon Card

Beacons are a cost-effective option compared to other technologies. They do consume a lot of power, so their range is less than 100 m (Hu & Ge, 2019).

The price per beacon is $3,000, and if you have an airport that needs an upgrade, they can be used (Yang &Oh et al., 2009). This Wi-Fi beacon-based system consists of three key components: the Android application (front end), the database (back-end), and the beacon (Wi-Fi).

The app connects to the server as soon as it's inside the airport using beacon technology as shown in Figure 13.11, so you don't need to waste time finding a Wi-Fi connection.

Once you are in the airport, make sure your phone is on and connected to the Wi-Fi beacon so that you can get notifications about flight progress, gate numbers, flight details, and various offers (Yang & Oh et al., 2009). There is also a map within the application which provides direction (Hu & Ge, 2019).

13.3.2.2 RFID Tags/Reader

13.3.2.2.1 Active RFID

Active RF Identification tags installed on the bracelets will respond to messages transmitted by a transceiver. There will be multiple antennae on each vessel installed for constant communication (Lou & Cai et al., 2019).

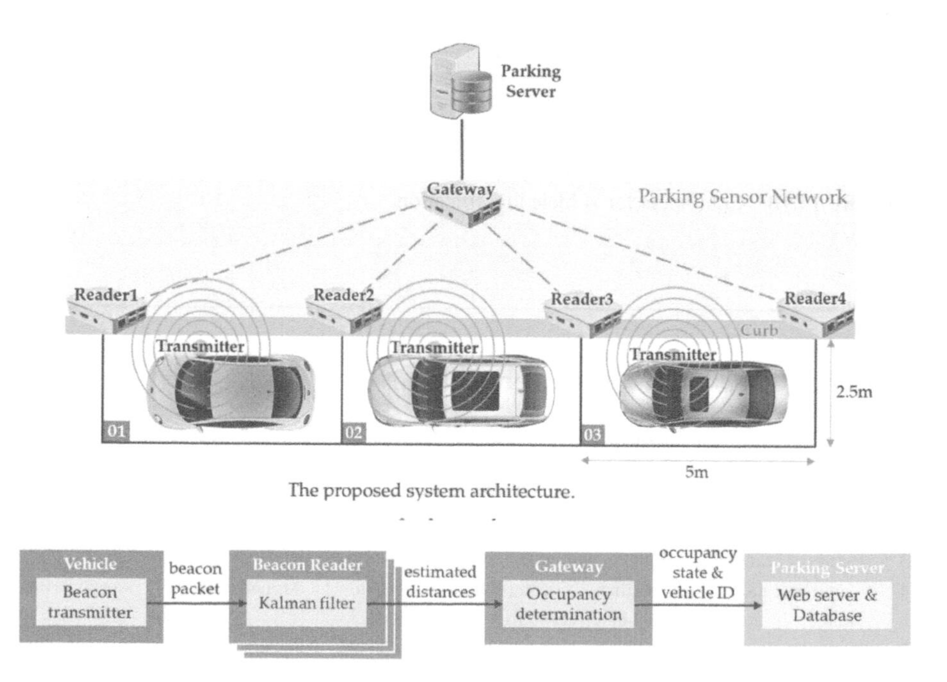

FIGURE 13.11 Beacon card for passenger identification.

These antennas will be able to track the location of the bracelet so that we'll always know where a person is on the ship. They'll work anywhere from the inside of a ship, to on deck, and in any different type of terrain.

We took privacy rights into consideration with the development of this system. As a result, you'll get an efficient missing passenger location system in case of emergencies.

13.3.2.2.2 Passive RFID

Passive RFID systems work by first sending a radio signal to the tag, and the tag then uses that energy to power on and reflect a signal back to the reader (Yu & Wang et al., 2016). Passive RFID tags are increasingly being used to identify a person's location in RFID-based transportation systems.

These tags send some kind of information once they come within the proximity of a reader, and then that reader sends this info to other readers.

13.3.2.2.3 UHF RFID

We developed an RFID detection system that searches for tags inserted in airport charging cards (Simutis and Vaitkus et al., 2013). Using card holders with RFID tags in different parts of their body, the researchers found that a loss in percentage of recognition occurred when cards with more than one tag (usually two) were placed on one part.

On the other hand, cards with only UHF tags have a recognition rate of 91% (Anu & Sarikha et al., 2015). The increasing number of reader antennas also improves recognition by an additional 2% per antenna. The authors say that antenna radiation and line of sight are aspects of the recognition system's performance which should not be overlooked.

13.3.2.3 Bluetooth

Bus companies often use Bluetooth in buses to diagnose problems after the bus returns to its garage. Scenarios include automatic download of reports and diagnostics for drivers, or Wi-Fi hot spot detection (Haroon Yousaf & Murtaza et al., 2019).

Further, Bluetooth has been considered a replacement for cables, which can run up to 4 km on a single bus. This will reduce weight and overall petrol consumption. In the future, a prototype of a system may consider using Bluetooth as a way to provide people with the internet while they're in their car.

Traveling with a smartphone? These clever systems may just make your trip easier as shown in Figure 13.12. This approach, however, requires custom software to run on passenger's mobile devices, and that can introduce considerable development costs as well as compatibility issues.

13.4 POLICIES

This information is intended to answer some questions about vehicle identification policies in the state.

FIGURE 13.12 Bluetooth equipment for passenger identification.

13.4.1 Policies for Vehicle Identification

1. Unless you're exempt from the law, you must always display a government-issued vehicle registration certificate on both your front doors.
   - A device substantially similar to the State seal or State emblem, or such other device approved by the DAS Division of Plant and Property Management pursuant to list item (2) (Langdon & Liang et al., 2014).
   - The agency that operates this vehicle is anonymous (Langdon & Liang et al., 2014).
2. All State vehicles operated on public ways will be registered with the Department of Safety, Division of Motor Vehicles, and will be assigned and display permanent government registration plates unless exempted or granted a waiver (Langdon & Liang et al., 2014).
3. This policy applies to all motor vehicles owned by or leased for 12 months or more by the agency (Anu & Sarikha et al., 2015).

13.4.2 Policies for Passenger Identification

The World Customs Organization's main goal is to simplify and harmonize customs formalities and make trade more efficient. This mandate only applies to passenger movements and the movement of commercial cargo across international boundaries (Miller & Habib et al., 2021).

As a result of the increased risk of transnational crime and international terrorism, customs have had to increase their security checks on passengers in order to apprehend offenders and minimize global risks.

With the need to strengthen borders and the growth in passenger traffic across borders, more resources are needed to regulate it all. Resulting in delays (quite severe in some instances) and increased pressure on airport facilities, many of which were designed to cater for much lower passenger numbers.

The WCO is interested in API because it allows its members to bolster what they can do and continues to improve the service they offer:

a. Providing members with information about API programs, development, and the benefits it can bring.
b. Ensuring that the constraints on API are discussed and resolved, for example.
c. Seeking to jointly agree on standards with the airline industry so that API does not develop and proliferate in an inconsistent or unstructured way.
d. The World Customs Organization has recognized that API is a powerful tool to enhance security while accommodating low-risk passengers, which benefits all stakeholders at the border (Anu & Sarikha et al., 2015).
e. We want to see a healthy, orderly, and disciplined development of the API market, and this is why we call for the establishment of standards and agreed principles. This would allow the sharing of data and make sure that developers are working together on shared goals.

13.5 PROTOCOLS

Some overviews of protocols used in both vehicle identification and passenger identification are listed as follows in Table 13.5.

13.5.1 Protocols for Vehicle Identification

13.5.1.1 Zigbee Protocol

Zigbee is based on the IEEE 802.15.4 standard for wireless personal area network (WPAN). It is widely used in commercial and research applications (Malinowski & Kwiatkowski et al., 2020). A Zigbee system consists of three Zigbee RF modules that are responsible for communication as shown in Figure 13.13.

- The first RF module is placed on the gate side to access vehicle information from the vehicle module. This module connects to the central database RF module for verification.

TABLE 13.5
Protocols Used in Vehicle and Passenger Identification

No	Characteristics	ZigBee	RFID	GPS
1.	Battery life (days)	100–1,000+	No battery(Passive tags)	1–7
2.	Range (meter)	1–75 m	1 m	1,000+
3.	Power consumption	Low	Low	High
4.	Data transfer rate	Low	High	Low

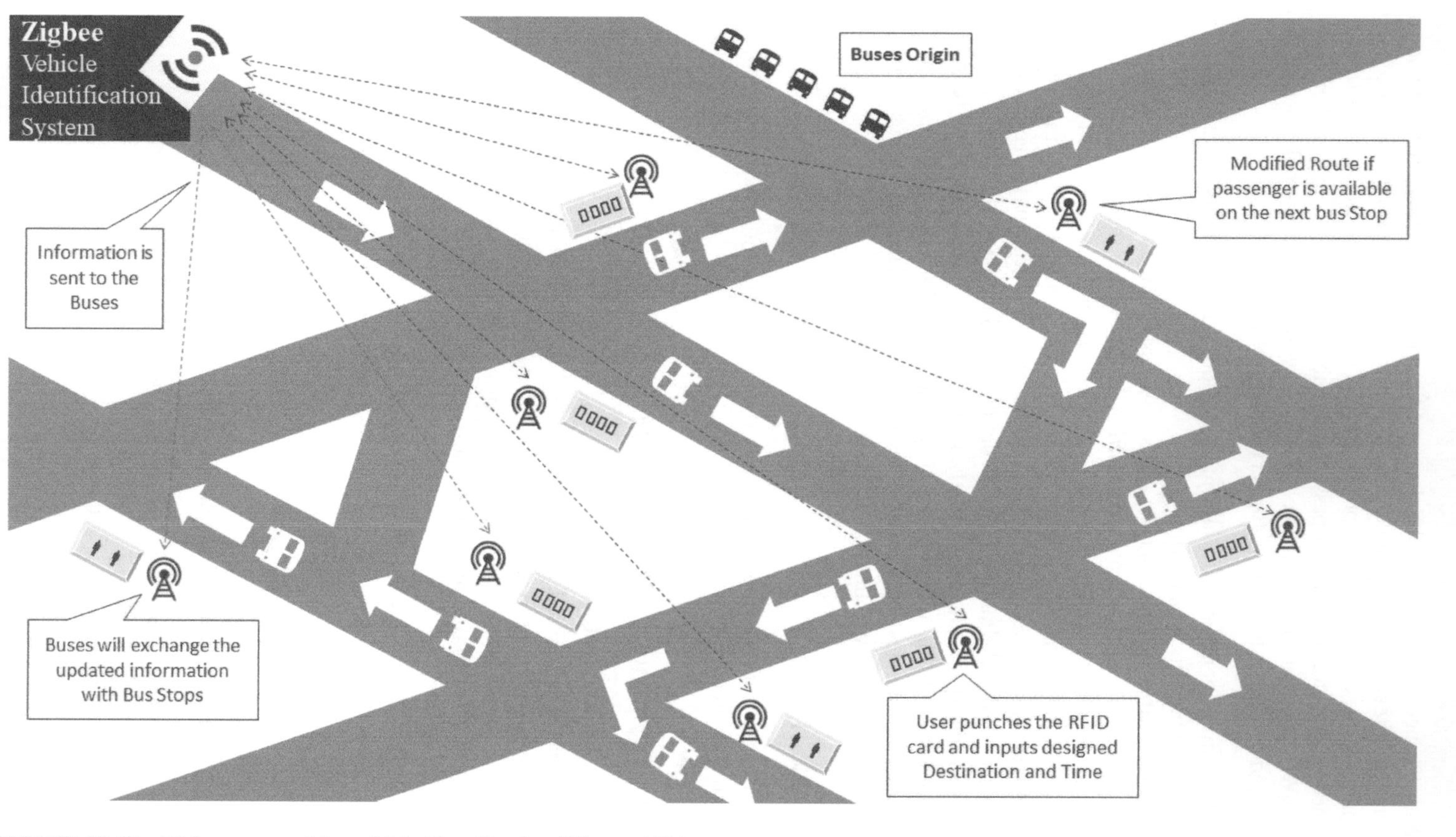

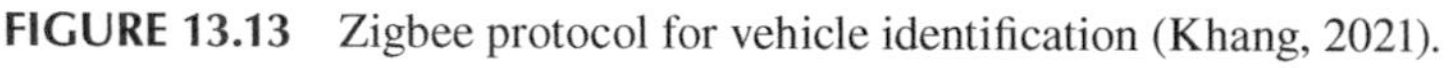

FIGURE 13.13 Zigbee protocol for vehicle identification (Khang, 2021).

- The second RF module is placed inside the vehicle. This module has a unique serial number, keypad, and fingerprint scanner.
- The third module is placed at the central computing device and contains an up-to-date list of authorized vehicles with a personal password for person authentication.
- ←→ Internet connection (3G, 4G, 5G)
- ←→ Internet connection wired
- Bus stop
- Passenger waiting at bus stop
- ▯ No passenger on bus stop
- Zigbee communication

13.5.1.2 RFID-Based Protocol

The protocols that are used to communicate between RFID tags and readers have a large impact on the efficiency of Intelligent Vehicles Identification Systems (IVIS). Using this information, it will be examined how different protocols can affect the system as well as how they work.

It is concluded that Electronic Product Code (EPC) Global and International Standards Organization/International Electrotechnical Commission (ISO/IEC) standards are more widely used than others (Raj & Ganesh et al., 2018).

The performance of the system can be measured by the rate of successful vehicle identification (Kasapovic & Hadzimehmedovic et al., 2016). We compared and analyzed several popular protocols and found the one suited for identifying vehicles by taking into account the characteristics of IVIS on an expressway combined with the collision-avoidance algorithm that corresponded most closely as shown in Figure 13.14.

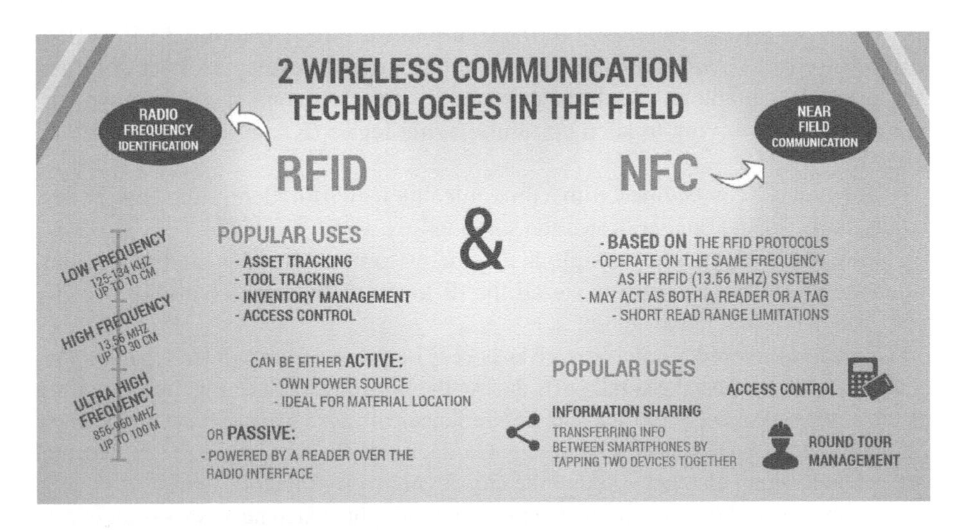

FIGURE 13.14 RFID-based protocols.

13.5.2 Protocols for Passenger Identification

13.5.2.1 Real-Time Identification (ReID) and Positioning of Passenger Using RFID Protocol

The security scanning channel is entered by passengers where security operatives monitor their progress (Chen & Lai et al., 2021) (Skirtic & Jelusic, 2014). Cameras in the corridors surrounding this area collect panoramic images of passengers and analyze ReID recognition results for similarity with an existing database of known faces (a composite sketch), using a mapping service to calculate the real-time locations of passengers and save them in the database.

13.5.2.2 Internet Protocol-Based Communication

The Intermodal Transport Information System provides mobile updates to passengers during their trips. With TRIAS (Travelers Real-time Information and Advisory Standard) technology, passengers can streamline the process with a single connection (Prakashkumar & Thiagarajan, 2020).

The TRIAS interface is very modular and easy to use, it offers a variety of services that can be accessed by anyone (Skirtic & Jelusic, 2014). Services could include the user's mobile application or their previous booking system.

ITIS helps passengers plan their trip and acquire the information they need, even when on the go. For example, the app can show potential adjustments to a planned trip depending on new information about disruptions.

13.6 VEHICLE IDENTIFICATION AND SMART CITY

Most smart cities are suffering from increased levels of traffic in the city center and shopping zones. Injuries to pedestrians as well as a general sense of unease seem inevitable given today's signs and symptoms of congestion (Iordache & Cormos et al., 2021).

Nedap can provide intelligent traffic solutions that are easy to use, highly effective, cost-effective, and safe. Nedap is a multinational developer of security and surveillance technologies with experiences in over 100 countries. Their systems are constantly evolving to keep up future technologies (Khang & Hahanov et al., 2022).

All entrances can be fitted with a mix of technologies for identifying cars, pedestrians, and cyclists. Barrier protection solutions are a case in point.

However, features like traffic lights, cameras to monitor areas, and voice communication devices can also make all the difference. With one controller per city entrance, it doesn't get any easier.

Certain authorized retailers are given access by using a transponder, whereas suppliers can only access the small parts during the morning and evening by using their license plates. Police, fire trucks, and ambulances always have swift access because of their specially designed vehicle tags (Iordache & Cormos et al., 2021). Normal access badges can also get you into the city if you're on official business.

However, you'll only be granted vehicle access rights when necessary – for example, when going through certain routes to specific destinations (Iordache & Cormos

FIGURE 13.15 Vehicle identification for smart city.

et al., 2021). At the core of our advanced access control system is an IP-connected central administration system as shown in Figure 13.15.

Our Nedap solutions help make a city safe and accessible and attractive – identity management, RFID, CCTV, fire and intrusion detection.

13.7 PASSENGER IDENTIFICATION AND SMART CITY

Public transportation is one of the most important aspects of any city as it meets the transport needs of its people. If a country doesn't offer easy and flexible public transportation, it will lead to an increase in people driving, which can have detrimental effects on the country's air quality and resources.

What better way to leverage technology and become a smart city than by investing in AI-based smart transport is just one example of this – the traffic problem that afflicts so many cities can be solved (Kostakos, 2008).

With the help of suitable transport providers, traveling through PT becomes more attractive for the public and reduces the traffic on the roads (Khanh & Khang, 2021).

This is one way to reduce greenhouse gas emissions from private cars. Sharing cars instead of owning one is proven to reduce air pollution and will likely improve the quality of life for the population. Several developed countries have implemented smart systems for PT that provide mobility services for citizens.

Most of these systems use special infrastructures to determine the location of citizens and public buses. The AI system has been implemented on a smartphone which uses the GPS technology to track the location of the public transportation drivers. It can easily calculate your arrival time for buses that are close to your bus stop (Khang & Rani et al., 2022).

This system monitors buses and calculates availability for seats by using a low-cost electronic device. A new smart mobility system has been proposed, and its working is being explained in this work. The work also contains practical

FIGURE 13.16 Passenger identification for smart city.

examples of its usefulness in practice, which you can try out yourself as shown in Figure 13.16.

13.8 CONCLUSION

In this chapter, a comprehensive study of vehicle and passenger identification is conducted. Also, the concept of smart city is also elaborated in this study. Moreover, the relationship between the identifications and smart city is presented. Basically, vehicle identification has a wide spectrum (Rani & Khang et al., 2022).

There are several technologies and protocols that evolved over the years for automation of vehicle identification. Of course, these technologies comply with the policies that are followed for social benefits (Khang & Hajimahmud et al., 2022).

Similarly, some of the techniques are available for passenger identification. However, it is not fully automated nor full-proof since it involves human beings. There is scope for improvement in this regard, and it needs to be integrated with the vehicle identification for an effective implementation.

Furthermore, mobility is an important indicator of smart city. So, smart mobility is achievable with the evolution of vehicle identification and passenger identification systems.

REFERENCES

Anu, M., Sarikha, D., Sai Keerthy, G., Jabez, J.V. "An RFID based system for bus location tracking and display", *in* 2015 International Conference on Innovation Information in Computing Technologies(ICIICT) (2015). Chennai. https://ieeexplore.ieee.org/abstract/document/7396080/

Bhambri, P., Rani, S., Gupta, G., Khang, A. *Cloud and Fog Computing Platforms for Internet of Things* (2022). CRC Press. https://doi.org/10.1201/9781032101507

Blumentritt, C.W. "Automatic Vehicle Identification Techniques," *Texas Transportation Institute the Texas A&M University System College Station*, Texas, 1982. https://static.tti.tamu.edu/tti.tamu.edu/documents/1779-1.pdf

Chen, Y.-Y., Lai, J.-H., Chen, T.-Z. "Estimating bus cross-sectional flow based on machine learning algorithm combined with Wi-Fi probe technology", *Sensors*, vol. 21 (Jan 2021). https://www.mdpi.com/975874

Guo, Z., Wu, J. et al. "Real-time vehicle detection based on improved YOLO v5", *Sustanability* (Sept 2022). https://www.mdpi.com/1853640

Hahanov, V., Khang, A., Litvinova, E., Chumachenko, S., Hajimahmud, V.A., Alyar, A.V. "The Key Assistant of Smart City – Sensors and Tools", *AI-Centric Smart City Ecosystems: Technologies, Design and Implementation* (1st Ed.) (2022). CRC Press. https://doi.org/10.1201/9781003252542-17

Haroon Yousaf, S., Murtaza, M., Velastin, F., Khan, H. "Passenger detection and counting for public transport system", *NED University Journal of Research* (2019). https://search.ebscohost.com/login.aspx?direct=true&profile=ehost&scope=site&authtype=crawler&jrnl=10233873&AN=142350202&h=e1aoyNyg1dnm0riHpd8TxwtI0PanV0IvYGt9YfA6k7%2FY6HE7ZvTrLIpNClHAR2t%2B19LQKGfggoIDJuq2wOG5xA%3D%3D&crl=c

Hu, Y., Ge, P. "Vehicle type classification based on improved HOPG_SVM", in 3rd International *C*onference on Mechatronics *E*ngineering and Information Technology (2019). China. https://www.atlantis-press.com/proceedings/icmeit-19/55917241

Iordache, V., Cormos, A., Gheorghiu, R. "Analysis of the possibility to detect road vehicles via bluetooth technology", *Sensors*, vol. 21 (Nov 2021). https://www.mdpi.com/1340546

Kasapovic, S., Hadzimehmedovic, A., Banjanovic-Mehmedovic, L., Jagodic, I. "Localization and monitoring of public transport services based on zigbee", *International Journal of Advanced Computer Science and Applications* (2016). Tuzla. https://pdfs.semanticscholar.org/3422/ac6533441998e4a3e18041862b5fed41d4f8.pdf

Khang, A., Abdullayev, V., Hahanov, V., Shah, V. *Advanced IoT Technologies and Applications in the Industry 4.0 Digital Economy* (1st Ed.) (2024). CRC Press. https://doi.org/10.1201/978-1-003-43426-9

Khang, A., Hahanov, V., Abbas, G.L., Hajimahmud, V.A. "Cyber-Physical-Social System and İncident Management", *AI-Centric Smart City Ecosystems: Technologies, Design and Implementation* (1st Ed.) (2022). CRC Press. https://doi.org/10.1201/9781003252542-2

Khang, A., (2021). Material4Studies, *Material of Computer Science, Artificial Intelligence, Data Science, IoT, Blockchain, Cloud, Metaverse, Cybersecurity for Studies*. https://www.researchgate.net/publication/370156102_Material4Studies

Khang, A., Ragimova, N.A., Hajimahmud, V.A., Alyar, A.V. "Advanced Technologies and Data Management in the Smart Healthcare System", *AI-Centric Smart City Ecosystems: Technologies, Design and Implementation* (1st Ed.) (2022). CRC Press. https://doi.org/10.1201/9781003252542-16

Khang, A., Rani, S., Sivaraman, A.K. *AI-Centric Smart City Ecosystems: Technologies, Design and Implementation* (1st Ed.) (2022). CRC Press. https://doi.org/10.1201/9781003252542

Khanh, H.H., Khang, A. "The Role of Artificial Intelligence in Blockchain Applications", *Reinventing Manufacturing and Business Processes through Artificial Intelligence*, pp. 20–40 (2021). CRC Press. https://doi.org/10.1201/9781003145011-2

Kostakos, V. "Using Bluetooth to capture passenger trips on public transport buses", CoRR, June 2008, *Phys. Rev.* vol. 47, pp. 777–780 (Jan 2008). https://www.researchgate.net/profile/Vassilis-Kostakos/publication/220484433_Using_Bluetooth_to_capture_passenger_trips_on_public_transport_buses/links/0046352266cbc780b7000000/Using-Bluetooth-to-capture-passenger-trips-on-public-transport-buses.pdf

Kubinaa, M., Bubelínya, O. "Impact of the concept Smart City on public transport", in *14th International Scientific Conference on Sustainable, Modern and Safe Transport* (2021). https://www.sciencedirect.com/science/article/pii/S2352146521005366

Kulakov, A., Tomikj, N. "Vehicle detection with HOG and linear SVM", *Journal of Emerging Computer Technologies*, vol. 1 (Jan 2021). https://dergipark.org.tr/en/pub/ject/issue/64437/980065

Kumari, N. et al. "Smart public transportation for smart cities", in *International Conference on Advances in Engineering Science Management and Technology 2019* (2019). Uttarakhand. https://papers.ssrn.com/sol3/papers.cfm?abstract_id=3404487

Langdon, P.M., Liang, J., Godsill, S.J., Delgado, M., Bashar, T.P., Ahmad, I. "Driver and Passenger Identification from smart phone data", *IEEE Transactions on Intelligent Transportation Systems* (2014). https://ieeexplore.ieee.org/abstract/document/8411173/

Liang, H., Li, H., Dai, Z., Yun, X., Song, H. "Vison-based vehicle detection and counting system using deep learning in highway scenes," European Transport Research Review, 2019. https://link.springer.com/article/10.1186/s12544-019-0390-4

Lou, X., Cai, Y., Li, Y., Chen, L., Wang, H. "Real-time vehicle detection algorithm based on vision and Lidar", *Journal of Sensors* (Apr 2019). https://www.hindawi.com/journals/js/2019/8473980/

Hao, L. Automatic Vehicle Detection and Identification using Visual. *Document* (2017). https://search.proquest.com/openview/664a37b2c35442afe9fd20bf277efcc2/1?pq-origsite=gscholar&cbl=18750

Malinowski, A., Kwiatkowski, S., Sniegula, A., Wieczorek, B., Mastalerz, M. "Passenger BIBO detection with IOT support and machine learning techniques for intelligent transport systems", in *24th International Conference on Knowledge Based and Intelligent Information &Engineering Systems, Poland* (2020). https://www.sciencedirect.com/science/article/pii/S1877050920319001

Miller, E.J., Habib, K.N., Liu, Y. "Detecting transportation modes using smartphone data and GIS information: evaluating alternative algorithms for an integrated smartphone-based travel diary imputation", *Transportation Letters the International Journal of Transportation Research* (Jul 2021). https://www.tandfonline.com/doi/abs/10.1080/19427867.2021.1958591

Muthukumar, V., Chintalacheruvu, N. "Video based vehicle detection and its application in intelligent transportation systems", *Journal of Transportation Technologies* (Aug 2012). https://www.scirp.org/html/23832.html

Nalwade, P. Jagtap, S., Kulkarni, D., Kulkarni, S. "Automatic Bus Transport System with bus location, identification" (Feb 2016). https://www.academia.edu/download/58525117/IJSRED-V2I1P37.pdf

Patni, J., Alshmrani, S., Chaudhari, V., Dumka, A., Singh, R., Rashid, M., Gehlot, A., Alghamdi, A., Rajput, S. "Automatic vehicle identification and classification model using the YOLOv3 algorithm for a toll management system", *Sustainability*, vol. 14 (Jul 2022). https://www.mdpi.com/1744542

Prakashkumar, S., Thiagarajan, R. "Identification of Passenger Demand in Public Transport Using Machine Learning," *Marudupandiyar College*, Tamil Nadu, ISSN 1735-188X, 2021 2020. https://www.webology.org/data-cms/articles/20210429121340pmWEB18068.pdf

Raj, M.P., Kumar, V.V., Ganesh, S. "Passenger identification system using IoT based transportation", *International Journal of Scientific Research in Science and Technology*, vol. 4, no. 5 (Mar 2018). https://ieeexplore.ieee.org/abstract/document/9087066/

Rana, G., Khang, A., Sharma, R., Goel, A.K., Dubey, A.K. *Reinventing Manufacturing and Business Processes through Artificial Intelligence* (2021). CRC Press. https://doi.org/10.1201/9781003145011

Rani, S., Bhambri, P., Kataria, A., Khang, A. "Smart City Ecosystem: Concept, Sustainability, Design Principles and Technologies", *AI-Centric Smart City Ecosystems: Technologies, Design and Implementation* (1st Ed.) (2022). CRC Press. https://doi.org/10.1201/9781003252542-1

Rani, S., Bhambri, P., Kataria, A., Khang, A., Sivaraman, A.K. *Big Data, Cloud Computing and IoT: Tools and Applications* (1st Ed.) (2023). Chapman and Hall/CRC. https://doi.org/10.1201/9781003298335

Rani, S., Chauhan, M., Kataria, A., Khang, A. "IoT Equipped Intelligent Distributed Framework for Smart Healthcare Systems", *Networking and Internet Architecture* (2021). CRC Press. https://doi.org/10.48550/arXiv.2110.04997

Ritika, H.J., Arora, A., Kejriwal, M.R. "Vehicle detection and counting using deep learning based YOLO and deep SORT algorithm for urban traffic management system", in *International Conference on Electrical, Electronics, Information and Communication Technologies (ICEEICT)* (2022). Bengaluru. https://ieeexplore.ieee.org/abstract/document/9768653/

Simutis, R., Vaitkus, V., Maskeliunas, R., Lengvenis, P. "Application of computer vision systems for passenger counting in public transport", *Electronika Ir Elechtrotechnika*, vol. 3, no. 3 (2013). https://epubl.ktu.edu/object/elaba:3259226/

Skirtic, M., Jelusic, N. "An RFID based vehicle identification system for public transport priority", in *22nd International Symposium on Electronics in Transport* (2014). Croatia. https://ietresearch.onlinelibrary.wiley.com/doi/abs/10.1049/smc2.12032

Tailor, R.K., Pareek, R., Khang, A. "Robot Process Automation in Blockchain", *The Data-Driven Blockchain Ecosystem: Fundamentals, Applications, and Emerging Technologies*, pp. 149–164 (2022). CRC Press. https://doi.org/10.1201/9781003269281-8

Vijayalaxmi, P., Mahadevan, S. "Design of algorithm for vehicle identification by number plate recognition", in *IEEE-Fourth International Conference on Advanced Computing, ICoAC* (2012). Chennai. https://ieeexplore.ieee.org/abstract/document/6416823/

Yang, S.-C., Oh, D.H., Kim, J.-D., Park, H.-S. "Design and implementation of WLAN-based automatic vehicle identification", in *2009 International Conference on Computational Science and Engineering,* (2009). South Korea. Phys. Rev. 47, 777–780. https://ieeexplore.ieee.org/abstract/document/8528677/

Yu, G., Wang, Y., Wu, X., Ma, Y., Xu, Y. "A hybrid vehicle detection method based on viola-jones and HOG + SVM from UAV images", *Sensors*, vol. 16, Phys. Rev. 47, pp. 777–780 (2016). https://www.mdpi.com/1424-8220/16/8/1325

14 5G-Assisted UDV Networks Based on Energy-Efficient Optimal Route Scheduling for Smart City

Parul Priya and Sushma S. Kamlu

14.1 INTRODUCTION

Unmanned drone vehicle (UDV)-based technologies have historically been researched and employed primarily to support military as well as rescue task forces. Such involve navigational and vigilance, instruments both on the ground and in the air for inspection, and mapping of unsafe places. There has been a significant increase in the use of UDVs as a consequence of the prevalence of cutting-edge approaches to developing complex UDVs that are prompted by several applications.

In accordance with the technology employed, different accessories like scanners, recorders, and many other sensing equipment can be used in UDVs. They are primarily helpful in instances where human persons would find them tedious (such as when monitoring agricultural land) or dangerous (such as when monitoring borders or nuclear power plants).

UDVs can only operate in a restricted region for their persistence and visual coverage restrictions. The previously mentioned problems of energy and exposure can be resolved by a multi-UDV approach (more than one UDVs is deployed concurrently) (Zhao & Lu et al., 2019; Ejaz & Ahmed et al., 2020; Priya & Kamlu, 2022). By collecting data from different points of view, it also increases the diversity of observations. Additionally, it improves the fault tolerance and reliability of UDV data.

The interactive features of FR that have been just highlighted draw clients and businesses from many different industries. Drone market is expected to grow at a 7.9% CAGR from 2022 to 2027, from $26.2 billion in the year 2022 to $38.3 billion in 2027 (Markets, 2018).

UDV communication is currently the primary study subject drawing scientists from all over the world and focusing on coverage scalability, increased throughput, and security (Khang & Hahanov et al., 2022). There are three different kinds of it:

a. UDV-to-UDV (U2U): UDVs can communicate data and collaborate in multi-UDV systems to boost performance in areas, including scalability, reliableness, and precision (Khang & Gujrati et al., 2023).

DOI: 10.1201/9781003376064-14

b. UDV2Base facility: A SB (SB) could further command or direct UDVs via wireless medium, and UDVs can transmit factual dataset to SB for the purpose of accurate and efficacious outcome.
c. UDV2Satellite: Satellites are used to monitor and control UDVs when UDV exits within line-of-sight, and SB transitioned to a satellite-control (Azmat & Kummer, 2020).

From the initial generation (1G) of communication platforms to the most current generation fifth-generation (5G) of communication technologies, there has been a phenomenal transition in this field (5G). Fourth-generation (4G) technology has been enhanced, but 5G technology ushers in a completely new age of wireless communication.

Several technologies, like cell densification, massive MIMO, and mm-wave communication, are being used to meet the objectives of 5G (M-MIMO). Cell densification is the process of incorporating more number of cells (relay, micro, pico, femto, nano, and Wi-Fi connectivity) into a wireless communication network's traditional big size, high-performance macro SB (BS).

The main goal is to maximize network capacity by repeatedly recycling the spectrum and to reduce latency by bringing end users closer to the network. Moreover, the UDV communication infrastructure may be more dependable, sunken delay, and high fault-tolerant. These could be accomplished by using the cutting-edge (5G) communication system.

This chapter has been organized as follows. Section 14.2 deals with the problem statement and proposed system description. Section 14.3 deals with the results and discussion and the last section consists of conclusion.

14.2 RELATED WORK

The execution and communication of UDVs are impacted by a number of factors, including meteorological conditions like temperature, direction and wind speed, and UDV height, among others (Liu & Dai et al., 2020). Energy that UDVs can receive from batteries is restricted, thus it's critical to limit how much energy it utilizes. By carefully planning their positions and flying routes, this can be accomplished.

Several earlier attempts took wind speed and direction into account while deciding where to put things, so that a UAV's flight path would be both efficient and safe (Ji & Meng et al., 2020).

The QoE and QoS for UDV communication have been enhanced by shrinking the route loss as well as time required to regulate the mobility of the UDV. The UDV has been placed in a way to minimize the possibility of an outage, which solved the problem of link failure in a confined environment. However, links have been broken since the repercussion of the weather upon UDV communication has been neglected.

Inevitably intertwined articles in Goh and Leow et al. (2023), Kodheli and Lagunas et al. (2020), and Li and Fei et al. (2018) examined the energy efficiency of UDV SBs but ignored the energy consumption of UDVs for propulsion. Goh and Leow et al. (2023) suggested lowering UDV transmission power to allow for energy-efficient deployment while still enabling wireless overage from ground users.

Kodheli and Lagunas et al. (2020) considered numerous UDV energy-efficient three-dimensional (3D) deployment scenarios by reducing the overall transmit power to the bare minimum and achieving the QoS requirements for end users.

Li and Fei et al. (2018) indicate that the energy efficiency of UDV-assisted 5G systems employing interference awareness has been optimized when device-to-device communication has been supported.

Despite the fact that UDV propulsion normally requires a lot more energy than communication does, optimizing the use of propulsion energy has been a great challenge to the service life of the UDV. As a result, the propulsion energy consumption cannot be disregarded when evaluating the UDV's energy efficiency.

This served as the motivation for a thorough analysis of the propellant energy requirements for fixed-wing and rotary-wing UDVs in Yu and Li et al. (2020) and Zeng and Xu et al. (2019), respectively. Moreover, models are given for how much energy these two UDVs need when communicating wirelessly.

Musavian and Ni et al. (2015) looked into how to optimize UDVs' energy efficiency while it has been utilized for secure transmission in accordance with the rotary-wing UDV's energy dissipation model (Zeng & Xu et al., 2019).

Some studies Azari and Rosas et al. (2019), Zhan and Zeng et al. (2019), Yu and Li et al. (2020) solely focus on the energy consumption of UDVs at a specific altitude, although taking into account the UDV's propulsion energy demand. Since there is a correlation between propulsion energy usage and flight altitude, the UDV's ability to alter flight altitude can significantly reduce energy consumption.

The inherent time-varying characteristics of the wireless channel make it extremely difficult and impractical to provide a deterministic delay in UDV-enabled wireless networks. In order to address this problem, UDV networks with statistical lag QoS provision have been substantially integrated with efficient capacity.

The number of Internet-of-Things (IoT) devices that can join up is maximized on the uplink of the UDV routing system, and the diverse statistical QoS requirements of IoT devices are addressed (Rani & Chauhan et al., 2021).

Li and Qingliang et al. (2022), Garcia-Rodriguez and Geraci et al. (2019), Zhan and Zeng et al. (2020) have maximized the system's overall efficacious capacity by simultaneously optimizing the 3D location of the UDV and allocating resources while addressing the statistical lag QoS for individual end users.

The functional capacity has been optimized while subject to average power and peak rate constraints and varied statistical QoS criteria. Sadly, none of the studies look for ways to deploy the UDV in an energy-efficient manner.

14.3 UDV COMMUNICATION

Unmanned vehicles will have a significant impact on smart cities for economical and safety objectives, especially integrated 5G and IoT technology (Khang & Rani et al., 2022). UDVs' integration with cellular networks, whether as ground users or transmission hubs, adds previously unseen visual possibilities, including challenges (Bhambri & Khang et al., 2022).

The high altitude and agility of UDVs, the potential for UDV-ground LoS networks, the unique communication service quality (QoS) prerequisites for CNPC versus operation payload data, the stringent environmental SWAP limitations of UDVs, and the modern design D-o-F by cohesively leveraging UDV agility control, both cellular-interrelated and UDV-aided transmission techniques, differ significantly from their geostationary counterpart.

The primary design potential and obstacles for cellular networking using UDVs are described below:

High altitude: UDV BSs and users frequently fly at altitudes that are significantly greater than those of traditional terrestrial BSs and users. A standard altitude for a geostationary BS is roughly 10 m and 25 m in Urban Micro and Macro implementation, respectively, in contrast to the current norm of 122 m.

The UDV-BS/relay can achieve greater ground coverage than their terrestrial equivalents in UDV-assisted communication systems because of the high UDV height (Paving the path to 5G, 2019; Cheng & Gui et al., 2019).

High LoS probability: UDVs' high altitude makes them different from terrestrial communication channels in terms of their air-ground channel characteristics as it has lesser path loss since obstructions here are minimized. Moreover, LoS predominant air-ground links make UDV communications more vulnerable to jamming/eavesdropping attacks by malicious ground hubs (Sun & Shen et al., 2019).

High 3D mobility: Unlike ground networks, where the BSs/relays are frequently in fixed locations and the users move sporadically and at random, UDVs can travel at high speeds in 3D space with partially or completely regulated mobility.

SWAP constraints: The SWAP constraints of UDVs provide significant barriers to their endurance and communication capabilities, in contrast to terrestrial communication systems, where ground BSs and users frequently have a stable power.

UDVs often require additional propulsion energy in addition to the regular energy used by communication transceivers, in order to stay in the air and fly freely (Xiao & Xu et al., 2019).

UDV environment with 5G enabled: 5G network integration has been facilitated by innovations in cellular technology in order to enhance UDV communication.

For UDV operation to be successful, reduced latency and reliable communication are essential components. This integration solution provides services with high throughput and minimal latency (Khang and Gupta et al., 2023).

Extreme mobile broadband (eMBB), massive machine-type communication (mMTC) (Deng & Zhao et al., 2020), and ultra-reliable low-latency communication (URLLC) are three types of services provided by the 5G-equipped UDV network as shown in Figure 14.1.

The terrestrial SB is thought to control and manage the cellular-connected UDVs (TCB). In heterogeneous contexts, the URLLC application addresses both latency and reliability objectives (McEnroe & Wang et al., 2022).

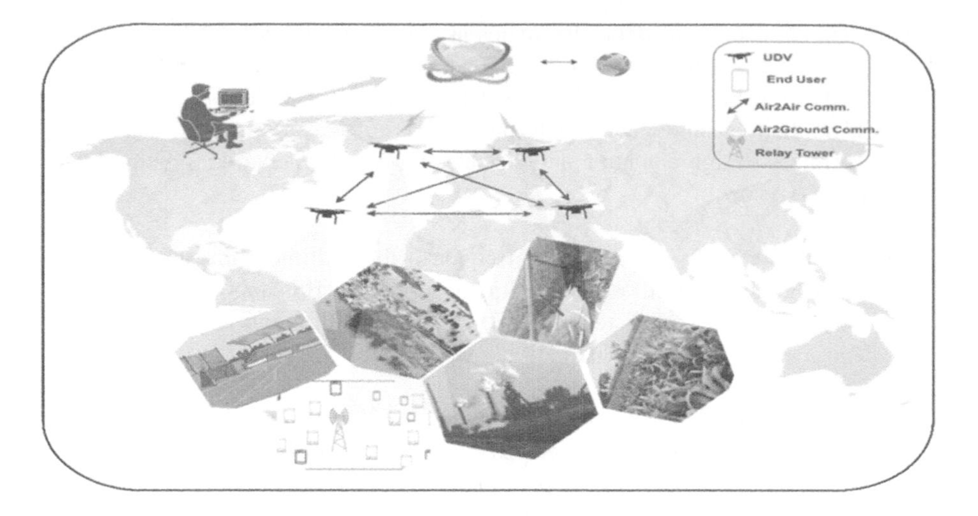

FIGURE 14.1 5G enabled UDV environment.

This method of UDV to TCB transmission is related to air-to-ground (A2G) communication. Wider transmission areas brought on by 5G control stations increase the amount of data needed for the 5G communication system (Jebaraj & Khang et al., 2024).

Table 14.1 lists the uses and functions of UDVs. The UDV and users share the cellular spectrum in A2A communication. A2G communication gathers environmental data so that action can be taken at a specific time. The 5G network provides the UDV with low latency and secure communication.

14.4 PROBLEM STATEMENT

In this method, current position and course planning of the UDV are jointly optimized while taking meteorological aspects into account so as to reduce the time as well as energy consumption for the UDVs.

TABLE 14.1
The Uses and Functions of UDVs

Field	Illustration
Agriculture	Soil analysis, crop monitoring, and spraying pesticides
Environmental	Information collection regarding climate change
Military	Search and rescue for injured or lost soldiers
Law enforcement	Search and rescue
Public safety communication	Communication in disaster condition, role like aerial SBs
Photography	Captures the events, like sports and competitions

The count of UEs in the exigency circumstance should be written as $UE_n = [1,2,3,.....k], n \in \{1,b\}$. No. of $UDV = [1,2,3,.....b]$ is used as BS to represent the number of UDVs in emergency scenarios. The variable "$\partial_{kb} \in \{0,1\}$" indicates whether or not the user "*k*" is sheathed by UDV "*b*". If $\partial_{kb} = 0$, then "*b*" does not cover "*k*", and if $\partial_{kb} = 1$, then "*b*"does cover "*k*".

The 3D orientation of the UDV is separated into two processes, the first of which is the 2D positioning of the UAV, followed by the definition of the UDV's altitude. This is simply the phenomena that altitude only influences the channel's pathway losses.

Once the UDV's coverage is arbitrated, the altitude is hence investigated to attain the lowest trajectory course loss. The UDV b's scope can be expressed as

$$\|a_b - \delta_{kb} c_k\| \leq CR_b + V(1 - \delta_{kb}) \tag{14.1}$$

where *b* is represented by a_b and *k* is represented by c_k in the horizontal plane. The covering radius of *b* is indicated by CR_b, and the enormous constant higher than the achievable horizontal stretch is indicated by V. The baud rate dispersion can be expressed as

$$\sum_{k=1}^{|W|} d_k \delta_{kb} \leq DC_b, b \in \{1,2,...|U|\} \tag{14.2}$$

where d_k stands for the data rate requirement required by *k* and DC_b for the data capacity required by *b*, respectively. The UDV's positioning can be expressed as

$$\max_{CR_b a_b} \sum_{b=1}^{|U|} \sum_{k=1}^{|W|} \delta_{kb} \tag{14.3}$$

This is possible under specific circumstances that constitute the UDV's ideal coverage. Equation (14.1) states that the end-user *k* being covered by the UDV *b* when the horizontal stretch amid *k* and *b* is smaller than CR_b. Equation (14.2) offers the second requirement for figuring out the data rate of the UDV *b*.

The following formula can be used to determine whether the user *k* is sheathed by more than one UDV as

$$\sum_{b=1}^{|U|} \delta_{kb} \leq 1, k \in \{1,2,...|W|\} \tag{14.4}$$

The following condition establishes that UDV coverage radii not being bigger than that of maxima of the coverage radius $CR_{\max}$, which is defined as

$$CR_b \leq CR_{\max}, b \in \{1,2,...|U|\} \tag{14.5}$$

The UDV's *b* height, which influences the trajectory course loss (PL) amid *k* and *b*, be expressed as

$$AL_b = CR_b \tan(\theta_{\max}) \tag{14.6}$$

The UDV flight time is defined as the amount of time spent moving the UAV as well as the time spent collecting dataset from that UE in an emergent situation. This can be calculated as

$$F_t = m_t + c_t \tag{14.7}$$

where c_t stands for data collecting time and m_t stands for motion time.

The energy consumed by the UDV is characterized into two brackets: energy consumption during flight and energy consumption for data assemblage. The total amount of energy used can be expounded as

$$EC_{tot} = m_t m_p + c_t c_p \tag{14.8}$$

where m_p and c_p are the energies used for flight and packet transmission, respectively.

So as to reduce the UDV's flying time as well as energy consumption, path planning is done. This can be illustrated by Equation 14.9:

$$Minimize\ F_t, EC_{tot} \tag{14.9}$$

In order to maximize coverage and use the least amount of energy, the UDVs have been positioned and their paths have been planned using a number of existing technologies, but these methods had significant shortcomings.

The deployment of mobile edges used swarm intelligence to place UDVs predicated on traffic fright and unpredictable weather (SWIM) (El-Sayed & Chaqfa et al., 2019). UDVs act as the bees in this approach, Bee-Swarm-Intelligence (BSI) implementation.

The BSI are used in the deployment phase for efficacious deployment and position update the reasons, leading in efficacious data transmission. So, to minimize alignment losses, various models have been presented predicated on the consideration of the effects of weather and wind changes on the location of the UDV.

14.5 SYSTEM MODEL

This section provides a detailed description of the proffered 5G-enabled UDV environment, which consists of three components: a 5G ground-predicated station (TCB), end-user equipment (UE), and an UDV as Figure 14.2.

Unfortunately, data transmission for timely event warning is impacted in disaster areas. The proffered work will be performed in a disaster-stricken setting with damaged or overburdened TCB.

The overall system model has been shown in Figure 14.2 and highlights the key processes that are engaged in this work and is discussed as follows:

- Weather forecast
- Cell partitioning with density awareness
- Placement for several UDVs
- Energy-efficient routes planning

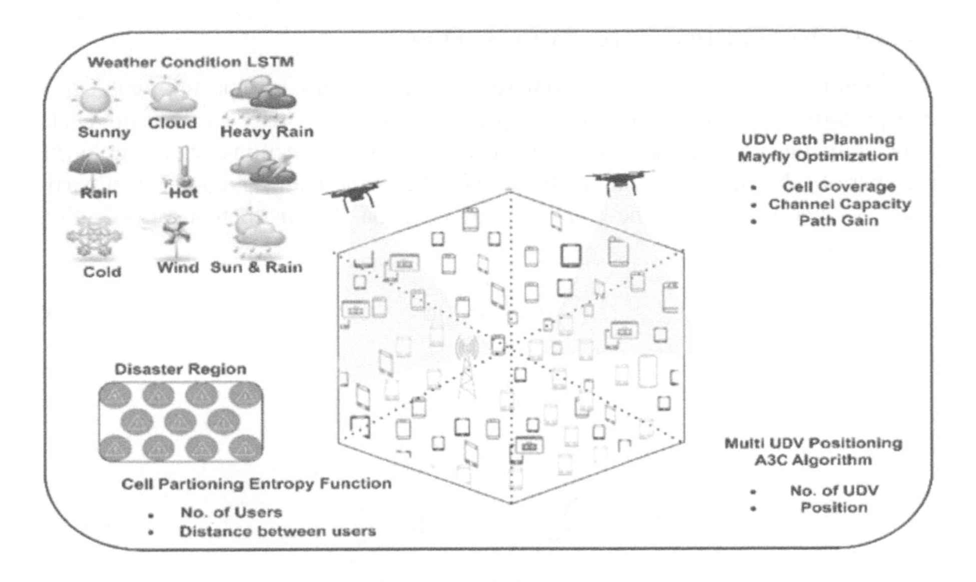

FIGURE 14.2 System model.

14.5.1 Weather Forecast

UDV communication is significantly impacted by the weather. Different weather conditions have an impact on the UDV's location as well as communication port in particular. Wind-related attributes, such as propeller vortex, wind gust-speed and shield, and turbulent flow, are what most perturb the UDV.

Moreover, the temperature has an impact on the UDV's energy usage. Severe weather makes it impossible to operate the UDV and damages it physically. Using Analytical Short Long-Term Memory (A-SLTM), found to have shrinkage on training loss than standard methods, allows for accurate weather prediction in SLTM system.

In comparison to regular LSTM, the suggested A-SLTM is an enhanced variant of LSTM that has edge to comprehend time series data. A-SLTM used a novel kind for recurrent cell consists of one secret state $H(t-1)$ and two cell states $C1(t)$ and $C2(t)$.

The forget gate, the initial component of the A-SLTM, requires input to contemporize, the candidate valuation is momentarily stowed. Every $H(t-1)$ and the input $I(t)$ for each time stamp (t), which are calculated on the basis of parameter weight values, are acknowledged as input and output. Here, the forget gates for the A-SLTM receives input in the form historical series data on climate conditions P_{wc} and present environment data C_{wd}.

A-SLTM generates output such as a climate prediction report based on the input. Wind speed (w), temperature (t), wind direction (wd), humidity (h), precipitation (p), dew-point (D), pressure (δ), and other climate parameters are predicted by this network. These measurements are taken into account when positioning UDVs and planning paths for effective communication.

The expected climate features are additionally used for appropriate channel modeling for data collection purposes.

14.5.2 Modularizing Cells Based on Density

Environment is made up of clustered hexagonal cells for effective partitioning. Users are positioned on the ground via cell partitioning so that they can communicate efficiently. Each cell in the hexagon is divided into equal sections according to the number of subscribers (n) and the distance (ε) between them. The number of partitions is determined using the graph entropy function. The mathematical definition of entropy is as

$$(\varphi) = (n, \varepsilon)\phi \tag{14.10}$$

$$\varphi = t(1) + t(2) \tag{14.11}$$

$$t(1) = -\sum_{i=1}^{n} \left(\frac{n_i}{\sum_{i=1}^{n} n_k} \right) \log \left(\frac{n_i}{\sum_{i=1}^{n} n_k} \right) \tag{14.12}$$

$$t(2) = -\sum_{i=1}^{n} \left(\frac{\varepsilon_i}{\sum_{i=1}^{n} \varepsilon_k} \right) \log \left(\frac{\varepsilon_i}{\sum_{i=1}^{n} \varepsilon_k} \right) \tag{14.13}$$

$$\varphi = \log(t1) + \log(t2) + \log(t3) \tag{14.14}$$

$$\log(t1) = \log \left(\sum_{i=1}^{n} n_k \right) - \frac{\sum_{i=1}^{n} n_i \log n_i}{\sum_{i=1}^{n} n_k} \tag{14.15}$$

$$\log(t2) = \log \left(\sum_{i=1}^{n} \varepsilon_k \right) - \frac{\sum_{i=1}^{n} \varepsilon_i \log \varepsilon_i}{\sum_{i=1}^{n} \varepsilon_k} \tag{14.16}$$

where φ represents the graph entropy threshold value. Proportional threshold partitions separate the hexagonal' cells. Due to the high number of individuals and the close proximity of the cells, the threshold (φ) value dictates that the cells be divided into many equal partitions. Effective UDV communication is made possible by such cellular partitioning.

14.5.3 Multi-UDV Positioning

Parameters such as weather, elevation, speed, entanglement, as well as the occurrence of blockages that prevent effective communication amongst ground users and UDV are taken into account during the dynamic positioning process. Thus, appropriate UDV positioning is required for QoS to be attained.

Each subdivision of the hexagonal cell's UDV and its corresponding position are calculated using the (A3C) algorithm. The forecasted weather, scope, entanglement sensitivity, baud rate, delayed sensitivity, elevation angle, route loss, and LOS/NLOS attributes all play a role in determining the optimal placement of UDVs and, in turn, the necessary number of UDVs.

Attributes of the state (ST), such as the quantity and location of UDV, are used to generate the Action (A). The actor network as well as the critic network makes up the A3C network. The actor network makes efforts to fulfill the guideline, whereas the critic network provides constructive feedback to help the actor network do a better job.

Specifically, the actor network is represented by (Al), the critic network by $\mathbb{C}\left(\frac{\pi}{\phi}\right)(ST)$, and the critic parameters by (ϕ). $\mathbb{Q}$ function is determined as

$$\mathbb{Q}(\mathrm{ST},\mathrm{A}) \approx \eta' + \beta\mathbb{C}^{\pi}(ST') \approx \eta' + \beta\mathbb{C}\frac{\pi}{\phi}(ST') \tag{14.17}$$

where β,η are the discounting function and reward function, respectively. $S\mathrm{T}'$ characterizes the transition to the following state. We specify the benefit function as

$$AD(S\mathrm{T},\mathrm{A}) \approx \mathbb{Q}^{\pi}(S\mathrm{T},\mathrm{A}) - \mathbb{C}^{\pi}(ST) \approx \eta' + \beta\mathbb{C}\frac{\pi}{\phi}(ST') - \mathbb{C}^{\pi}(ST) \tag{14.18}$$

The critic network is trained using an optimized approach that is perhaps expressed as

$$\mathbb{C}(ST) = \mathrm{E}_{A-\pi(A|ST)}\,\mathrm{E}_{ST'-P(ST'|ST,A)} \tag{14.19}$$

The method by which the updated target value is determined using the most recent estimate is graphically represented as

$$y = n' + \beta\mathbb{C}\frac{\pi}{\phi}(ST') \tag{14.20}$$

14.5.4 UDV Path Planning

Path planning is a predominant step for efficient data exchange for which the route of the UDV throughout its' flight is optimized to reduce the UDV's energy consumption in different environmental conditions.

Intelligent Mayfly-Based Optimization (IMO) is being accustomed to plan the UDV's path optimally so as to maximize cell extent coverage, channel competence, and path-gain. Wind properties (w), air temp (T), energy (e), distance (D), speed (s),

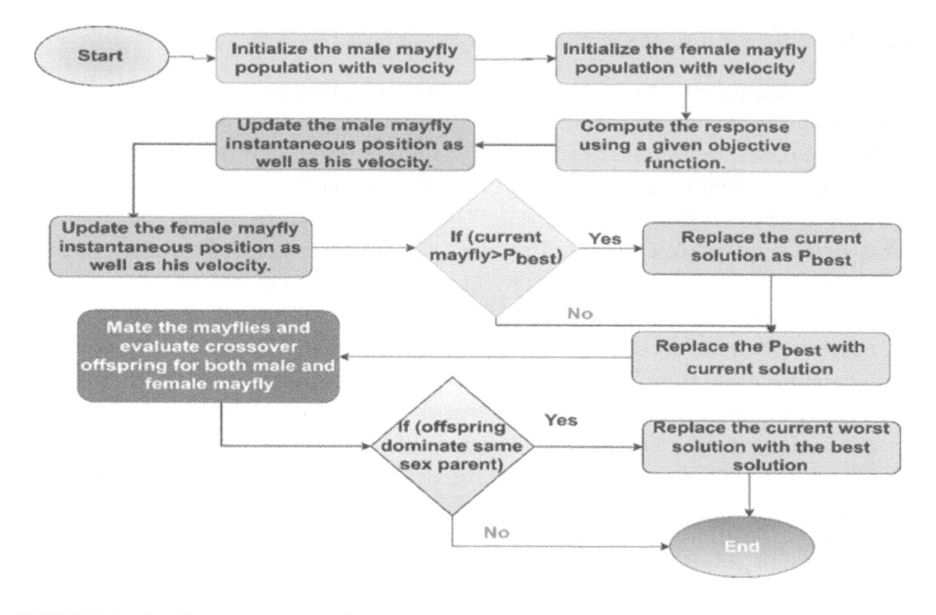

FIGURE 14.3 Flowchart of IMO algorithm.

obstructions (o), LOS/NLOS, optimality, cost efficacy (ce), and collision evasion are used to plan the path.

The proffered IMO combines the benefits of PSO, GA, and firefly optimization technique. The position of each for aging behavior depicts the likelihood of a successful outcome that has been presented in Figure 14.3.

In this scenario, UDV is represented as a mayfly that selects the optimal course to lower energy use. The mayfly cluster, which is defined in Equation 14.21, is used to determine the mayfly's first location along the search area.

$$X = [x1, x2, \ldots, xn] \tag{14.21}$$

where X represents the male mayfly's (mf_1) position. The objective function evaluates the performance of the position vector. Mayfly velocities are defined as

$$V = [v1, v2, \ldots, vn] \tag{14.22}$$

where V represents the mayfly's velocity, which can be used to update its position. The search agent's position is updated based on its own proper position, which is expressed as P_{best}, and the acceptable position is modified by distinct mayflies, which is expressed as g_{best}. A swarm of mayflies (mf_1) demonstrates how their locations are revised based on individual and social perspectives. The mf_1 mayfly's position expressed as

$$\mathrm{X}_i(t+1) - V_i(t+1) + \mathrm{X}_i(t) i_{th} \tag{14.23}$$

where $\mathrm{X}_i(t)$ indicates the i_{th} mayfly's current position. The values $\mathrm{X}_i(t+1)$ and $V_i(t+1)$ indicate the velocity and position of the i_{th} mayfly in the succeeding stage.

The mayfly dance is perpetrated at an elevation of a few meters, and also its speed is measured as

$$v_{id}(t+1) = v_{id}(t) + a1 \times \exp(-\varepsilon d_p^2) \times (P_{bestdd} - x_{id}(t)) \\ + a2 \times \exp(-\varepsilon d_g^2) \times (g_{bestd} - x_{id}(t)) \tag{14.24}$$

where $a1$ and $a2$ are the significant positive allurement constants, which portray the mayfly visibility limits for one another. d_p indicate the space between P_{best} and g_{best}, and x_{id} indicate speed as well position, with d dimensional aspect, valuation of $d = [1,2,3...,D]$ be defined as

$$P_{best} = \begin{cases} \mathrm{P}_i(t+1), F(\mathrm{P}_i(t+1) < F(\mathrm{P}_{besti})) \\ \mathrm{P}_{besti}, F(\mathrm{P}_i(t+1) \geq F(\mathrm{P}_{besti}) \end{cases} \tag{14.25}$$

where $f(.)$ represents the fitness function that determines the solution quality. The formulas for calculating d_p^2 and d_g^2 are expressed as

$$d_p^2 = \left(\sum_{j=1}^{d\max} (P_{id} - P_{besti}) \right)^{0.5} \tag{14.26}$$

$$d_g^2 = \left(\sum_{d=1}^{d\max} (P_{id} - g_{besti}) \right)^{0.5} \tag{14.27}$$

$$F = \sum_{d=1}^{d\max} |w, T, e, s, D, \mu| \tag{14.28}$$

where F presents the fitness value measured by taking into account w, e, T, s, μ, D where the optimal route is obtained by the UDV it reaches τ, c, c_e, α. To secure optimal route planning, a bunch of relevant mayflies linger over their nuptial dance with vertical move. As a result, the optimal mayfly maintains its speed change while adding a spontaneous features defined as

$$v_i(t+1) = v_{id}(t) + nd \times \delta \tag{14.29}$$

where nd represents the connubial dance and δ represents an arbitrary value between $[-1 and 1]$. The female then fly apropos the male mayflies to begin the mating step. Female mayfly mf_2 population initialization can be defined mathematically as

$$Y = [y1, y2, \ldots, yn] \tag{14.30}$$

where Y represents mf_2's position, and mf_2's velocity is defined as

$$U = [u1, u2, \ldots, un] \tag{14.31}$$

The latest updated local position for mf_2 Mayfly denoted as

$$Y_i(t+1)-V_i(t+1)+Y_i(t) \tag{14.32}$$

The infatuation process is carried out in which the relevant mf_2 is allured to the relevant mf_1. This process is repeated until the end. The mf_2 mayfly's velocity calculation is expressed as

$$V_{id}(t+1)=\begin{cases} V_{id}(t)+a.2\times\exp(-\varepsilon d_f^2)\times X_{id}(t)-Y_{id}(t) \\ F(Y_i)>F(X_i) \\ u_{id}+a_n\times\delta, F(Y_i)\leq F(X_i) \end{cases} \tag{14.33}$$

where Y_{id} and u_{id} denote mf_2 mayfly's location along speed in the d_{th} dimensional aspect. d_f^2 represents the length among mf_1 as well as mf_2 mayflies, and a_n portrays the arbitrary coefficient.

The mf_2 and mf_1 optimal matching method is dependent upon objective function as well as fitness function. The offspring of a crossover are calculated as

$$\alpha_1=\sigma\times m f+(1-\sigma)\times mf_1 \tag{14.34}$$

$$\alpha_2=\sigma\times f f+(1-\sigma)\times mf_2 \tag{14.35}$$

where α_1 and α_2 portray the offspring 1 and 2, and σ portray an arbitrary value with a specified range. Thus, robust trajectory course planning is attained, enhancing the system's energy efficacy.

14.6 EXPERIMENTAL AND COMPARATIVE STUDY

This section examines the planned ICROS-UDV and does a comparative analysis. In order to maximize coverage, reduce flight time, and conserve energy, weather forecasting is first done using historical weather reports, and then UDV positioning as well as path planning are done using those results.

The best positioning as well as path routing of UDVS are made easier by taking meteorological conditions into account. The UDVs obtain a higher coverage ratio, allowing the UEs to reliably gather data without any delays. The ground surveillance center gets the UDVs' collected data and makes it easier to report emergencies.

By doing this, UDV communication achieves QoS, and energy efficacy and reliability. Three dissimilar perpetration metrics, that is coverage extent ratio, cell extent coverage, and lag, can be used to determine the QoS for UDV positioning along trajectory course planning as Figure 14.4.

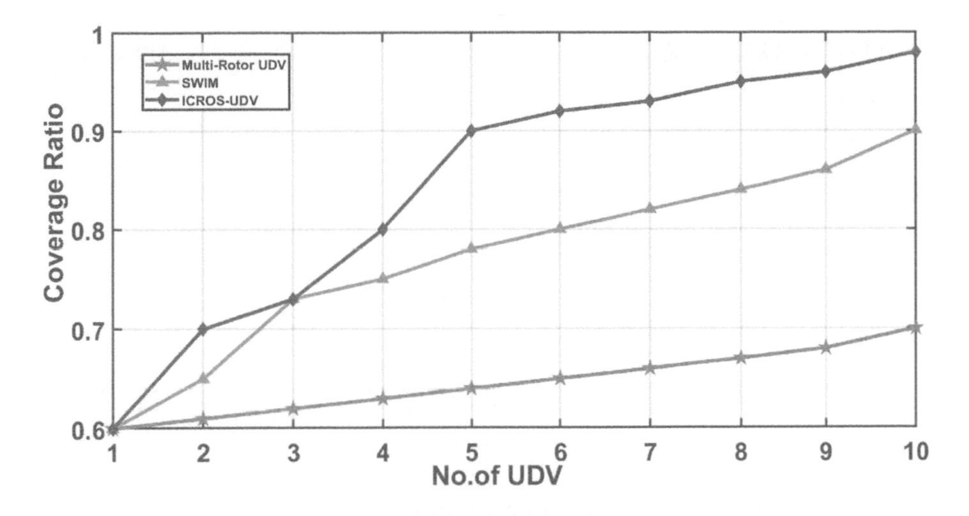

FIGURE 14.4 Coverage extent ratio vs UDV counts.

The ratio of UEs that are covered by the UDV to all UEs present during the emergency is known as the coverage ratio. To attain higher QoS value, an approach's coverage ratio needs to be high as Figure 14.5.

Due to the selection of the necessary number of UDVS and their positioning, the proposed strategy has a high coverage ratio. With the increasing order of position updates, the proposed method shows better enhancement in cell coverage as Figure 14.6.

The suggested mayfly optimization takes into account the climatic conditions estimated in the prevailing process, as well as the residue energy and the

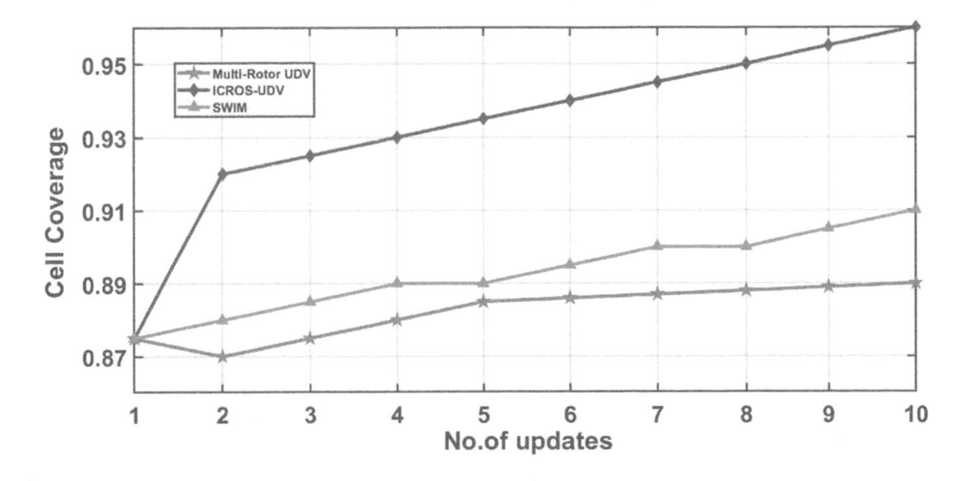

FIGURE 14.5 Cell extent coverage vs updates counts.

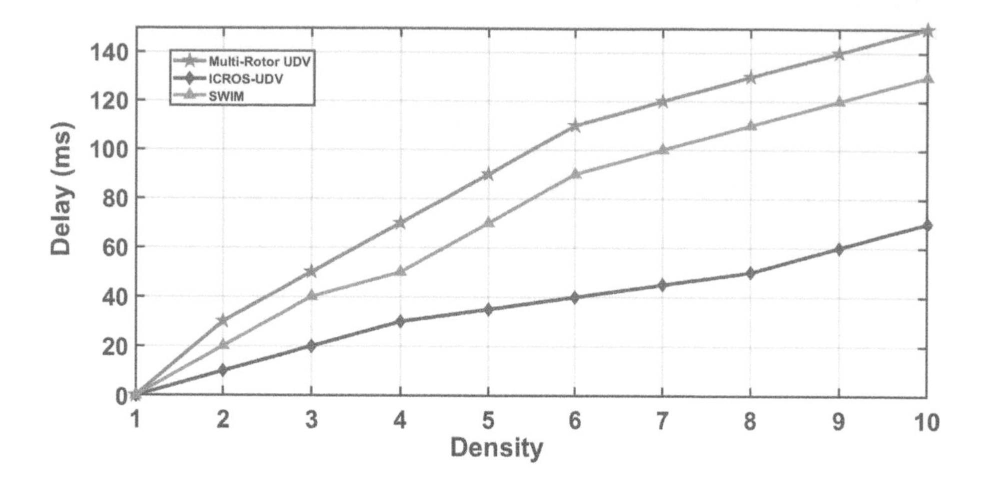

FIGURE 14.6 Coverage delay vs density of end user.

occurrence of clogging in environment, consequent to realistic UE coverage in for the cells.

The QoS is significantly impacted by the latency in collecting data from the UEs. The lag should be extremely minimal, especially in emergency cases as Figure 14.7.

While assessing an approach's effectiveness, it is important to consider its reliability as a quantitative metric. In order to establish a reliable data collection, a high route gain is regarded as a positive criterion as Figure 14.8.

The figure indicates that the UDV's increased speed has an impact on packet collecting. Because of the enhanced coverage ratio it has attained, the suggested solution has more packets gathered as Figure 14.9.

The dependability of the suggested IWPOP-UDV technique as well as the existing approaches is numerically analyzed in Table 14.2.

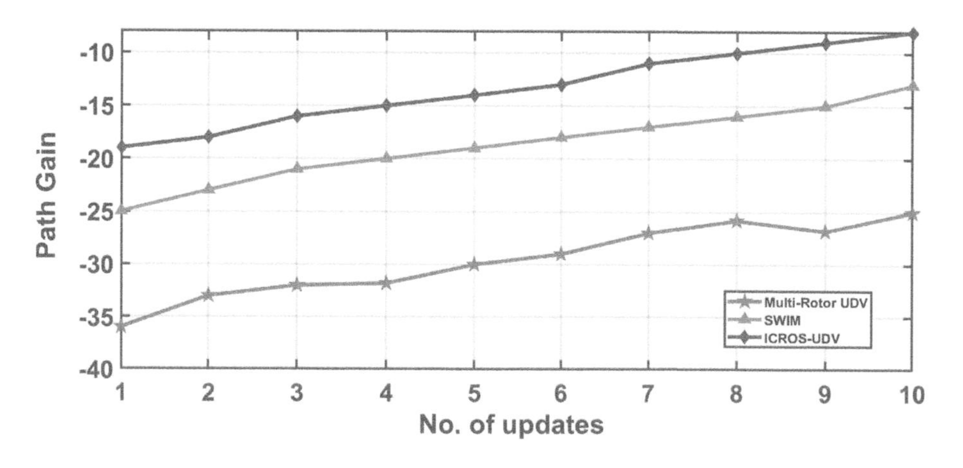

FIGURE 14.7 Path gain vs updates counts.

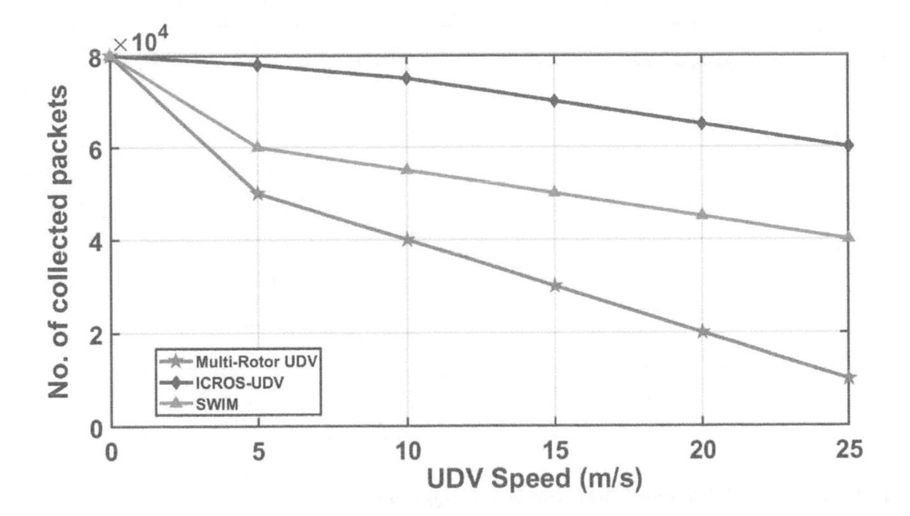

FIGURE 14.8 Updates counts vs UDV speed.

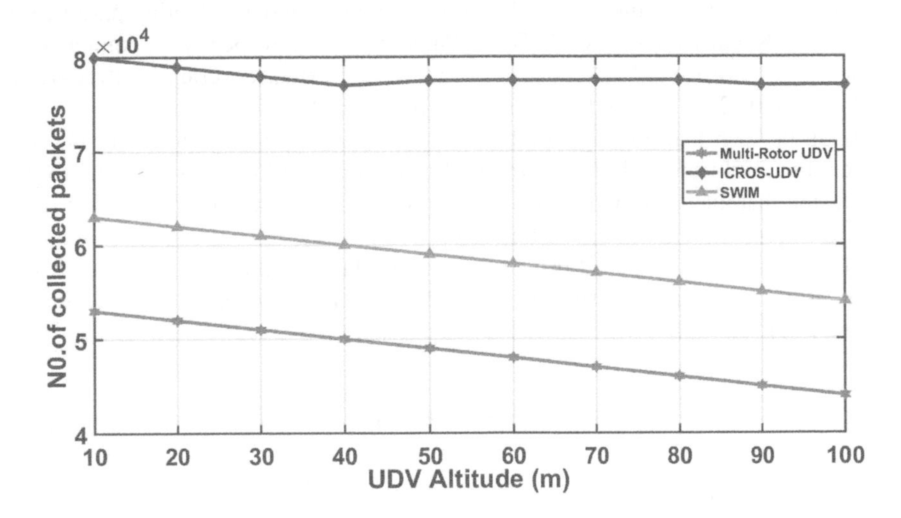

FIGURE 14.9 Updates counts vs UDV's altitude.

TABLE 14.2
QoS Analysis

	Coverage Extent Ratio	Cell Extent Coverage	Coverage Delay (*ms*)
Methods	**Updates Counts**	**UDV Counts**	**Density of End User**
Multirotor UDV	0.657 ± 0.45	0.792 ± 0.4	88 ± 0.5
SWIM	0.782 ± 0.34	0.983 ± 0.3	74.6 ± 0.4
ICROS-UDV	0.856 ± 0.31	0.894 ± 0.2	37.4 ± 0.2

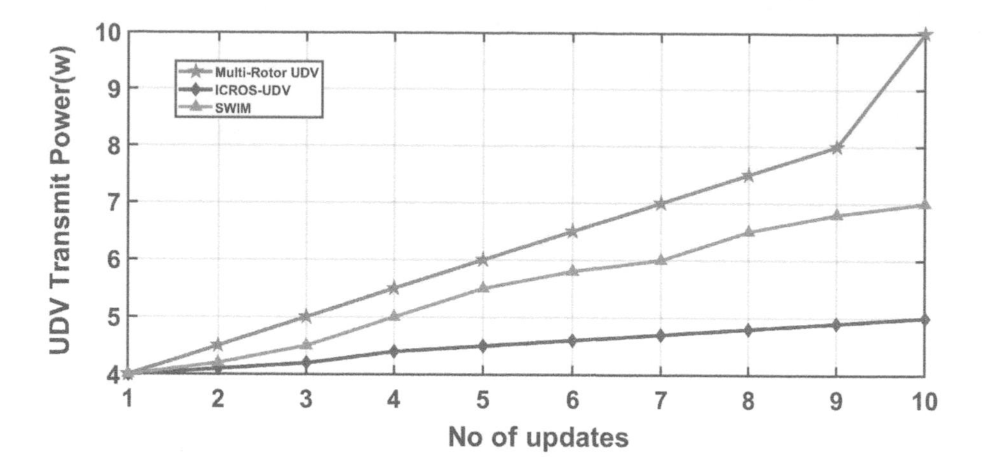

FIGURE 14.10 UDV power transmit vs updates counts.

Path gain and the quantity of packets collected are two dependability parameters that are analyzed, with a lag of 0.856, 0.894, and 37.4 *ms*, respectively, compared to the previous approaches' coverage extent ratio, cell extent coverage, with a lag of 0.65 to 0.78, 0.82 to 0.89, and 73.5 to 89 *ms*, respectively.

It can be seen that the suggested strategy achieves a route gain that is 14.6 dB higher than multirotor UDV and 5.5 dB higher than SWIM. This leads us to the conclusion that the suggested method is trustworthy for gathering information from UEs in emergent scenarios as Figure 14.10.

The suggested IMO-based trajectory course planning approach consumes less energy than other alternatives in terms of the number of updates as shown in Figure 14.11.

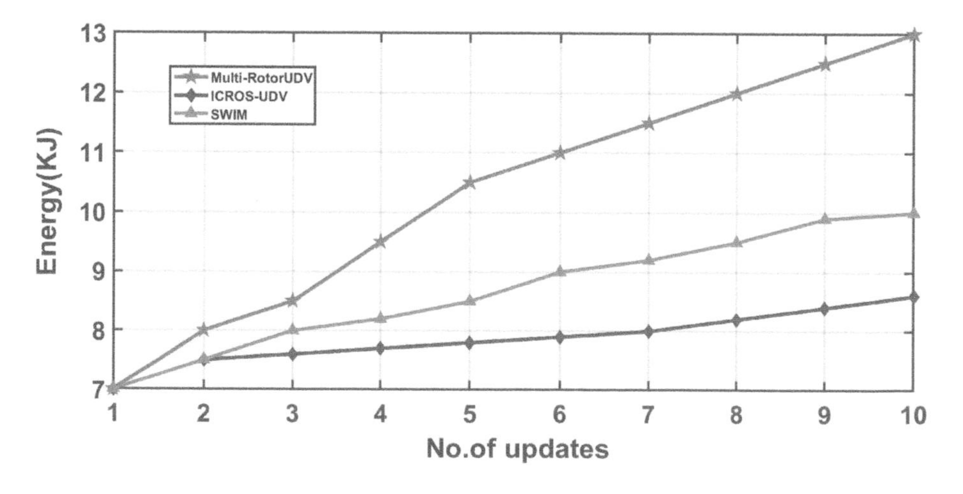

FIGURE 14.11 Energy vs updates counts.

TABLE 14.3
Analysis of Reliability and Energy Efficiency

	Path Gain	Collected Packets Counts		UDV Power Transmit	Energy
Methods	**Updates Counts**	**UDV Speed**	**Updates Counts**	**Updates Counts**	
Multirotor UDV	-28.9 ±0.5	39432.2± 0.49	48123.3 ± 0.4	5.9 ± 0.06	11.37 ± 0.05
SWIM	-19.9 ± 0.49	54000.5± 0.3	58300.4± 0.3	5.626 ± 0.04	7.98 ± 0.04
ICROS-UDV	-14.3 ± 0.09	72332.2± 0.2	79576.9± 0.2	3.96 ± 0.02	8.05 ± 0.02

A plan should be used to successfully manage the UDVs due to their restricted energy. Energy efficiency serves as a measure of the UDV's endurance throughout a task given. The two performance metrics used for the energy study are UDV transmit energy and power use.

From Table 14.3, one can conclude that proposed method is more efficient in terms of energy consumption and enhanced data collection with regard to the number of updates.

UDV various parameters were calculated based on the sample data as shown in Table 14.4.

TABLE 14.4
Sample Parameters

Sample Parameters	Value
UDV counts [a1]	10
Count of UEs	110
Speed of wind	5.5 m/s
Loss constant in path	2.9
UDV power transmit	5W
UEs power transmit	10mW
Data rate	310 kbps
Size of packet	120 Bytes
Width of path	13m
Density of air	1.325 Kg/ cubic meter
GP	-14.6
Cell extent coverage	0.89
QoS	1.02
Rate of learning, actor network, critic network	0.002

14.7 CONCLUSION

To address the problems in UDV communication, formulation was performed on the basis of climatic conditions-sensitive positioning as well as path planning of UDVs. In order to increase QoS, dependability, and energy efficiency, the fundamental objective of this method is to enhance UDV coverage while optimizing its energy consumption (Rana & Khang et al., 2021).

The forecasting of weather conditions is performed using A-SLTM based on obtained data to enable effective communication. The proposed method is assessed in QoS, energy analysis, and reliability. The suggested strategy will be developed in the future to incorporate energy harvesting to lengthen UDV's lifespan (Khanh & Khang, 2021).

REFERENCES

Azari, M.M., Rosas, F., Pollin, S. "Cellular connectivity for UAVs: network modeling, performance analysis, and design guidelines", *IEEE Transactions on Wireless Communications*, vol. 18, no. 7, pp. 3366–3381 (2019). https://ui.adsabs.harvard.edu/abs/2018arXiv180408121M/abstract

Azmat, M., Kummer, S. "Potential applications of unmanned ground and aerial vehicles to mitigate challenges of transport and logistics-related critical success factors in the humanitarian supply chain", *Asian Journal of Sustainability and Social Responsibility*, vol. 5, pp. 1–22 (2020). https://ajssr.springeropen.com/articles/10.1186/s41180-020-0033-7?w=745&h=430

Bhambri, P., Rani, S., Gupta, G., Khang, A. *Cloud and Fog Computing Platforms for Internet of Things* (2022). CRC Press. https://doi.org/10.1201/9781032101507

Cheng, F., Gui, G., Zhao, N., Chen, Y., Tang, J., Sari, H. "UAV-relaying assisted secure transmission with caching", *IEEE Transactions on Communications*, vol. 67, no. 5, pp. 3140–3153 (2019). https://ieeexplore.ieee.org/abstract/document/8626132/

Deng, S., Zhao, H., Fang, W., Yin, J., Dustdar, S., Zomaya, A.Y. "Edge intelligence: the confluence of edge computing and artificial intelligence", *IEEE Internet of Things Journal*, pp. 7457–7469 (2020). https://ieeexplore.ieee.org/abstract/document/9052677/

Ejaz, W., Ahmed, A., Mushtaq, A., Ibnkahla, M. Energy-efficient task scheduling and physiological assessment in disaster management using UAV-assisted networks, 2020, *Computer Communications*, https://www.sciencedirect.com/science/article/pii/S0140366419318638

El-Sayed, H., Chaqfa, M., Zeadally, S., Puthal, D. "A traffic-aware approach for enabling unmanned aerial vehicles (UAVs) in smart city scenarios", *IEEE Access*, vol. 7, pp. 86297–86305 (2019). https://ieeexplore.ieee.org/abstract/document/8735690/

Garcia-Rodriguez, A., Geraci, G., Lopez-Perezp, D., Giordano, L.G., Ding, M., Bjornson, E. "The essential guide to realizing 5Gconnected UAVs with massive MIMO", *IEEE Communications Magazine*, vol. 57, no. 12, pp. 84–90 (2019). https://ieeexplore.ieee.org/abstract/document/8869706/

Goh, C.Y., Leow, C.Y., Nordin, R. "Energy efficiency of unmanned aerial vehicle with reconfigurable intelligent surfaces: a comparative study", *Drones*, vol. 98 (2023). https://www.mdpi.com/2504-446X/7/2/98

Jebaraj, L., Khang, A., Chandrasekar, V., Pravin, A.R., Sriram, K. (Eds.). "Smart City Concepts, Models, Technologies and Applications", *Smart Cities: IoT Technologies, Big Data Solutions, Cloud Platforms, and Cybersecurity Techniques* (1st Ed.) (2024). CRC Press. https://doi.org/10.1201/9781003376064-1

Ji, X., Meng, X., Wang, A., Hua, Q., Wang, F., Chen, R., Zhang, J., Fang, D. E2PP: An energy-efficient path planning method for UAV-assisted data collection, 2020, *Security and Communication Networks* 8850505:1–8850505:14.

Khang, A., Gupta, S.K., Rani, S., Karras, D.A. *Smart Cities: IoT Technologies, Big Data Solutions, Cloud Platforms, and Cybersecurity Techniques* (2023). CRC Press. https://doi.org/10.1201/9781003376064

Khang, A., Hahanov, V., Abbas, G.L., Hajimahmud, V.A. "Cyber-Physical-Social System and İncident Management", *AI-Centric Smart City Ecosystems: Technologies, Design and Implementation* (1st Ed.) (2022). CRC Press. https://doi.org/10.1201/9781003252542-2

Khang, A., Rani, S., Gujrati, R., Uygun, H., Gupta, S.K. *Designing Workforce Management Systems for Industry 4.0: Data-Centric and AI-Enabled Approaches* (2023). CRC Press. https://doi.org/10.1201/99781003357070

Khang, A., Rani, S., Sivaraman, A.K. *AI-Centric Smart City Ecosystems: Technologies, Design and Implementation* (1st Ed.) (2022). CRC Press. https://doi.org/10.1201/9781003252542

Khanh, H.H., Khang, A. "The Role of Artificial Intelligence in Blockchain Applications", *Reinventing Manufacturing and Business Processes through Artificial Intelligence*, pp. 20–40 (2021). CRC Press. https://doi.org/10.1201/9781003145011-2

Kodheli, O., Lagunas, E., Maturo, N., Sharma, S.K., Shankar, B., Montoya, J.F., Duncan, J.C., Spano, D., Chatzinotas, S., Kisseleff, S., Querol, J. "Satellite communications in the new space era: a survey and future challenges", *IEEE Communications Surveys & Tutorials*, pp. 70–109 (2020).

Li, B., Fei, Z., Zhang, Y. "UAV communications for 5g and beyond: recent advances and future trends",, *IEEE Internet Things Journal*, vol. 6, pp. 2241–2263 (2018). https://ieeexplore.ieee.org/abstract/document/8579209/

Li, B., Li, Q., Zeng, Y., Rong, Y., Zhang, R. "3D trajectory optimization for energy-efficient UAV communication: a control design perspective", *IEEE Transactions on Wireless Communications*, vol. 21, no. 6, p. 4579 (2022). https://ieeexplore.ieee.org/abstract/document/9652043/

Liu, Y., Dai, H.-N., Wang, H., Imran, M., Wang, X., Shoaib, M. UAV-Enabled data acquisition scheme with directional wireless energy transfer for internet of things, 2020, *Computer Communications* https://www.sciencedirect.com/science/article/pii/S0140366419304852

Markets and Markets, Unmanned aerial vehicle (UAV) market, 2018, https://www.marketsandmarkets.com/Market-Reports/unmanned-aerialvehicles-uav-market-662.html.

McEnroe, P., Wang, S., Liyanage, M. "A survey on the convergence of edge computing and AI for UAVs: opportunities and challenges", *IEEE Internet of Things Journal* (2022). https://ieeexplore.ieee.org/abstract/document/9778241/

Musavian, L., Ni, Q. "Effective capacity maximization with statistical delay and effective energy efficiency requirements", *IEEE Transactions on Wireless Communications*, vol. 14, pp. 3824–3835 (2015). https://ieeexplore.ieee.org/abstract/document/7061966/

Paving the path to 5G: optimizing commercial LTE networks for drone communication." https://www.qualcomm.com/news/onq/2016/09/06/paving-path-5g-optimizing-commercial-lte-networks-drone-communication, accessed: 2019.

Priya, P., Kamlu, S. "Improved GA-PI technique for non-linear dynamic modelling of a UAV", in *International Conference on Connected Systems & Intelligence (CSI)*, pp. 1–6 (2022). Trivandrum, India. https://ieeexplore.ieee.org/abstract/document/9924088/

Rana, G., Khang, A., Sharma, R., Goel, A.K., Dubey, A.K. *Reinventing Manufacturing and Business Processes through Artificial Intelligence* (2021). CRC Press. https://doi.org/10.1201/9781003145011

Rani, S., Chauhan, M., Kataria, A., Khang, A. "IoT Equipped Intelligent Distributed Framework for Smart Healthcare Systems", *Networking and Internet Architecture* (2021). CRC Press. https://doi.org/10.48550/arXiv.2110.04997

Sun, X., Shen, C., Ng, D.W.K., Zhong, Z. "Robust Trajectory and Resource Allocation Design for Secure UAV-Aided Communications", in *IEEE International Conference on Communications Workshops, Workshops (ICC Workshops)*, pp. 1–6 (2019). https://ieeexplore.ieee.org/abstract/document/8756815/

Xiao, L., Xu, Y., Yang, D., Zeng, Y. "Secrecy energy efficiency maximization for UAV-enabled mobile relaying", *IEEE Transactions on Green Communications and Networking*, pp. 1–1 (2019). https://ieeexplore.ieee.org/abstract/document/8884126/

Yu, X., Li, C., Zhou, J. "A constrained differential evolution algorithm to solve UAV path planning in disaster scenarios", *Knowledge-Based Systems*, vol. 204, p. 106209 (2020). https://www.sciencedirect.com/science/article/pii/S0950705120304263

Yu, X., Yang, J., Li, S. Finite-time path following control for small scale fixed-wing UAVs under wind disturbances, 2020, Journal of The Franklin Institute. https://www.sciencedirect.com/science/article/pii/S001600322030435X

Zeng, Y., Xu, J., Zhang, R. "Energy minimization for wireless communication with rotary-wing UAV", *IEEE Transactions on Wireless Communications*, vol. 18, pp. 2329–2345 (2019). https://ieeexplore.ieee.org/abstract/document/8663615/

Zhan, C., Zeng, Y. "Completion time minimization for multi-UAVenabled data collection", *IEEE Transactions on Wireless Communications*, vol. 18, no. 10, pp. 4859–4872 (2019). https://ieeexplore.ieee.org/abstract/document/8779596/

Zhan, C., Zeng, Y. "Aerial–ground cost tradeoff for multi-UAVenabled data collection in wireless sensor networks", *IEEE Transactions on Communications*, vol. 68, no. 3, pp. 1937–1950 (2020). https://ieeexplore.ieee.org/abstract/document/8943326/

Zhao, N., Lu, W., Sheng, M. et al. "UAV-assisted emergency networks in disasters", *IEEE Wireless Communications*, vol. 26, no. 1, pp. 45–51 (2019). https://ieeexplore.ieee.org/abstract/document/8641424/

15 Role of Vehicular Ad Hoc Network-Aided Traffic Management in Smart City Development

Joseph Wheeder and Dillip Rout

15.1 INTRODUCTION

A smart city is continually being defined and has changed throughout time. However, some important turning points in the development of smart cities include the following: Early 1990s: The National Research Council of Canada initially used the phrase "smart city" to refer to a city that employs information and communication technology (ICT) to raise the standard of living for its inhabitants (Khang & Rani et al., 2022).

Late 1990s and early 2000s: As under its Intelligent Energy Europe initiative, which sought to leverage ICT to improve energy efficiency and lower greenhouse gas emissions, the European Union began to make investments in the creation of smart cities (Rani & Bhambri et al., 2022).

Around this time in the 2000s, cities all over the world started to use the phrase "smart city" to characterize their efforts to employ technology and data analysis to enhance both local services and the quality of life for residents (Khang & Hahanov et al., 2022).

The usage of smart city technology, such as sensors and data analytics, started to grow beyond merely energy efficiency and environmental concerns in the late 2000s and early 2010s. These technologies are now being employed for transportation, public safety, and other municipal operations (Hahanov & Khang et al., 2022).

Early 2010s and even beyond: The concept of a "smart city" has evolved further, with an emphasis on utilizing data and technology to build communities that are more comfortable, economic, and equitable. The COVID-19 pandemic has also brought attention to the significance of combining data and technology to control human safety and support remote learning and employment.

A smart city is an urban region that makes use of numerous electronic data gathering sensors to provide data that is utilized to effectively manage resources and assets.

To monitor and manage traffic and transportation systems, power plants, utilities, water supply networks, waste management, crime detection, information systems, schools, libraries, hospitals, and other community services, data that is gathered from persons, gadgets, and investments is analyzed. By increasing the effectiveness of municipal operations and services and empowering residents to make better decisions, a smart city aims to enhance the quality of life for its inhabitants.

DOI: 10.1201/9781003376064-15

An important piece of technology used in smart cities is a vehicular ad hoc network (VANET). VANET is a wireless network that allows vehicle-to-vehicle (V2V) and infrastructure-to-vehicle communications. This communication can be utilized for a number of things, including enhancing the efficiency of transportation systems, decreasing traffic congestion, and promoting road safety. One of the key elements of a smart city is traffic control, which is made possible via VANETs.

VANET operates by putting the ideas of mobile ad hoc networks (MANETs), a decentralized kind of wireless network, into practice. Because it doesn't rely on a pre-existing infrastructure, like routers in wired networks or access points (APs) in wireless networks, this network is ad hoc. Instead, each node takes part in routing by sending data from one node to another; as a result, the choice of which node will forward data is determined dynamically based on the connectivity of the network and the chosen routing algorithm.

VANETs can be used in a smart city to collect real-time information regarding traffic patterns, road conditions, and the positioning and movement of vehicles. This information may be examined and utilized to route cars more effectively, enhance traffic flow, and raise the general effectiveness of the transportation system. Drivers can get real-time information from VANETs regarding traffic patterns, collisions, and other occurrences that could influence their path.

VANETs can be utilized to enhance public safety in addition to enhancing transportation. For instance, VANETs may be used to coordinate disaster response efforts as well as to warn vehicles of possible risks like accidents or construction. Therefore, the usefulness and quality of life for residents might be greatly enhanced by the integration of VANETs into smart cities. Smart cities are able to allocate resources wisely and provide for the requirements of its residents by gathering and examining real-time data.

In this chapter, usability of VANET in smart city development is presented. Majorly, the means of traffic management and smart mobility are focused. So, a comprehensive review is articulated considering the research in this field for the last two decades, large-scale practical implementation still requires some time.

The remainder of this chapter is organized as follows. The next section describes the overview of smart city and its components. Then, the functioning of VANET is discussed in Section 15.3. In Section 15.4, the roles of VANET are presented, which are helpful in developing a smart city.

In the next section, particularly smart mobility is widely discussed with the help of VANET. Lastly, the concluding remarks are given in Section 15.6 consisting of the current trend and future directions.

15.2 SMART CITY

A smart city is an urban region that makes use of cutting-edge data and technology analysis to promote sustainability, improve livability, and simplify urban services (SCC-WSC, 2023).

Smart cities collect information on anything from traffic patterns to energy use using Internet of Things (IoT) sensors, big data analytics, and other modern devices

(Rani & Chauhan et al., 2021). Then, using this data, city processes are optimized, and data-driven choices are made that can boost effectiveness, decrease waste, and improve citizen happiness.

Smart cities strive to enhance their people's quality of life while simultaneously enhancing the city's infrastructure, transportation, energy, and public services. This may entail taking steps to lower crime, increase access to healthcare and learning, and encourage environmentally friendly behaviors (EC-SC, 2018) as Figure 15.1.

In general, smart cities provide a perspective on the future of urban planning that aims to use data and technology to build more comfortable, effective, and sustainable communities (Bhambri & Rani et al., 2022).

15.2.1 Smart City Definition

A smart city is characterized as a metropolis that employs cutting-edge technology and data processing to boost sustainability, improve livability, and simplify urban services (SCC-WSC, 2023).

Smart cities use big data analytics, IoT sensors, and other digital technologies to gather and analyze data on a variety of municipal functions, including energy use, public services, and traffic patterns (Khang & Gupta et al., 2023).

The knowledge gathered from this data is utilized to make data-driven choices, optimize city operations, and enhance resident life satisfaction (Smart Cities Council, what-smart-city).

15.2.2 Smart City Categories

The components of a smart city include smart governance, smart mobility, smart energy, smart building, smart environment, smart healthcare, and smart public participation (Jebaraj & Khang et al., 2024).

The details of each one are described in the following paragraphs. Based on the numerous components that make up the smart city idea, different kinds of smart cities can be identified. Among which the most popular categories are depicted in the following sections.

15.2.2.1 Smart Governance

This category is concerned with using data and technology to enhance government decision-making and advance transparency and accountability.

In order to enhance the efficacy and efficiency of urban administration and decision-making processes, smart governance refers to the application of cutting-edge technology and data analytics. To improve transparency, accountability, and public engagement in governance processes, it entails the integration of digital platforms and intelligent technologies.

Open data portals, public interaction platforms, and real-time monitoring systems are examples of smart governance technologies that enable cities to collect and analyze data to influence policy choices and improve service delivery (Rani & Bhambri et al., 2023).

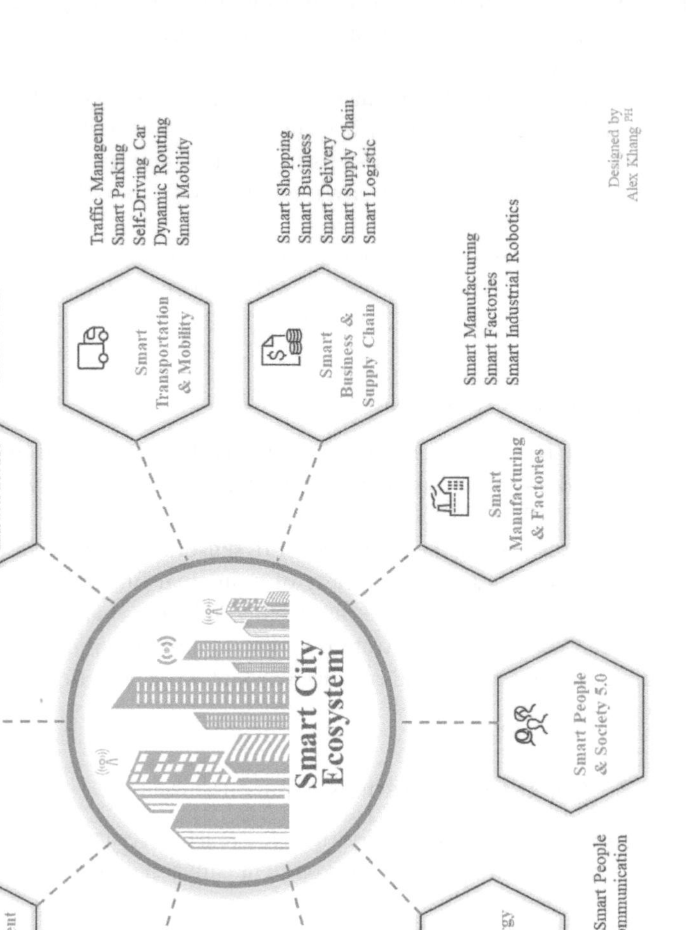

FIGURE 15.1 Showing the overview of a smart city (Khang, 2021).

Smart governance attempts to foster more inclusive, responsive, and effective urban government that matches the changing demands of residents and encourages sustainable urban growth by utilizing technology and data (EC-SC, 2018).

15.2.2.2 Smart Mobility

This area focuses on the application of technology to enhance transportation infrastructure, lessen gridlock, and encourage sustainable mobility. The term "smart mobility" describes the use of cutting-edge technology in urban transportation networks for the purpose of enhancing accessibility and maximizing mobility for inhabitants.

In order to lessen traffic congestion, enhance public transportation, and promote sustainable modes of transportation like cycling and walking, it makes use of intelligent traffic management systems, real-time transit information, and shared mobility services (Khang & Chowdhury et al., 2022).

The goal of smart mobility solutions is to increase the mobility alternatives available to individuals, shorten travel times, and lessen the negative environmental effects of transportation, such as greenhouse gas emissions and air pollution.

Smart mobility seeks to advance sustainable urban development and raise the standard of living for inhabitants by utilizing technology to better transportation networks (SCC-SMB, 2023).

15.2.2.3 Smart Energy

This is a subcategory that focuses on using technology to eliminate waste, maximize energy consumption, and encourage the use of renewable energy sources using cutting-edge technology like smart grids, power storage, and energy-efficient structures; smart energy is a component of a smart city that strives to improve energy efficiency and encourage the use of renewable energy sources.

Reducing energy use, reducing greenhouse gas emissions, and boosting energy security are the three main objectives of smart energy in a smart city. By combining data from multiple sources, such as weather forecasts, energy demand, and supply trends, smart energy solutions enable cities to optimize energy consumption and make choices that increase energy efficiency in real-time.

Smart energy solutions can, for example, be implemented in the form of sensors that are installed in houses and buildings and that regulate heating and cooling depending on occupancy patterns as well as energy storage systems that enable the use of renewable sources such as solar and wind power. Smart cities may lower their carbon footprint and promote sustainable urban growth by using smart energy solutions (SCC-SEN, 2023).

15.2.2.4 Smart Buildings

This category focuses on the application of technology to improve the planning, constructing, and maintenance of structures in order to increase their sustainability and energy efficiency.

A smart building is a part of a smart city that employs cutting-edge technology like sensors, automation, and data analytics to enhance the sustainability and energy efficiency of buildings. In a smart city, energy consumption reduction, indoor air quality improvement, and occupant comfort and safety enhancement are the main objectives of smart buildings.

By combining data from numerous sources, such as occupancy trends, weather predictions, and energy consumption, smart building solutions enable communities to optimize building operations while also making decisions that increase energy efficiency in real-time.

For instance, installing sensors that track interior air quality and modify ventilation systems as necessary as well as installing automation systems that regulate temperature and lighting depending on occupancy patterns are examples of smart building solutions.

Smart cities may lower their energy consumption and environmental impact while also enhancing the quality of life for building occupants by deploying smart building technologies (SCC-SBD, 2023).

15.2.2.5 Smart Environment

The goal of this category is to employ technology to protect and enhance the environment's long-term sustainability.

The goal of a smart environment is to enhance the quality of the urban environment by incorporating cutting-edge technology like sensors, data analytics, and real-time monitoring systems.

In a smart city, the aim of a smart environment is to deal with problems like air pollution, water management, waste management, and urban greenery.

Cities can monitor and control environmental conditions in real-time with the help of smart environment solutions that gather and analyze data from a variety of sources, including weather sensors, air quality monitors, and waste management systems.

One example of a smart environment solution is the installation of smart trash cans that send alerts to the appropriate authorities when they are full.

Another example is the deployment of green infrastructure initiatives that increase urban flora and lessen the impact of the urban heat island. Smart cities may improve the quality of life for residents by providing a clean and healthy urban environment and promoting sustainable urban growth via the use of smart environment solutions (SCC-SEN, 2023).

15.2.2.6 Smart Healthcare

This subcategory is centered on the application of technology to raise both the accessibility and caliber of healthcare services. By using cutting-edge technology like telemedicine, wearables, and electronic health records, smart healthcare works to increase access to healthcare services and enhance public health in smart cities.

Enhancing healthcare service quality while lowering healthcare costs and enhancing citizen health outcomes are the objectives of smart healthcare in a smart city.

By utilizing data analytics to track and manage health concerns in real-time, smart healthcare systems allow communities to offer individualized and effective healthcare services. For instance, telemedicine services that offer remote consultation and diagnosis as well as the usage of wearables that monitor vital signs and give early identification of health concerns are examples of smart healthcare solutions.

Smart cities may improve citizen quality of life, public health and wellness, and healthcare expenses by deploying smart healthcare solutions. Moreover, smart healthcare systems may promote equitable healthcare services throughout the city and increase healthcare access for underprivileged people (SCC-SHC, 2023).

15.2.2.7 Smart Public Participation

It is the use of technology to increase citizen involvement and engagement in decision-making processes. The goal of smart public involvement is to increase citizen engagement and participation in municipal government processes by utilizing cutting-edge technology like digital and social media platforms.

By giving residents the chance to engage in decision-making processes and offer input on municipal services, smart public participation in a smart city aims to improve openness, accountability, and inclusion in urban government.

By offering digital platforms for residents to access information, offer input, and participate in urban planning and policy-making processes, smart public participation solutions help cities increase civic engagement.

The adoption of citizen engagement platforms that allow people to report problems and offer input on municipal services is one example of a smart public participation solution.

Another is the use of social media platforms to gauge public opinion on policies and programs for cities. Smart cities may support more inclusive, responsive, and effective urban government that meets the changing requirements of residents and encourages sustainable urban growth by putting in place smart public involvement solutions (SCC-SCE, 2023).

Depending on the requirements and objectives of the city, more components may be added to a smart city. These subcategories are not all-inclusive as shown in Figure 15.2.

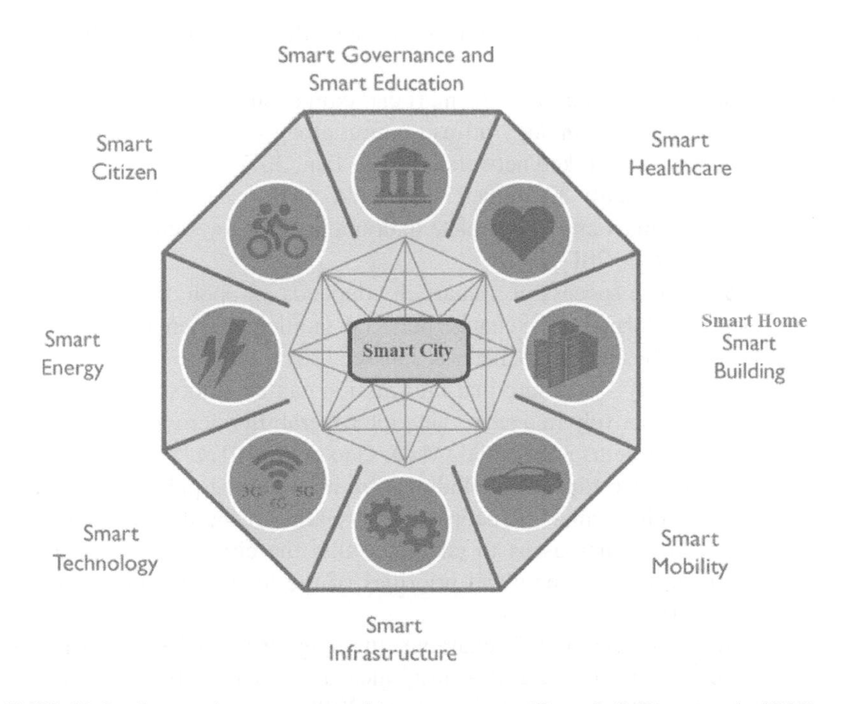

FIGURE 15.2 Smart city concepts and its components (Frost & Sullivan et al., 2020).

15.2.3 Smart City Importance

Smart cities offer a range of benefits and advantages, including:

- **Better quality of life**: Smart cities leverage technology and data to increase inhabitants' overall quality of life by improving the delivery of key services like healthcare, education, and transportation.
- **Enhanced efficiency and cost savings**: Smart cities may cut waste and save money by simplifying city processes and maximizing the use of resources (SCC-BSC, 2023).
- **Enhanced sustainability**: To encourage sustainability, lower carbon emissions, and safeguard the environment, smart cities use technology and data (SCC-SEN, 2023).
- **Greater participation of citizens** in decision-making processes thanks to the use of technology in smart cities, which also fosters a feeling of community and enhances governance.
- **Enhanced security and safety**. Smart cities leverage technology and data to enhance protection and reliability (Khang & Hahanov et al., 2022), for example, through quicker emergency response times and camera systems (SCC-SPS, 2023).
- **Enhanced economic growth**: Smart cities may attract new enterprises, entrepreneurs, and talented people, enhancing economic growth (SCC-EDV, 2023), by enhancing quality of life and fostering sustainability.

To be deemed "smart," a city generally needs the following elements as Figure 15.3:

- **Technology infrastructure**: A smart city cannot succeed without a strong and cutting-edge technology infrastructure, along with the access to high-speed internet and linked networks (Lai & Tsai, 2017).
- **Data management**: To gather, store, and analyze data from multiple sources in a smart city, a complete data management system is necessary (Lu & Ratti et al., 2017).
- **Open data**: The success of a smart city depends on open data regulations that permit information exchange between municipal agencies, residents, and for-profit businesses (Scherer & Janssen, 2015).
- **Interoperability**: To ensure that information can be smoothly transferred and used productively in a smart city, interoperability across various technologies and systems is essential (Ma & Gao et al., 2017).
- **Effective partnerships and collaboration** between the public and private sectors, as well as amongst people, are essential to the development of a smart city since they assist to guarantee that the city's objectives are in line with the requirements and priorities of the local population (Lee & Klysubun, 2015).
- **Resilience and security**: To guarantee the long-term viability and success of a smart city, resilience and security measures, such as disaster recovery and cybersecurity, are essential (O'Brien, 2018).

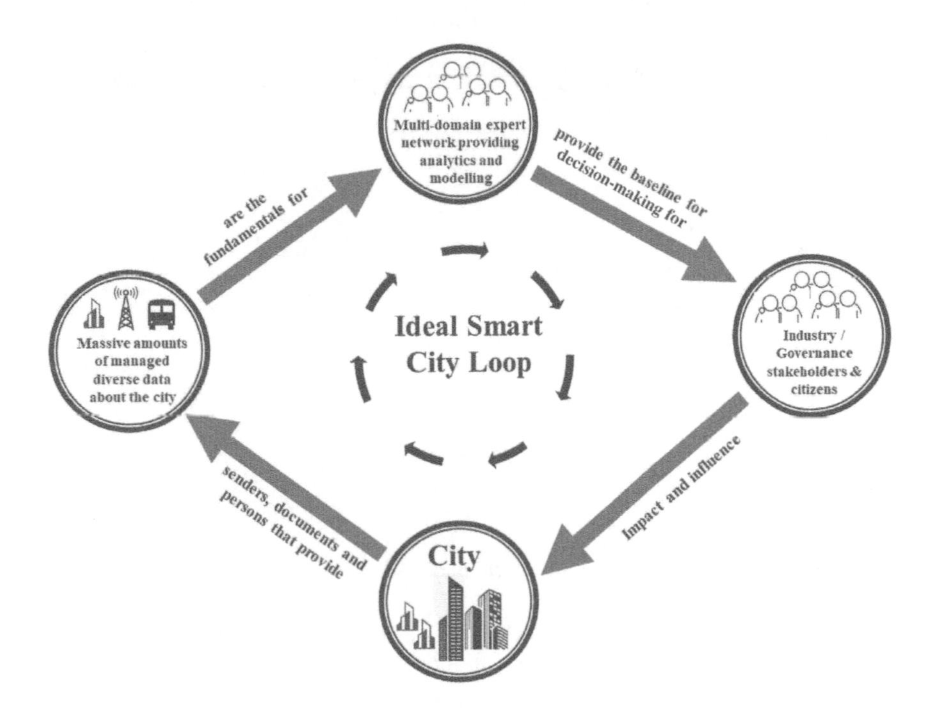

FIGURE 15.3 The ideal smart city loop. A conceptual model for a reactive system that addresses the challenges of today's smart city applications (Khang, 2021).

15.2.4 Smart City Architecture

The technical framework and mechanisms that enable a city to enhance its functionality, quality of life, and sustainability via the use of ICT are referred to as smart city architecture (ICTs) as in Figure 15.4.

The core elements of a smart city architecture typically include the following:

- **Sensors and networks:** To collect real-time data on numerous elements of city operations, such as flow of traffic, energy usage, trash management, and air quality, sensors are installed all over a city. These sensors are linked to networks that provide the data for analysis and processing to a central location. The sensors can be connected to existing communications infrastructure, like the internet, or they can use specialized networks, such as Zigbee or LoRaWAN, which are designed specifically for the transmission of IoT data.
- **Data management:** A centralized site must be used to store, handle, and analyze the vast volumes of data produced by the sensors. A strong data management system that can handle the volume, velocity, and diversity of data is necessary for this. In addition to databases, data warehouses, and big data platforms like Hadoop for storing and processing data, the information management system may also incorporate analytics tools like algorithms for machine learning for data analysis and insight generation.

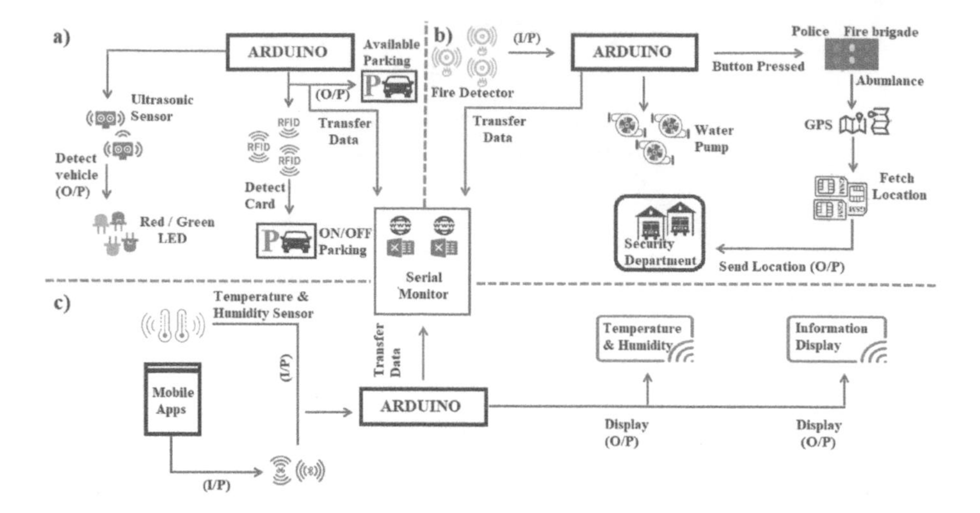

FIGURE 15.4 The smart city's architectural plan.

- **Applications and services:** On top of the data management system, applications and services are developed to give residents and city authority's simple-to-use tools for accessing and utilizing the data. Web portals, smartphone applications, and Chatbots are some examples of these that offer up-to-date data and insights on many elements of city operations. A citizen may use a smartphone app, for instance, to monitor real-time traffic data, locate the closest recycling facility, or report a damaged street light.
- **Interoperability:** The capacity of various elements of the smart city design to easily and effectively communicate and exchange data with one another is known as interoperability. This is essential to make sure that the different apps and services can get the data they require and that the data can be used consistently and cohesively. Common data formats, APIs, and protocols that enable data interchange and interoperability between various systems can be used to accomplish interoperability.
- **Security:** To guarantee the privacy and confidentiality of the data as well as to prevent unwanted access to the data and systems, the security of the data generated by the sensors and networks is essential. This necessitates the use of access control technologies, such as role-based access control, to manage who really has access to the data and what they can do with it, as well as encryption and authentication technologies, such as SSL/TLS, to safeguard the data as it is transported over the networks.
- **User Engagement:** Participating citizens in the creation and use of smart city solutions is one way to ensure that they can take advantage of the technology. This might entail developing user-friendly software and services, giving citizens a chance to offer comments and recommendations, and carrying out outreach and education initiatives to assist people comprehend technology and how it can be applied to better their lives.

In general, a smart city design is a sophisticated, integrated system that uses technology to improve urban livability, productivity, and sustainability. Cities may enhance quality of life, cut down on waste and inefficiencies, and better understand and respond to the needs of their residents by gathering and analyzing data in real-time (Kaya & Yilmaz et al., 2019).

15.2.5 Connectivity Infrastructure

The IoT is one of the key technologies for implementing a smart city, requiring the deployment of a variety of urban areas with an ever-increasing number of heterogeneous connected devices.

This is illustrated in Figure 15.5, which also provides an overview of the estimated number of connected IoT devices (also known as Smart Objects) in the period from 2015 to 2025, according to IoT Analytics.

A smart city will need connectivity anytime (day or night), anywhere (inside, outside, or while moving), and between highly heterogeneous entities (such as PCs, smartphones, tablets, and battery-powered IoT devices) characterized by different communication needs.

Based on the increasing trend shown in Figure 15.5, it is obvious that the communication infrastructure is the first important factor to deal with in future smart cities, and interactions between Machine-to-Machine (M2M), Human-to-Human (H2H), Human-to-Thing, and Thing-to-Thing.

Wireless networks are typically used to ensure connectivity among IoT devices, making information exchange possible with flexible and affordable installations. For instance, when end-to-end communications are not possible due to power restrictions or impediments, mesh networks and short-range communication technologies are preferred.

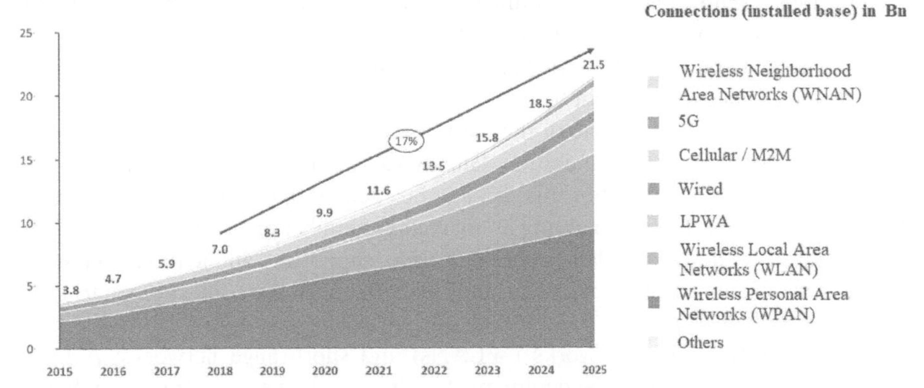

FIGURE 15.5 Trend on the global number of connected IoT devices in the period 2015/2025.

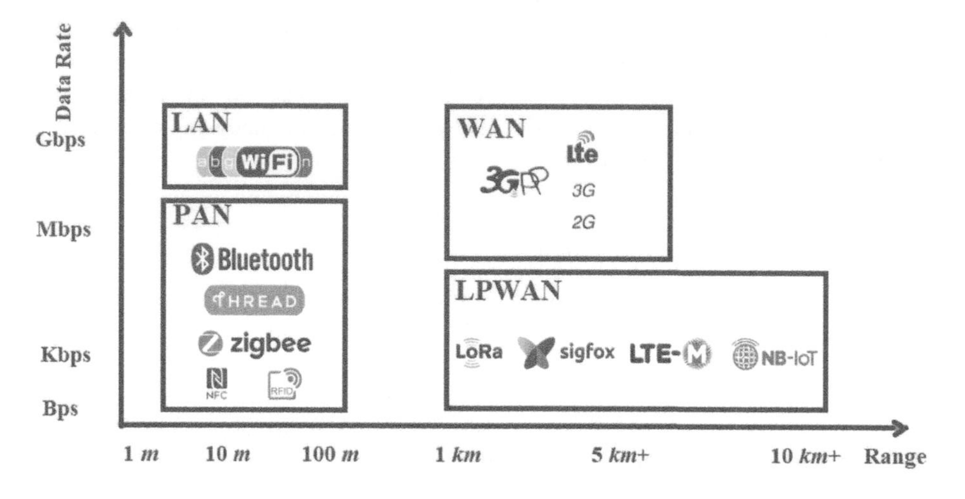

FIGURE 15.6 Network communication technologies.

In this scenario, data may be sent from IoT nodes (data sources) to data consumers (e.g., servers and border routers) and reversed via hop-by-hop connections between devices. In situations when there are direct and reliable communication links available between a central "hub" and all IoT nodes within its coverage, star networks and long-range communication technologies are preferred.

The following is a summary of the most important communication technologies that may be applied in the context of smart cities as in Figure 15.6.

Low-Power Wide-Area Networks (LPWANs): These networks are specifically made to link battery-operated devices across large distances at low bit rates. LPWANs are being used in a variety of applications where just a small amount of data has to be transferred because of their low cost, extensive coverage, and simple setup.

The most significant LPWANs are LTE-M (Gozalvez, 2016), Narrowband IoT (NB-IoT) (Feltrin & Tsoukaneri et al., 2019), and LoRa (Magrin & Centenaro et al., 2017), which operate on both unlicensed and licensed frequency bands and incorporate various standards (that can be open or proprietary).

IoT is quickly becoming one of the main use cases for 5G due to its cutting-edge capabilities, such as its extremely low latency (less than 1 ms) and extremely high bandwidth (10 Gb/s). In light of this, 5G can be seen as an enabling communication technology for smart cities, enabling an increasing number of IoT devices to connect to the internet regardless of their location or time, supporting applications like smart traffic systems, public safety, security, and surveillance in the context of smart cities (Rao & Prasad, 2018).

Wireless Local Area Networks (WLANs) and short-range networks: A lot of smart city use cases call for the deployment of "regional" (i.e., covering a small geographical area) and, in some circumstances, "individual" networks (such as Personal Area Networks, PANs).

The communication methods that are accessible in this context are incredibly diverse, ranging from Bluetooth and Bluetooth Low Energy (BLE) through protocols based on IEEE 802.15 (Yaqoob & Hashem et al., 2017).

15.2.6 Smart City Framework

Intelligent traffic lights (ITLs) are placed in a few of the intersections in the smart city framework that we have created. These ITLs gather real-time traffic information from passing cars and compute statistics on traffic, such as the amount of traffic in the streets nearby (between consecutive crossroads). These ITLs may notify passing cars on traffic conditions and warn them in the event of accidents at the same time.

The ITLs in this sub-network may exchange the data they have gathered and provide statistics for the entire city. Vehicles are therefore well informed about the city's traffic condition. The design of this smart city framework and the purposes for which the ITL will be used are covered in the following sections.

The blocks in the smart city idea are square in shape and include structures on all four sides. ITLs are in charge of controlling the flow of the vehicles that make up a VANET. It is not necessary to place these ITLs at every junction. Only a small portion of the city's traditional traffic lights will be switched out for ITLs. This is so because each ITL includes all four streets and an entire junction as in Figure 15.7.

As a result, any ITL within a covered range receives data from all passing cars (the four streets and the intersection). By putting this paradigm into practice, it is more cost-effective to not have an ITL at every junction.

Vehicles are supposed to have a global positioning system (GPS) device, a driving assistance device, and complete city map information, including the location of the ITLs. Vehicles may choose the closest ITL with ease as a result. Ad hoc On-Demand Distance Vector (AODV) routing protocol was configured for each ad hoc node (i.e., ITLs and automobiles) included in the scenario. AODV was chosen because of its ease of use.

Although it is commonly recognized that AODV is not ideal for use as a general-purpose routing protocol in VANETs, there are specific situations where AODV may perform admirably. The ease and extensive use of AODV are advantages.

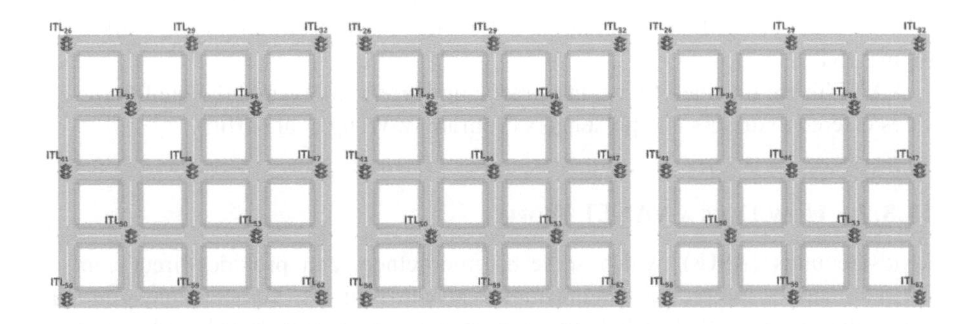

FIGURE 15.7 Intelligent traffic lights distribution.

The fundamental disadvantage of AODV is that end-to-end routes are required for data forwarding, which is challenging to manage because with VANETs end-to-end paths endure very briefly owing to the fast vehicle speeds.

15.2.7 Smart City Issues

While smart cities offer many benefits and opportunities, they also face various issues, limitations, and problems, including:

- Privacy and security: As personal data is susceptible to hacking and exploitation, the collecting and use of data in smart cities raises issues about privacy and security (Wang & Chen, 2016).
- Equity and social justice: There are worries that smart city programs may exacerbate existing disparities rather than try to alleviate them. This is because not everyone may have equal access to technology and services.
- Technical complexity: Implementing smart city technology can be difficult and technically demanding, necessitating hefty investments in staff, infrastructure, and training (Jain, 2017).
- Resistance to change: Implementation of new technology and practices in a smart city may encounter resistance from stakeholders, including residents, city agencies, and other stakeholders (Mehmood & Venkatesh, 2015).
- Interoperability and integration: It can be challenging to ensure that various systems and technologies are interoperable and can share data in a smooth manner, which can result in data silos and inefficiencies (Ma & Gao, 2017).
- Sustainability: There are questions regarding the long-term viability of these solutions due to the potential energy consumption and environmental effect of smart city technology (Scherer & Janssen, 2015).

15.3 VEHICULAR AD HOC NETWORKS (VANETs)

15.3.1 Overview

A MANET of vehicles that are outfitted with wireless communication devices is known as a VANET. These vehicles can speak with one another directly or via infrastructure along the road, creating a dynamic and self-organizing network as in Figure 15.8.

VANETs are intended to increase traffic flow, road safety, and the quality of services offered to drivers and passengers (Sohrabi & Wong et al., 2016).

15.3.2 How Does a VANET Work?

Roadside units (RSUs), which serve as middlemen, can provide direct contact between cars or indirect communication between vehicles. Vehicles may communicate via messages that include position, speed, direction of travel, and any other pertinent information like traffic conditions, road dangers, and road conditions.

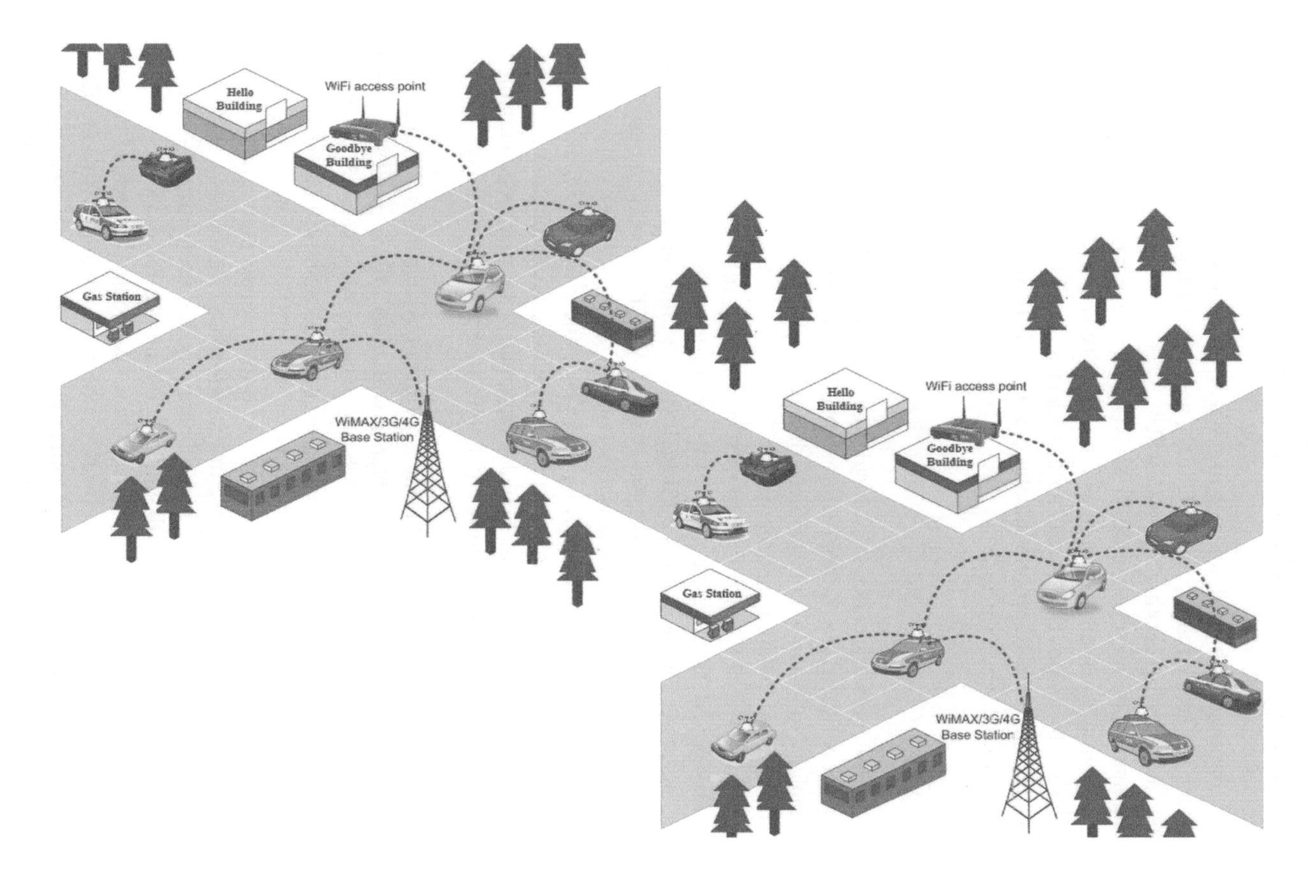

FIGURE 15.8 Illustration of working of a VANET.

Based on the existence or lack of RSUs, VANETs can either be infrastructure-based or infrastructure-less. While in infrastructure-less VANETs, cars connect with one another directly utilizing a peer-to-peer communication method, RSUs are employed in infrastructure-based VANETs to create a more dependable and stable communication network.

Wireless communication technologies like Wi-Fi, cellular networks (3G/4G/5G), dedicated short-range communications (DSRC), and others make it possible for cars to communicate with one another and with RSUs.

The communication protocols used in VANETs can vary based on the application and the communication requirements, but they typically adhere to the standards established by groups like the Institute of Electrical and Electronics Engineers (IEEE) and the European Telecommunications Standards Institute (ETSI) (Munir & Gani et al., 2018).

15.3.3 VANET Components

A sort of wireless network called a "vehicular ad hoc network" (VANET) permits communication between cars, infrastructure, and other road users. VANETs employ DSRC technology to instantly communicate data with other cars and infrastructure, such as vehicle speed, position, and status.

The major objectives of VANET are to increase driver comfort, traffic efficiency, and road safety. VANETs are made up of a variety of parts that cooperate to accomplish these objectives. Onboard units (OBUs), RSUs, communication protocols, and applications are some of the essential elements of a VANET. We will go deeper into these elements in this post and learn how they contribute to a VANET's functioning.

Vehicles with wireless communication devices: Since vehicles make up the majority of nodes in a VANET, they must have wireless communication devices in order to function as network nodes. Vehicles equipped with wireless communication devices are essential for facilitating communication between other vehicles and infrastructure in a VANET.

OBUs on these vehicles use DSRC technology to communicate in real-time with RSUs and other vehicles. The OBUs can interact with other OBUs and RSUs nearby thanks to their usual connection to the CAN bus (Controller Area Network) of the car.

Vehicle speed, position, direction, acceleration, and road and traffic conditions are just a few of the details that may be sent between cars and infrastructure. Applications like collision warning, junction management, and traffic management can all be supported by this data. In order to increase traffic efficiency, road safety, and driver comfort in a VANET, communication and collaboration between infrastructure and cars are made possible by vehicles with wireless communication devices.

Roadside infrastructure: Roadside infrastructure may be used to extend the communication range of cars and boost network connection. Examples include RSUs and APs. A VANET that facilitates communication between automobiles and between vehicles and the roadside infrastructure is made up in large part of roadside infrastructure.

RSUs, which are placed throughout the road network to facilitate communication with cars, make up the roadside infrastructure. In order to connect with OBUs deployed in vehicles, RSUs are furnished with antennae, wireless communication tools, and processing units.

RSUs are able to gather data on the flow of traffic and the state of the roads and provide it instantly to the moving cars. This data may include things like traffic jams, road closures, weather reports, and other pertinent information.

Applications like junction management, emergency vehicle preemption, and traffic signal priority may all be supported by RSUs. The success of a VANET depends on roadside infrastructure since it offers the necessary communication infrastructure to facilitate real-time communication between infrastructure and cars, enabling a variety of protection and traffic management applications.

Communication protocols: Information is sent and received between automobiles and roadside infrastructure via communication protocols like Dedicated Short Range Communications (DSRC) or Long-Term Evolution (LTE).

Communication protocols are a key element that regulates the information transmission between cars and infrastructure in a VANET. The rules and guidelines known as communication protocols specify how information is sent, received, and processed between various networked devices.

The high-speed mobility of vehicles and the quickly changing topology of the network present issues that are intended to be addressed by VANET communication protocols.

They are also made to enhance network performance and guarantee data protection, privacy, and dependability. The Wireless Access in Vehicular Environments (WAVE) protocol suite, which includes the IEEE 1609 family of standards, as well as the DSRC protocol, are two of the most widely used communication protocols in VANETs.

These protocols facilitate V2V and vehicle-to-infrastructure (V2I) communication, enable real-time communication between cars and infrastructure, and offer the required security and privacy measures to guarantee safe and secure information exchange in a VANET.

Applications: A broad range of applications, including those for traffic control, public safety, and entertainment services, may be created to function on the VANET.

Apps are a crucial part of a VANET, which offers customers value-added services and improves the network's operation. Three types of applications may be found in a VANET: security, mobility, and comfort.

By giving drivers immediate warnings and notifications in life-or-death circumstances, such as accident avoidance and emergency vehicle preemption, safety applications are created to improve road safety.

By offering real-time traffic updates and the most efficient route suggestions, mobility applications are made to ease congestion and enhance traffic flow. The purpose of comfort apps is to offer drivers and passengers more services and conveniences, such as entertainment, information, and communications.

Management systems: The VANET's cars, roadside infrastructure, and communication protocols are all under the supervision and observation of management systems. A vehicular ad-hoc program's management mechanisms are crucial.

An integral part of a VANET that enables effective network administration and control is management systems. Systems for monitoring, analyzing, and controlling network activities are made up of both hardware and software components.

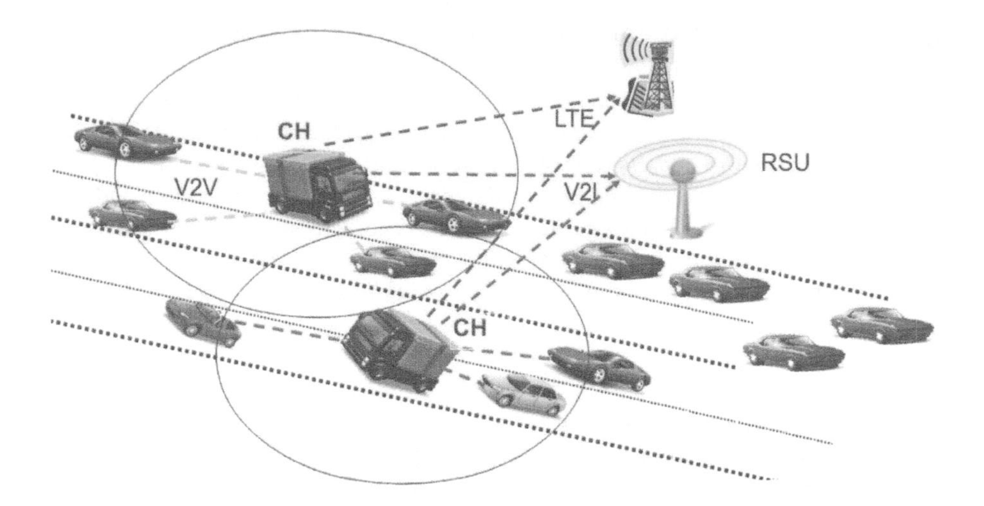

FIGURE 15.9 Communications in VANET.

Network resource management, performance optimization, and real-time problem detection and repair are all possible with management systems. Network management systems, security management systems, and traffic management systems are a few of the important management systems utilized in VANETs. Network performance is monitored, network problems are found and fixed, and network resources like bandwidth, power, and frequency are managed using network management tools.

To guarantee the security and privacy of the data transmitted between infrastructure and cars, security management solutions are employed. By delivering real-time traffic information and enabling dynamic route planning, traffic management systems are employed to enhance traffic flow and alleviate traffic congestion as Figure 15.9.

15.4 ROLES OF VANETs IN SMART CITIES

The concept of MANETs, which spontaneously form a wireless network for data transmission, is applied to the realm of automobiles to produce VANETs. They are an essential part of systems for intelligent transportation system (ITS). While VANETs were initially thought of as only a one-to-one application of MANET concepts, they have subsequently grown into a distinct area of study.

Although the emphasis continues to be on the feature of spontaneous networking and much less on the usage of infrastructure like Road Side Units (RSUs) or cellular networks, the word VANET mostly became synonymous with the more general term inter-vehicle communication (IVC) by the year 2015.

Due to their potential to serve a variety of services and applications for smart cities, VANETs have attracted a lot of attention in recent years. Using wireless communication devices, cars may connect with one another and with roadside infrastructure like RSUs and APs as part of mobile networks known as VANETs (APs). These communication networks may be utilized to offer a range of services, including as improved services for drivers and passengers, traffic efficiency, and road safety.

Increasing road safety is one of the main functions of VANETs in smart cities. Vehicles may communicate in real-time via VANETs on the state of the roads, traffic congestion, accidents, and road closures. Vehicles may utilize this information to make educated judgments and steer clear of hazardous circumstances on the road. For instance, if a car sees an accident up ahead, it may alert other vehicles in the network so that they can take a different path and stay out of harm's way. This can contribute to fewer accidents on the road and increased road safety for all users.

The improvement of traffic efficiency is another crucial function of VANETs in smart cities. In order to route cars more effectively and relieve traffic congestion, VANETs can be used to communicate data regarding traffic congestion and road conditions.

For instance, if a vehicle detects a road closure up ahead, it may alert other cars in the network so that they can choose an alternative route to escape the traffic. This may shorten travel times and enhance city traffic flow.

Additional services for drivers and passengers can be offered using VANETs. These services may include real-time traffic updates, entertainment, and navigation. For instance, a VANET may provide drivers with real-time traffic data so they can pick the optimal route and steer clear of busy regions. VANETs can also be utilized to offer drivers and passengers entertainment services like streaming music and videos.

Supporting intelligent transportation management is another important use of VANETs in smart cities. Real-time data regarding the whereabouts and motion of cars may be provided through VANETs, enabling better resource management and traffic flow.

For instance, a VANET may be used to coordinate the movement of emergency vehicles, including ambulances and fire engines, making it possible for them to get there more quickly and effectively. In emergency situations, this can speed up reaction times and save lives.

VANETs may be used to provide a variety of different services and applications in addition to these advantages. In order to help vehicles identify available parking spots more quickly and spend less time looking for a place, smart parking systems can be supported by VANETs, for instance.

Additionally, smart charging systems for electric vehicles may be supported by VANETs, enabling faster charging and less waiting time for the vehicle.

VANETs can help with environmental management and monitoring as well. For instance, VANETs may be used to track noise and air pollution in real-time, enabling better control of these problems. In order to manage weather-related problems, including flash floods and snowstorms, more effectively, VANETs may also be utilized to help weather management and monitoring.

Some of the examples of applications of VANETs are as follows. Electrical brake lights, which enable a driver (or an autonomous vehicle) to respond to braking cars even when they may be concealed (e.g., by other vehicles).

Platooning, which enables cars to follow a leading vehicle closely (down to a few inches), establishes electrically connected "road trains" after wirelessly receiving acceleration and steering information.

Traffic information systems, which employ VANET connectivity to send satellite navigation systems in a vehicle the most recent updates on obstacles. Of course, the application ranges are beyond the mentioned ones.

15.4.1 An Overview of VANET's Role

An overview of VANET's role in smart cities for other researchers

a. *Carolina Tripp Barba and Miguel Angel Mateos' study on a smart city for vanets that uses traffic statistics, intelligent traffic signals, and warning messages.*
   In the past 20 years, governments and automakers have made road safety a top priority. The improvement of road safety has become a top priority for businesses, researchers, and institutions as a result of the development of new vehicle technology.

   Researchers have been able to construct communication systems where automobiles engage in the communication networks thanks to the development of wireless technology over the past few decades.

   As a result, new kinds of networks, such as VANETS, have been developed to enable communication between cars and infrastructure. In recent years, new ideas like "smart cities" and "living labs" that heavily rely on vehicle networks have emerged.

   Intelligent traffic management, which allows data from the infrastructures of TICs (traffic information centers), is a feature of smart cities.

   Living laboratories (cities where newly developed systems may be evaluated in real-world settings) have been built all around Europe to test the viability of these future cities. The framework being employed aims to send data regarding traffic conditions to assist the driver (or the vehicle itself) in making wise judgments.

   The construction of a warning system made up of ITLs that inform drivers about the volume of traffic and the weather on city streets is suggested and assessed through simulations (Kaur & Singh, 2016).
b. *An efficient privacy-preserving authentication protocol in VANET research by Zhang J., Zhen W., Xu M. (2013).*
   They demonstrate how VANETs, a crucial element of ITSs, may give drivers safer and more comfortable driving conditions. To the best of his knowledge, the two most popular methods in VANET for providing privacy-preserving of the vehicle's identity were group-oriented signatures and pseudonym certificates.

   The two techniques did, however, have a number of efficiency issues that limit their use. These resolve the aforementioned issues that were present in the first two procedures. In that study, they suggested a revolutionary self-certified signature-based authentication mechanism that protects privacy.

   Moreover, they demonstrate how his plan may accomplish conditional privacy preservation and how its security was demonstrated by the random oracle.

The short length of the signature and cheap calculation costs of the technique were further benefits (Kaur & Singh, 2016).

c. *A smart city frame for intelligent traffic system using VANET by Ganesh S. Khekare; Apeksha V. Sakhare (2013)*
The overall number of cars worldwide has increased significantly, increasing traffic congestion and the number of accidents that arise from it. So, rather than improving the quality of the roads, manufacturers, researchers, and the government redirected their attention to improving on-road safety.
A variety of new types of networks, such as the VANET, which allowed for communication between vehicles and between vehicles and roadside equipment, were created as a result of the successful development of wireless technology.
In recent years, a number of novel ideas, including smart cities and living labs, have been developed. His proposed plan is compared to a review of several ITSs and routing technologies.
It also established a brand-new program made up of a smart city framework that transmits data about traffic conditions to assist drivers in making wise selections. They included a warning message module made up of ITLs, which informed the driver of the traffic situation at the time (Kaur & Singh, 2016).

d. *Attacks on security goals (Confidentiality, Integrity, Availability) in VANET: A survey by Irshad Ahmed Sumra, Halabi Bin Hasbullah and Jamalul-lail Bin Ab Manan (2014)*
Because of its potential safety applications and non-safety applications, researchers in academia and industry have been paying more attention to the VANET recently. One form of adversary in the VANET that causes security issues is malicious users.
Security objectives include confidentiality, integrity, and availability (CIA) as key elements. Reviewing the attacks on security objectives is necessary because of the growing research interest, prospective uses, and security issues in VANET.
The purpose of this description is to give a review of assaults on security objectives and to describe in detail the types of attacks and the behavior of attackers using various network situations. Also, they improved our understanding of security objectives, and in the end, they offered an analysis and categorized assaults according to security objectives into several danger categories that may aid in the practical use of VANET.
They demonstrated how his plan might accomplish conditional privacy preservation and demonstrated how safe it was using a random oracle. Also, the suggested approach has the following benefits: a short signature and little processing (Kaur & Singh, 2016).

15.4.2 A Smart City Framework

ITLs are located in some of the intersections in the framework for the smart city that has been created. These ITLs gather real-time traffic information from passing cars and compute statistics on traffic, such as the amount of traffic in the streets nearby (between consecutive crossroads).

ITLs are set up. An omnidirectional propagation pattern is the aerial pattern utilized to cover the whole region. As a result, each ITL on its covered range receives data from all passing cars. Vehicles are supposed to have a GPS unit, a driving assistance device, and a complete map of the city that includes the locations of the ITLs (Rana & Khang et al., 2021).

Vehicles may choose the closest ITL with ease as a result. AODV routing protocol was built up for each ad hoc node (including the cars and ITLs) included in the scenario. The ease and extensive use of AODV are advantages (Khanh & Khang, 2021).

The fundamental disadvantage is that AODV requires end-to-end pathways for data forwarding, which is problematic since with VANETs end-to-end paths endure for a short time due to high vehicle speeds (Kaur & Singh, 2016).

But the deployment and application of VANETs in smart cities also come with a number of difficulties. Ensuring the security and privacy of the data carried over the network is one of the major concerns. VANETs must be built to guard user privacy and stop illegal access to sensitive data (Sohrabi & Wong et al., 2014).

15.5 TRAFFIC MANAGEMENT WITH VANETs

Through routing protocols, VANET enables communication between vehicles. It relates to V2I and V2V, which is shown in Figure 15.10.

The following topics will be covered in this section: the intersection-based algorithm, the greedy curve metric-based routing algorithm, the beam-forming technique,

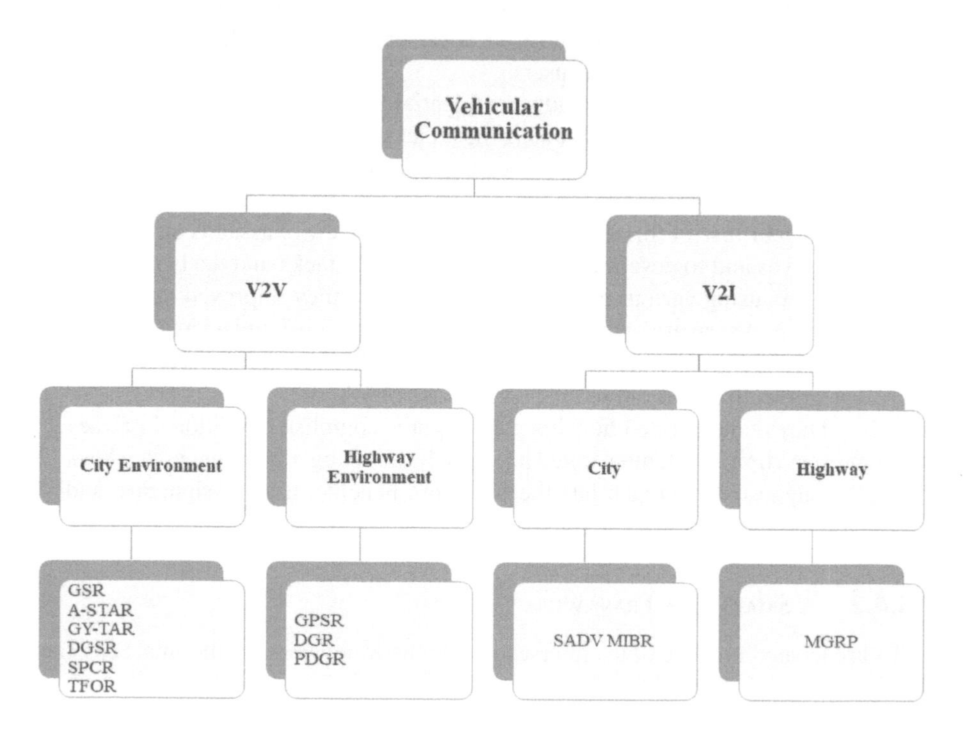

FIGURE 15.10 Categories of position-based routing protocol (Baber & Rizwan, 2019).

and the connectivity-oriented routing protocol. We will receive the ideal packet-forwarding path from each of the aforementioned algorithms (Baber & Rizwan, 2019).

GPSR is a merger of greedy forwarding. The position of the destination is included in GPSR packets. Only those nodes that are closest to the destination receive data from an intermediary node via this information source. Each node employs greedy forwarding and calculates the shortest path to the following node using Dijkstra's algorithm as Figure 15.11.

Only when a local minimum occurs is the perimeter forwarding technique applied. The drawback of GPSR is that it needs the geographic coordinates of nearby nodes. With GPSR, the farther away the node, the faster the speed change, which results in packet loss and poor service quality.

Supporting the urban environment is the main goal of anchor-based street and traffic aware routing (A-STAR). The protocol can even function in areas with low traffic intensity. This algorithm's goal is to send the most packets possible to the target by leveraging information about city bus routes to identify anchor pathways.

The recovery mode of this approach is quite efficient. It computes a new anchor path from the local maximum when there is no node nearby to send packets to.

The next hop is selected via a vector-based technique using a partition model and vector computation. The GPSR algorithm has been enhanced (Smart Cities Council, smart-mobility).

By fixing nodes at intersections, this protocol avoids local optimum and chooses more authenticated nodes. It breaks the forwarding approach into two pieces and concentrates on that. (I) Nodes that are intersecting employ the partition model algorithm to determine the routing direction, and then the intersection greedy forwarding strategy is used to move on to the next node.

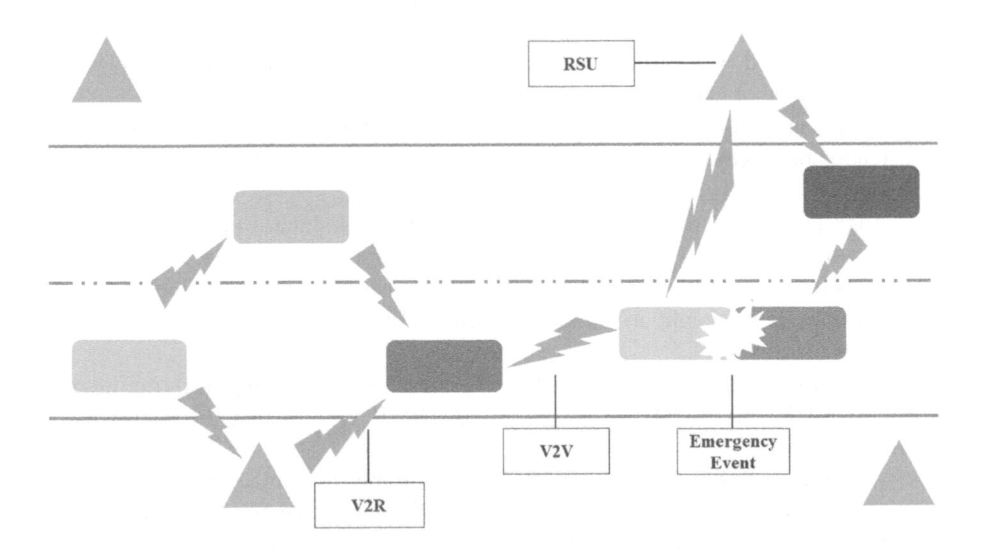

FIGURE 15.11 VANET showing types of vehicular communication.

The deployment of a (II) forwarding routing technique will enable nodes to forward data packets. The inability of vector-based algorithms to deal with environmental constraints is one of their main drawbacks.

Greedy Curve Metric-Based Protocol is the second routing protocol covered in this article from EC-SC (2018). In a metropolitan setting, there are many obstructions, such as trees and buildings that might reduce packet reception and degrade signal quality when using VANET with a greedy routing strategy.

A new routing protocol was developed that employed curve metric distance rather than Euclidean distance to locate the shortest path in order to prevent problems of this nature.

The nearest node to the destination is chosen using curve metric distance using geographic information about the neighbors, a digital city map, and the destination coordinates. Its rehabilitation plan is not very dependable. That should be enhanced. Failure in packet delivery results in local issues that are most costly and create end-to-end delays.

Intersection-based distance and traffic aware routing (IDTAR) is a more effective method than Greedy Curve Metric based protocol. The main determinant of communication between vehicles and between vehicles and infrastructure is routing.

Throughout the city, IDTAR performs efficiently. To determine the ideal way, distance and current traffic density are taken into consideration. One of the most flexible routing protocols is this one. As it chooses the path dynamically, IDTAR outperforms all other position-based protocols in terms of packet delivery ratio and end-to-end latency.

Unicast routing protocol based on attractor choosing is another routing strategy that may be adjusted to the environment (URAS). It offers a configurable environment through packet routing feedback. It utilizes a fresh algorithm known as approach for order preference by resemblance to an ideal response (TOPSIS).

By doing so, fewer nodes must be chosen in order to hop. Performance is enhanced by URAS utilizing the TOPSIS approach. By keeping the present path in place and assessing the next way through the existing one, it analyses itself to determine the optimum course of action.

Cellular attractor selection models are used in this routing system to provide resilience and flexibility while making decisions based on feedback data (Baber & Rizwan, 2019).

Geographic protocol is the greatest routing protocol since it prevents overhead, bulky storage, delays, interruption of nodes, and bandwidth waste. Geographical protocols rely on positions where each node is aware of its own position as well as the positions of its neighbors and surroundings. For this purpose, it makes use of resources like the GPS.

In this work, GPSR, A-STAR, the Greedy curve metric geographical routing protocol, IDTAR, and URAS are all discussed. GPSR verifies that the packet came from the source.

It is not easy to build a protocol due to the number of vehicles moving around, especially in urban areas. In urban areas, speed of vehicles is different from the highway as there are so many obstacles while driving like traffic lights, zebra crossing, and vehicle density is more which weakens signal strength (Baber & Rizwan, 2019).

15.6 PROBLEM WITH GEOGRAPHIC ROUTING PROTOCOL

Local maximum is one of the major challenges that geographic routing protocol faces. When the current node cannot find any nodes closer to its destination to hop into, this is referred to as a local maximum.

Because there is no neighboring node to destination, it results in a dead end. Many methods have been proposed for local maximum issues, like GPSR, A-STAR, IDTAR, URAS, and vector-based routing protocol. Through these routing protocols, it gives us the shortest path. Major issue is that this protocol uses Dijkstra's distance, which is a straight line distance.

Our environment is full of obstacles. While moving around, we get hindrances, so shortest distance is not always the best path. Sometimes we need a full path to get the shortest distance. Table 15.1 gives a brief comparison of VANET routing

TABLE 15.1
Comparison of Different Publications

Publications	Year	Strength	Weakness
Nebbou et al., "Greedy Curvemetric-based Routing Protocol for VANETs"	2018	This strategy uses destination coordinates to deliver a packet and selects the closest node to the destination through curve metric distance to avoid obstacles like building and trees.	Its recovery strategy is not that reliable. It needs to be improved. Packet delivery failure causes local maximum problems causing high cost and end-to-end delay.
Zhang et al. "A Vector based-Improved Geographic Information Routing Protocol"	2017	This strategy chooses the next hop through calculation of the vector and partition model. Through this protocol, local optimum is avoided by fixing nodes at intersection by which it selects more authenticated nodes.	One of the major problems faced using vector-based algorithms is that they cannot tackle obstacles in the environment.
Tian et al. "A Microbial Inspired Routing Protocol for VANET"	2017	It self-evaluates to find the best path by maintaining the current path and evaluating the next path through the current one. This routing protocol uses cellular attractor selecting models to give robustness and adaptability to take decisions through feedback data.	Position services may fail in tunnels or obstacles (missing satellite signal) Unnecessary flooding
Ahmed et al. "Intersection-based Distance and Traffic-Aware Routing (IDTAR) Protocol for Smart Vehicular Communication"	2017	This strategy takes distance and real-time density of traffic into account to get the best path. It gives the highest packet delivery ratio and lowest end-to-end delay as it determines the path dynamically.	It has security issues such as authentication and trust. Position services may fail in tunnels or obstacles (missing satellite signal)

protocols as discussed in this chapter. It shows how each routing protocol is better than the other.

Smart cities must effectively manage their traffic, and VANETs can help by enhancing traffic flow and easing congestion. The foundation of VANETs is the idea of real-time vehicle communication with RSUs and other vehicles.

This real-time connection may be utilized to give drivers important information about traffic, accidents, and road conditions, which will assist to improve traffic flow and ease congestion (Jayapal & Roy, 2016).

Traffic management on VANETs is mostly accomplished by ITSs. Advanced traffic control techniques, such as routing and scheduling, can be used with these systems to improve traffic flow and lessen congestion. For instance, VANETs can give drivers real-time traffic data so they can choose the optimal routes and steer clear of busy areas (Jayapal & Roy, 2016).

Delivering real-time information regarding traffic flow and road conditions is another important function of VANETs. City officials may utilize this data to change traffic lights, ease congestion, and enhance road safety. For instance, if a road is clogged, VANETs can offer real-time information regarding the location and reason for the clog, enabling municipal officials to take the necessary steps to lessen the effect on traffic flow (Jayapal & Roy, 2016).

VANETs can be utilized to increase traffic safety in addition to streamlining traffic and easing congestion. For instance, VANETs can alert drivers in real-time about possible road dangers, including accidents, road closures, and other situations. Drivers may use this knowledge to make educated judgments and stay out of potentially hazardous situations.

Additionally, the deployment of autonomous cars can be supported by VANETs. In order to function safely and effectively, autonomous cars depend on real-time coordination and communication, and VANETs offer the communication infrastructure required to enable these systems.

The use of VANETs in smart cities is accompanied by a number of difficulties, though. Making sure the VANET network is dependable and secure is one of the major difficulties. In order to protect the network from assaults and data breaches, this calls for strong communication protocols and security measures.

Making sure the VANET network is compatible with other transportation systems, such as public transit and roadside equipment is another difficulty. To accomplish this, universal data interchange formats and standardized communication protocols must be created.

Making sure that the VANET network is scalable and can support growing numbers of cars and devices is another problem. To do this, infrastructure and effective communication protocols must be developed, and large-scale deployments must be supported (Jayapal & Roy, 2016).

Despite these difficulties, VANETs have a lot to offer in terms of traffic management benefits. The effectiveness and safety of road transportation in smart cities might be greatly enhanced by the capacity to offer real-time information and manage traffic flow.

To sum up, by enabling real-time communication between cars and roadside devices, maximizing traffic flow, enhancing road safety, and assisting the deployment of autonomous vehicles, VANETs may significantly improve traffic management in smart cities.

The potential advantages of VANETs in smart cities are substantial and call for more study and development, notwithstanding the difficulties connected with their deployment.

15.7 CONCLUSION

In conclusion, VANETs are crucial for the growth of smart cities, particularly for traffic control. Traffic flow may be optimized, traffic congestion can be decreased, and road safety can be increased via VANETs, which offer a real-time communication infrastructure between cars and roadside equipment.

An essential part of VANETs for managing traffic is ITS, which enables the use of sophisticated traffic control algorithms to improve traffic flow and lessen congestion. In order for autonomous cars to function safely and effectively, real-time communication and coordination are essential, and VANETs are a key component in enabling their adoption.

The adoption of VANETs in smart cities, however, is fraught with difficulties, such as maintaining network security and stability, compatibility with other transportation systems, and scalability to handle an expanding fleet of cars and gadgets. Despite these difficulties, VANETs in smart cities have significant potential advantages that call for more study and development.

In conclusion, by enabling real-time communication between cars and roadside devices, maximizing traffic flow, enhancing road safety, and assisting the deployment of autonomous vehicles, VANETs can significantly improve traffic management in smart cities.

Future research and development should focus on the use of VANETs in smart cities, which has the potential to dramatically increase the effectiveness and security of road transportation.

VANETs are an essential technology for the growth of smart cities, especially in the field of traffic control. They have the potential to transform the way cities manage traffic and enhance the quality of life for their residents through their capacity to deliver real-time communication, optimize traffic flow, increase road safety, and assist the deployment of autonomous cars.

To fully exploit the promise of VANETs in smart cities and to handle the deployment issues, more study and development in this field are required.

REFERENCES

Ahmed AIA, A Gani, SH Ab Hamid, S Khan, N Guizani, K Ko. Intersection-based distance and traffic-aware routing (IDTAR) protocol for smart vehicular communication. 2017 13th international wireless communications and mobile. 2017. https://ieeexplore.ieee.org/abstract/document/7986334/

Baber, F., Rizwan, M. "Traffic management through VANET", *International Journal of Scientific and Research Publications (IJSRP)*, vol. 9, p. 8647 (2019). https://doi.org/10.29322/IJSRP.9.02.2019.p8647

Bhambri, P., Rani, S., Gupta, G., Khang, A. "*Cloud and Fog Computing Platforms for Internet of Things*" (2022). CRC Press. https://doi.org/10.1201/9781032101507

EC-SC, 2018. European Commission (smart-cities). Smart Cities and Communities. Retrieved on (2023) from https://commission.europa.eu/eu-regional-and-urban-development/topics/cities-and-urban-development/city-initiatives/smart-cities_en

Feltrin, L., Tsoukaneri, G., Condoluci, M., Buratti, C., Mahmoodi, T., Dohler, M., Verdone, R. "Narrowband IoT: a survey on downlink and uplink perspectives", *IEEE Wireless Communications*, vol. 26, pp. 78–86 (2019). https://ieeexplore.ieee.org/abstract/document/8641430/

Frost & Sullivan, www.frost.com. (Source: Frost and Sullivan). (April 2020).

Gazette, Smart-Building-Market-Increased-Demand. (2020). https://designsmartcity.com/smart-building-market-increased-demand/

Ganesh S. Khekare; Apeksha V. Sakhare, International Multi-Conference on Automation, Computing, Communication, Control and Compressed Sensing (iMac4s). March 2013. https://ieeexplore.ieee.org/abstract/document/6526427/

Gozalvez, J. "New 3GPP standard for IoT [mobile radio]", *IEEE Vehicular Technology Magazine*, vol. 11, pp. 14–20 (2016). https://ieeexplore.ieee.org/abstract/document/7419993/

Hahanov, V., Khang, A., Litvinova, E., Chumachenko, S., Hajimahmud, V.A., Alyar, A.V. "The Key Assistant of Smart City – Sensors and Tools", *AI-Centric Smart City Ecosystems: Technologies, Design and Implementation* (1st Ed.) (2022). CRC Press. https://doi.org/10.1201/9781003252542-17

Jain, R. "Technical challenges in implementing smart city initiatives", *Journal of King Saud University – Computer and Information Sciences*, vol. 29, no. 4, pp. 337–341 (2017). https://www.igi-global.com/chapter/cyber-security-challenges-for-smart-cities/226916

IA Sumra, HB Hasbullah, JB AbManan, Vehicular Ad-hoc Networks for Smart Cities: First International Workshop. 2014 Springer, https://link.springer.com/chapter/10.1007/978-981-287-158-9_5

Jayapal, C., Roy, S. Road traffic congestion management using VANET. pp. 1–7 (2016). https://doi.org/10.1109/HMI.2016.7449188

Jebaraj, L., Khang, A., Vadivelraju, C., Antony, R.P., Kumar, S. (Eds.). "Smart City Concepts, Models, Technologies and Applications", *Smart Cities: IoT Technologies, Big Data Solutions, Cloud Platforms, and Cybersecurity Techniques* (1st Ed.) (2024). CRC Press. https://doi.org/10.1201/9781003376064-1

Kaur, A., Singh, E.P. "Role of VANET in smart city", *International Journal for Research in Applied Science & Engineering Technology (IJRASET)*, vol. 4, no. VIII, pp. 1–5 (2016). ISSN: 2321–9653. http://jnc.digitallibrary.co.in/bitstream/123456789/475/1/Publication.pdf

Kaya, A., Yilmaz, A., Kalemci, A. "Smart City Architecture: A Comprehensive Review", *Proceedings of the 10th International Conference on Ambient Systems, Networks and Technologies* (ANT 2019), pp. 97–104 (2019). Springer. https://www.hindawi.com/journals/amse/2020/8167402/

Khang, A. "Material4Studies", *Material of Computer Science, Artificial Intelligence, Data Science, IoT, Blockchain, Cloud, Metaverse, Cybersecurity for Studies* (2023). https://www.researchgate.net/publication/370156102_AlexKhangMaterial4Studies

Khang, A., Chowdhury, S., Sharma, S. "*The Data-Driven Blockchain Ecosystem: Fundamentals, Applications, and Emerging Technologies*" (1st Ed.) (2022). CRC Press. https://doi.org/10.1201/9781003269281

Khang, A., Gupta, S.K., Rani, S., Karras, D.A. "*Smart Cities: IoT Technologies, Big Data Solutions, Cloud Platforms, and Cybersecurity Techniques*" (2023). CRC Press. https://doi.org/10.1201/9781003376064

Khang, A., Hahanov, V., Abbas, G.L., Hajimahmud, V.A. "Cyber-Physical-Social System and İncident Management", *AI-Centric Smart City Ecosystems: Technologies, Design and Implementation* (1st Ed.) (2022). CRC Press. https://doi.org/10.1201/9781003252542-2

Khang, A., Ragimova, N.A., Hajimahmud, V.A., Alyar, A.V. "Advanced Technologies and Data Management in the Smart Healthcare System", *AI-Centric Smart City Ecosystems: Technologies, Design and Implementation* (1st Ed.) (2022). CRC Press. https://doi.org/10.1201/9781003252542-16

Khang, A., Rani, S., Sivaraman, A.K. "*AI-Centric Smart City Ecosystems: Technologies, Design and Implementation*" (1st Ed.) (2022). CRC Press. https://doi.org/10.1201/9781003252542

Khanh, H.H., Khang, A. "The Role of Artificial Intelligence in Blockchain Applications", *Reinventing Manufacturing and Business Processes Through Artificial Intelligence*, pp. 20–40 (2021). CRC Press. https://doi.org/10.1201/9781003145011-2

Lai, Y., Tsai, T. "A review of smart city development", *Sustainability*, vol. 9, no. 4, 636 (2017). https://www.mdpi.com/2076-328X/13/2/134

Lee, Y.K., Klysubun, W. "Collaborative governance and smart city: a comparative case study of Amsterdam and Seoul", *Proceedings of the 33rd International Conference on Information Systems, 15–17*, 2015, Dublin, Ireland (pp. 1–12). Association for Information Systems. https://aisel.aisnet.org/icis2015/proceedings/InnovationandEntrepreneurship/7/

Lu, J., Ratti, C., Aiello, L.M. "Big data for urban informatics: a review of research", *ACM Transactions on Intelligent Systems and Technology (TIST)*, vol. 8, no. 6, pp. 1–22 (2017). https://www.tandfonline.com/doi/abs/10.1080/19475683.2018.1471518

Ma, Q., Gao, J. "Interoperability in smart city development: a review of the state-of-the-art and future directions", *Future Generation Computer Systems*, vol. 74, pp. 319–333 (2017). https://www.sciencedirect.com/science/article/pii/S0167739X18300025

Magrin, D., Centenaro, M., Vangelista, L. "Performance evaluation of LoRa networks in a smart city scenario", *Proceedings of the 2017 IEEE International Conference on Communications (ICC), Paris, France*, 21–25 May 2017; pp. 1–7. https://ieeexplore.ieee.org/abstract/document/7996384/

Mehmood, A., Venkatesh, V. "Technological innovations in smart cities: a review of the literature", *Journal of Management Information Systems*, vol. 32, no. 1, pp. 213–246 (2015). https://www.academia.edu/download/52879719/6.pdf

Munir, M., Gani, A., Imran, M. "A comprehensive survey of vehicular ad hoc networks: from evolution to recent trends", *Journal of Network and Computer Applications*, vol. 109, pp. 81–103 (2018). https://www.sciencedirect.com/science/article/pii/S108480451300074X

Nebbou T., Lehsaini M., Greedy curvemetric-based routing protocol for VANETs. 2018 International Conference on Selected Topics in Mobile. 2018. https://ieeexplore.ieee.org/abstract/document/8428952/

O'Brien, A. "Smart city resilience: challenges and opportunities", *Smart Cities*, vol. 1, no. 3, pp. 97–102 (2018). https://link.springer.com/chapter/10.1007/978-3-319-46131-1_23

Rana, G., Khang, A., Sharma, R., Goel, A.K., Dubey, A.K. "*Reinventing Manufacturing and Business Processes Through Artificial Intelligence*" (2021). CRC Press. https://doi.org/10.1201/9781003145011

Rani, S., Bhambri, P., Kataria, A., Khang, A. "Smart City Ecosystem: Concept, Sustainability, Design Principles and Technologies", *AI-Centric Smart City Ecosystems: Technologies, Design and Implementation* (1st Ed.) (2022). CRC Press. https://doi.org/10.1201/9781003252542-1

Rani, S., Bhambri, P., Kataria, A., Khang, A., Sivaraman, A.K. "*Big Data, Cloud Computing and IoT: Tools and Applications*" (1st Ed.) (2023). Chapman and Hall/CRC. https://doi.org/10.1201/9781003298335

Rani, S., Chauhan, M., Kataria, A., Khang, A. "IoT Equipped Intelligent Distributed Framework for Smart Healthcare Systems", *Networking and Internet Architecture* (2021). CRC Press. https://doi.org/10.48550/arXiv.2110.04997

Rao, S.K., Prasad, R. "Impact of 5G technologies on smart city implementation", *Wireless Personal Communications*, vol. 100, pp. 161–176 (2018). https://link.springer.com/article/10.1007/s11277-018-5618-4

Scherer, M., Janssen, P. "A research framework for smart city indicators", *Journal of Ambient Intelligence and Smart Environments*, vol. 7, no. 3, pp. 225–244 (2015). https://www.sciencedirect.com/science/article/pii/S1747938X16300252

SCC-BSC, 2023. Smart Cities Council (benefits-smart-cities). Benefits of Smart Cities. Retrieved on (2023) from https://www.plantemoran.com/explore-our-thinking/insight/2018/04/thinking-about-becoming-a-smart-city-10-benefits-of-smart-cities

SCC-EDV, 2023. Smart Cities Council (economic-development). Economic Development. Retrieved on (2023) from https://www.cocoflo.com/resources/driving-economic-growth-with-smart-city-innovations

SCC-SBD, 2023. Smart Cities Council (smart-buildings). Smart Buildings. Retrieved on (2023) from https://www.axians.com/innovation-technology/smart-building-cities/

SCC-SCE, 2023. Smart Cities Council (smart-citizen-engagement). Smart Citizen Engagement. Retrieved on (2023) from https://www.beesmart.city/en/strategy/how-smart-cities-boost-citizen-engagement

SCC-SEN, 2023. Smart Cities Council (smart-energy). Smart Energy. Retrieved on (2023) from https://smartcities.ieee.org/newsletter/february-2022/smart-energy-for-smart-cities

SCC-SEV, 2023. Smart Cities Council (smart-environment). Smart Environment. (2023). Retrieved from https://nexusintegra.io/smart-city-environments/

SCC-SHC, 2023. Smart Cities Council (smart-healthcare). Smart Healthcare. Retrieved on (2023) from https://www.orange-business.com/en/magazine/smarter-healthcare-in-smart-cities

SCC-SMB, 2023. Smart Cities Council (smart-mobility). Smart Mobility. Retrieved on (2023) from https://rideamigos.com/smart-mobility-in-smart-cities

SCC-SPD, 2023. Smart Cities Council (smart-public-safety). Smart Public Safety. Retrieved on (2023) from https://www.digi.com/solutions/by-industry/smart-cities

SCC-WSC, 2023. Smart Cities Council (what-smart-city). What is a Smart City?. Retrieved on (2023) from https://commission.europa.eu/eu-regional-and-urban-development/topics/cities-and-urban-development/city-initiatives/smart-cities_en

Sohrabi, F., Wong, W., Gao, J. "Protocols for secure vehicular ad hoc networks", *Proceedings of the IEEE*, vol. 104, no. 9, pp. 1727–1740 (2014). https://doi.org/10.1016/j.vehcom.2014.01.001

Tian D, K Zheng, J Zhou, X Duan, Y Wang, Z Sheng, Q Ni. A microbial inspired routing protocol for VANETs. IEEE Internet of Things Journal, 2017. https://ieeexplore.ieee.org/abstract/document/8004418/

Wang, H., Chen, J. "Big data privacy in the era of smart cities", *Journal of Ambient Intelligence and Humanized Computing*, vol. 7, no. 4, pp. 427–441 (2016). https://link.springer.com/article/10.1007/s12652-017-0466-8

Yaqoob, I., Hashem, I.A.T., Mehmood, Y., Gani, A., Mokhtar, S., Guizani, S. "Enabling communication technologies for smart cities", *IEEE Communications Magazine*, vol. 55, pp. 112–120 (2017). https://ieeexplore.ieee.org/abstract/document/7823347/

Zhang J., Zhen W., Xu M., 2013 IEEE 9th International Conference on Mobile Ad-hoc and Sensor, 2013. https://ieeexplore.ieee.org/abstract/document/6726342/

Zhang L., Guo J., A vector-based improved geographic information routing protocol. 2017 7th IEEE International Conference on Electronics Information. https://ieeexplore.ieee.org/abstract/document/8076596/

16 Big Data Analytics Tools, Challenges and Its Applications

Shashi Kant Gupta, Olena Hrybiuk, NL Sowjanya Cherukupalli, and Arvind Kumar Shukla

16.1 INTRODUCTION

Data is classified as the significance of resolution-making process that includes the accountability of materials. Due to lack of high-value data, which provides significant material at required time, observing, analyzing and estimating significant regulations are highly impossible (Oguine & Oguine et al., 2022).

In this scenario, a present consideration of data-driven and data methods from various professionals and academics exist, as knowledge drown from the process of data visualization that deals with the enhancement of advanced activity, organizational transformation, national data storage, records of public services and information protection of enterprise corporations in Industry 4.0 economy.

Presently, in the industrial revolution, the focus of various organizations and administrations is on the enhancement capabilities that give extracted knowledge from huge and complicated datasets, significantly called "big data". It is a noteworthy topic in the field of finance and business; later, it plays an enormous part in economic strategy and has improved its significance in forming economic range by making significant ways (Lutfi et al., 2023). Various enterprise that includes huge organizations attempts to get data-based culture stressed for reasonable engagement amongst rivals.

Enterprises focus to influence the generated data within organizations through various procedures to gain insight information for faster, better and more significant decisions in essential business problems. The beginning of Web 2.0 permits users to associate with every person on social platforms, making enterprises get access to large amounts of information better and faster. In significance to that, the involvement of Web 3.0 gives enhanced chances for collecting external data sources.

Precise information can be accurately gathered using mobile devices like tablets and smart phones; since these devices are enabled with internet connectivity, they provide personalized transaction and location-based applications. This proficiency will continuously offer significant research problems and challenges through the years (Kreiser & Wright et al., 2022).

Organizations like Amazon, Google and other social network notice the importance of big data, identifying different methods that can be used for customer's satisfaction that can enhance the aids of these organizations. Various organizations have

DOI: 10.1201/9781003376064-16

initiated to gain from those prospects enlighten by large improvement in big data methodologies.

Every industries and enterprises rely on data-driven analytics rather than leader-based knowledge (Dorn & Khailaie et al., 2022). Nevertheless, exploration of data requires people with expertise and skillset who will be useful to analyze the scope of providing data efficient knowledge to make decisions.

16.2 BIG DATA

The enormous cohort of data which will grow past 180ZB in the year 2025, gives an upper hand to the data that can change the 21st century into a data-driven world along with the revolution of business and markets. Digital revolution from heterogeneous and complex data gathered from any point in time forming a new age, the period of big data (Sivarajah & Kamal et al., 2017).

Big data illustrates huge a dataset that is unable to be gathered, achieved and analyzed by traditional methods. Data are not particularly huge in size but also complex in nature that make the decision making very difficult. These data normally include data from transactional, operational, marketing and sales. This unstructured form of data rapidly generates a structured form of data and has bagged 90% of all data around the world. Figure 16.1 illustrates the overall lifecycle of the data and its challenges.

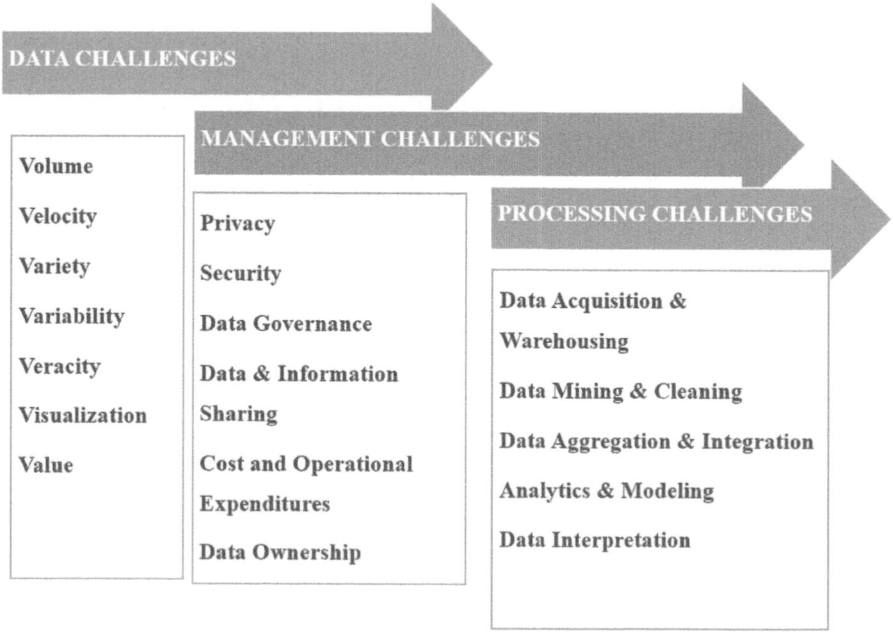

DATA CHALLENGES
Volume
Velocity
Variety
Variability
Veracity
Visualization
Value

MANAGEMENT CHALLENGES
Privacy
Security
Data Governance
Data & Information Sharing
Cost and Operational Expenditures
Data Ownership

PROCESSING CHALLENGES
Data Acquisition & Warehousing
Data Mining & Cleaning
Data Aggregation & Integration
Analytics & Modeling
Data Interpretation

FIGURE 16.1 Data lifecycle and management challenges.

Therefore, a unique form of handling competences is needed for receiving the insight information that has enhanced decision abilities. Process of the data can be categorized into various types like process, data and controlling issues. Experiments in data depict the classification of big data, consisting of velocity, variety, volume and veracity.

Process issues are linked with methodologies required for acquisition of data, integration, analysis and conversion in order to acquire knowledge from larger data. The administration issues consist of data privacy, expenditure of cost and governance.

16.2.1 Volume

It depicts a large portion of datasets. It is a concept that increases the connectivity using sensors, and smart devices, in association with fast-developing communication and information consisting of Artificial Intelligence (AI), have credited to the enormous data formation (transaction, files, tablets and records). The data speed rapidly increases such that it overcomes Moore's regulations and the amount of data formed incorporating new estimation for storage of data (Rana & Khang et al., 2021).

16.2.2 Variety

It illustrates the enhancing diversity of the generated data sources and various formats of data. Web 3.0 indicates the enhancement of social and web systems which lead to the formation of various data types.

Photos, updates, messages and videos are depicted in social media systems such as Twitter or Facebook, GPS, SMS and various other forms of signals, transaction of customers in banks and data obtained from business and retail markets. Various sources of data are significantly unique, consisting of various computing devices that supply large data that connect with the behavior of human and their location.

Normally, larger data also insists on forms of data that are formed, thus consisting of structured data, semi-structured data and unstructured data and various other formats of data that are hard to categorize derived from video and various appliances.

16.2.3 Variability

It is normally tangled with different variety of data which relates to enormous deviation of denotation. For example, words in a sentence can have various meanings based on the context of the sentence; thus, for an accurate analysis of sentiment, algorithm requires to identify the meaning of every word taking context into account.

16.2.4 Velocity

A larger source of data is classified by the higher generation of data, for instance generated data by associating web-based devices inward an organization in real time. This amount of pace is largely substantial for organizations, thus considering different actions that enable them to become more significant amongst competitors, in spite of fact that certain organizations have explored big data to provide their customers with purchase suggestions.

16.2.5 Veracity

Data denotes statistics accuracy and reliability. Collection of data has statistics that aren't smooth and correct, and for this reason, information veracity denotes back to the facts uncertain and the phases of correlated reliability with certain forms of facts.

16.2.6 Visualization

It is the technology of visual illustration of larger data in day-to-day life. It offers qualitative and quantitative statistics in certain forms considering schematic patterns, trends and versions, in approaches that are unable to be provided in various kinds like textual content.

The influence of large information can provide precious data, and for that reason, the fee provided with the aid of the records analysis method can gain firms, businesses, communities and purchasers as shown in Figure 16.2.

Organizations that overcome the issues and explore big facts correctly have greater precise data and are capable of creating new information with the aid of which they enhance their method and commercial enterprise operations concerning nicely described targets like productiveness, monetary overall performance and

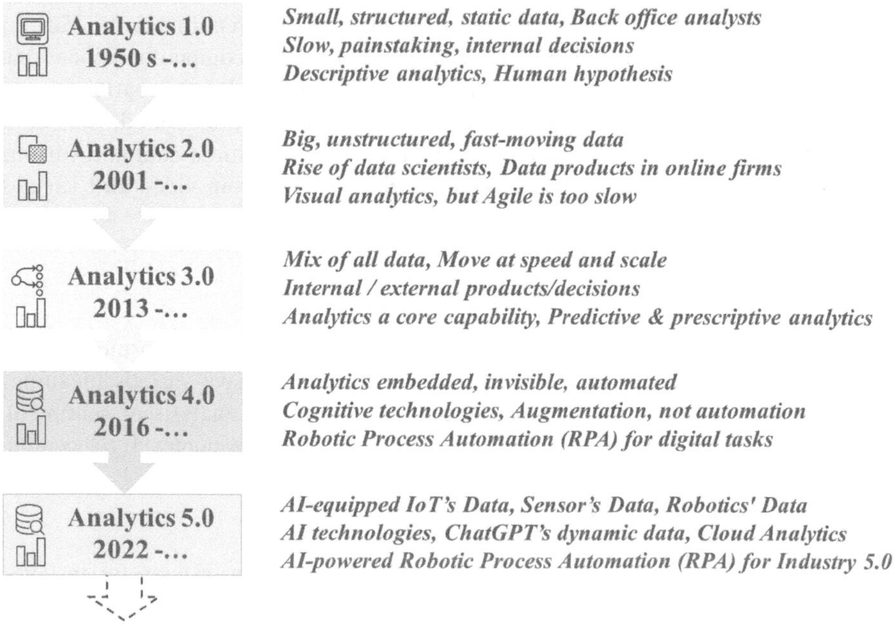

FIGURE 16.2 The advancement of analytics eras (Khang, 2021).

marketplace cost (Dai & Rosenberg et al., 2022), even though massive information plays a main function in the digital conversion of organizations introducing innovations. Therefore, a growing hobby in the manipulation of huge records amongst corporations and businesses exists (Figure 16.2).

16.3 BIG DATA ANALYTICS

It includes the evaluation of big statistics group in corporations, the time period of huge records analytics related to data science, and commercial enterprise analytics. Data technology is described as a group of essential concepts that encourages taking records and understanding statistics (Gopalakrishnan et al., 2022). Since past years, records driven procedures like Business Intelligence (BI) and business analytics are classified as crucial to working businesses.

BI is described because the techniques, networks and packages for gathering, getting ready and analyzing statistics to offer records supporting choice takers. In different terms, BI structures are records-based decision system (Power, 2008), whilst based on technologies related to business and programs which are used to research important enterprise information for assisting them to apprehend their enterprise on time.

The decision of business analytics is of good size quantities of statistics to improve its score, while BI especially distillates ancient data in charts and records table reviews as a manner to offer queries without records and improve its cost.

Business analysis started defining the fundamental analytical detail in BI in the overdue 2000s. Subsequently, the standings of massive data and large information analysis are used to explain strategies for facts – sets which might be so massive and composite, wanting advanced records storage, control, and evaluation and visualization equipment.

In that unexpectedly developing surroundings, the significance of facts makes the deviation of records into valued understanding quickly a need. The variations between traditional analysis and rapid analytics consisting of huge records are in analytics traits (type, goal and technique), records features (type, age/glide, extent) and primary independence as shown in Table 16.1 (Helbing, 2015).

TABLE 16.1
Big Data Analytics and Conservative Models

Type	Conventional Analytics	Big Data Analytics
Type of analysis	Predictive and descriptive model	Prescriptive and predictive model
Method of analysis	Based on hypothesis	Based on machine learning (Khanh & Khang, 2021)
Primary scope	Performance management and decision support	Data-driven amenities
Type of data	Defined and structured	Undefined and unstructured
Flow data	>24 h static data	<Min constant data flow
Volume of data	Less than terabytes	100 terabytes to petabytes

Web 2.0 structures, such as the creation of social media networks like Facebook, offer establishments extra facts with statistics approximately organizations products. The present is due to enhancement of cell gadgets towards the variety of devices added a unique generation of analytics like user generated information using social channels.

Computing strategies have greater promotion, e.g., notably cellular, region-conscious and person-centered methods and dealings. Therefore, data-pushed decisions take on records gathered from all the assets of corporations, even as forecasting and device on the basis of the traditional facts and new modern resources like Internet of Things (IoT).

Data evaluation is the technique of examining, cleansing, converting and modeling facts gaining beneficial facts for pointers and aid in selection making. It has a couple of sides and methods, encompassing diverse strategies below a sort of details of different businesses, technological know-how and social technological know-how schemes, while "Big Data Analytics" (BDA) depicts the enhanced methods, considering huge and several styles of datasets to observe and gather knowledge from huge records, substituting a sub-procedure in obtaining the records process.

Using superior methods, BDA consists information organization, open-supply platforms like Hadoop, and statistical evaluation, such as sentimental based data collection evaluation, visualization equipment that help shape and connect information to find backdrop behavior and unidentified patterns along with insights.

The system of BDA is an aid for decision-making technique for extensive enhancements in strategic functionality, unique streams and importance. In this particular information, the procedure of having insights from huge facts can be classified into stages: information control and facts analysis.

Data control is associated with the strategies and skills for statistics technology, loading, mining and guidance for analysis, and at the same time, fact evaluation refers back to the strategies and strategies for observing the insights and information (Larson & Chang et al., 2016, figure 3). Data analytics can be classified into four divisions, ranging from diagnostic to analytics-based enhanced prescriptive and predictive analytics.

16.3.1 Descriptive Analytics

It is on the basis of present and historical data which significantly have insights of sources about past scenarios and the correlation amongst different pattern identification of determinants using the statistical measures such as range, mean and standard deviation.

This type of analytics is used such as online analytical processing (OLAP) which explores the knowledge from various experiences obtained from previous data of significant analysis which consists of data conception, reports, graphs and charts illustrating different metrics of organization comprising orders and sales.

16.3.2 Diagnostic Analysis

Additionally, historical facts give insights approximately into the root cause of a few results of the beyond. Thus, businesses can take higher choices fending off errors and terrible outcomes of the beyond.

16.3.3 Predictive Analytic

It's about providing information regarding chance of data estimation and end analysis using opportunities. Using different strategies which includes facts mining, facts modelling and gadget studying, the operation of predictive analysis is based on big for any organization's phase (Gupta & Khang et al., 2023).

Most acknowledged packages of that sort of analytics is the extrapolation of patron conduct, figuring out operations, advertising and avoiding threat. Using historic and different available facts, predictive assessments are able to discover styles and find the relationships in statistics that can be used for prediction (Dubey & Bryde et al., 2022).

Predictive analytics inside the digital generation is a big weapon for corporations in the aggressive race. Therefore, corporations exploiting predictive assessment can perceive future developments and styles, presenting revolutionary products/offerings and novelties in their enterprise models.

16.4 BIG DATA ANALYTICS TOOL

The data is developing rapidly and getting extra complicated. Commercial entities (businesses) use large statistics equipment to improve the internal in addition to external performances via reading the records. It requires the right equipment that suits their necessities for analyzing the amassed records. Interestingly, among the high-quality acknowledged huge statistics gear available are open-supply tasks.

Various tools have been evolved and lots of are still in the procedure of improvement. Out of absolutely or partly advanced gear and framework, this section drives out the salient capabilities of a few famous equipment.

16.4.1 Apache Cassandra

It is an open-supply distributed DBMS. It is able to deal with a good sized quantity of records unfolding over a huge set of machines and attempts to offer records availability with zero failure (Thiess & Fiol et al., 2022). Apache Cassandra offers aid for a group of servers which spreads over many facts' centers, with graspless replication.

This means that client operation will take less time. Highest throughput may be performed for an optimum number of nodes, but it creates a put-off for read and write operations. It introduced CQL as an opportunity for RPC interface. The nodes are decentralized which supports asynchronous grasp with much less replication and multi-statistics center replication. Fault tolerance is completed through more than one node replication.

16.4.2 Apache Mahout

It is used to produce open-source algorithm implementation (Vidhya & Shanmugalakshmi et al., 2020). It is an implementation in progress; certain algorithms are still lacking. It mainly focuses on collaborative analysis, classification and clustering. It also gives various libraries for normal operation and primitive collection.

16.4.3. Apache Avro

It is the most significant data serialization and procedural call procedure. Avro is vitally used to give format serialization for persistent data. This association between nodes of Hadoop to client is also given by Avro, and for this reason, wire is based on format.

JSON is used by this tool for definition of data, binary and format of every data. It consists of substitute description language syntax known as Avro IDL.

16.4.4 Sqoop

It is an application-oriented tool which is used for the transfer of data associated with relational form of data to Hadoop. Sqoop involves different types of jobs such single content load or query using SQL and some different of saved jobs.

16.4.5 MapReduce

It includes two different associated function. Initial function includes map, which filters the initial data and sort it in some necessary form, and second one is used to reduce that summary that data received from the map function.

16.4.6 Apache Hive

This provides different services which consist of normalization of large dataset, query using data and analysis. It can be thought as structural warehouse infrastructure. Hive includes its own query design called HiveQL. It is transparent divided queries into different types of schedules.

YARN is used to run these execution engines, whereas Hive is run on the basis of decisions on different types of indexes to move the queries.

16.4.7 Apache Hbase

This tool is a significant schema, associated database implemented to provide fast accessibility to huge total structured database. It ropes random search on huge data. It is used as hash tables for managing internal data and also used by organizations like social media for real-time monitoring and searching.

16.5 APPLICATION OF BIG DATA ANALYTICS

From the retrieved primary research, this review also highlights numerous application domains for industrial BDA. Industry 4.0 application development, which involves transformation of intensive information and other significant industries, uses BDA as part of a power play in a networked setting.

The Industry 4.0 framework will be used by numerous enterprises. Sufficient ability to foresee problems, schedule maintenance, and adapt to new requirements, unanticipated adjustments are made to the production procedures.

The results of the investigations demonstrate that BDA actionable data and information have been used in a variety of contexts, including Smart cities, healthcare and medical services, financial fraud detection, etc. That is the most recent and important application under subsection (Khang & Rani, 2022).

16.5.1 Education Sector

Through the use of big data and BDA techniques, educational enhancement and implementation lead to experienced tremendous success. The information gained by BDA used to address issues in education includes institutional management, quality management, online educational platforms and the prediction of student dropouts.

As an illustration (Wei & Karuppiah et al., 2022), The PABED (Project – Analyzing Big Education Data) tool for analyzing massive educational data has been applied to issues with analysis. The tools enable users to generate tables using data from Google Big Query as their data source. However, Google-based accounts are still in the incubation stage and do not have the most fundamental features.

Using US higher education, a sample offers a conceptual model for the efficient application of big data analysis in education. They pointed out that educational organizations can use a framework to change institutional activities such as student profiles, attendance records, performance ratings, athletic participation, evaluations of extracurricular activities, alumni, etc.

16.5.2 Healthcare Domain

This domain makes use of some computing devices like ECG, PPG, and EEG, among others; BDA aids in capturing and analyzing various recovery outcomes and further prediction analysis process of patients. Clinic analysis, laboratories, pharmaceutical companies, management of hospital and sensors of all produce heterogeneous data (Jones & Collier et al., 2020).

Large datasets are also produced by hospitals, clinics, and medical care; these datasets include electronic medical records (EMRs), imaging data, pharmaceutical EMRs, personal practices and preferences, and financial information (Vrushank & Vidhi et al., 2023).

Consider the improved effectiveness and utility of outpatient medical rehabilitation as evidenced by data obtained by healthcare in medical records unit (Firouzi & Rahmani et al., 2018). However, the preprocessed and appropriately harnessed generated data aid in the formulation and execution of an intervention program that would improve data analytics process.

For occurrence, Manogaran and Varatharajan et al. (2017) studied the use of the IoT and the big data ecosystem to deliver high-quality healthcare while protecting patient privacy and data security (Khang & Hahanov et al., 2022).

The authors pointed out that BDA technologies offer meaningful and superior-based computing devices and mechanisms to gather stream-based signals and data like temperature, heart rate (HR), blood glucose, blood pressure, cardiovascular sensors like PPG, ECG, etc.

Additionally, BDA processes services for healthcare provision, illness exploration, and system support. BDA is also used in the healthcare industry to tailor healthcare services, react to public health programs and disease evolution, and share patient data with doctors (Vrushank & Khang, 2023).

16.5.3 Transaction in Financial Sector

Financial scandals, money laundering, and financial fraud are all on the rise in the industry. Financial domains like computer systems, security agencies and various commodities have generated a significant amount of data that has revealed an increase in crime rates as well as other nefarious business practices in the financial institutions (Aggarwal & Khang, 2023).

As a result, BDA technology has demonstrated a significant impact in addressing significant financial abnormalities. For instance, a BDA approach to assess the credit risk platform of e-business (Rani & Chauhan et al., 2021).

A recent study by Yang and Yu et al. (2020) also suggested a BDA model (like SVM) for forecasting financial market volatility.

To help regulators and investors to succeed in the market capital, key factors, including modeling both high frequency and short storage, Barbaglia and Consoli et al. (2021) also used Syferlock and MapReduce in HDFS to stop fraudulent happenings in financial institutions involving credit and debit logs.

The big data technologies (like MapReduce) logic in HDFS increased significantly for speedy execution response and analysis of credit/debit card data.

16.5.4 National Security

The evaluation of the chosen studies demonstrates the value of BDA in enhancing national security. For instance, Gurlev and Yemelyanova et al. (2019) used the Russian Far North as a case study to create a transformation of services to sustain data-driven security. This equipment enhanced the Northern Sea Route's high-speed infrastructure management and telecommunication services.

In order to foresee future approaches to managing this dreadful issue that threatens the peace of the populace, data gathered in that regard must be effectively connected with appropriate high-tech intelligence and learned technologies. Therefore, to analyze security data, BDA methodologies like clustering methods, machine learning, etc. (Bhambri & Khang et al., 2022).

16.5.5 Logistics and Transportation

BDA were used in logistics and transportation for operational research, management, revenue optimization and other intelligent transit. BDA's experience in the logistics and intelligent transportation sectors aids in addressing issues, including high fuel costs, high carbon dioxide emissions, traffic congestion, vehicle management and enhancing road safety.

Additionally, it raises global economic performance, e-commerce and individual mobility (Khang & Gupta et al., 2023).

16.5.6 Prediction of Power of Transportation

BDA has also made great progress in the prediction of transient power. To increase the power stability and operation, for instance used multidimensional data created, hostile support vector machine (ASVM), and traditional support vector machine (CSVM) approaches.

BDA also aids in the fourth industrial revolutions planning, manufacture, operation, and handling the power networks. Machine learning, one big data technology, aids in the forecast of the power system for the best possible output, sales and market usage (Rani & Khang et al., 2023).

16.6 CONCLUSION

With the advent of the Web 2.0 period, internet usage rose, making it easier and less expensive for businesses to acquire large volumes of data. With the advent of the Web 3.0, options for external data collecting have even grown (Khang & Gupta et al., 2023).

Nowadays, practically every firm has quietly transitioned to the big data era as a result of the realization that data-driven decisions are frequently smarter and more accurate (Khang & Chowdhury et al., 2022).

However, just because a lot of businesses across a variety of industries are using business analytics, including BDA, doesn't mean that they are all reaping the rewards by gaining insightful knowledge and practical business value from the available data (Jebaraj & Khang et al., 2024).

Using analytical methods and technologies is only one aspect of being a data-driven organization. To ensure the effectiveness of data-driven decision-making, businesses must employ individuals with systematic thinking skills (Subhashini & Khang, 2024).

Today's data-driven business climate requires the ability to think analytically while making decisions. Domain knowledge and analysis cannot be viewed as separate fields because the amount of data is always increasing (Shah & Jani et al., 2024).

REFERENCES

Aggarwal, P., Khang, A. "A Study on the Impact of the Industry 4.0 on the Employees Performance in Banking Sector", *Designing Workforce Management Systems for Industry 4.0: Data-Centric and AI-Enabled Approaches* (1st Ed.), pp. 384–400 (2023). CRC Press. https://doi.org/10.1201/9781003357070-20

Barbaglia, L., Consoli, S., Manzan, S., Recupero, D.R., Saisana, M., Pezzoli, L.T. "Data Science Technologies in Economics and Finance: A Gentle Walk-in", *Data Science for Economics and Finance*, pp. 1–17 (2021). Springer. https://doi.org/10.1007/978-3-030-66891-4

Bhambri, P., Rani, S., Gupta, G., Khang, A. "*Cloud and Fog Computing Platforms for Internet of Things*" (2022). CRC Press. https://doi.org/10.1201/9781032101507

Dai, T., Rosenberg, J.M., Lawson, M. "Data representations and visualizations in educational research", *Routledge*, 2022. https://doi.org/10.4324/9781138609877-REE148-1

Dorn, F., Khailaie, S., Stoeckli, M., Binder, S.C., Mitra, T., Lange, B., Lautenbacher, S. et al. "The common interests of health protection and the economy: evidence from scenario calculations of COVID-19 containment policies", *The European Journal of Health Economics*, p. 18 (2022). https://doi.org/10.1007/s10198-022-01452-y

Dubey, R., Bryde, D.J., Dwivedi, Y.K., Graham, G., Foropon, C. "Impact of artificial intelligence driven big data analytics culture on agility and resilience in humanitarian supply chain: a practice-based view", *International Journal of Production Economics*, p. 108618 (2022). https://doi.org/10.1016/j.ijpe.2022.108618

Firouzi, F., Rahmani, A.M., Mankodiya, K., Badaroglu, M., Merrett, G.V., Wong, P., Farahani, B. "Internet-of-things and big data for smarter healthcare: from device to architecture, applications and analytics", *Future Generation Computer Systems*, vol. 78, pp. 583–586 (2018). https://doi.org/10.1016/j.future.2017.09.016

Gopalakrishnan, et al. "Data-driven machine criticality assessment–maintenance decision support for increased productivity", *Production Planning & Control* vol. 33, no. 1, pp. 1–19 (2022). https://doi.org/10.1080/09537287.2020.1817601

Gupta, S.K., Khang, A., Somani, P., Dixit, C.K., Pathak, A. "Data Mining Processes and Decision-Making Models in Personnel Management System", *Designing Workforce Management Systems for Industry 4.0: Data-Centric and AI-Enabled Approaches* (1st Ed.), pp. 89–112 (2023). CRC Press. https://doi.org/10.1201/9781003357070-6

Gurlev, I., Yemelyanova, E., Kilmashkina, T. "Development of Communication as a Tool for Ensuring National Security in Data-Driven World (Russian Far North Case-Study)", *Big Data-Driven World: Legislation Issues and Control Technologies*, pp. 237–248 (2019). Springer,.

Helbing, D. "*Thinking Ahead-Essays on Big Data, Digital Revolution, and Participatory Market Society*" Vol. 10 (2016). Springer. https://doi.org/10.1007/978-3-319-15078-9

Jebaraj, L., Khang, A., Chandrasekar, V., Pravin, A.R., Sriram, K. (Eds.). "Smart City Concepts, Models, Technologies and Applications", *Smart Cities: IoT Technologies, Big Data Solutions, Cloud Platforms, and Cybersecurity Techniques* (1st Ed.) (2024). CRC Press. https://doi.org/10.1201/9781003376064-1

Jones, M., Collier, G., Reinkensmeyer, D.J., DeRuyter, F., Dzivak, J., Zondervan, D., Morris, J. "Big data analytics and sensor-enhanced activity management to improve effectiveness and efficiency of outpatient medical rehabilitation", *International Journal of Environmental Research and Public Health*, vol. 17, no. 3, p. 748 (2020). https://doi.org/10.3390/ijerph17030748

Khang, A. "Material4Studies", *Material of Computer Science, Artificial Intelligence, Data Science, IoT, Blockchain, Cloud, Metaverse, Cybersecurity for Studies* (2023). https://www.researchgate.net/publication/370156102_AlexKhangMaterial4Studies

Khang, A., Chowdhury, S., Sharma, S. "*The Data-Driven Blockchain Ecosystem: Fundamentals, Applications, and Emerging Technologies*" (2022). CRC Press. https://doi.org/10.1201/9781003269281

Khang, A., Gupta, S.K., Rani, S., Karras, D.A. "*Smart Cities: IoT Technologies, Big Data Solutions, Cloud Platforms, and Cybersecurity Techniques*" (1st Ed.) (2023). CRC Press. https://doi.org/10.1201/9781003376064

Khang, A., Gupta, S.K., Shah, V., Misra, A. "*AI-Aided IoT Technologies and Applications in the Smart Business and Production*" (1st Ed.) (2023). CRC Press. https://doi.org/10.1201/9781003392224

Khang, A., Hahanov, V., Abbas, G.L., Hajimahmud, V.A. "Cyber-Physical-Social System and İncident Management", *AI-Centric Smart City Ecosystems: Technologies, Design and Implementation* (1st Ed.) (2022). CRC Press. https://doi.org/10.1201/9781003252542-2

Khang, A., Rana, G., Tailor, R.K., Hajimahmud, V.A. "*Data-Centric AI Solutions and Emerging Technologies in the Healthcare Ecosystem*" (1st Ed.) (2023). CRC Press. https://doi.org/10.1201/9781003356189

Khang, A., Rani, S., Sivaraman, A.K. "*AI-Centric Smart City Ecosystems: Technologies, Design and Implementation*" (1st Ed.) (2022). CRC Press. https://doi.org/10.1201/9781003252542

Khanh, H.H., Khang, A. "The Role of Artificial Intelligence in Blockchain Applications", *Reinventing Manufacturing and Business Processes through Artificial Intelligence*, pp. 20–40 (2021). CRC Press. https://doi.org/10.1201/9781003145011-2

Kreiser, R.P., Wright, A.K., McKenzie, T.L., Albright, J.A., Mowles, E.D., Hollows, J.E., Limbocker, S., Eslinger, M., Limbocker, R., Nguyen, L.T. "Utilization of standardized college entrance metrics to predict undergraduate student success in chemistry", *Journal of Chemical Education*, vol. 99, no. 4, pp. 1725–1733 (2022). https://doi.org/10.1021/acs.jchemed.1c00719

Larson, D., Chang, V. "A review and future direction of agile, business intelligence, analytics and data science", *International Journal of Information Management*, vol. 36, no. 5, pp. 700–710 (2016). https://doi.org/10.1016/j.ijinfomgt.2016.04.013

Lutfi, A., et al. "Drivers and impact of big data analytic adoption in the retail industry: a quantitative investigation applying structural equation modeling", *Journal of Retailing and Consumer Services*, vol. 70, no. 103129 (2023). https://doi.org/10.1016/j.jretconser.2022.103129

Manogaran, G., Varatharajan, R., Lopez, D., Malarvizhi, P., Sundarasekar, R., Thota, C. "A new architecture of internet of things and big data ecosystem for secured smart healthcare monitoring and alerting system", *Future Generation Computer Systems*, vol. 82, pp. 375–387 (2017).

Oguine, O.C., Oguine, K.J., Bisallah, H.I. "*Big Data and Analytics Implementation in Tertiary Institutions to Predict Students Performance in Nigeria*", 2022. 2207.14677. https://doi.org/10.48550/arXiv.2207.14677

Power, D.J. "Understanding data-driven decision support systems", *Information Systems Management*, vol. 25, no. 2, pp. 149–154 (2008). https://doi.org/10.1080/10580530801941124

Rana, G., Khang, A., Sharma, R., Goel, A.K., Dubey, A.K. "*Reinventing Manufacturing and Business Processes through Artificial Intelligence*" (2021). CRC Press. https://doi.org/10.1201/9781003145011

Rani, S., Bhambri, P., Kataria, A., Khang, A., Sivaraman, A.K. "*Big Data, Cloud Computing and IoT: Tools and Applications*" (1st Ed.) (2023). Chapman and Hall/CRC. https://doi.org/10.1201/9781003298335

Rani, S., Chauhan, M., Kataria, A., Khang, A. "IoT Equipped Intelligent Distributed Framework for Smart Healthcare Systems", *Networking and Internet Architecture* (2021). CRC Press. https://doi.org/10.48550/arXiv.2110.04997

Shah, V., Jani, S., Khang, A. "Automotive IoT: Accelerating the Automobile Industry's Long-Term Sustainability in Smart City Development Strategy", *Smart Cities: IoT Technologies, Big Data Solutions, Cloud Platforms, and Cybersecurity Techniques* (1st Ed.) (2024). CRC Press. https://doi.org/10.1201/9781003376064-9

Sivarajah, U., Kamal, M.M., Irani, Z., Weerakkody, V. "Critical analysis of big data challenges and analytical methods", *Journal of Business Research*, vol. 70, pp. 263–286 (2017). https://doi.org/10.1016/j.jbusres.2016.08.001

Subhashini, R., Khang, A. "The Role of Internet of Things (IoT) in Smart City Framework", *Smart Cities: IoT Technologies, Big Data Solutions, Cloud Platforms, and Cybersecurity Techniques* (1st Ed.) (2024). CRC Press. https://doi.org/10.1201/9781003376064-3

Thiess, H., Fiol, G.D., Malone, D.C., Cornia, R., Sibilla, M., Rhodes, B., Boyce, R.D., Kawamoto, K., Reese, T. "Coordinated use of health level 7 standards to support clinical decision support: case study with shared decision making and drug-drug interactions", *International Journal of Medical Informatics*, vol. 162, p. 104749 (2022). https://doi.org/10.1016/j.ijmedinf.2022.104749

Vidhya, K., Shanmugalakshmi, R. "Modified adaptive neuro-fuzzy inference system (M-ANFIS) based multi-disease analysis of healthcare Big Data", *The Journal of Supercomputing*, vol. 76, no. 11, pp. 8657–8678 (2020). https://doi.org/10.1007/s11227-019-03132-w

Vrushank, S., Khang, A. "Internet of Medical Things (IoMT) Driving the Digital Transformation of the Healthcare Sector", *Data-Centric AI Solutions and Emerging Technologies in the Healthcare Ecosystem* (1st Ed.), p. 1 (2023). CRC Press. https://doi.org/10.1201/9781003356189-2

Vrushank, S., Vidhi, T., Khang, A. "Electronic Health Records Security and Privacy Enhancement Using Blockchain Technology", *Data-Centric AI Solutions and Emerging Technologies in the Healthcare Ecosystem* (1st Ed.), p. 1 (2023). CRC Press. https://doi.org/10.1201/9781003356189-1

Wei, J., Karuppiah, M., Prathik, A. "College music education and teaching based on AI techniques", *Computers and Electrical Engineering*, vol. 100 (2022). https://doi.org/10.1016/j.compeleceng.2022.107851

Yang, R., Yu, L., Zhao, Y. et al. "Big data analytics for financial market volatility forecast based on support vector machine", *International Journal of Information Management*, vol. 50, pp. 452–462 (2020). https://doi.org/10.1016/j.ijinfomgt.2019.05.027

17 Dual Access Control for Cloud-Based Data Storage and Sharing

Geetha C., Neduncheliyan S., and Alex Khang

17.1 INTRODUCTION

17.1.1 What Is Cloud Computing?

Resource excretion of computers (hardware and software) provided as a service across a network is known as cloud computing (typically the internet). The name is derived from the widespread use of a cloud-shaped symbol in system diagrams as a metaphor for the intricate infrastructure it holds.

Cloud computing entrusts the data, software, and processing of a user to remote services. Hardware and software resources are made accessible via the internet as managed third-party services in cloud computing. These services often give users access to cutting-edge server networks and sophisticated software programmers.

17.1.2 How Cloud Computing Works?

Cloud computing aims to apply conventional supercomputing, or high-performance computing power, typically used by military and research facilities, to perform tens of trillions of computations per second in consumer-oriented applications like financial portfolios, deliver personalized information, provide data storage, or power massively multi-player computer games (Rani & Chauhan et al., 2021).

Networks of enormous clusters of computers, often running low-cost consumer PC technology with specialized connections, are used in cloud computing to distribute data processing tasks among them. Large networks of interconnected systems make up this shared IT infrastructure. Virtualization methods are frequently employed to increase the power of cloud computing (Bhambri & Khang et al., 2022).

17.1.3 Characteristics and Services Models

The characteristics of salient cloud computing based on the initial explanation provided by the National Institute of Standards and Terminology (NIST) are outlined (Figure 17.1) as follows:

On-demand self-service: Consumers can autonomously provision computing resources, including server time and network storage, as needed automatically without involving the providers of those services in any direct communication.

DOI: 10.1201/9781003376064-17

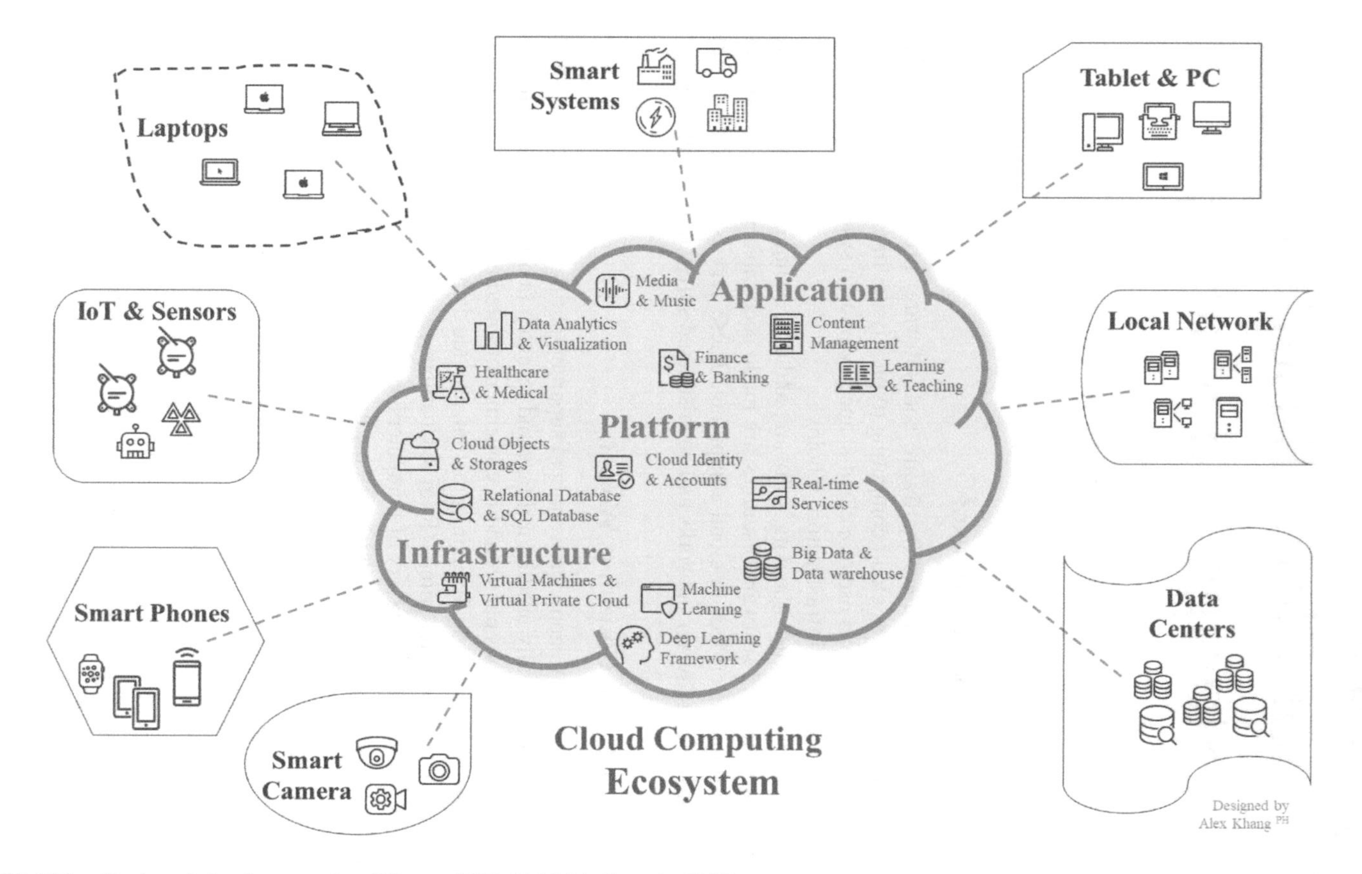

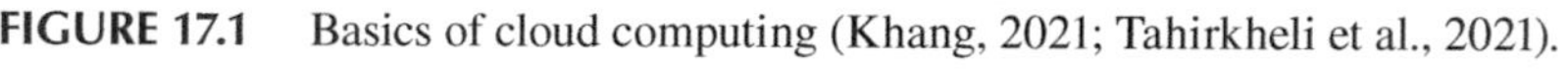
FIGURE 17.1 Basics of cloud computing (Khang, 2021; Tahirkheli et al., 2021).

Broad network access: Capabilities are available over the network and penetrate through caliber mechanisms to be promoted by heterogeneous platforms with a thin or thick client (e.g., TaB phones, laptops, and PDAs).

Resource pooling: Using a multi-tenant approach, the provider pools computing resources to serve several customers to be overcome with various physical and virtual resources that are dynamically assigned and reassigned in response to customer demand. The customer typically has no control or knowledge over the precise location of the resources offered, although they might be able to specify location at a higher level of abstraction, giving the impression that the resources are location-independent (e.g., country, state, or data center). Storage, computation, memory, network bandwidth, and virtual machines (VMs) are a few examples of resources.

Rapid elasticity: For quick scale out and quick scale in, capabilities can be swiftly and elastically provisioned, sometimes automatically. Consumers frequently perceive the provisioning capabilities as being limitless and able to be ordered in any quantity at any time.

Measured service: By utilizing a metering capability at an abstraction level relevant to the type of service, cloud systems automatically manage and optimize resource utilization (e.g., storage, processing, bandwidth, and active user accounts). The management, control, and reporting of resource utilization can ensure transparency for both the service provider and the service user as Figure 17.2.

17.2 CHARACTERISTICS OF CLOUD COMPUTING

17.2.1 Services Models

Three main service models are included in cloud computing: Infrastructure-as-a-Service (IaaS), Platform-as-a-Service (PaaS), and Software-as-a-Service (SaaS). An end-user layer that encompasses the end-user perspective on cloud services completes the three service models or layers.

A cloud user can run her applications on the resources of a cloud infrastructure if she accesses services at the infrastructure layer, for example, and is still in charge of the support, upkeep, and security of these apps (Khang & Hahanov et al., 2022). These responsibilities are usually handled by the cloud service provider (CSP) if she uses a service that is accessible at the application layer as Figure 17.3.

17.2.2 Benefits of Cloud Computing

- Achieve economies of scale: Boost output or productivity with fewer workers. Your cost per unit, project fall dramatically.
- Achieve economies of scale: Boost the executed result or productivity with fewer workers or good.
- Reduce spending on technology infrastructure: Maintain simple access to your data with little initial outlay. Depending on demand, pay as you go (weekly, quarterly, or annually).

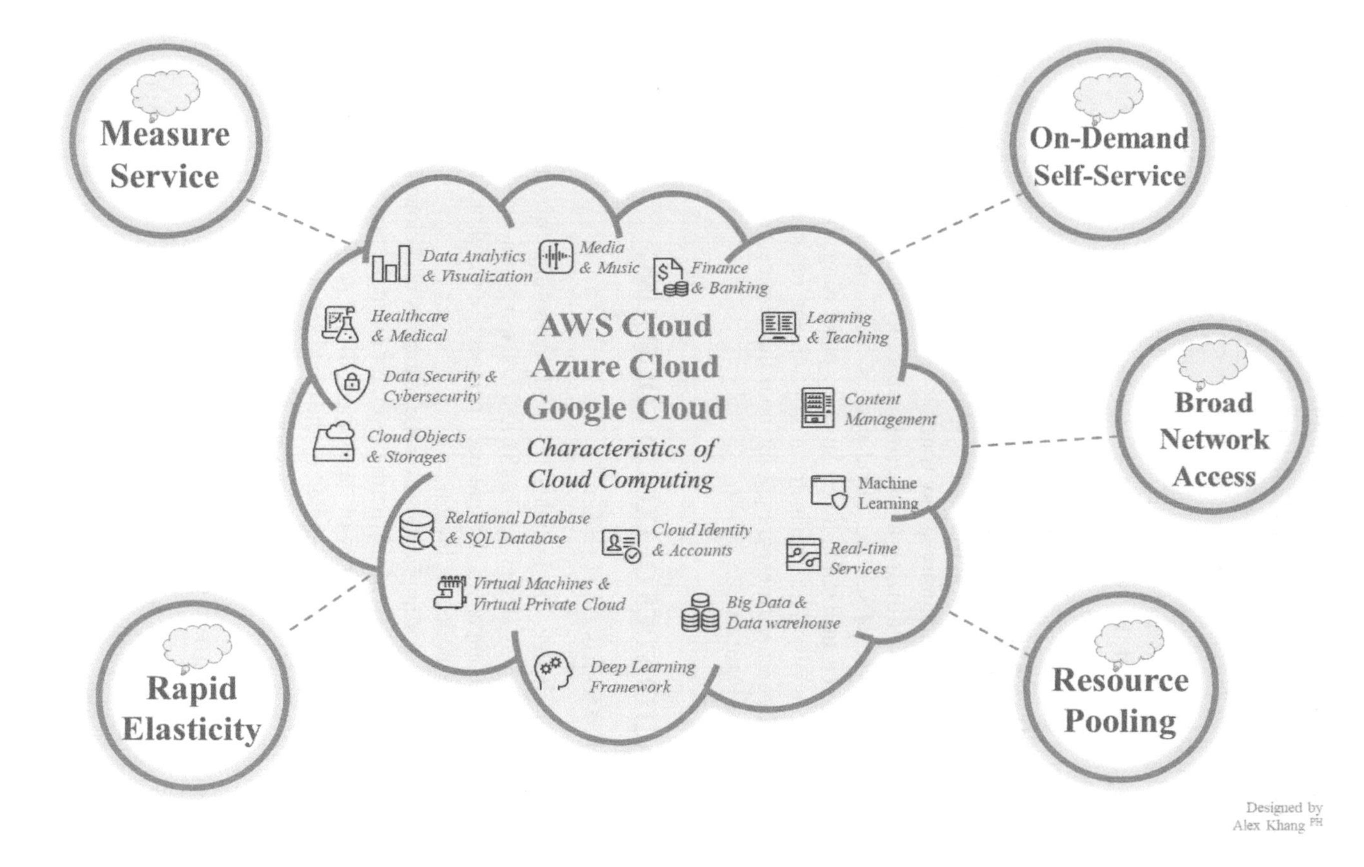

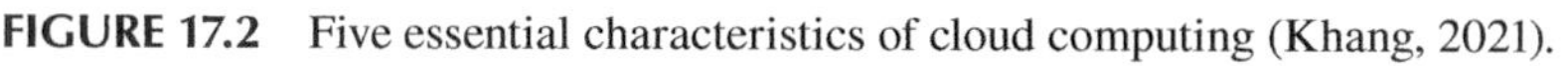
FIGURE 17.2 Five essential characteristics of cloud computing (Khang, 2021).

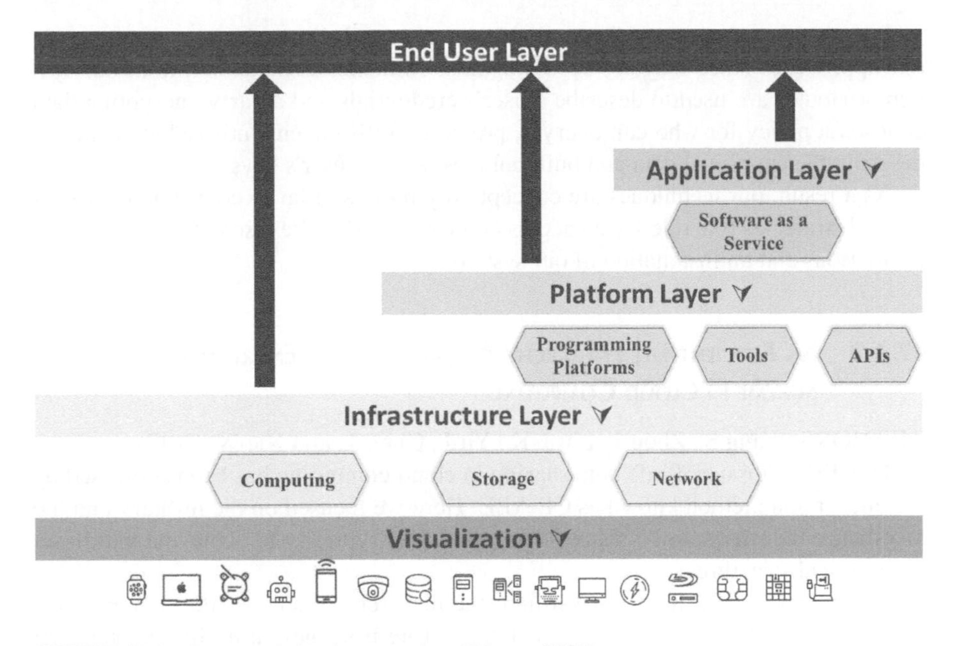

FIGURE 17.3 Layers of cloud computing (Khang, 2021).

- Globalize your workforce cheaply: As long as they have an internet connection, anyone can access the cloud.
- Simplify processes: To do more in less time with fewer personnel.
- Lower the cost of capital: There is no need to invest a lot of money on license costs, hardware, or software.
- Increasing access: Your life is made so much easier by having access whenever and wherever you want!
- Improved project monitoring: Keep costs under control and ahead of the completion cycle times.

17.3 LITERATURE SURVEY

17.3.1 Cipher text-policy Attribute-Based Encryption

AUTHOR: Waters B. (2011)

A user should only be permitted access to data in some distributed systems if they possess a specific set of credentials or qualities. Currently, using a trusted server to store the data and handle access control is the only way to enforce such regulations.

The confidentiality of the data will, however, is jeopardized if any server hosting the data is compromised. In this research, we introduce cipher text-policy attribute-based encryption (CP-ABE), a system for implementing complicated access control on encrypted data.

Even if the storage server is unreliable, encrypted data may be kept private using our approaches since they are secure against collusion assaults. While in our system attributes are used to describe a user's credentials and a party encrypting data chooses a policy for who can decrypt, previous ABE systems utilized attributes to describe the encrypted data and built policies into the user's keys.

As a result, our techniques are conceptually more similar to conventional access control strategies like role-based access control (RBAC). We also offer performance evaluations and an installation of our system.

17.3.2 An Efficient File Hierarchy Attribute-Based Encryption Scheme in Cloud Computing

AUTHORS: Wang S., Zhou J., Liu J. K., Yu J., Chen J., and Xie W. (2017)

The difficult issue of safe data sharing in cloud computing has been addressed by the encryption method known as CP-ABE. Here we focused on the military and the healthcare industries, to be shared these data files typically to achieve a multilevel hierarchical structure.

However, CP-ABE hasn't looked into the hierarchy of shared files. An effective file hierarchy ABE technique for cloud computing is suggested in this chapter. The layered access structures are combined into a single access structure, and the combined access structure is then used to encrypt the hierarchical files.

The files might share the attributes-related cipher text components. Finally, the result shows of both decryption storage and encryption time have been reduced. Additionally, it is demonstrated that the suggested system is secure under the common assumption. The proposed technique is extremely effective in terms of encryption and decryption, according to experimental simulation. The benefits of our plan become more and more obvious as the quantity of files grows.

17.3.3 CP-ABE with Constant-Size Keys for Lightweight Devices

AUTHORS: Guo F., Mu Y., Susilo W., Wong D. S., and Varadharajan V. (2014)

A limitation for many applications, particularly security applications, is the limited storage capacity of lightweight devices like radio frequency identification tags. A promising cryptographic technique called CP-ABE allows the encryptor to choose the access structure that will be utilized to safeguard the sensitive data.

The problem with current CP-ABE techniques is that their decryption keys are lengthy and inversely proportional to the number of characteristics. Due to this flaw, it is not practical to use portable devices to store the CP-ABE users' decryption keys. In this study, we offer a solution to the long-standing problem mentioned above, which will make the CP-ABE very useful.

With constant-size decryption keys that are independent of the number of characteristics, we suggest a unique CP-ABE technique. The size can be as little as 672 bits, as we discovered. The suggested system is the sole CP-ABE with expressive access structures when compared to other schemes in the literature, making it ideal for CP-ABE key storage in portable devices.

17.3.4 JPBC: Java Pairing-Based Cryptography

AUTHORS: De Caro A. and Iovino V. (2011)

A specific bilinear pairing map that adds extra structure to some cyclic groups that could be utilized in cryptography has just been shown admitted. Bilinear pairing maps were initially used to crack cryptosystems (see, for instance), but it was later discovered that the additional structure might be utilized to create cryptosystems with more advanced features.

The identity-based encryption method developed by Boneh and Franklin (2001) is the most well-known early illustration of what may be done with bilinear maps. Following that, many different cryptosystems were created employing bilinear maps. Up until this effort, there was no complete, open-source implementation of pairing-based cryptography.

Recent approaches either lack the necessary source code or only support a small variety of elliptic curves, which prevents them from reaching this objective. Furthermore, none of them uses preprocessing, which is essential to speed up calculation. In this work, we introduce Java pairing-based cryptography (JPBC), a Java port of the C-written PBC library.

Even for a non-cryptographer, JPBC offers a complete ecosystem of interfaces and classes to make using the bilinear maps simple. JPBC is prepared for mobile use and supports a variety of elliptic curve types. Preprocessing can greatly speed up computation. To gauge the difference between the two libraries, a benchmark comparison between JPBC and PBC has been conducted. Additionally, JPBC has been evaluated on several Android mobile platforms.

17.3.5 Hierarchical Attribute-Based Encryption and Scalable User Revocation for Sharing Data in Cloud Server

AUTHORS: Wang G., Liu Q., Wu J., and Guo M. (2010)

As cloud computing expands quickly, more and more businesses will outsource the sharing of their sensitive data online. It seems sensible to only store encrypted data in the cloud to protect shared data from unreliable CSPs.

Establishing access control for the encrypted data and canceling users' access rights when they are no longer permitted to view the encrypted data are the main issues with this approach. This essay attempts to address both issues.

To provide not only fine-grained access control but also full delegation and high performance, we first propose a Hierarchical ABE (HABE) scheme by combining a Hierarchical Identity-Based Encryption (HIBE) system and a CP-ABE system. Then, to effectively remove users' access rights, we provide a scalable revocation strategy that applies Proxy Re-Encryption (PRE) and Secure Re-Encryption (SRE) to the HABE scheme.

17.4 PROPOSED WORK

Twin Access Management is a novel methodology that combines Attribute-based Encryption (ABE) and Cipher text-policy ABE (CP-ABE) to address the issues with the current system of securing data in cloud-based storage services.

This methodology offers both confidentiality of outsourced data and fine-grained control over the access to that data.

ABE is a powerful technique that allows data to be encrypted and decrypted based on certain attributes or properties, rather than using traditional cryptographic keys. CP-ABE takes this one step further by allowing access policies to be defined for authorized encrypted data, specifying the access privilege of potential data receivers.

By combining these two techniques, Twin Access Management provides a reliable method of data encryption and access control for cloud-based storage services. We have also compared our findings with simulations of two other systems under identical circumstances, highlighting the advantages of our approach.

Overall, Twin Access Management offers a promising solution to the challenges of securing data in cloud-based storage services and may have implications for a wide range of industries and applications.

17.5 IMPLEMENTATION

Cloud computing architecture is a combination of service-oriented architecture and event-driven architecture, and it consists of two main parts – the front end and the back end.

The front end is the part of the cloud computing architecture that is visible to the user, and it includes the user's device or computer, the web browser or software application used to access the cloud, and the network that connects the user's device to the cloud.

The back end is the part of the cloud computing architecture that is responsible for the storage, processing, and delivery of data and services to the front end. It includes servers, storage systems, databases, and other components that are responsible for managing the data and services in the cloud.

The back end can be further divided into different layers, including the infrastructure layer, platform layer, and software layer, which are responsible for different aspects of the cloud computing environment. These layers work together to provide scalable, on-demand computing resources to users, making cloud computing a popular choice for businesses of all sizes as Figure 17.4.

17.5.1 Application

Front End: The front end is the user-facing part of a software application or website that interacts with the user and enables them to perform actions or access information. In cloud computing, the front end may consist of a web-based graphical user interface (GUI) or an application programming interface (API) that allows clients to access cloud resources, such as computing power, storage, and applications.

Back End: The back end is the server-side of an application that processes and stores data, performs computations, and manages the resources needed to provide services to the front end. In cloud computing, the back end may include virtualized servers, databases, storage systems, load balancers, and other infrastructure components.

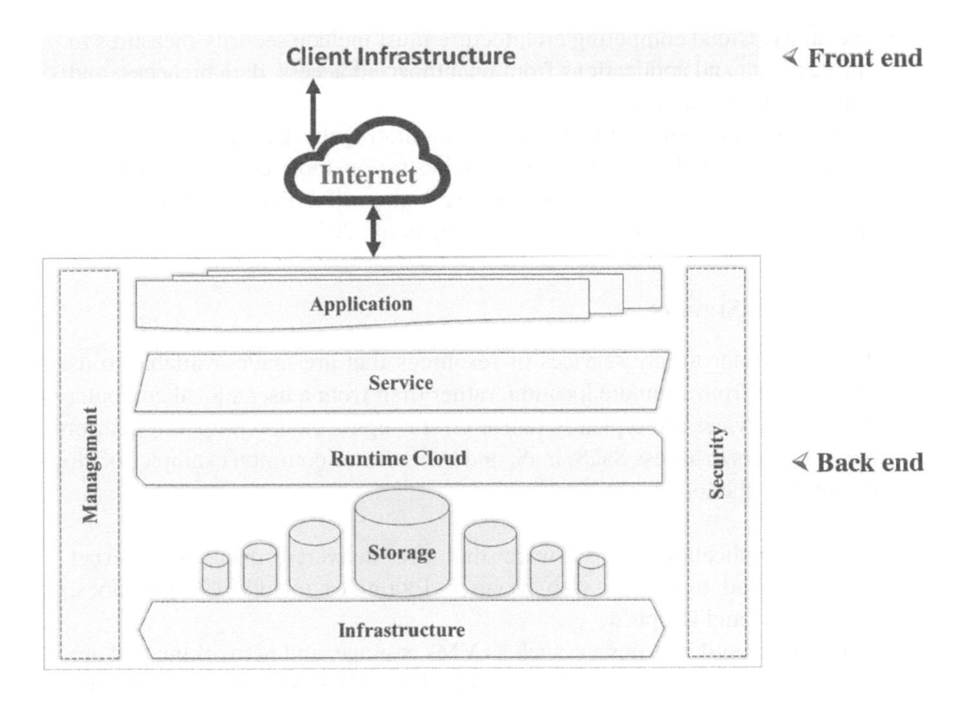

FIGURE 17.4 Architecture of cloud computing. (Khang, 2021)

It's worth noting that the front end and back end are not limited to cloud computing but are common concepts in software development and web development. Additionally, in the context of web development, the front end usually refers to the client-side code that runs in a user's browser, while the back end refers to the server-side code that runs on a web server.

Cloud computing architecture typically consists of several key components, including the following:

- **Front end:** This is the user interface that allows users to interact with the cloud-based services. It typically consists of web browsers or mobile applications.
- **Back end:** This is the cloud infrastructure that provides the necessary resources to run the cloud-based services. It includes servers, storage devices, databases, and network infrastructure.
- **Cloud service provider:** This is the company that provides the cloud-based services to customers. Examples of CSPs include Amazon Web Services (AWS), Microsoft Azure, and Google Cloud Platform.
- **APIs:** APIs are the communication protocols that enable interaction between different software components. They enable developers to create custom applications that can integrate with cloud-based services.
- **Virtualization:** This is the process of creating virtual versions of hardware, software, and storage resources. Virtualization enables the CSP to optimize resource utilization and provide on-demand scalability to customers.

- **Security:** Cloud computing architecture must include security measures to protect data and applications from unauthorized access, data breaches, and other security threats.
- **Orchestration:** This is the process of automating the deployment and management of cloud-based services. Orchestration tools enable the CSP to manage resources efficiently and ensure high availability and scalability of cloud-based services (Hahanov & Khang et al., 2022).

17.5.2 Cloud Services

Cloud services refer to any services or resources that are made available to users over the internet from a remote location, rather than from a user's local computer or server. Cloud services are typically delivered through a cloud computing platform, which allows users to access SaaS, IaaS, and PaaS. Some common examples of cloud services include the following:

- **SaaS:** Applications or software that are delivered over the internet and accessed through a web browser. Examples include Google Docs, Salesforce, and Dropbox.
- **IaaS:** Computing resources, such as VMs, storage, and networking that are provided over the internet. Examples include AWS, Microsoft Azure, and Google Cloud Platform.
- **PaaS:** A platform for developing, running, and managing applications without the complexity of building and maintaining the underlying infrastructure. Examples include Heroku, Google App Engine, and Microsoft Azure.

Cloud services offer several benefits over traditional on-premise solutions, including cost savings, scalability, and flexibility. Users can access cloud services from anywhere with an internet connection and can easily scale up or down based on their needs. Additionally, cloud services often offer built-in security features and automatic backups, making them a popular choice for businesses and organizations.

17.5.3 Runtime Cloud

Runtime cloud refers to the environment in which applications and services are executed in the cloud. In this environment, the CSP manages the infrastructure, servers, and other resources necessary to run the application, while the user only needs to manage the application itself.

Runtime cloud services are often offered as part of a PaaS offering. They allow developers to deploy and run their applications in the cloud without worrying about the underlying infrastructure. The CSP takes care of scaling, load balancing, security, and other aspects of running the application.

Examples of runtime cloud services include AWS Elastic Beanstalk, Microsoft Azure App Service, and Google App Engine. These services offer support for a wide variety of programming languages and frameworks, making it easy for developers to deploy their applications in the cloud.

Using a runtime cloud service can be beneficial for businesses and organizations because it allows them to focus on developing and improving their applications without having to worry about the infrastructure. It also provides the flexibility to scale up or down as needed, reducing costs and improving performance.

17.5.4 Cloud Storage

Cloud storage refers to the practice of storing data on remote servers that are accessible over the internet. Instead of storing files locally on a device or server, cloud storage allows users to store and access files from anywhere with an internet connection.

Cloud storage is typically offered as a service by cloud providers, such as AWS, Microsoft Azure, and Google Cloud Platform. These providers offer a variety of storage options, including object storage, file storage, and block storage.

Object storage is designed for unstructured data, such as photos, videos, and documents, and allows users to store and access files as objects. Examples of object storage services include Amazon S3, Microsoft Azure Blob Storage, and Google Cloud Storage.

File storage is designed for structured data, such as databases and files that need to be accessed through a network file system. Examples of file storage services include Amazon EFS, Azure Files, and Google Cloud Filestore.

Block storage is designed for storing and accessing data as blocks, which are typically used for VMs and databases. Examples of block storage services include Amazon EBS, Azure Disk Storage, and Google Cloud Persistent Disk.

Cloud storage offers several benefits over traditional on-premise storage solutions, including cost savings, scalability, and flexibility. Users can easily scale up or down based on their storage needs and can access their files from anywhere with an internet connection. Additionally, cloud storage often offers built-in security features and automatic backups, making it a popular choice for businesses and organizations (Jebaraj & Khang et al., 2023).

17.5.5 Infrastructure

Cloud infrastructure refers to the underlying hardware and software components that make up a cloud computing environment. This includes the servers, storage devices, networking equipment, and software necessary to support cloud computing services (Khang & Rani et al., 2022).

Cloud infrastructure is typically provided by CSPs, such as AWS, Microsoft Azure, and Google Cloud Platform. These providers offer a range of infrastructure services, including compute, storage, networking, and security, which can be configured and deployed to meet specific needs.

The compute services offered by cloud providers typically include VMs, containers, and serverless computing. VMs are virtualized instances of a computer that can run any operating system and software, while containers allow multiple applications to run on a single host operating system. Serverless computing allows developers to deploy code as functions, without having to manage the underlying infrastructure.

Storage services offered by cloud providers typically include object storage, block storage, and file storage. Object storage is designed for unstructured data, such as photos and videos, while block storage is used for structured data, such as databases. File storage is used for network file sharing.

Networking services offered by cloud providers typically include virtual private clouds (VPCs), load balancers, and content delivery networks (CDNs). VPCs allow users to create a private network within the cloud, while load balancers distribute incoming traffic across multiple instances of an application. CDNs cache content at various locations around the world, making it faster for users to access content.

Cloud infrastructure provides several benefits over traditional on-premise infrastructure, including cost savings, scalability, and flexibility. Users can easily scale up or down based on their needs and can access their resources from anywhere with an internet connection. Additionally, cloud infrastructure often offers built-in security features and automatic backups, making it a popular choice for businesses and organizations (Rani & Khang et al., 2022).

17.5.6 Cloud Management

Cloud management refers to the process of managing and monitoring cloud resources and services. Cloud management is essential for ensuring that cloud services are available, secure, and optimized for performance.

Cloud management involves several key tasks, including the following:

- **Provisioning and Deployment:** The process of creating and deploying cloud resources, such as VMs, storage, and networking.
- **Monitoring and Optimization:** The process of monitoring cloud resources to ensure they are performing as expected, and optimizing them to improve performance and reduce costs.
- **Security and Compliance:** The process of ensuring that cloud resources are secure and compliant with relevant regulations and standards.
- **Cost Management:** The process of monitoring and optimizing cloud costs to ensure that resources are being used efficiently and cost-effectively.

Cloud management is typically performed using specialized tools and platforms that provide visibility into cloud resources and services. These tools can be provided by CSPs, such as AWS, Microsoft Azure, and Google Cloud Platform, or by third-party vendors that specialize in cloud management.

Cloud management platforms typically provide a range of features and capabilities, including automation, monitoring, reporting, and analytics. These platforms allow users to manage cloud resources from a single dashboard, making it easier to monitor and optimize cloud services.

Effective cloud management is essential for businesses and organizations that rely on cloud services. It can help improve performance, reduce costs, and ensure that cloud resources are secure and compliant with relevant regulations and standards.

17.5.7 Cloud Security

Cloud security refers to the set of policies, technologies, and controls used to protect data, applications, and infrastructure hosted on cloud computing platforms. Cloud security is a critical aspect of using cloud services as it helps to ensure that data and assets are protected from unauthorized access, data breaches, and other cyber threats.

Cloud security involves several layers of security measures, including physical security, network security, data security, and access management. Some of the key techniques used in cloud security include encryption, multi-factor authentication, firewalls, intrusion detection and prevention systems, and security monitoring.

CSPs also play a critical role in ensuring cloud security by implementing security measures at the infrastructure level, such as secure data centers, redundant data storage, and disaster recovery plans. Additionally, cloud providers offer various security services, such as Identity and Access Management (IAM), threat detection, and compliance monitoring.

It is essential to understand that cloud security is a shared responsibility between the cloud provider and the customer. Cloud providers are responsible for securing the infrastructure, while the customer is responsible for securing the data and applications they deploy on the cloud. As such, it is crucial for customers to understand their responsibilities and take necessary measures to secure their data and applications in the cloud.

17.5.8 Dataflow Diagram

The bubble chart is another name for the dataflow diagram (DFD). It's an easy graphical formalism that will be wont to depict a system in terms of the information that's fed into it, the various operations that square measure performed thereon, and therefore the information that's made as results of those operations.

The most important modeling tool is the information flow sheet (DFD). The system's element models square measure created victimization it. These components embrace the system's operation, the information it uses, a 3rd party that engages with it, and therefore the approach data moves through it.

DFD demonstrates the information's flow through the system and therefore the varied changes that have an effect on it. It's a graphical methodology for representing data flow and therefore the changes created to information because it travels from input to output.

DFD is another name for a bubble chart. Any degree of abstraction for a system is drawn by a DFD. DFD is divided into stages that correspond to escalating purpose full complexness and knowledge flow.

17.5.9 Functionality

Dual Access Control (DAC) is a methodology that combines ABE and Attribute-based Access Control (ABAC) to provide secure storage and sharing of data in cloud-based environments. Here's how it works.

- **Data Encryption:** First, the data is encrypted using ABE techniques. In this process, data is encrypted based on certain attributes or properties of the data, and the encryption key is also based on these attributes. This ensures that only users who have the appropriate attributes can decrypt and access the data.
- **Access Control:** Next, ABAC is used to control access to the encrypted data. This involves defining policies that specify which users are allowed to access the data based on their attributes. For example, a policy might state that only users with a certain job title or in a certain department can access the data.

- **DAC:** In DAC, there are two levels of access control. The first level controls access to the encrypted data based on the attributes of the user. The second level controls access to the decryption key based on the attributes of the user. This ensures that only authorized users can access both the encrypted data and the decryption key.
- **Secure Sharing:** DAC also allows for secure sharing of data. When a user wants to share encrypted data with another user, they can generate a new decryption key based on the attributes of the user they want to share with. This ensures that the recipient can decrypt and access the data, but no one else can.

Overall, DAC provides a robust and flexible approach to securing data in cloud-based environments. It ensures that only authorized users can access the data and decryption key and allows for secure sharing of data between authorized users.

17.6 RESULTS AND DISCUSSION

The DAC is a security mechanism that provides two layers of access control for cloud-based data storage and sharing. The DAC model combines RBAC and ABAC to provide fine-grained access control.

The implementation of DAC involves integrating RBAC and ABAC policies into the cloud-based data storage and sharing system. RBAC policies define roles and permissions, while ABAC policies define access based on attributes.

The implementation of DAC in a cloud-based data storage and sharing system has several benefits, including the following:

- **Improved security:** DAC provides two layers of access control, making it more difficult for unauthorized users to gain access to sensitive data.
- **Fine-grained access control:** DAC enables the specification of access control policies based on user attributes and roles, providing more granular control over data access.
- **Flexibility:** DAC allows for dynamic changes to access control policies based on changes in user attributes or roles.

DAC is a security mechanism used for cloud-based data storage and sharing. It provides an extra layer of security by allowing data owners to grant access to their data based on two factors: user identity and user attributes. The purpose of DAC is to ensure that only authorized users can access the data stored in the cloud and prevent unauthorized access to sensitive information.

The result of implementing DAC for cloud-based data storage and sharing is improved security for the data stored in the cloud. DAC provides a more sophisticated access control mechanism that allows data owners to control who has access to their data and under what conditions.

By incorporating user identity and user attributes, DAC ensures that only authorized users can access the data stored in the cloud. This reduces the risk of data breaches and unauthorized access to sensitive information. Additionally, DAC allows data owners to specify different levels of access for different users, which further enhances the security of the stored data.

DAC also provides an audit trail that allows data owners to track who has accessed their data and when. This feature helps identify potential security breaches and allows data owners to take corrective action to prevent further unauthorized access.

In summary, the implementation of DAC for cloud-based data storage and sharing provides improved security for the stored data, reduces the risk of data breaches, and allows data owners to control who has access to their data and under what conditions.

17.7 CONCLUSION

We showed twice access management systems associated with self-addressing an intriguing and pervasive issue with cloud-based information sharing. DDoS/Economic Denial of Sustainability (EDoS) assaults cannot be used against the prompt systems (Khanh & Khang, 2021).

We have a tendency to claim that completely different CP-ABE constructions will "trans-plant" the tactic used to get the feature with network self-activated management on transfer requests.

The projected solutions do not impose an outsized procedure or communication overhead, per the findings of our experiments (compared to its underlying CP-ABE building block). We have a tendency to use the shortcoming to extract the key data keep within the territory in our improved system (Rana & Khang et al., 2021).

The access patterns (Waters, 2011) or alternative comparable side-channel attacks (Gentry & Silverberg, 2002) reveal that territory might, however, leak a number of its secret(s) to a malicious host. Thus, Li and Wang et al. (2011) introduce the clear territory execution model.

Associate intriguing challenge is making a twin access management theme for cloud information sharing from a clear territory. We'll take under consideration the relevant application for the guidance of problem-solving approach in our coming work (Rani & Khang et al., 2023).

REFERENCES

Bhambri, P., Rani, S., Gupta, G., Khang, A. *Cloud and Fog Computing Platforms for Internet of Things* (2022). CRC Press. https://doi.org/10.1201/9781032101507

Boneh-franklin Identity-Based Cryptosystem (2001). The Boneh–Franklin scheme is an identity-based encryption system proposed by Dan Boneh and Matthew K. Franklin in 2001. https://en.wikipedia.org/wiki/Boneh%E2%80%93Franklin_scheme

De Caro, A., Iovino, V. "JPBC: Java pairing based crypto graphy", in *IEEE Symposium on Computers and Communications (ISCC)*, pp. 850–855 (2011). IEEE. https://ieeexplore.ieee.org/abstract/document/5983948/

Gentry, C., Silverberg, A. "Hierarchical id-based cryptography", in *International Conference on the Theory and Application of Cryptology and Information Security*, pp. 548–566 (2002). Springer. https://link.springer.com/chapter/10.1007/3-540-36178-2_34

Guo, F., Mu, Y., Susilo, W., Wong, D.S., Varadharajan, V. "CP-ABE with constant-size keys for lightweight devices", *IEEE Transactions on Information Forensics and Security*, vol. 9, no. 5, pp. 763–771 (2014). https://ieeexplore.ieee.org/abstract/document/6757025/

Hahanov, V., Khang, A., Litvinova, E., Chumachenko, S., Hajimahmud, V.A., Alyar, A.V. "The Key Assistant of Smart City – Sensors and Tools", *AI-Centric Smart City Ecosystems: Technologies, Design and Implementation* (1st Ed.) (2022). CRC Press. https://doi.org/10.1201/9781003252542-17

Jebaraj, L., Khang, A., Vadivelraju, C., Antony, R.P., Kumar, S. (Eds.). "Smart City Concepts, Models, Technologies and Applications", *Smart Cities: IoT Technologies, Big Data Solutions, Cloud Platforms, and Cybersecurity Techniques* (1st Ed.) (2023). CRC Press. https://doi.org/10.1201/9781003376064-1

Khang, A., (2021). Material4Studies, *Material of Computer Science, Artificial Intelligence, Data Science, IoT, Blockchain, Cloud, Metaverse, Cybersecurity for Studies.* https://www.researchgate.net/publication/370156102_Material4Studies

Khang, A., Hahanov, V., Abbas, G.L., Hajimahmud, V.A. "Cyber-Physical-Social System and İncident Management", *AI-Centric Smart City Ecosystems: Technologies, Design and Implementation* (1st Ed.) (2022). CRC Press. https://doi.org/10.1201/9781003252542-2

Khang, A., Rani, S., Sivaraman, A.K. *AI-Centric Smart City Ecosystems: Technologies, Design and Implementation* (1st Ed.) (2022). CRC Press. https://doi.org/10.1201/9781003252542

Khanh, H.H., Khang, A. "The Role of Artificial Intelligence in Blockchain Applications", *Reinventing Manufacturing and Business Processes through Artificial Intelligence*, pp. 20–40 (2021). CRC Press. https://doi.org/10.1201/9781003145011-2

Li, J., Wang, Q., Wang, C., Ren, K. "Enhancing attribute-basedencryption with attribute hierarchy", *Mobile Networks and Applications*, vol. 16, no. 5, pp. 553–561 (2011). https://link.springer.com/article/10.1007/s11036-010-0233-y

Rana, G., Khang, A., Sharma, R., Goel, A.K., Dubey, A.K. *Reinventing Manufacturing and Business Processes through Artificial Intelligence* (2021). CRC Press. https://doi.org/10.1201/9781003145011

Rani, S., Bhambri, P., Kataria, A., Khang, A. "Smart City Ecosystem: Concept, Sustainability, Design Principles and Technologies", *AI-Centric Smart City Ecosystems: Technologies, Design and Implementation* (1st Ed.) (2022). CRC Press. https://doi.org/10.1201/9781003252542-1

Rani, S., Bhambri, P., Kataria, A., Khang, A., Sivaraman, A.K. *Big Data, Cloud Computing and IoT: Tools and Applications* (1st Ed.) (2023). Chapman and Hall/CRC. https://doi.org/10.1201/9781003298335

Rani, S., Chauhan, M., Kataria, A., Khang, A. "IoT equipped intelligent distributed framework for smart healthcare systems", *Networking and Internet Architecture* (2021). CRC Press. https://doi.org/10.48550/arXiv.2110.04997

Tahirkheli, A.I., et al. "A survey on modern cloud computing security over smart city networks: threats, vulnerabilities, consequences, countermeasures, and challenges", *Electronics*, vol. 10, no. 15, p. 1811 (2021). https://doi.org/10.3390/electronics10151811

Wang, G., Liu, Q., Wu, J. "Hierarchical attribute-based encryption for fine-grained access control in cloud storage services", in *Proceedings of the 17th ACM Conference on Computer and Communications Security*, pp. 735–737 (2010). ACM. https://dl.acm.org/doi/abs/10.1145/1866307.1866414

Wang, S., Zhou, J., Liu, J.K., Yu, J., Chen, J., Xie, W. "An efficient file hierarchy attribute based encryption scheme in cloud computing", *IEEE Transactions on Information Forensics and Security*, vol. 11, no. 6, pp. 1265–1277 (2017). https://www.tandfonline.com/doi/abs/10.1080/19361610.2019.1649534

Waters, B. "Ciphertext-policy attribute-based encryption: An expressive, efficient, and provably secure realization", *International Workshop on Public Key Cryptography*, pp. 53–70 (2011). Springer. https://link.springer.com/chapter/10.1007/978-3-642-19379-8_4

18 An Analysis of Securing Internet of Things (IoT) Devices from Man-in-the-Middle (MIMA) and Denial of Service (DoS)

Vicky Tyagi, Amar Saraswat, and Shweta Bansal

18.1 INTRODUCTION

The Internet of Things (IoT), a connection that connects numerous things, including tangible objects, has emerged consequently by combining the notion of the Internet alongside numerous technology (wired and wireless) to handle and monitor multiple activities in our surroundings (Rizzardi & Sicari et al., 2016).

IoT includes cloud-based platforms, surveillance systems, smart parking, entertainment-related activities, smart cities, machine-to-machine communications, and smart electricity grids. The Internet of Everything (IoE), which could also simplify all data generated and shared through these IoT devices and sensors, will revolutionize our civilization (Saxena & Grijalva et al., 2016).

Because of its capacity to deliver a wide range of applications, the IoT has now become a constantly evolving innovation with a significant impact on community life and corporate environments (Khang & Rani et al., 2022).

The IoT is becoming pervasive in many aspects of modern life, particularly education, healthcare, and business, and includes the storage of private information about corporations and individuals, financial data exchanges, product creation, and marketing (Abomhara & Køien et al., 2015).

The massive proliferation of connected devices in the IoT has created a large demand for better protection to satisfy the growing requirements of millions, if not billions, of devices connected and services throughout the world (Khang & Hahanov et al., 2022).

Nearly every day, the frequency of attacks increases, and attacks become more dynamic and varied. Not just is the number of possible hackers growing in lockstep with network scale, but the tools in their arsenal have also become more sophisticated, efficient, and impactful. As a consequence, IoT must also be secured from dangers and vulnerabilities in order to accomplish its maximum capabilities (Rani & Chauhan et al., 2021).

Security is defined as a way of protecting an item against physical harm, security breaches, stealing, or damage by preserving the confidentiality and integrity of data

DOI: 10.1201/9781003376064-18

about the item and making that available information whenever needed (Bhambri & Rani et al., 2022).

According to some researchers, there really is no such thing as a safe mode for just anything, physical or digital, because no element can ever be highly secured while still being effective (Bertino & Islam, 2017). A method is deemed to be safe if it can maintain its maximum essential nature in a variety of scenarios.

IoT security standards are the same as they are for any other ICT system. As a consequence, maintaining the highest intrinsic value of both tangible and intangible goods (devices) is critical to ensuring IoT security (services, information, and data).

18.2 ARCHITECTURE OF IoT

The IoT middleware is a layer of software that sits between both the technical and application levels and is made up of a number of sub-layers (Patil & Chaudhari, 2016). The essential virtue of concealing diverse technical aspects in middleware allows a programmer to concentrate on the creation of the particular applications allowed by IoT infrastructures, rather than on difficulties that are not specifically linked to his attention (AlFuqaha & Guizani et al., 2020).

The Service Oriented Architecture (SOA) notion is used in IoT middleware systems. The implementation of SOA principles allows composite systems to be decomposed into services with smaller and very well elements. Because an SOA method does not necessitate a set technology for service implementation, it allows for the reuse of both hardware and software (Tahir & Kanwer et al., 2016).

The IoT must be capable of connecting billions or trillions of different objects, which necessitates a vibrant distributed architecture (Habibzadeh & Nussbaum et al., 2019). The ever-increasing number of possible designs has still yet to yield a basic framework.

Some initiatives, like IoT-A, are seeking to create a standard infrastructure based on a better knowledge of researcher and industrial needs.

The core model, which was picked from a collection of proposed approach, is a three-layer architecture consisting of the Application, Network, and Perception Layers.

In the literature, further approaches that add additional sophistication to the IoT architecture have recently been suggested. One of the popular designs depicted in Figure 18.1 is the 5-layer variant, which has been used. Following that, we'll take a quick look at all these five layers.

18.2.1 Application Layer

Users make requests for services, which the application layer fulfills. For instance, the application layer is responsible for providing temperature and pressure information to a customer who requests them (Rani & Bhambri et al., 2023).

This layer is important for the IoT although it may be capable of providing high-quality advance intelligent services that fulfill user's requirements. Smart home, building automation, commuting, factory automation, and intelligent healthcare are just a few of the various sectors covered by the application layer (AlFuqaha & Guizani et al., 2020).

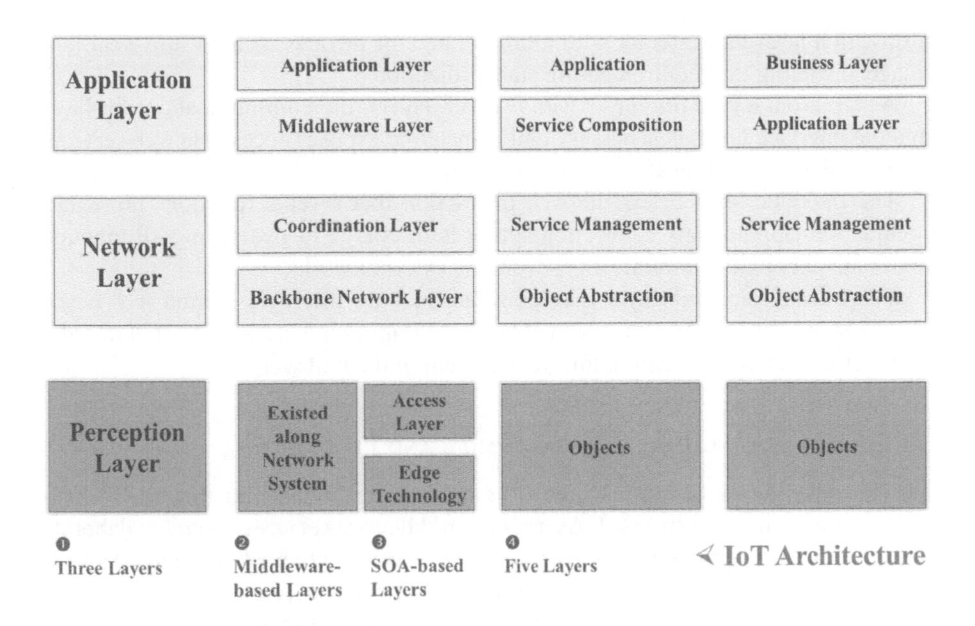

FIGURE 18.1 IoT architecture (Khang, 2021).

18.2.2 Service Composition Layer

The SOA middleware's higher layer is usually the service composition layer. The basic purpose of this layer is to give a simple configuration of specific network object services in order to construct specific implications.

The flow of business processes, and also the design and maintenance of complex processing, can be represented in this layer using workflow languages such as Business Process Execution Language (BPEL) and Jolie.

18.2.3 Service Management Layer

The service management layer provides critical functionalities which should be accessible to all objects and enables each object to be managed in an IoT framework. A basic range of services includes status monitoring, service configuration, and object dynamic detection.

This layer permits remotely the use of new services during processing times to satisfy the requirements of the application. The service management layer also offers a more advanced set of features such as quality of service management, lock management, and so on. A storage mechanism is provided at this layer to keep track of the records of services associated with each object in the network.

18.2.4 Object Abstraction Layer

The IoT is based on a principle of many different and complicated items, including one that must perform specific functions in its own language. As a consequence, an abstraction layer can interact with a broad range of devices using a universal language, obviating the requirement for standardization.

As a reason, a cladding layer with two sub-layers, the communication sub-layer and the interface layer, is necessary until a machine exposes accessible web services over an IP network (Lawal & Shaikh et al., 2020).

The interface layer offers the web interaction that reveals functions through a regular web applications connection, and it is in charge of overseeing all inbound mail and outbound communications from the external world.

The transmission sub-layer is responsible for both the logic behind web-based service strategies and the conversion of these techniques into a specific set of instructions for the machine to communicate with item in the real world.

18.2.5 Managing Integrity, Anonymity, and Protection

The introduction of communication over the Internet items into their daily lives poses a great danger to survival. As a result, middleware services should collaborate to manage data interchange trust, secrecy, and security (Khanh & Khang, 2021).

The appropriate operations can be on a specific tier of the preceding, or (as is generally the case) dispersed all from object simplification to service selection, all the way up the hierarchy, without degrading system performance or introducing unnecessary overheads.

18.3 IoT APPLICATIONS AREAS

The IoT has a lot of ramifications. IoT has a broad array of potential applications, ranging from controlling a single network for a small company to controlling an entire apartment. IoT can also be used in practically anything that would connect to the network.

IoT is used in a variety of ways by individuals all over the globe. One of the most popular applications is the home automation, which includes smart equipment such as light bulbs, televisions, and refrigerators shown in Figure 18.2.

Self-driving cars with many sensors have been already released by tech giants such as Tesla. These sensors send and receive information in real time at a really high speed, and they also make choices on the road at the same time based on data collected.

Despite the fact that the experiment was a massive success, the car's self-driving feature is still very much in beta since there are tons of aspects the automobile can't perform. A few of the most common IoT applications are listed below.

18.3.1 Smart Traffic

The basic goal of the IoT is to link "things" instantaneously so that all traffic data is added at the same time and efficiently. It can play a very important role by forwarding and gathering data from many sources such as sensors, camera systems, and geolocation (Hahanov & Khang et al., 2022).

All of this information is being used to detect driving patterns and then use a data-driven approach, and to train data sets applying AI algorithms (Rana & Khang

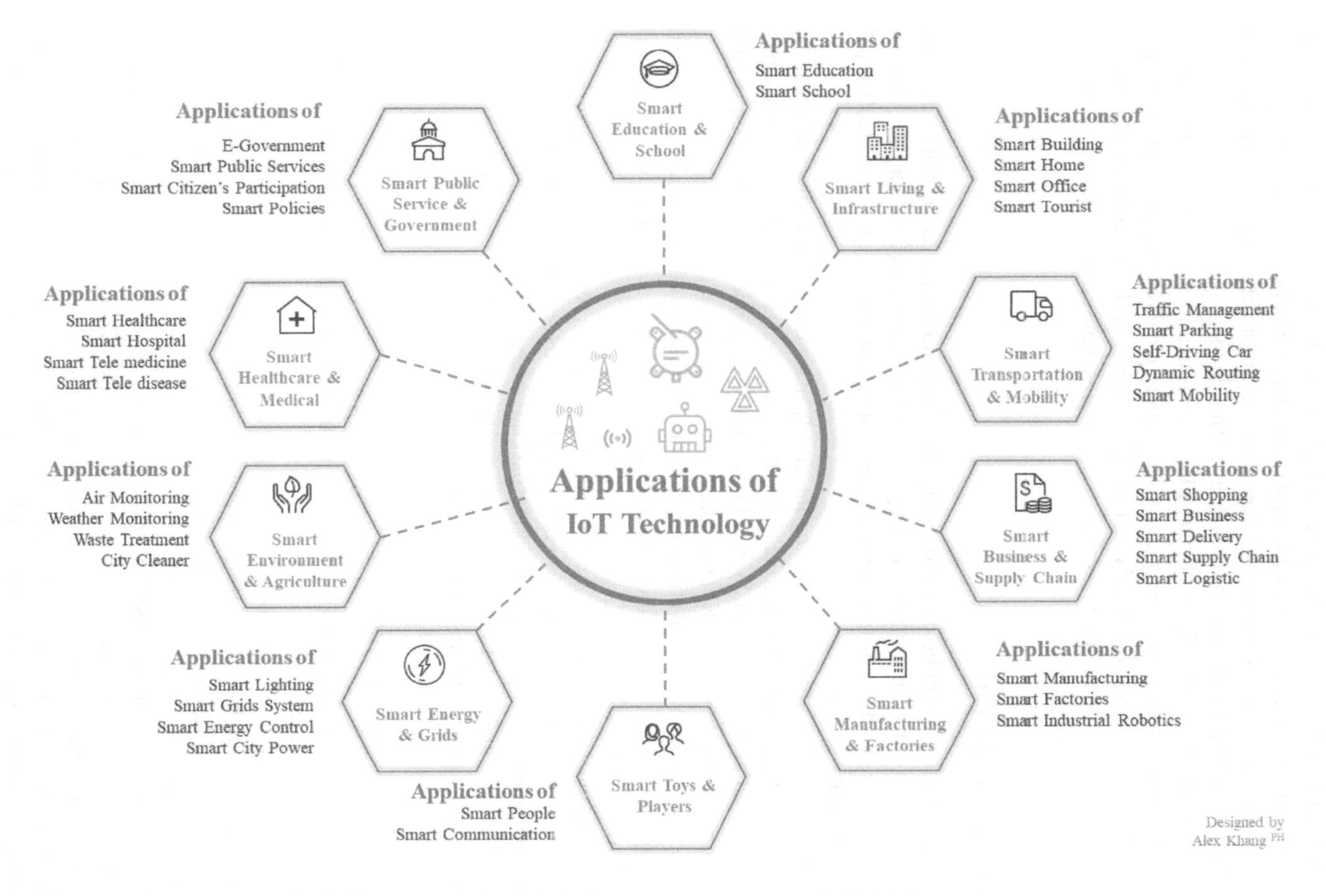
Applications of
IoT Technology
Smart Education & School
Applications of
Smart Education
Smart School
Smart Public Service & Government
Applications of
E-Government
Smart Public Services
Smart Citizen's Participation
Smart Policies
Smart Living & Infrastructure
Applications of
Smart Building
Smart Home
Smart Office
Smart Tourist
Smart Healthcare & Medical
Applications of
Smart Healthcare
Smart Hospital
Smart Tele medicine
Smart Tele disease
Smart Transportation & Mobility
Applications of
Traffic Management
Smart Parking
Self-Driving Car
Dynamic Routing
Smart Mobility
Smart Environment & Agriculture
Applications of
Air Monitoring
Weather Monitoring
Waste Treatment
City Cleaner
Smart Business & Supply Chain
Applications of
Smart Shopping
Smart Business
Smart Delivery
Smart Supply Chain
Smart Logistic
Smart Energy & Grids
Applications of
Smart Lighting
Smart Grids System
Smart Energy Control
Smart City Power
Smart Manufacturing & Factories
Applications of
Smart Manufacturing
Smart Factories
Smart Industrial Robotics
Smart Toys & Players
Applications of
Smart People
Smart Communication
Designed by
Alex Khang PH

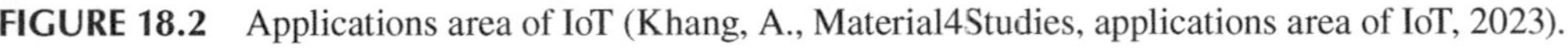
FIGURE 18.2 Applications area of IoT (Khang, A., Material4Studies, applications area of IoT, 2023).

et al., 2021) to optimize the prediction model for improved performance and future investigation possibilities (Rabby & Islam et al., 2019).

It is critical to manage transportation in cities; alternatively, there will be huge traffic congestion in popular locations such as Delhi, Noida, etc., and entirely empty roadways elsewhere. This is indeed dependent on the street's structure and planning, but it can still be handled with intelligent traffic signals.

For example, traffic lights should adapt dynamically depending on the flow of vehicles, with green lights going for longer if there is more traffic and shorter whenever the roads are vacant. Roads and bridges could also have sensors embedded in them to evaluate their status so that they're being renovated if there is significant physical damage (Naveen, 2016).

18.3.2 Smart City

From commercial applications to emergency responders, public transit and mobility, public security, municipal lighting, as well as other smart urban uses, the IoT is now present in each and every business and government industry (Jebaraj & Khang et al., 2024).

Municipalities are now becoming intimately associated as a result of advances in IoT technology in an attempt to enhance the effectiveness of building deployments, improve reliability and availability of emergency responders, reduce expenses, and much more (Subhashini & Khang, 2024).

And the concept of innovation is never-ending. In the coming years, we anticipate seeing even smarter city initiatives using IoT technology for this industry (Deepak & Prakshi et al., 2016).

IoT technologies could be used to increase the efficiency of societies on a bigger level. The purpose of smart cities is to use the IoT to enhance inhabitants' lives by enhancing congestion control, tracking parking space accessibility, analyzing quality of the air, and sometimes even sending out notifications when garbage cans are overflowing (Nur & Khang et al., 2024).

18.3.3 Smart Education

Education is perhaps the most effective means of ensuring our continuing creative and technical development.

We lay the groundwork for ultimate prosperity by providing that next generations have access to high-quality information and resources available. The IoT can help us enhance education in measurable ways over time without spending a lot of money (Byrne & O'Sullivan et al., 2017).

In terms of location, position, and capability, the IoT can assist us in making education more affordable. There are countless possibilities for integrating IoT technology into educational settings (Javaid & Sufian et al., 2018). They will serve as a strong platform for building a more detailed understanding of IoT learning in higher education. It may take a bit of time for IoT to make its way into conventional teaching.

The goal of this series on IoT applications in education is to highlight the numerous advantages of incorporating IoT into school environments (Kaur & Maheshwari, 2016).

18.3.4 Smart Healthcare

The wearable IoT device for overall health is primarily used for remote health monitoring, therapy, and in certain circumstances, rehabilitation. Before sending the user/health patient's information over the Internet for some further research, the sensors are used to collect health-related information, and the equipment may conduct limited computation (Khang & Rana et al., 2023).

Information may also be retrieved by the gadget, allowing the individual to create additional judgments. In many instances, usable gadgets that are connected to cell phones evaluate recorded information before transmitting it to a cloud computing based architecture like MS Azure or Amazon Web Services (AWS) for storage, processing, and analysis (Dian & Vahidnia et al., 2020).

The scientists developed a smart power chair that enables a disabled person to engage well with power chair using a smartphone application that evaluates information from various sensors and visualizes the findings for caretakers to assess the condition of patient virtually (Ghorbel & Bouguerra et al., 2018).

18.3.5 Smart Grid

The most crucial building block within the advancement of technology is smart electricity networks. Networks allow low-budget and low-cost access to the highest quality possible power generation. The deployment of smart grids ensures efficiency.

Smart grids, economic, and social aspects all contribute to growth. The creation of micro-networks, which enable more decentralized generation and power conservation at the end-user level, is among the major aims of smart networks (Kadar & Kalim et al., 2018).

With just the rise of energy output from micro energy systems and renewable power, systems and power storage systems that will function as essential energy reserves were becoming increasingly important to meet the rising demand for electricity consumption. For the energy grid, faster and better telecommunication mechanisms and technologies are required to prevent many electric connections and voltage fluctuations (Yang & Yue et al., 2011).

18.4 ATTACKS ON IoT

IoTs communicate over the Internet, which is a publicly accessible network. As a result, it is vulnerable to a variety of assaults that disrupt the smooth operation of IoTs. Below are a few examples of attacks (Bhardwaj & Kumar et al., 2017).

Denial of service (DoS) attack: This attack brings the services party to a halt. It's a risky assault because the offender impersonates the actual source. The attacker makes the receiver believe they are still receiving a valid communication. RFID technology is the most vulnerable to this form of threat (Heartfield & Loukas et al., 2018).

18.4.1 Eavesdropping

The intruder gains access to the data getting exchanged between both the sender and recipient, resulting in a loss of confidentiality. Other devices might closely monitor data from the infected smartphone and send bogus messages to collect personal information from it.

18.4.2 Alteration

Attackers can modify or modify information processed by IoT devices, posing a danger to the authenticity needs of the IoT system. Attackers are doing this to fool the data transmission protocol.

18.4.3 Fabrication

The attacker in this situation adds or edits data into IoT system without authorization. Because the sender is unaware that the system has been infiltrated, the authentication of the system is jeopardized.

18.5 MIMA ATTACKS IN IoT

Man-in-the-Middle (MITM, referred to as MIM, MiM, MitM, or MITMA in the literature) is a kind of strike in which an attacker steals a telecommunication connection between two or even more destinations invisibly. The MITM attacker can obtain, update, or alter the targeted users' data transmission (this distinguishes a MITM from a simple eavesdropper).

Furthermore, because the victims are uninformed of the invader, they believe the communication connection is secure (Conti & Dragoni et al., 2016).

MITM attacks can be carried out across a variety of communication channels, including GSM, UMTS, LTE, Bluetooth, Near-Field Communication (NFC), and Wi-Fi. The cybercriminals are interested not only in the data transmitted across destinations but rather in the data's security. MITM attacks, specifically, are designed to expose information (Efe & Aksöz et al., 2018).

- By listening in on conversations, confidentiality is maintained.
- Eavesdropping and manipulating conversations are used to ensure the integrity.
- By monitoring and wiping communication or by altering communications to trigger one of its participants to cease message transmission, availability could be achieved.

There is a minimum of three alternative approaches to characterizing MITM cyber-attacks, which consist of three distinct classifications:

- MITM is based on the use of impersonation tactics.
- MITM relies on the route of communication used to carry out the assault.
- MITM varies on the attacker's and victim's network locations.

The first division (MITM based on impersonation techniques) separates tactics predicated on the manner in which perpetrators convince victims that they are genuine endpoints. Furthermore, collecting the fragile and powerful sides of algorithms through various communication channels is beneficial.

As a result, researchers may see which approaches are more resistant to MITM assaults, and then use those ways to upgrade weak protocols and develop new MITM-resistant techniques. MITM attacks can be classified into the following categories:

- Spoofing-based MITM is a spoofing attack where the assaulter hijacks genuine interactions between different hosts and controls the sent information while the sufferers are unaware of an intermediary. In certain circumstances (such as DNS spoofing), the perpetrator spoofs equipment among targets, while in others (such as ARP spoofing), the assaulter spoofs the target's systems very much straightforward.
- SSL/TLS MITM is an operational circuit eavesdropping method in which the assaulter enters the interaction link between different targets (typically the victim's browser and the web server). The attacker then opens two distinct SSL connections with every target, relaying communications among them in such a manner that neither party is informed of the intermediary. This configuration allows an attacker to capture all data sent across the network and perhaps even change the contents sent arbitrarily (Bastos & Shackleton et al., 2018).
- BGP MITM is an assault that is set on IP takeover, yet where the perpetrator directs intercepted communication to the intended target. As a result, traffic passes within the Autonomous Station (AS) of the assaulter, where it could be altered.
- Based on a fictitious base station (FBS-based), MITM is a type of attack in which a third party compels a target to connect to a false Base Transceiver Station (BTS), which the assailant then uses to alter the victim's traffic. Additional method to distinguish MITM attacks is to look at the channel of communication within which they are carried out.

MITM attacks are listed in Table 18.1 throughout OSI layers and cellular networks. To provide protection, each level takes on a different method. Despite this, none of them would be immune to MITM assaults.

TABLE 18.1
MITM Attacks on Different Layers of OSI Model

OSI Layer	MITM Attacks
Application layer	BGP MITM, DHCP spoofing-based MITM, DNS spoofing-based MITM
Presentation layer	SSL/TLS MITM
Transport layer	IP Spoofing-based MITM
Network layer	
Data link layer	ARP Spoofing-based MITM

18.6 DoS ATTACKS IN IoT

In modern environment, cyber threats are a common occurrence on the Internet. A DoS attack is an approach to sabotage the proper functioning of a system in just about any way. The attack is known as distributed, or DDoS, if it comprises numerous systems hitting at the very same time (typically one lakh). A DDoS attack can target a particular workstation, function, or perhaps even the company's infrastructures (Huraj & Šimon et al., 2020).

An effective assault disrupts a service or network's normal operation and causes damage to the owners, which is measurable depending on the system's architecture. A DDoS assault can be carried out in a variety of ways (Sharma & Khang, 2024).

Following that, the several types of DDoS assaults used in the case study are detailed. During one of these attacks, individuals' capacity to use IoT sensors was evaluated. The effectiveness of DDoS attacks was tested in real time using several situations (Fadele & Othmana et al., 2017).

DoS attempts to make IoT devices unreachable to their target purposes for a short or long period of time. Jamming, collision, and malicious inside assaults are some of the numerous forms of DoS attacks that can be conducted opposed by the IoT; the very last kind can wreak greater damage since it manages part of the infrastructure (Sicari & Rizzardi et al., 2018).

To develop an efficient solution capable of recognizing and recovering an IoT system in the event of a DoS attack, we must first define an appropriate IoT architecture, as well as the entities involved. The authors defined this architecture and it is a versatile and cross-domain middleware called Networked Smart object (NOS) that is customized to the IoT context (Huraj & Šimon, 2019).

NOSs are able to manage information from diverse sources in a distributed way and analyze the protection and privacy quality of information using suitable algorithm, allowing consumers to be informed of the degrees of truthfulness of the facilities gathered by NOSs itself (Sahu & Khare, 2020). NOSs offer a lightweight and safe exchange of information methodology derived from an authorized post and subscription technique using the MQTT protocol.

Furthermore, an enforcement framework analyses NOS behavior to ensure that policies are implemented correctly.

18.7 MIMA TARGETS AN IoT NETWORK

In a traditional MITM attack scenario, both victims (the two endpoints) and attackers are involved (a third party). After gaining access to the communication channel, the attacker attempts to alter the data going back and forth between two locations.

As a consequence, a hostile third-party attacker can gain, modify, replace, or manipulate data delivered over the transmission medium between both the two endpoints in MITM attacks. Because the victims are uninformed of the invaders, they feel the communications connection is secure. MITM assaults can be carried out through a wide range of communication modes, including GSM, LTE, UMTS, Bluetooth, and Wi-Fi (Bhushan & Sahoo et al., 2018).

The hackers are after the information that travels within destinations, and also the integrity and confidentiality of that data. An intruder can breach legitimacy and security by listening in on conversations and altering messages via communications takeover.

Malicious users can also eavesdrop, modify, or destroy communications in order to prohibit one of the participants from interacting, causing a breach of service concern.

MITM attacks can take four multiple variations. To begin with, there are spoofing-based MITM attacks, in which an adversary uses a spoofing attack to monitor lawful communication and handle data sent alone without targets being aware of the intruder's existence.

The attacker spoofs hardware between both endpoints in DNS spoofing, whereas in ARP spoofing, the assaulter explicitly spoofs these endpoints or the perpetrator's devices.

Second, SSL/TSL MITM attacks, wherein the intruder gains access to the communication connection in between endpoints or targets. Invading force establishes two separate SSL connections and uses those to deliver a message back and forth (Singh & Singh et al., 2017). As a result, the attacker can capture all conversations and deliberately alter the data.

Next, BGP MITM attacks, in which the attacker distributes, trying to disrupt connections to the target. IP hijacking occurs when traffic flows through to an attacker's independent unit, where it is possible for traffic to be controlled.

Finally, there's also the MITM fake ground station attack, wherein the attacker constructs a fake transmitter station and afterward uses it to tamper with the victim's personal information.

18.7.1 ARP's Attack Method Spoofing

A spoofing attack includes an attacker or hostile party impersonating any user or devices that are connected over a network. This is a type of spoofing-based MITM attack. Network addresses are translated into MAC addresses using ARP protocols. ARP is a stable and necessary protocol for LAN connectivity.

Malicious users alter the local ARP cache database by linking the owner's MAC address to the victim's IP address. An MITM attack is designed to gain entry to an individual's private details.

There are two kinds of ARP spoofing attacks: those that trick the host and those that cheat the internal network gateway. ARP requests are sent throughout the network whenever a host wants to communicate with some other computer within the same network with an undisclosed MAC address.

Since there are no proper authentication mechanisms, it is simple to fake cache data. By avoiding transmissions, the originating computer could save IP to a MAC entry to speed things up communications in the future. In the caching system, ARP, a stateless rule of engagement, has less security.

18.7.2 TSL/SSL Attack Methodology MITM (Man-in-the-Middle) Attacks

The encrypting technologies Safe Socket Layer (SSL) and Transport Layer Security (TLS) offer secure information transfer and interaction over the web. These develop a safe channel of two-way communication (client and server). These guidelines necessitate the use of four more protocols in order to create a session.

Record Protocols assure dependability and anonymity, whereas Handshaking Protocol is being used to negotiate session parameters, Cipher Spec Protocol is being used to install newly negotiated links, and Alert Protocol has been used to notify almost any form of breach.

The procedure of certificate validation is crucial to protection. If the hacker has a valid certificate, he or she has hacked a CA or compelled it to issue one. If the attacker has an incorrect certification and the target overlooks the security restrictions, they may prevail.

18.7.3 BGP-Based MITM Attacks Have the Following Attack Methodology

The network reachability information is communicated using the Border Gateway Protocol, an inter-autonomous systems procedure. This data is used to create an AS (Autonomous System) connection network, which is then used to enforce specific policy judgments at the AS level.

BGP is a basic Internet routing system that allows AS to choose the quickest route for transmitting data. This normally chooses the shortest AS path, which has the fewest amount of AS names because BGP does not conduct peer-to-peer verification and does not provide security over MITM attacks.

Note: AS Path Attribute is one of the most used BGP Path Attribute by Service Providers. In AS Path Attribute mechanism, whenever a route passes an AS, it adds the number of AS it passed. So, AS Path Attribute is a list of AS numbers that the router traverse.

18.7.4 FBS-Based MITM Attack

A malicious hacker outsmarts a transceiver base station in a fake base station (FBS) assault. By sending BTS signals over the air, a fake BTS can impersonate a legitimate BTS, letting neighboring cell devices communicate with it.

It does have a real-time jamming technology that shuts down all active connections instantaneously. Since the enemy may flood legitimate BTS, the jammer strategy compels sufferers to link to fraudulent BTS.

Devices which does not validate network identity or display one-way identification are susceptible to FBS-based MITM attacks. A third party can trick a victim into connecting to a fake BTS, allowing their data transmission on a network to be altered. The attacker may use phony BTSs with different protocols to pull out the MITM attack.

18.8 DoS TARGETS AN IoT NETWORK

An attacker can send many packets in a DDoS assault by altering the id and location of each one repeatedly.

The attacker cars coordinate themselves so that they do not exceed the threshold value while still being capable of attacking. As a result, an attacker can simply design an attack without exceeding the threshold value, making methods for timely identification of this attack impractical.

DDoS attacks, wherein an intruder uses a network of adversely affected IoT sensors or machines to overwhelm and overflow the target, forcing the victim's activities to become inaccessible, are common.

The case study, on different aspect, portrays the IoT device as a sufferer in a smart home system in an effort to stop the client from talking with and operating the IoT sensors of these devices. DDoS assault exploit was inspected in real time corresponding to different situations.

18.8.1 SYN Flood Attack

In the classic server-client architecture, the server must obtain a SYN (synchronize) signal from the client throughout TCP connection establishment. As a result, the server allocates some resources for such a half-open TCP link and replies to the client with an SYN-ACK message.

Because server resources are insufficient, if a user never responds over an ACK (acknowledgment) packet and a large number of SYN packets from several further hostile clients are received, the server's resources can be exhausted, and the server will be unable to link to any new clients, as seen in Figure 18.3.

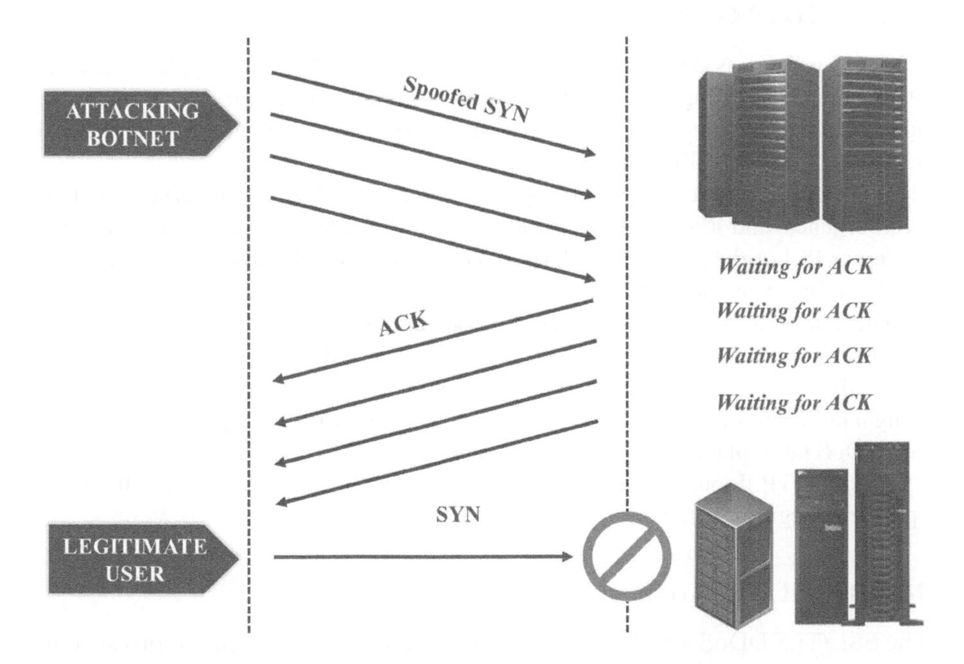

FIGURE 18.3 SYN flood distributed denial of service (DDoS) attack (Khang, A., Material4Studies, SYN flood distributed denial of service (DDoS) attack, 2023).

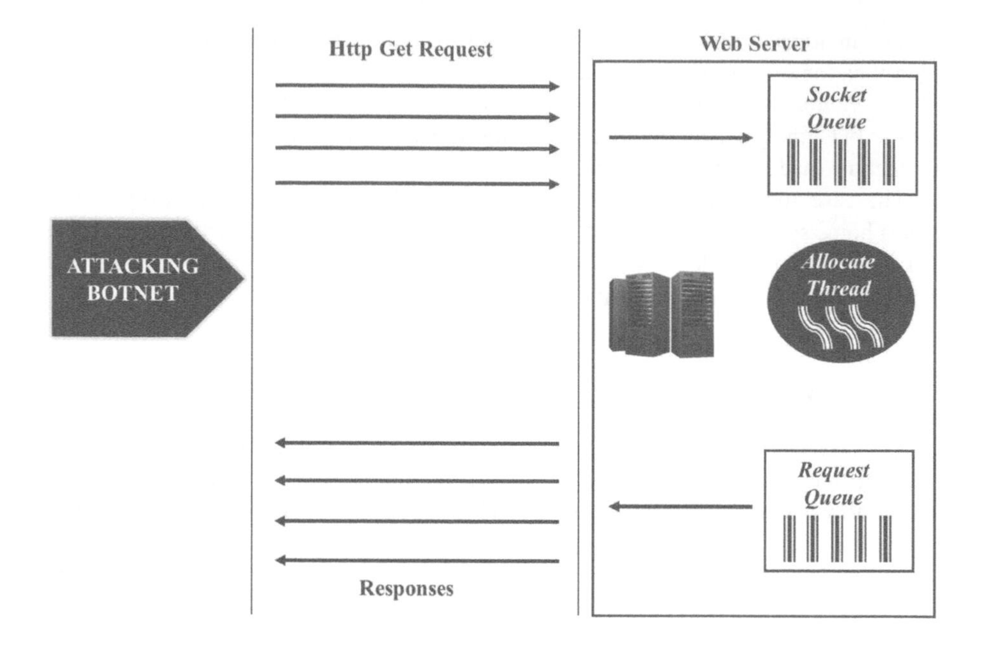

FIGURE 18.4 HTTP get flood DDoS attack (Khang, 2021).

18.8.2 HTTP Get Flood Attack

HTTP Get flood assaults are among the most common types of application layer DDoS attacks. During an HTTP Get flood attack, an offender uses genuine IP addresses that appear to be valid sources, causing the web server to repeatedly accept and execute HTTP Get requests.

When a significant number of requests are submitted, the web server becomes overburdened, and it is unable to handle any additional HTTP Get requests, as seen in Figure 18.4. A detection method based on spotting faked IP addresses or blacklisting IP addresses is ineffective.

There seem to be two forms of HTTP flood attacks: one which has already been discussed, while the other is known as HTTP POST flood attack. This approach takes advantage of the mismatch in proportional consumption of resources by generating a large number of post responses to a single server till its capacity is exceeded and a DoS takes place.

Since HTTP flood assaults use regular URL queries, it's tough to predict them apart from legitimate requests.

18.8.3 SSL/TLS Flood

The SSL/TLS DDoS attack took leverage of the server's necessity to invest computational power when establishing a secured TLS connection. By delivering a huge quantity of junk to the server or repeatedly attempting to re-establish the

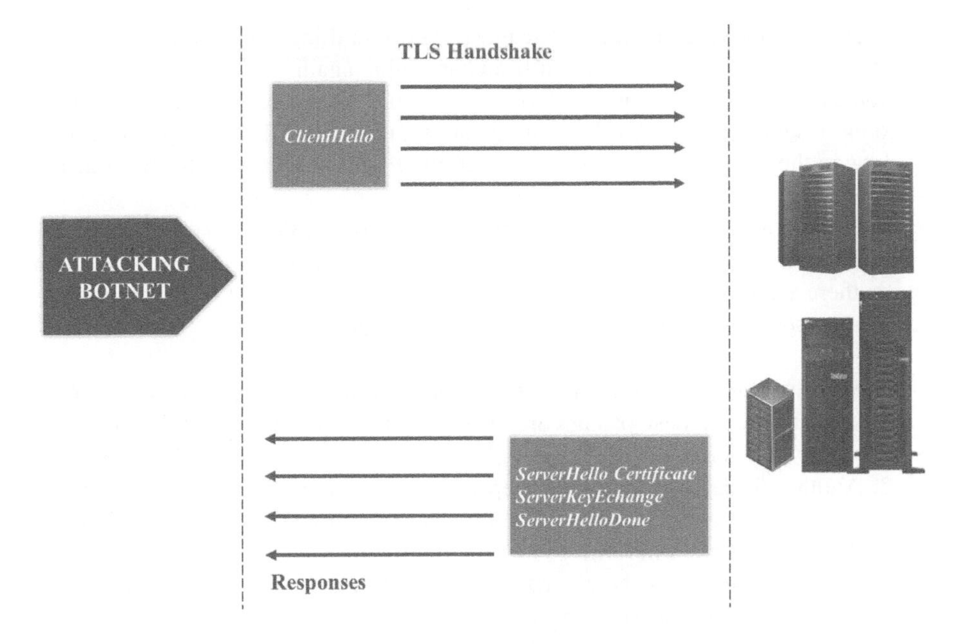

FIGURE 18.5 SSL/TLS flood DDoS attack (Khang, 2021).

link, the hacker overloads the server's resources and shuts it down throughout TLS negotiation as Figure 18.5.

18.9 SOME POSSIBLE SOLUTIONS TO AVOID/PREVENT CYBER ATTACKS

There are some most possible solutions to cyber-attacks as most people in the world already know but due to complexities of the discussed attacks, that is MITM and DoS attacks, are kind of hard to handle once triggered.

To be realistic, there really is nothing particularly unique about cyber-attacks; each has its own level of sophistication. However, while evaluating these cyber-attacks, they can be difficult to tackle and sometimes even dangerous if it's not handled properly.

A range of hacking attempts must be protected against those in order to avoid a break of corporate network and its devices. To eliminate the threat from exploiting the vulnerabilities or flaw, the proper contingency plan must always be implemented.

Any firm's initial defense mechanism is to examine and adopt security protocols. Even this is much more fruitful in the case of deep learning and machine learning applications (Saraswat & Sharma, 2022) when launched through IoT devices.

From launching an application-specific assault on web servers to delivering spam scams with harmful attachments or links, cyber threats come in various kinds and sizes. While determining the goal of a cyber-assault can indeed be useful, it was not the most important consideration. The most important thing is to figure out how the incident happened and how to stop it from happening again.

Some most possible solutions are to simply destroy the device on which the attack is going. It's not the most feasible solution at that time since there are possibly a lot of data on the device and also some or possibly all are in danger until the attack is stopped.

For any organization, data is one of the most important things. Recently hackers stole the source code for software for Samsung Galaxy devices. This is just single news; there are a lot of attacks on the biggest tech companies that one cannot imagine. Some most possible preventive measures are given below:

1. Security patches are one of the best things for any security system, and in security attack, these patches are the best thing. It isn't the best solution but it gets the job done.
2. Another thing is antivirus to keep the devices safe until a certain limit. Since any antivirus is still a waste for a zero day not always but apart from that this is another best thing.
3. Regular security check-ups might save a company from a big catastrophe. Most companies did and suggest it as well.
4. Regular searching for vulnerabilities also might be a good way to ensure the security of systems.
5. It is most often considered non-serious threat in security concerns and that is social engineering. It sounds stupid that but it is way more dangerous than any attack out there present in our world and the only way to deal with these attacks is to simply encourage employees of any organization about this attack and also help them manage it.
6. Using cloud base platform is also a great way to get protection from the attacks but since any security is not full proof so it can't be considered as a permanent solution, but also at the same time with further research and development the platform will become in future.
7. Training the employees of an organization can also play a significant role in security of firm data. As employees are the actual people that tend to engage with the day-to-day data and information and if they know that by doing something they can put the organization into a huge attack they will tend to report it and may also delay the attack or maybe finish it off even before it happens.
8. Endpoint security safeguards connections that are connected to machines via a virtual gateway. Mobile devices, tablets, and laptops linked to company network provide security risks with entry points. Endpoint security application is needed to safeguard those pathways.

18.10 CONCLUSIONS

Across both wired and wireless IoT scenarios, machine-to-machine interaction is necessary. Machines in the IoT are embedded with sensors and have limited computational resources due to energy limitations. These devices communicate with one another using open communication channels. As a consequence, these broadcasts are exposed to a wide range of hazards.

The main problem with such devices is that they do not jeopardize user privacy or permit information to be sent in its original form. Therefore, as response, maintaining data authenticity and integrity is crucial. Since these devices are powered by batteries and must operate for lengthy stretches of time, lightweight and secure cryptographic methods are essential.

We additionally looked at MITM attacks and came up with a full taxonomy centered on impersonating tactics. We have included a list of MITM detection systems, as well as explanations of each. The aim of this journal was not just to provide a comprehensive appraisal of suggestions for future research. Certainly, this would force an extreme investigation of future interaction technology trends as well as a technical auditing of technological advances.

DDoS attacks used two kinds of assaults, SYN flood and HTTP Get flood, that flooded the smart devices [IoT enabled] having a huge amount of data packets in order to crush it and restrict any interactions with smart sensors. The best method for managing smart sensors has been to use smart home digital assistant software (like Google Assistant, Alexa, and Siri) to interface with client IoT sensors.

The conducted pilot survey can be used as a basis for future study, and this can be expanded in the long term by looking into the behavior of additional IoT protocols under diverse DDoS attacks in similar real-world situations. In addition, the case of cloud connection among IoT sensors and personal assistant machines must be investigated extensively, and a robust framework immune to DDoS assault in the IoT-enabled home might be presented.

Even when this study is over, there is still plenty to debate because the area is continuously constantly being updated, and it is critical to keep up with this now, at least in terms of cyber security (Khang & Hahanov et al., 2022). It is a large field that will require more research in order to provide more correct and reliable approaches and outcomes. IoT also was regarded to have been in the initial phases of development because there is much to learn about it and many more applications are still being tested or beta tested.

18.11 RESULTS AND DEMONSTRATIONS

As previously said, cyber-attacks are a huge source of concern because they generate significant problems for businesses. There seem to be undoubtedly some solutions that have been discussed, but they may not be sufficient for the major technological corporations.

So, with all of this in consideration, some experts have developed a system called REACTO that stands for REActing TO DoS attacks within Internet of Things middleware called NOS.

The program can now be used to defend a network from DoS attacks. The program is capable of detecting and counteracting attacks on IoT devices. The suggested solution, which was adapted to NOS structure, was tested using a real test-bed, which consisted of an NOS mock-up running on a Raspberry Pi and receiving freely accessible data services instantaneously from a wide range of sources.

The program seems to be in beta because it has only been tested on one main architecture. With more research and development, the program may be able to defend against DoS attacks against any architecture in future.

Furthermore, on an IoT-based mechanical arm, certain researchers demonstrate a MITM attack. This chapter delves into every stage in great depth to ensure a thorough comprehension of the intrusion.

Using all of the knowledge available, a similar incident can likewise be avoided. A precautionary method to counteract the strike is also discussed, as well as a working footage of the intrusion is also provided.

The study emphasizes that, at the very least, source-authentication (authenticity) of the data is required to prevent or protect the network; this allows for the detection of communications that have been altered by a private entity.

Even though the invasion in the demonstration appears to be straightforward, it illustrates just what would actually occur in much more large-scale cloud-IoT architectures in which a false sense of control can be presented.

The researcher has complex encryption HTTPS-TLS security from the cloud to an IoT gateway, but they rely on whichever security is presented at Layer two (IEEE 802.15.4, Sigfox, LoRaWAN Network Session Key) inside the IoT.

From the above results, we can say that further research is still required to have much more accurate preventive measures as based on the above results which are tested in a controlled environment and cannot be considered a solution to the attacks.

But these solutions can become as a base for the solution needed to counter the attacks. Since the attacks can be countered in particular scenarios, there is a huge possibility that with further research and development, these attacks can be handled easily in future.

REFERENCES

Abomhara, M., Køien, G.M. "Cyber security and the Internet of things: vulnerabilities, threats, intruders and attacks", *Journal of Cyber Security and Mobility*, vol. 4, pp. 65–88 (2015). https://link.springer.com/chapter/10.1007/978-981-16-7182-1_31

AlFuqaha, A., Guizani, M., Mohammadi, M., Aledhari, M., Ayyash, M. "*Internet of Things: A Survey on Enabling Technologies, Protocols and Applications*" (2020). doi: 10.1109/comst.2015.2444095

Bastos, D., Shackleton, M., El-Moussa, F. "Internet of things: A survey of technologies and security risks in smart home and city environments", *Proceedings of the Living in the Internet of Things: Cybersecurity of the IoT*, London, UK, 28–29 March 2018. https://digital-library.theiet.org/content/conferences/10.1049/cp.2018.0030

Bertino, E., Islam, N. "Botnets and internet of things security", *Computer*, vol. 50, no. 2, pp. 76–79 (2018). https://ieeexplore.ieee.org/abstract/document/7842850/

Bhambri, P., Rani, S., Gupta, G., Khang, A. *Cloud and Fog Computing Platforms for Internet of Things* (2022). CRC Press. https://doi.org/10.1201/9781032101507

Bhardwaj, I., Kumar, A., Bansal, M. "A review on lightweight cryptography algorithms for data security and authentication in IoTs", *2017 4th International Conference on Signal Processing, Computing and Control (ISPCC)*, pp. 504–509 (2017). doi: 10.1109/ISPCC.2017.8269731.

Bhushan, B., Sahoo, G., Rai, A.K. "Man-in-the-middle attack in wireless and computer networking — A review", *2017 3rd International Conference on Advances in Computing, Communication & Automation (ICACCA) (Fall)*, pp. 1–6 (2018). doi: 10.1109/ICACCAF.2017.8344724.

Byrne, J.R., O'Sullivan, K., Sullivan, K. "An IoT and wearable technology hackathon for promoting careers in computer science", *IEEE Transactions on Education*, vol. 60, no. 1, pp. 50–58 (Feb. 2017). doi: 10.1109/TE.2016.2626252

Conti, M., Dragoni, N., Lesyk, V. "*A Survey of Man in the Middle Attacks*" (2016). Vol 18, Issue 3. https://doi.org/10.1109/COMST.2016.2548426

Deepak Kumar, L., Prakshi, V., Addala, V., Prashant, S., Ritakshi, G., Kumar, P.M., Rajat, B. "Smart electronic wheelchair using arduino and bluetooth module", *International Journal of Computer Science and Mobile Computing*, vol. 5, no. 5, pp. 433–438 (2016). https://bitelectronicsystem.com/wp-content/uploads/2020/12/Smart-Electronic-Wheelchair-Using.pdf

Dian, F., Vahidnia, R., Rahmati, A. "Wearables and the internet of things (IoT), applications, opportunities, and challenges: a survey", *IEEE Access*, pp. 1–1 (2020). 10.1109/ACCESS.2020.2986329

Efe, A., Aksöz, E., Hanecioğlu, N., Yalman, S.N. "Smart security of IoT against to DDoS attacks". *International Journal of Innovative Engineering Applications*. vol. 2, pp. 35–43 (2018). https://dergipark.org.tr/en/pub/ijiea/issue/42454/487630

Fadele, A., Othmana, M., Hashem, I., Alotaibi, F. "Internet of things security: a survey", *Journal of Network and Computer Applications* (2017). https://doi.org/10.1016/j.jnca.2017.04.002

Ghorbel, A., Bouguerra, S., Amor, N.B., Jallouli, M. "Cloud based mobile application for remote control of intelligent wheelchair", *2018 14th International Wireless Communications & Mobile Computing Conference (IWCMC)*, pp. 1249–1254 (2018). doi: 10.1109/IWCMC.2018.8450366

Habibzadeh, H., Nussbaum, B.H., Anjomshoa, F., Kantarci, B., Soyata, T. "A survey on cybersecurity, data privacy, and policy issues in cyber-physical system deployments in smart cities", *Sustainable Cities and Society*, vol. 50, p. 101660 (2019). https://www.sciencedirect.com/science/article/pii/S2210670718316883

Hahanov, V., Khang, A., Litvinova, E., Chumachenko, S., Hajimahmud, V.A., Alyar, A.V. "The Key Assistant of Smart City – Sensors and Tools", *AI-Centric Smart City Ecosystems: Technologies, Design and Implementation* (1st Ed.) (2022). CRC Press. https://doi.org/10.1201/9781003252542-17

Heartfield, R., Loukas, G., Budimir, S., Bezemskij, A., Fontaine, J.R., Filippoupolitis, A., Roesch, E. "A taxonomy of cyber-physical threats and impact in the smart home", *Computers and Security*, vol. 78, pp. 398–428 (2018). https://www.sciencedirect.com/science/article/pii/S0167404818304875

Huraj, L., Šimon, M. "Realtime attack environment for DDoS experimentation", *Proceedings of the 2019 IEEE 15th International Scientific Conference on Informatics, Poprad, Slovakia*, 20–22 November 2019; pp. 165–170. https://ieeexplore.ieee.org/abstract/document/9119271/

Huraj, L., Šimon, M., Horák, T. "Resistance of IoT sensors against DDoS attack in smart home environment", *Sensors (Basel, Switzerland)* (2020). https://doi.org/10.3390/s20185298

Javaid, S., Sufian, A., Pervaiz, S., Tanveer,"MehakSmart traffic management system using Internet of Things" pp.393–398 (2018). https://doi.org/10.23919/ICACT.2018.8323770

Jebaraj, L., Khang, A., Vadivelraju, C., Antony, R.P., Kumar, S. (Eds.). "Smart City Concepts, Models, Technologies and Applications", *Smart Cities: IoT Technologies, Big Data Solutions, Cloud Platforms, and Cybersecurity Techniques* (1st Ed.) (2024). CRC Press. https://doi.org/10.1201/9781003376064-1

Kadar Muhammad Masum, A., Kalim Amzad Chy, M., Rahman, I., Nazim Uddin, M., Islam Azam, K. "An Internet of things (IoT) based smart traffic management system: A context of Bangladesh", *2018 International Conference on Innovations in Science, Engineering and Technology (ICISET)*, pp. 418–422 (2018). doi: 10.1109/ICISET.2018.8745611

Kaur, M.J., Maheshwari, P. "Building smart cities applications using IoT and cloudbased architectures", *2016 International Conference on Industrial Informatics and Computer Systems (CIICS)*, pp. 1–5 (2016). doi: 10.1109/ICCSII.2016.7462433

Khang, A. (2021). "Material4Studies", *Material of Computer Science, Artificial Intelligence, Data Science, IoT, Blockchain, Cloud, Metaverse, Cybersecurity for Studies.* https://www.researchgate.net/publication/370156102_Material4Studies

Khang, A., Gupta, S.K., Rani, S., Karras, D.A. "*Smart Cities: IoT Technologies, Big Data Solutions, Cloud Platforms, and Cybersecurity Techniques*" (1st Ed.) (2023). CRC Press. https://doi.org/10.1201/9781003376064

Khang, A., Hahanov, V., Abbas, G.L., Hajimahmud, V.A. "Cyber-Physical-Social System and İncident Management", *AI-Centric Smart City Ecosystems: Technologies, Design and Implementation* (1st Ed.) (2022). CRC Press. https://doi.org/10.1201/9781003252542-2

Khang, A., Rana, G., Tailor, R.K., Hajimahmud, V.A. "*Data-Centric AI Solutions and Emerging Technologies in the Healthcare Ecosystem*" (1st Ed.) (2023). CRC Press. https://doi.org/10.1201/9781003356189

Khang, A., Rani, S., Sivaraman, A.K. "*AI-Centric Smart City Ecosystems: Technologies, Design and Implementation*" (1st Ed.) (2022). CRC Press. https://doi.org/10.1201/9781003252542

Khanh, H.H., Khang, A. "The Role of Artificial Intelligence in Blockchain Applications", *Reinventing Manufacturing and Business Processes through Artificial Intelligence*, pp. 20–40 (2021). CRC Press. https://doi.org/10.1201/9781003145011-2

Lawal, M.A., Shaikh, R.A., Hassan, S.R. "Security analysis of network anomalies mitigation schemes in IoT networks", *IEEE Access*, vol. 8, pp. 43355–43374 (2020). https://ieeexplore.ieee.org/abstract/document/9016241/

Naveen, S. "*Study of IoT: Understanding IoT Architecture, Applications, Issues and Challenges*" (2016). https://www.researchgate.net/profile/Soumyalatha-Naveen/publication/330501274_Study_of_IoT_Understanding_IoT_Architecture_Applications_Issues_and_Challenges/links/5c434fea458515a4c731d4bb/Study-of-IoT-Understanding-IoT-Architecture-Applications-Issues-and-Challenges.pdf

Nur Aeni, M., Khang, A., Yakin, A.A., Yunus, M., Cardoso, L. "Revolutionized Teaching by Incorporating AI-Chatbot for Higher Education", *Smart Cities: IoT Technologies, Big Data Solutions, Cloud Platforms, and Cybersecurity Techniques* (1st Ed.) (2023). CRC Press. https://doi.org/10.1201/9781003376064-16

Patil, S., Chaudhari, S. "DOS attack prevention technique in wireless sensor networks", *Procedia Computer Science*, vol. 79, pp. 715–721 (2016). https://www.sciencedirect.com/science/article/pii/S1877050916002258

Rabby, M.K.M., Islam, M.M., Imon, S.M. "A review of IoT application in a smart traffic management system", *2019 5th International Conference on Advances in Electrical Engineering (ICAEE)*, pp. 280–285 (2019). doi: 10.1109/ICAEE48663.2019.8975582

Rana, G., Khang, A., Sharma, R., Goel, A.K., Dubey, A.K. "*Reinventing Manufacturing and Business Processes through Artificial Intelligence*" (2021). CRC Press. https://doi.org/10.1201/9781003145011

Rani, S., Bhambri, P., Kataria, A., Khang, A., Sivaraman, A.K. "*Big Data, Cloud Computing and IoT: Tools and Applications*" (1st Ed.) (2023). Chapman and Hall/CRC. https://doi.org/10.1201/9781003298335

Rani, S., Chauhan, M., Kataria, A., Khang, A. "IoT Equipped Intelligent Distributed Framework for Smart Healthcare Systems", *Networking and Internet Architecture* (2021). CRC Press. https://doi.org/10.48550/arXiv.2110.04997

Rizzardi, A., Sicari, S., Miorandi, D., Coen-Porisini, A. "Aups: an open source authenticated publish/subscribe system for the Internet of things", *Information Systems*, vol. 62, pp. 29–41 (2016). https://www.sciencedirect.com/science/article/pii/S030643791630237X

Sahu, S.K., Khare, R.K. "DDOS attacks and mitigation techniques in cloud computing environments", *Gedrag & Organisatie Review*, vol. 33, pp. 2426–2435 (2020). https://ieeexplore.ieee.org/abstract/document/8528677/

Saraswat, A., Sharma, N. "Bypassing confines of feature extraction in brain tumor retrieval via MR images by CBIR", *ECS Transactions*, vol. 107, no. 1, pp. 3675–3682 (Apr. 2022). doi: 10.1149/10701.3675ecst

Saxena, N., Grijalva, S., Chaudhari, N.S. "Authentication protocol for an IoTenabled LTE network", *ACM Transactions on Internet Technology*, vol. 16, no. 4, 20 Pages (2016). Article 25 (December 2016). http://dx.doi.org/10.1145/2981547

Sharma, A., Khang, A. "Exploratory Analysis on Current Application-Layer HTTP-Payload Based DDoS Attack on IoT Devices", *Smart Cities: IoT Technologies, Big Data Solutions, Cloud Platforms, and Cybersecurity Techniques* (1st Ed.) (2024). CRC Press. https://doi.org/10.1201/9781003376064-21

Sicari, S., Rizzardi, A., Miorandi, D., Coen-Porisini, A. "REATO: REActing TO denial of service attacks in the Internet of things", *Computer Networks* (2018). doi: 10.1016/j.comnet.2018.03.020

Singh, K., Singh, P., Kumar, K. "Application layer HTTP-GET flood DDoS attacks: research landscape and challenges", *Computers and Security*, vol. 65, pp. 344–372 (2017). https://www.sciencedirect.com/science/article/pii/S0167404816301365

Subhashini, R., Khang, A. "The Role of Internet of Things (IoT) in Smart City Framework", *Smart Cities: IoT Technologies, Big Data Solutions, Cloud Platforms, and Cybersecurity Techniques* (1st Ed.) (2024). CRC Press. https://doi.org/10.1201/9781003376064-3

H. Tahir, A. Kanwer and M. Junaid. "*Internet of Things (IoT): An Overview of Applications and Security Issues Regarding Implementation*" (2016). http://www.ijmse.org/Volume7/Issue1/paper3.pdf

Yang, Z., Yue, Y., Yang, Y., Peng, Y., Wang, X., Liu, W. "Study and application on the architecture and key technologies for IoT", *2011 International Conference on Multimedia Technology*, pp. 747–751 (2011). doi: 10.1109/ICMT.2011.6002149

19 Exploratory Analysis on Current Application-Layer HTTP-Payload-Based DDoS Attack on Internet of Things (IoT) Devices

Ankita Sharma and Alex Khang

19.1 INTRODUCTION

Internet was intended to fulfill the essential condition of transferring data from one device to another with a very less focus on the security of data. Hence, attackers are a threat to both customers and service providers who use the facility of internet. Among, many attacks, Payload-based Distributed Denial of Service (DoS) attack is one of the eminent threats to the internet.

Nowadays, 80% of businesses are dependent on internet facility to grow their business by using website, applications, web portals, etc. Therefore, on Internet of Things (IoT) Devices, DoS attack increases from 2017 to 2018 by 45%–80%. There could be more reasons like increased size of network and data/information contained in the network. The words with which we are very familiar are "smart" like smart watches, smart cities, healthcare sector, and smart home (Rani and Chauhan et al., 2021).

All that are related to IoT are vast network-connected servers, computers, tablets, mobiles that are governed by a standard protocol for connected system. According to the latest statistics, 7 billion devices were connected to internet in 2018, and by 2020, estimate was 14 million.

The global market worth in 2018 is $150 billion and is expected to be $1.6 trillion by 2025. Therefore, we first need security at application layer because collection data start from end layer (Bhambri & Khang et al., 2022).

According to industry statistics, industries invest 75% currency for securing the IoT devices, although attacks are still increasing. In May 2018, CISCO told that Russia-linked botnets affecting at least 500,000 vulnerable routers and network access storage (NAS). So, hackers compromise the user's data by snooping traffic and steal some confidence data in December, 2018. Ku Leuven in Belgium found

 DOI: 10.1201/9781003376064-19

the vulnerability in the luxury cars and unlocked the cars' door lock without the customer authentication (Rani & Khang et al., 2023).

In Jan 2019, DDoS attacks were made with 145 thousand IoT compatible cameras (web cameras, security cameras) captured by the attackers. Recently, Mirai botnets by DYN (Dyn is a DNS provider) attack.

Typically, DDoS attacks come in three categories: application layer, perception, and the one based on volume. Volumetric DDoS attack on infrastructure between users in the internet and a data center by sending huge quantity of unwanted traffic near the victim. In 2017, 80% of attack is experienced by DDoS attack.

Perception layer is on RFID (radio frequency identification), wireless sensor network (WSN) for sending data from sensor without human interaction and touch, such as jamming, killing command attack. At the end, attack abuses the configuration and functionality of various application layers.

In the internet environment, a web application firewall (WAF) is used to protect web applications from a variety of application-layer attacks such as cross-site scripting (XSS), Structure Query Language (SQL) injection and cookie poisoning, among others. Attacks to business applications are the leading cause of breaches that they are the gateway to the valuable data of organization.

This includes the HTTP-get (Hypertext Transfer Protocol) flood etc. In 2018 and 2019, 95% attack is done on application layers. Dos attack is implemented by attackers on IoT devices for few seconds daily so that service provider faces the huge financial losses (Subhashini & Khang, 2023).

The motive behind these types of attacks generally occur among the business competitors in the market as Figure 19.1.

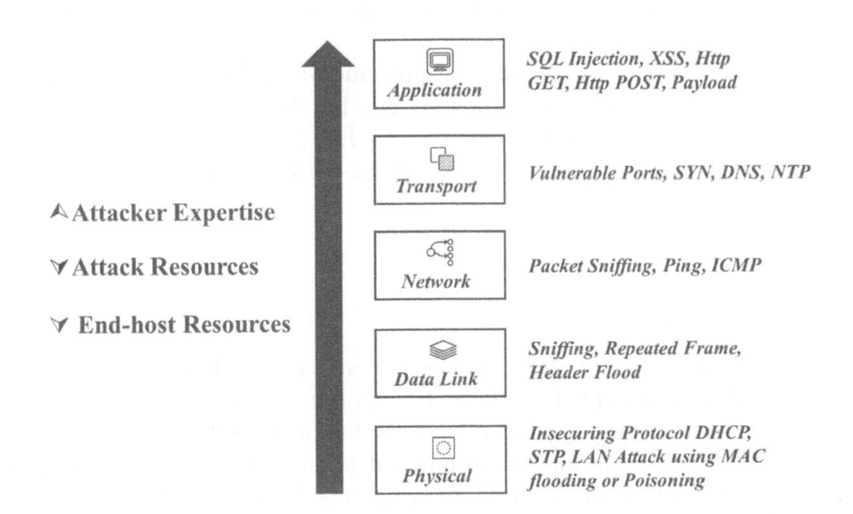

FIGURE 19.1 Common denial of services on IoT devices (Khang, 2021).

19.1.1 Motivation

Various organizations and industries take advantage of the increase of internet connections with the abundance of IoT devices. The IoT devices include smart cities, smart watches, healthcare sector, electric vehicle, agriculture and in-store contextualized market (Shah, & Jani et al., 2023).

Most of these services use the end layer (top-most layer of model) protocols like HTTP, DNS (Domain Name System), SMTP (Simple Mail Transfer Protocol) and NTP (Network Time Protocol).

Today attackers increase their level and attack on application layers. Such kind of attack is application-layer DDoS attack on IoT devices. According to the A10 Report (2019), we need more secure devices.

Hence, with an increasing number of IoT devices being offered these days, the likelihood of such attacks also increases. According to the report A10 Report (2019), DDoS attacks occur in the first quarter of 2019.

This level of data generates the payloads of packets; hence, payload analysis is required in layer 7 for such attacks. Payloads are encapsulated in frames of data; for example, content of packet without HTTP headers.

Analyzing payload is an effective way to detect attacks for the reason that network attack generates network traffic that contains the payload. So the recent application-layer IoT DoS attack evades existing attack detection (Khang & Hahanov et al., 2022).

In network traffic, payload header contains data without the HTTP header. Although analyzing payload is an effective method, in application layer, it is not easy to identify the payloads in HTTP header. Application-specific and targeted attacks are on increasing trend.

The security issue differs depending on the industry and environment, but application layer has some specific attack such as buffer overflow, command injection attacks and scripting attacks. This kind of attack is not detected by intrusion detection system.

To detect these kinds of data, we need application-level data and content-analysis-based data, such as intrusion detection system, but data is originated from attack traffic, which is a legitimate IP address of an IoT device. This review makes an attempt to expansively investigate the state-of-the-art defense solution against HTTP Payload flood IoT DDoS attack.

19.1.2 Contribution

The main spotlight of an HTTP Payload injector attack is just before inventing legitimate traffic that strictly simulates the authority of an individual users. Therefore, it becomes harder to identify legitimate users and legitimate traffic.

Our objective is not to find the minimal detection attack. This research aims at giving solutions to given problems of mitigating HTTP Payload injector DDoS attacks. This harmful attack has been increasing very rapidly.

This chapter is trying to find a methodical way to deal with building up a significant and total database of state-of-art text focused on HTTP Payload detection injector DDoS attacks. This survey contributes and is summarized as follows:

- cover all rich quality work in the area of HTTP Payload detection injector DDoS attacks;
- define the attack strategies;
- model method followed by detection techniques;
- dataset and various software tools;
- suggesting future investigative direction during a suspicious analysis of a variety of limitations and challenges.

This chapter is organized as follows: it discusses the linked work in Section 19.2. The detailed conditions on HTTP Payload injector DDoS attacks are presented in Section 19.3. Section 19.4 maps the results along with the answers to the research questions. A variety of limitations and challenges in the area are discussed in Section 19.5. Section 19.6 concludes the chapter.

19.2 RELATED WORK

After an in-depth research of the existing literature (A10 Report, 2019), this chapter represents an analysis of current IoT security tool modelers and computational analysis. We found that a very few chapters are dealing with defense-related work against HTTP Payload injector DDoS attacks.

Alaba and Othman et al. (2017) focuses on the state-of-the-art IoT security threats and vulnerabilities of the existing work in the area of IoT security. Moreover, the survey only deals with the overview of each layer's vulnerability.

Yu and Kim (2019) compares and analyzes security elements of international and domestic IoT platforms. This chapter only works on the domestic and international IoT security.

Grabovica and Pezer et al. (2016) explores protocols in security provided by IoT devices in different communication technologies. This chapter only studies about the security protocols of various IoT devices.

Tabanc and Zuva (2016) sets a challenge range from IoT security ethics privacy and threats. This only works on the security of RFID or network layer.

Chen and Thombre et al. (2017) shows strength privacy and security in location-based service for future IoT devices and suggests the strategies of different security and privacy aspects at location-based IoT.

Moreover, this chapter only discusses the privacy of location-based services. Ammar and Russello et al. (2019) surveys on the security of the main eight IoT frameworks with an emphasis on security. We explain the planned design, the development of basic third-party smart apps. This chapter only works on the IoT frameworks.

Our work is different from the above-mentioned existing survey in which we follow a methodical way to carry out an in-depth study and to offer more details in the literature on the detection technique, next to HTTP Payload injector for DDoS attacks.

An appropriate methodology for studying the writing at first picked up ubiquity in the field of brilliant application. In software engineering area, such reviews are normal in the field of IoT gadget.

Various studies additionally exist in the territory of IoT using distributed computing. In the late work, we directed a legitimate survey on plans that secured different methodologies, practical classes and assessment measurements of the procedures, alongside the difficulties and believable future.

19.3 HTTP PAYLOAD INJECTOR DDoS ATTACKS

The internet is highly dominant with World Wide Web, and all devices are connected through cloud called IoT devices that allow users to connect to their devices at any time and for the communication server using the HTTP protocol that manages all the transmissions of request from one application to another.

Each high-end commercial and non-commercial industry is adopting these devices for providing services to end-users today as they cover all fields such as sports, homes, automobiles, entertainments, healthcare, etc. The total number of IoT devices that are active globally will soon cross the two billion mark.

Therefore, against expanding the base of these devices, the attack targets these services. Of all applications, the most targeted protocol is HTTP as it is common between servers and users for communication.

HTTP-Payload-based attack is more vulnerable as it targets the area in the application layer. DDoS attack differs from lower layer flooding attack in the following way as shown in Figure 19.2:

- **Malicious code**: Lack of designing in web application security such as adding malicious code in HTTP Header.
- **Software modification**: Because the attackers use this method to put the malicious code in applications, for example, changing in HTTP header and updating the application.

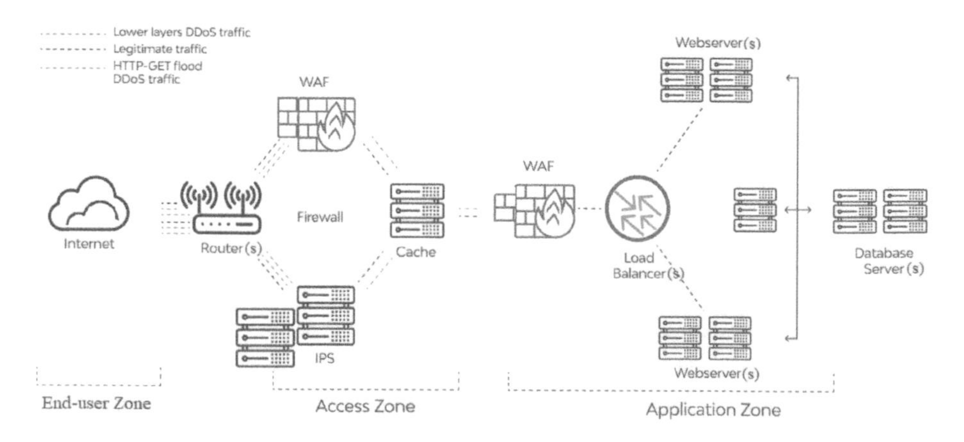

FIGURE 19.2 A typical scenario of receiving legitimate packet through end-user app's to server.

- **Mimicking humans**: The access behavior of the legitimate user due to which it becomes infeasible to filter the attack traffic.
- **Legitimate IP address**: The attacker sends valid IP address to compromised IoT devices to launch an HTTP DDoS Payload-based attack.

The end-user layer zone is the first level to inject such malicious code in different IoT devices such as smart cities, vehicle smart device, and many more. They all are connected to sensors and applications of end-users.

Access zone enters in various networking equipment such as router, firewall, Intrusion Prevention System (IPS), etc., whereas application zone contains resources such as mobile applications store in web and database servers.

Nowadays, attackers start working with payload-based strategies by adding malicious code in HTTP header's or they show user to update their mobile app and at that time they insert some new permission to the following app. Some malicious code acts as bot's (DDoS) attack.

The traffic to the interior system is at first directed to a firewall, IPS or other security hardware. The mentioned assets are conveyed to the client if present on web store, generally the heap balancer forward the solicitation to web server.

The authentic solicitation is assessed and permit to go through system, though lower layer assault traffic is obstructed at the entrance zone. Because of the authentic method used for correspondence, HTTP-GET flood assault traffic can break security premises ensured using system and transport layer protection instruments.

To verify the application benefits on the internet from such assaults, the security firms give WAFs whose nonattendance empowers the application-layer DDoS assault traffic (counting HTTP-GET flood) to easily arrive at the objective server. WAF can either be introduced as a strengthening module on the current server as shown in Figure 19.2.

For verifying web server, there are various strategies or systems accessible. This chapter gives a methodically overview, and some exploratory takes a shot at HTTP-GET Payload-based DDoS assault.

19.4 SURVEY PROTOCOL

The scheme is followed by different surveys for large-scale gain accepting some problem in hand. Elicit from the field of security (Chen and Karanpreet et al., 2017), it helps to provide a successful means to gather the literature.

The basic conceptual view of this survey is shown in Figure 19.3 that depicts the most prominent way to represent our organized survey that is defined in research questions on the basis of different articles we found in the system. This feature is used to conduct a complete search of literature that gives basic answers of the research questions.

There are many steps involved for conducting this organized survey. The result of this study will help to face the various challenges linked to field and some experimental work on http payloads which may inspiring the researcher to perform extra investigation.

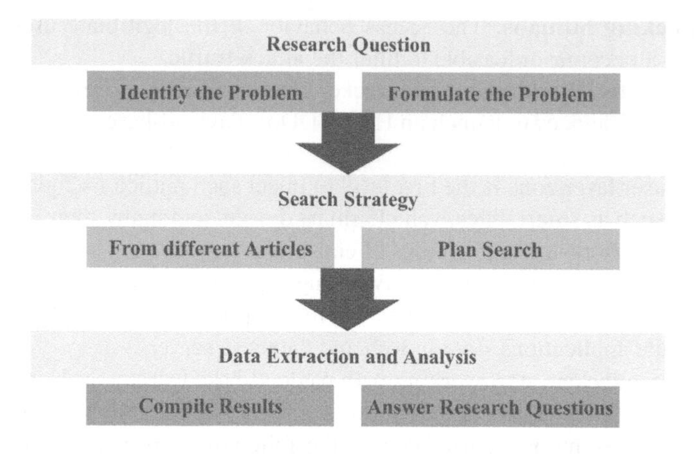

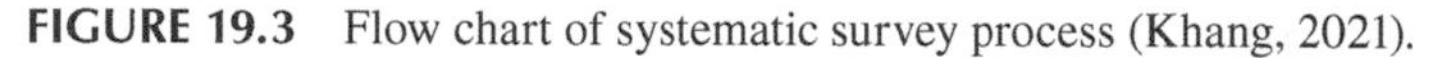
FIGURE 19.3 Flow chart of systematic survey process (Khang, 2021).

The following subsection contains all phases related to research questions, strategies, categorization methods and experimental results.

In this section, we have to understand that latest problems that are phased by researchers and those who are directly or indirectly connected to the IoT devices. To fulfill the goal, we create the four research questions that are answered by analyzing the different data.

This problem is facing by most of the researchers. First, we discussed that how much research has been performed on IoT security since 2019.

- **RQ1**. What are different attack strategies used by the attackers to launch the Http DDoS payload attack?
- **RQ2**. Which user and traffic attribute have been utilized for the detection of HTTP-DDoS Payload attack?
- **RQ3**. What are the various approaches and modeling in literature?
- **RQ4**. What kinds of dataset and software tools used for the evaluation of various attack detection techniques?

We mainly focus on the answer of the research questions, examine a literature survey on the detection of HTTP-DDoS Payload-based attack from multiple perspectives. In RQ1, we discover the distinctive attack strategies that have been explored in literature; in RQ2, we endeavor to recognize the detection attributes.

The classification of detection techniques is based on approaches. The order of discovery systems is dependent on approaches followed by the demonstrating techniques used to cover in RQ3. The wide scope of devices and dataset is used to evaluate these detection systems in RQ4.

19.5 RESULTS AND DISCUSSIONS

The area is separated into two different sections that give different outcomes which were obtained from the last set of main studies. First, discussion on the research questions is required, then we will proceed to some experimental works.

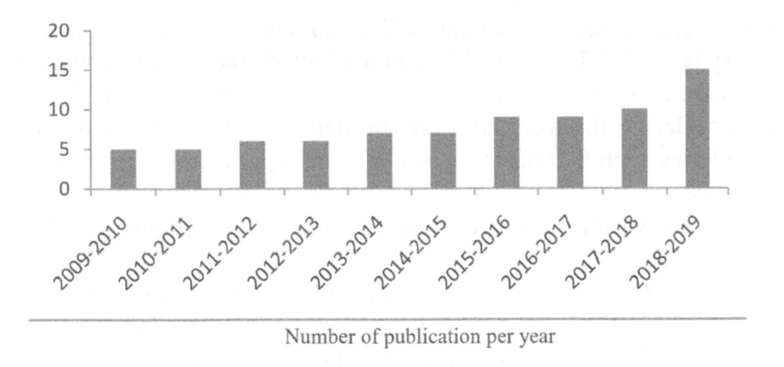

FIGURE 19.4 Number of publication per year.

To analyze how much research is going on the IoT security, we need some systematic review that shows the number of publications done from 2010 to 2019 as shown in Figure 19.4.

We found that from primary search, 30 out of 20 researcher studies have been working separately on every smart devices, and very few researchers address IoT security problem on application-layer platform.

RQ1. What are the various attack strategies that are used to launch the payload-based DDoS attack on the application layer?

In an attempt to escape attack detection, some technique is used to copy human activities. The attacker uses many strategies to perform an attack such as flood of data, malicious code, patterns, etc., to launch payload-based HTTP DDoS attack strategies, which have been studied in literature. These strategies are broadly classified into two categories shown in Figure 19.5.

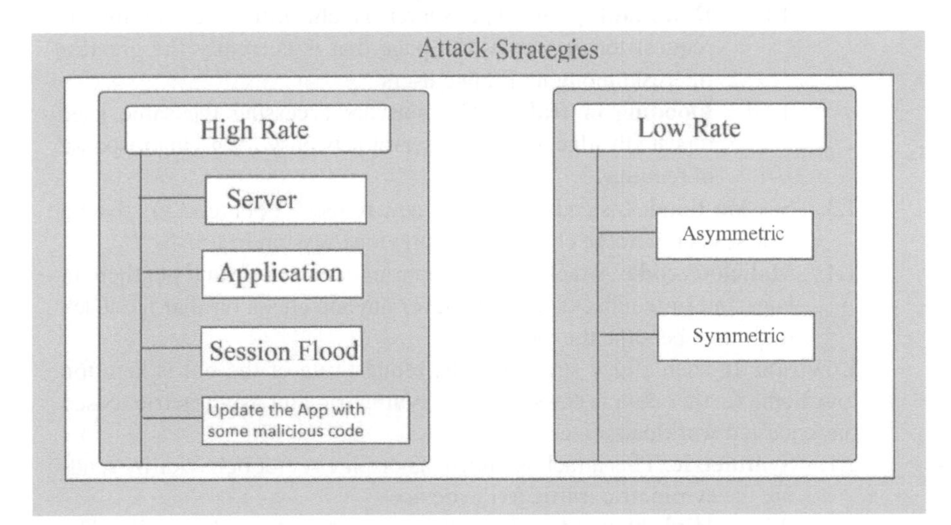

FIGURE 19.5 HTTP Payload-based attack strategies.

Due to a high demand of attack rate, HTTP targets DDoS attack at a high and low rate at every layer of IoT section. The attackers attack the sensors at the lower layer that is high and attack the user layer by sending a huge number of packets.

Only considering the low and high rate data flow, we can classify it into different categories such as reflection, volume, and applications of DDoS attack on smart devices.

In this survey, various possible attack strategies are discussed as follows:

1. **High rate**: Now the day's attacker becomes intelligent as they start using reflection of DDoS attack, that is, creating numerous bots through which attackers hide itself by using bots. The bots attack the server with his full load by sending a huge number of traffic so that the server gets crashed or attacker can get access of all smart devices during operation. Further, it is categorized into two: high request load and high session load.
   - 1.1. **Server**: The bots spread a huge request to the server due to this server resources run out and bot gets access. The buffer capacity of the request queue is quickly used and due to this, legitimate users are dropped by the server.
   - 1.2 **Application**: Most of the smart devices have their applications. The attackers start attacking the user-end by changing the HTTP header. They add some malicious codes in the file. A number of strategies attackers are using such as clicking on the comment, clicking on application window.
     - 1.1.1. **Lack of bad programming**: Developers don't have any idea regarding the secure code, secure programming.
     - 1.1.2. **Social engineering**: Attackers use this strategy to add new payloads to the API. Payload is added to submitting button in any form.
     - 1.1.3. **Dominant page**: The server is checking the continuous request for the same web page that is currently the greatest interest among legitimate users.
     - 1.1.4. **Flooding of traffic**: The bots are accessing the same page repeatedly; due to this, the server gets busy for giving response of requests.
   - 1.3. **Session flood**: Instead of creating new request, bots tend to create a new session without checking the previous session to finish.
   - 1.4. **Malicious code**: Attackers crate their malicious code and put them in bogs and in feedbacks. So, whenever anyone clicks on that feedback they must become the victim.
2. **Low rate**: In such attack strategies, the request rate of the bot is kept too low. Some further data is classified into symmetric and asymmetric, based on requested workloads size.
   - 2.1. **Symmetric**: This attack is known as display secret behavior by sending the symmetric traffic to the devices.
     - 2.1.1. **High blowout**: A request is generated at regular basics. The rate of attack only disturbs the server for some time.

2.1.2. **Non-recurrent**: The peaks' attacks produce the irregular intervals, that is, without any similarity among the bots. However, the peak is very high, and due to this, it degrades the server's performances.

2.1.3. **Slowloris**: Normally a software program, which is main for attacking the server. In this attack strategy, bot creates multiple connections with the server and simultaneously sends the HTTP request so that the connection becomes alive.

3. **Asymmetric**: This bots create a connection with the server and keep server busy by requesting the same data repeatedly. Such attacks are categorized in asymmetric attack because the download rate is much higher than the upload rate.
4. **Simultaneously**: The bots simultaneously send the request for the content and server becomes busy. A single request may help to operate multiple operations on the server.

Figure 19.5 describes the type of attack strategies done by the attacker on different smart devices such as smart watch, smart cities, etc. It's a high rate attack, because the impact of these attacks is very powerful, and researchers must focus on this area as compared to low-rate HTTP-DDoS Payload-based attacks.

The attack strategy is classified based on the different categories such as malicious code, API, Update of software, sending flood of data using HTTP Payloads. We found that the bots are the problem phase at a high rate. The problem of attack increases in the case of bots that attempts to send a huge number of requests.

RQ2. Which user and traffic attribute have been used for the detection of HTTP-DDoS Payload attack?

There are numerous ways to detect the process of application layer and lower layer DDoS attack. HTTP-Payload-based attack is highly dependent on identifying the behavior of attacking bots. The detection of HTTP–Payload-based DDoS attacks is highly dependent on identifying the activities of bots.

The detection of HTTP- Payload-based DDoS attack mainly relies on mobility and scalability, resources constraints, data interchange, connectivity, and we mainly consider the traffic-related characteristics.

A detection attribute is identified in different aspects such as API used by the users. A variety of detection attributes have been used independently in the lower layer and the application layer. The elementary detection attribute along with the short descriptions are presented in Table 19.1. Each of the attributes is assigned with a unique identifier.

Most of the attributes follow the same class. The attribute value is identified into less and more different for human and bots, which make the detection process easier and effective. We identify some new attributes that will contribute to secure our applications in terms of those attributes

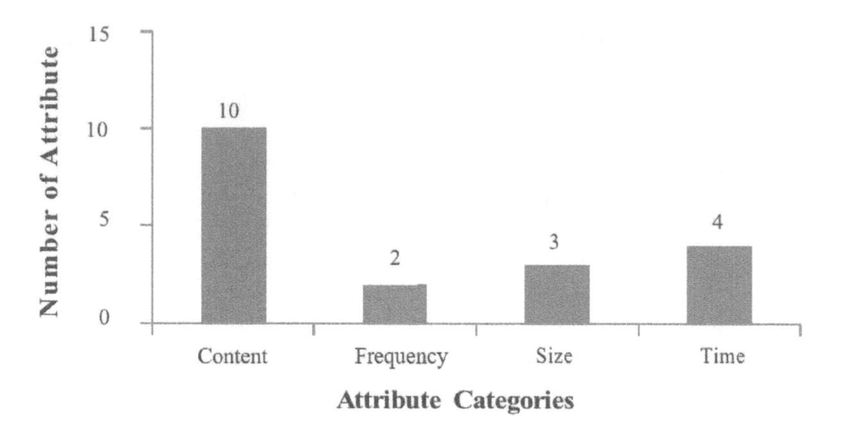

FIGURE 19.6 Distribution of a number of attributes per category.

The detection attribute is divided into four categories: frequency, size, time and content. These classes were decided only on the basis of literature and articles based on IoT security. The primary aim of attack is to take a control on the smart devices. Therefore, the large portion of these attributes come from content, time classes, as shown in Figure 19.6.

Besides, other classes like size is used by the attacker when they follow the asymmetric attack strategies. The class "content" recognizes what sort of data is putting away in the application. The "frequency" class identifies continuous requests coming from the legitimate users or not. The class time identifies the interval of time between two successive.

RQ3. What are the various modeling methods in literature?

Answer of the question is exploring a variety of modeling methods used to detect the HTTP-based Payload DDoS attacks. Machine learning, statistics and neural network are three main methods that form the basic majority of detection techniques in the present day, as shown in Table 19.1.

Modeling: There are different methods used by the researchers for detecting the attack on the application layer. Majority of the researchers are using the machine learning techniques and neural network for making their research more efficient and helpful.

- Machine learning techniques: Researchers are using different parameters to train the devices in that way. We map the method with different algorithms.
- Neural network: Nowadays, this field is in trend, peoples are using their algorithm and train device, and the efficiency of the device is better than others.

TABLE 19.1
Detection of Attributes for HTTP-Payload-Based DDoS Attack

ID	Attribute	Description	Typical Human Value	Typical Bot Value	Papers	Class
A (A Proposed Framework, 2016)	Length of queue	Request filled by the length of queue	–	High	17	Size
A (Kaspersky Q3 2018)	Request size	Huge number of Request in smart devices	Average	High	16, 26, 43	Size
A (A10 Report, 2019)	Application Front page Size	Total no. of request on pages	–	–	24, 46, 54	Size
A (Xie & Yu, 2005)	Downtime	Time gap after user reconnects to server	High	Average	16	Time
A (Alaba & Othman et al., 2017)	Timestamp	Request is received at particular time	–	–	23, 45, 46, 47	Time
A (Yu & Kim, 2019)	Interval of Packet arrival	Interval between successive packets	High	Average	30	Time
A (Grabovica & Pezer et al., 2016)	Time-based SQL	Number of injection using SQL query	High	Average		Time
A (Tabane & Zuva, 2016)	TCP packets receiving	Number of TCP packets received	High	Average	37, 55	Frequency
A (Chen & Thombre et al., 2017)	HTTP responds every packet	When HTTP server responds to error packets	High	Average	34, 39, 40, 45	Frequency
A (Ammar & Russello et al., 2019)	HTTP Verb Tampering	Bypass the authorization on this directory	High	Average		Content
A (Tang, 2012)	HTTP Parameter Tampering	Data entered by user are changed without user authorization	High	Average		Content
A (Beitollahi & Deconinck, 2013)	HTTP Response splitting	Entrusted source, requested by the legitimate users	High	Average		Content
A (Xie & Tang et al., 2013)	Header injection in HTTP	Dynamically generated based on user input	High	Average		Content
A (Kim & Kim et al., 2012)	Cookies without using HTTP flag set	Cookie value cannot read or set by client-side JavaScript	High	Average		Content
A (Xie & Yu, 2005)	Cookies without secure flag set	Browsers will not submit the cookie in any request	High	Average		Content
A (Wang & Yang et al., 2010)	Unwanted HTTP methods Implementation	Unwanted HTTP Methods implementations	High	Average		Content
A (Huang & Wang et al., 2014)	Check for X-XSS protection	Header to protect cross-site scripting	High	Average		Content
A (Umarani & Sharmila, 2014)	Check X-frame of HTTP header	Response header can be used	High	High		Content
A (Maciá-Fernández & Rodríguez-Gómez et al., 2010)	To Identify the X-content type-option	Acts as a marker which indicate the MIME	High	High		Content

Table 19.2 provides the classification on the basis of literature. In the literature survey, clustering is considered a part of machine learning. There are a number of categorical studies which do not fall in these categories, so they are classified into other categories.

Table 19.3 describes that detections as per category based on their functionalities. There are challenge-based techniques for the practical applications but failed to capture the researchers focus due to their hostile nature.

However, the majority of researchers are creating their encryption techniques for securing the IoT devices Depending on the apps types, user have different access patterns that can be captured using different models and tools.

RQ4. What kinds of dataset and software tools used for the evaluation of various attack detection techniques?

TABLE 19.2
Various Recent Detection Techniques for HTTP-Payload Based DDoS

	Modeling					
Research Papers	**ML Techniques**	**Clustering**	**Neural Networks**	**Statistical**	**Others**	**Algorithm**
1					✓	HADEC-Hadoop-based detection (Stevanovic & An et al., 2012)
2					✓	D-face anomaly detection (Stevanovic & Vlajic et al., 2013)
3	✓					User Behavior using detection (Hameed & Ali, 2019)
4			✓			Bat Algorithm (Behal & Kumar et al., 2018)
5				✓		Covariance matrix approach (Hameed & Ali, 2019)
6			✓			MLP-GA based (Sreeram and Vuppala, 2017)
7					✓	FPGA for detection (Aborujilah & Musa, 2017)
8			✓			Entropy based (Kaspersky Q3 2018)
9						HTTP solider (Hoque & Kashyap et al., 2017)
10		✓				Sparse vector decomposition rhythm matching (Johnson Singh & Thongam et al., 2016)
11					✓	FHSD (Hoque & Kashyap et al., 2017)
12			✓		✓	SIT secure IoT lightweight encryption technique (A10 Report, 2019)
13		✓				Entropy based clustering (Singh & De, 2017)
14	✓					SVM support vector machine (Shiaeles & Papadaki, 2014)
15		✓				*K*-mean clustering (Liao & Li et al., 2015)
16	✓					Navies Bayes (Wang & Yang et al., 2014)
17	✓					Principle Component Analysis (PCA)
18				✓		Connection: Score
19	✓					OC-SVM One class support vector machine (Shiaeles & Papadaki, 2014)
20				✓		Fuzzy test

TABLE 19.3
Available Standard Datasets

No.	Dataset	Online Link	Year
1	DARPA LLDOS	https://www.ll.mit.edu/r-d/datasets/2000-darpa-intrusiondetection-scenario-specific-datasets	2000
2	MIT Lincoln FIFA	https://www.kaggle.com/datasets/bryanb/fifa-player-stats-database	2019
3	CLARK Net HTTP	https://ita.ee.lbl.gov/html/contrib/NASA-HTTP.html	2019
4	Standard EPA -HTTP	https://catalog.data.gov/dataset	2018
5	DARPA_2009_malware DDoS_attack-20091104	https://ant.isi.z/datasets/all.html	2014
6	UNINA traffic traces	http://traffic.comics.unina.it/Traces/ttraces.php	2007
7	NTP DDoS dataset	https://ant.isi.edu/datasets/readmes/DARPA_2009_DDoS_attack20091105.README.txt	2009
8	IoT dataset	https://www.kaggle.com/datasets/atulanandjha/temperature-readings-iot-devices/discussion/120916	2010
9	UK Domestic Appliance-Level	http://www.doc.ic.ac.uk/~dk3810/data/	2014
10	Road Traffic Data, Pollution	http://iot.ee.surrey.ac.uk:8080/datasets.html	2013
11	CASAS datasets for activities of daily living	casas.wsu.edu/datasets.html	2010

The traffic follow-ups utilized for assessing a barrier strategy assume a significant job in building up performance standards. Some researchers have some standard which analyze the results with existing techniques.

An enormous variety of freely accessible datasets are generally used as traffic sources in the assessment procedure on different detection strategies. These datasets can additionally be merged with traffic follow-ups produced by various devices so as to structure numerous approval situations. These datasets have been classified into three categories such as educational, commercial and standard.

- Educational: These datasets incorporate web get to logs of college, organizations and division level sites. Such datasets can't be legitimately downloaded, as they are not freely accessible on the internet. A number studies have used educational datasets (Khang & Abdullayev et al., 2024).
- Standard: The datasets that are openly accessible online for aggregate use are referred to as standard datasets. The researchers from round the globe can easily download these datasets to calculate and compare their proposed techniques. Due to the absence of real HTTP-Payload-based DDoS attack datasets, the standard datasets, used as a source of background traffic, are commonly blended in with the traffic generally using different software tools in order to evaluate a detection approach (Khang & Vrushank et al., 2023).
- Commercial: Several commercial websites are considered on such commercial datasets. These web-logs are provided from various internet services providers (ISPs).

Such dataset is not freely available for the individual's use. In the given studies, (Grabovica & Pezer et al., 2016), Sreeram and Vuppala (2017) used the commercial dataset. Table 19.3 enrolls standard dataset along with online links available with the published year of the datasets.

Most of the researchers are used these datasets in their work. Hence, the standard dataset is listed, rather than educational and commercial datasets are not easily available in public.

Out of rest, 16 studies used the standard dataset for evaluating their work. The rest 2 studies used the educational dataset, and Wang and Yang et al. (2014) used the educational dataset in their work.

A large diversity of programming tools helps the researchers in creating and implementing their future techniques. Various software tools are used in literature along with their objectives, domains and online links presented in Table 19.4. Various software tools classified into three categories:

- **Identification tools:** These are the software tools that help to create their own environment by using different platforms. Due to this, researcher determine the performance of work, such as network simulation NS2.
- **Access control:** It can identify the trust calculations to validate the trust value of IoT devices. Only 20% of access control methods are used for the evaluation process. Most of the researchers are using the AVISPA, Test bed, Ban-logic, Matlab tools in the domain of smart card, smart-home, free authentication, Biometric authentication key agreement.
- **Server software**: When working in a copied domain, the victim individual that typically is a server can implement on the client of a specific server.

In this chapter, some broadly utilized server programming in the essential examinations are viable. The open source server software, Apache HTTP server, has been seen ordinarily used by various investigations as delineated in Table 19.4.

TABLE 19.4
Summary of Diverse Software Tools

Software Tool	Objectives	Domain	Online Link
NS2	Identification	Use of NFC, suitable for mobile environment	http://www.isi.edu/nsnam/ns
AVISPA	Access control	Biometric authentication, key agreement	http://www.avispa-project.org/
Test bed	Access control	For smart home, medical	https://www.softwaretestinghelp.com/test-bed-test-environment-management-best-practices/
Ban-logic	Access control	Smart card	–
Matlab	Access control	Free authentication	http://in.mathworks.com/products/matlab
Apache	Server	Server connection with app	https://httpd.apache.org

19.6 LIMITATIONS AND CHALLENGES

The developing interest for the field of detection of HTTP-Payload-based DDoS attack is observing numerous issues and difficulties arising out of the ongoing exploration (Khang & Rani et al., 2022). We discuss different challenges, either incomplete address or awaiting, while moving the direction of countering HTTP-Payload-based DDoS attacks.

- **Bot**: A big portion of the literature used different mathematical models to filter out legitimate users particularly bots. Such detection techniques use parameters such as behavior of the users. Trust issues come due the bots. How device can identify the legitimate user and real users is discussed.
- **Encryption**: IoT system needs the heterogeneity with end-to-end security. Thus, encryption is needed at every level of smart devices. So, most of the researchers had focused on implementing the lightweight encryption techniques.
- **Inaccessibility of attack datasets**: Many examinations have practiced their proposed systems on 10-year-old datasets.

Most analysts have used their college/association web logs rather than standard datasets to assess their proposed techniques. As these datasets are gathered from various destinations, the semantics of perusing practices fluctuate drastically (Khanh & Khang, 2021).

19.7 CONCLUSION

This survey intentional gives the IoT security current trends. The IoT is a rising technology that has considerably attracted the interest of both industry and academia. Researchers start working on the security aspects. The detection of payload-based DDoS attack is a rather challenging task. Due to many forms, devices are affected, such as Botnets.

We have conducted the systematic survey on the HTTP-Payload-based DDoS attack on the IoT devices and provide the current state-of-the-art supported classification of research landscape and challenge and future aspects.

The research question defines this aim of work. These sections include the attack strategy, detection attribute, methods, datasets used and software tools (Khang & Gupta et al., 2023).

We expect that the reader understands the various problems related to the HTTP-Payload based DDoS attacks in the application layer of smart devices. At last, we identify the future scope and challenges listed, which provides researchers an optimistic future (Tyagi & Saraswat et al., 2023).

REFERENCES

A Proposed Framework. *Securing the Internet of Things* (2016) http://www.cisco.com/c/en/us/about/security-center/secureiot-proposed-framework.html.

A10 Report. *DDoS attack Report* (2019) [online]; Available: https://securelist.com/ddos-report-in-q1-2018/85373

Aborujilah, A., Musa, S., "Cloud-based DDoS HTTP attack detection using covariance matrix approach", *Journal of Computer Networks and Communications*, vol. 2017, Article ID 7674594, 8 pages (2017). https://www.hindawi.com/journals/jcnc/2017/7674594/

Alaba, F.A., Othman, M., Hashem, I.A.T., "Internet of things security: a survey", *Journal of Network and Computer Application* (2017). https://www.sciencedirect.com/science/article/pii/S1084804517301455

Ammar, M., Russello, G., Crispo, B., "Internet of Things: A survey on the security of IoT frameworks", *Journal of Information Security and Application* (2019). https://www.sciencedirect.com/science/article/pii/S2214212617302934

Behal, S., Kumar, K., Sachdeva, M., "D-FACE: an anomaly based distributed approach for early detection of DDoS attacks and flash events", *Journal of Network and Computer Applications*, vol. 111, pp. 49–63 (2019). https://www.sciencedirect.com/science/article/pii/S1084804518301115

Beitollahi, H., Deconinck, G., "Connection score: a statistical technique to resist application-layer DDoS attacks", *Journal of Ambient Intelligence and Humanized Computing*, vol. 5, no. 3, pp. 425–442 (2013). https://link.springer.com/article/10.1007/s12652-013-0196-5

Bhambri, P., Rani, S., Gupta, G., Khang, A., *Cloud and Fog Computing Platforms for Internet of Things* (2022). CRC Press. https://doi.org/10.1201/9781032101507

Chen, L., Thombre, S., Karanpreet Jarvinen, E.S., Lohan, "Robustness, Security and Privacy in Location-Based Services for Future IoT: A Survey", *Security and Privacy in Applications and Services for Future Internet of Things* (2017). http://hbrppublication.com/OJS/index.php/RRAR/article/view/307

Grabovica, M., Pezer, D., Popic, S., Knezevic, V., "Provided security measures of enabling technologies in Internet of Things (IoT): a survey", *Zooming Innovation in Consumer Electronics International Conference (ZINC)* (2016). https://ieeexplore.ieee.org/abstract/document/7513647/

Hameed, S., Ali, U., "HADEC: hadoop-based live DDoS detection framework", *EURASIP Journal on Information Security*, vol. 2018, no. 1, p. 11 (2019). https://jis-eurasipjournals.springeropen.com/articles/10.1186/s13635-018-0081-z

Hoque, N., Kashyap, H., Bhattacharyya, D.K., "Real-time DDoS attack detection using FPGA", *Computer Communications*, vol. 110, pp. 48–58 (2017). https://www.sciencedirect.com/science/article/pii/S0140366416306442

Huang, C., Wang, J., Wu, G., Chen, J., "Mining web user behaviors to detect application layer DDoS attacks", *Journal of Software*, vol. 9, no. 4, pp. 985–990 (2014). https://citeseerx.ist.psu.edu/document?repid=rep1&type=pdf&doi=436e870303009a2395600155a5d1a1cc7670189a#page=197

Johnson Singh, K., Thongam, K., De, T., "Entropy-based application layer DDoS attack detection using artificial neural networks", *Entropy*, vol. 18, no. 10, p. 350 (2016). https://www.mdpi.com/159460

Kaspersky Q3. *DDoS Attack Report* (2018) [Online]: Available From: https://securelist.com/kaspersky-ddos-intelligencereport-for-q3-2016/76464/

Khang, A., Abdullayev, V., Hahanov, V., Shah, V., *Advanced IoT Technologies and Applications in the Industry 4.0 Digital Economy* (1st Ed.) (2024). CRC Press. doi: 10.1201/9781003434269

Khang, A., Gupta, S.K., Rani, S., Karras, D.A., *Smart Cities: IoT Technologies, Big Data Solutions, Cloud Platforms, and Cybersecurity Techniques* (1st Ed.) (2023). CRC Press. https://doi.org/10.1201/9781003376064

Khang, A., Hahanov, V., Abbas, G.L., Hajimahmud, V.A., "Cyber-Physical-Social System and İncident Management", *AI-Centric Smart City Ecosystems: Technologies, Design and Implementation* (1st Ed.) (2022). CRC Press. https://doi.org/10.1201/9781003252542-2

Khang, A. (2021), Material4Studies, *Material of Computer Science, Artificial Intelligence, Data Science, IoT, Blockchain, Cloud, Metaverse, Cybersecurity for Studies.* https://www.researchgate.net/publication/370156102_Material4Studies

Khang, A., Rani, S., Sivaraman, A.K., *AI-Centric Smart City Ecosystems: Technologies, Design and Implementation* (1st Ed.) (2022). CRC Press. https://doi.org/10.1201/9781003252542

Khang, A., Vrushank, S., Rani, S., *AI-Based Technologies and Applications in the Era of the Metaverse* (1st Ed.) (2023). IGI Global Press. https://doi.org/10.4018/9781668488515

Khanh, H.H., Khang, A., "The Role of Artificial Intelligence in Blockchain Applications", *Reinventing Manufacturing and Business Processes through Artificial Intelligence*, pp. 20–40 (2021). CRC Press. https://doi.org/10.1201/9781003145011-2

Kim, H., Kim, B., Kim, D., Kim, I.-K., Chung, T.-M., "Implementation of GESNIC for web server protection against HTTP GET flooding attacks", in: Proc. Int. Workshop Inform. Security Applicant, pp. 285–295 (2012). https://link.springer.com/chapter/10.1007/978-3-642-35416-8_20

Liao, Q., Li, H., Kang, S., Liu, C., "Application layer DDoS attack detection using cluster with label based on sparse vector decomposition and rhythm matching", *Security and Communication Networks*, vol. 8, no. 17, pp. 3111–3120 (2015). https://onlinelibrary.wiley.com/doi/abs/10.1002/sec.1236

Maciá-Fernández, G., Rodríguez-Gómez, R.A., DíazVerdejo, J.E., "Defense techniques for low-rate DoS attacks against application servers", *Computer Networks,* vol. 54, no. 15, pp. 2711–2727 (2010). https://www.sciencedirect.com/science/article/pii/S1389128610001386

Rani, S., Bhambri, P., Kataria, A., Khang, A., Sivaraman, A.K., *Big Data, Cloud Computing and IoT: Tools and Applications* (1st Ed.) (2023). Chapman and Hall/CRC. https://doi.org/10.1201/9781003298335

Rani, S., Chauhan, M., Kataria, A., Khang, A., "IoT equipped intelligent distributed framework for smart healthcare systems", *Networking and Internet Architecture* (2021). CRC Press. https://doi.org/10.48550/arXiv.2110.04997

Shah, V., Jani, S., Khang, A., "Automotive IoT: Accelerating the Automobile Industry's Long-Term Sustainability in Smart City Development Strategy", *Smart Cities: IoT Technologies, Big Data Solutions, Cloud Platforms, and Cybersecurity Techniques* (1st Ed.) (2023). CRC Press. https://doi.org/10.1201/9781003376064-9

Shiaeles, S.N., Papadaki, M., "FHSD: an improved IP spoof detection method for web DDoS attacks", *Computer Journal*, vol. 58, no. 4, pp. 892–903 (2014). https://academic.oup.com/comjnl/article-abstract/58/4/892/336014

Singh, K.J., Dc, T., "MLP-GA based algorithm to detect application layer DDoS attack", *Journal of Information Security and Applications*, vol. 36, pp. 145–153 (2017). https://www.sciencedirect.com/science/article/pii/S2214212616302162

Sreeram, I., Vuppala, V.P.K., "HTTP flood attack detection in application layer using machine learning metrics and bio inspired bat algorithm", *Applied Computing and Informatics*, in press (2017). https://www.sciencedirect.com/science/article/pii/S2210832717301655

Stevanovic, D., An, A., Vlajic, N., "Feature evaluation for web crawler detection with data mining techniques", *Expert Systems with Applications*, vol. 39, no. 10, pp. 8707–8717 (2012). https://www.sciencedirect.com/science/article/pii/S0957417412002382

Stevanovic, D., Vlajic, N., An, A., "Detection of malicious and non-malicious website visitors using unsupervised neural network learning", *Applied Soft Computing*, vol. 13, no. 1, pp. 698–708 (2013). https://www.sciencedirect.com/science/article/pii/S1568494612003778

Subhashini, R., Khang, A., "The Role of Internet of Things (IoT) in Smart City Framework," *Smart Cities: IoT Technologies, Big Data Solutions, Cloud Platforms, and Cybersecurity Techniques* (1st Ed.) (2023). CRC Press. https://doi.org/10.1201/9781003376064-3

Tabane, E., Zuva, T., "Is there a room for security and privacy in IoT?" in: *International Conference on Advances in Computing and Communication Engineering* (ICACCE) (2016). https://ieeexplore.ieee.org/abstract/document/8073758/

Tang, Y., "Countermeasures on application level low-rate denial-of-service attack", *in:* Proc. Int. Conf. Inform. Commun., pp. 70–80 (2012). https://link.springer.com/chapter/10.1007/978-3-642-34129-8_7

Tyagi, V., Saraswat, A., Bansal, S., Khang, A., "An Analysis of Securing IoT Devices from Man-in-the-Middle (MIMA) and Denial of Service (DoS)", *Smart Cities: IoT Technologies, Big Data Solutions, Cloud Platforms, and Cybersecurity Techniques* (1st Ed.) (2023). CRC Press. https://doi.org/10.1201/9781003376064-20

Umarani, S., Sharmila, D., "Predicting application layer DDoS attacks using machine learning algorithms", *International Journal of Electrical and Computer*, vol. 8, no. 10 (2014). http://publications.waset.org/10000388/predicting-application-layer-ddos-attacks-using-machine-learning-algorithms

Wang, J., Yang, X., Long, K., "A new relative entropy based app-DDoS detection method", in: Proc. IEEE Symp. Comput. Commun. pp. 966–968 (2010). https://ieeexplore.ieee.org/abstract/document/5546587/

Wang, J., Yang, X., Zhang, M., Long, K., Xu, J., "HTTP soldier: an HTTP-flooding attack detection scheme with the large deviation principle", *Science China Information Sciences*, vol. 57, no. 10, pp. 1–15 (2014). https://link.springer.com/article/10.1007/s11432-013-5015-2

Xie, Y., Tang, S., Huang, X., Tang, C., Liu, X., "Detecting latent attack behavior from aggregated web traffic", *Computer Communications*, vol. 36, no. 8, pp. 895–907 (2013). https://www.sciencedirect.com/science/article/pii/S0140366413000406

Xie, Y., Yu, S.-Z., "A detection approach of user behaviors based on HsMM", *in: Proc. Int. Teletraffic Congr. Perform. Challenges Efficient Next Generation Netwt.*, pp. 461–450 (2005). https://www.sciencedirect.com/science/article/pii/S0167404816301365

Yu, J.-Y., Kim, Y.G., "Analysis of IoT Platform Security: A survey", *IEEE International Conference on Platform Technology and Service (PlatCon)* (2019). https://ieeexplore.ieee.org/abstract/document/8669423/

20 Detection of Cyber Attacks in IoT-Based Smart Cities Using Integrated Chain Based Multiclass Support Vector Machine

Shobhna Jeet, Shashi Kant Gupta, Olena Hrybiuk, and Nupur Soni

20.1 INTRODUCTION

Internet of Things (IoT) is a network of networked devices that encourages easy information sharing between them (Chowdhury & Karmakar et al., 2019). With 27 billion connected IoT devices as of 2017 and an estimated 125 billion by 2030 as shown in Figure 20.1, the IoT's recent growth has considerably boosted its usage in communities and services throughout the globe (Sharma & Khang, 2024).

The services, technologies, and protocols that IoT devices employ vary. Therefore, maintaining future IoT infrastructures will become very difficult, which inevitably results in the system being vulnerable in an unfavorable way (Pahl & Aubet et al., 2018).

IoT devices are used in applications; therefore, cyber assaults may illegally acquire information about inhabitants' daily activities or change devices to a dangerous state (Tyagi & Saraswat et al., 2024).

There are a number of security concerns with smart city applications. To begin, there is the risk of zero-day attacks due to the use of unpatched protocols in smart city applications (Khang & Hahanov et al., 2022).

Due to resource (e.g., memory) limits, smart city IoT devices have limited on-board security capabilities and transport recorded data to the cloud server for processing. IoT data is stored on the cloud, which has powerful CPUs and processor (Restuccia & D'Oro et al., 2018).

Recent growth in IoT devices has increased the quantity of data transferred from the "IoT terminal layer" to the cloud, increasing delay and congestion (Rani & Chauhan et al., 2021).

Fog computing might solve these issues. Fog layer devices may share more of the cloud's computing load. This decreases internet traffic, delay, and data storage and

DOI: 10.1201/9781003376064-20

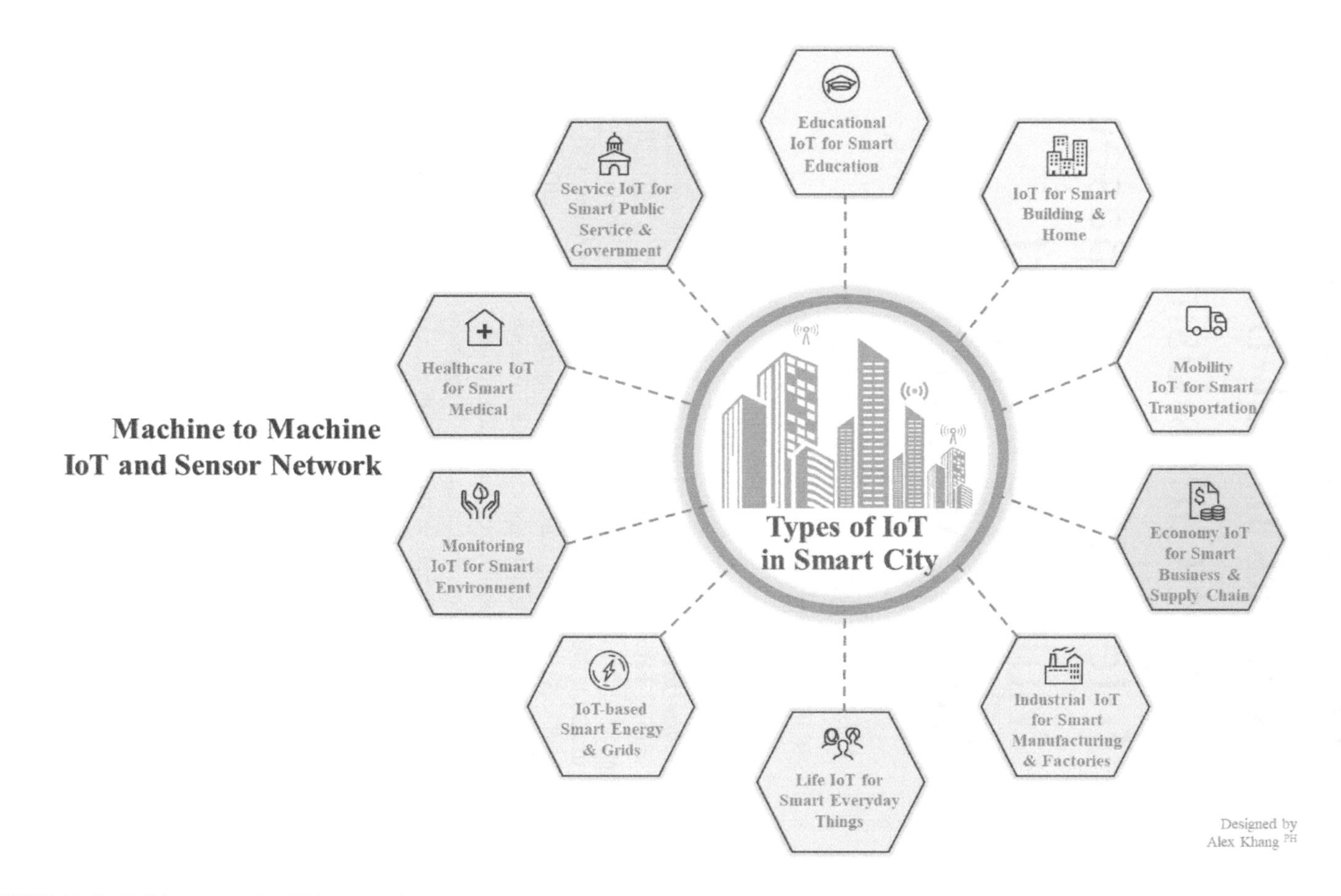

FIGURE 20.1 IoT in smart city (Khang, 2021).

transfer difficulties (Pu & Choo, 2022). It also intends to move the processing process closer to the edge device, allowing for a faster response to applications.

The advantages of detecting cyber-attacks in the fog layer are twofold. To begin, if assaults are detected early in this layer, the network administrator may take the required precautions to avoid significant harm. Second, it will not disrupt the regular flow of city life (Mihoub & Fredj et al., 2022).

We explore approach based on an Integrated Chain based Multiclass Support Vector Machine (IC-MSVM) to prevent and reduce IoT cyber-security risks in a smart city in this study (Gaber & El-Ghamry et al., 2022). This approach can detect affected sensor gadgets, which is significantly difficult in the cloud computing environment. This approach can detect affected IoT devices, which is a significant difficulty in the cloud environment (Elsisi & Tran, 2021).

A single classifier is often inadequate for developing efficient anomaly detection, prompting researchers to construct an IC-MSVM classifiers.

Taking into consideration a variety of models, the Integrated Chain based Multiclass model combines several models to build one finishing model. According to research, the Integrated Chain based Multiclass model outperforms the single classifier.

Contribution of this chapter:

- Data can be preprocessed using z-score and "min-max normalization" techniques.
- Principal Component Analysis (PCA) is a method for transforming raw data into numerical features that can be used to analyze the data without losing any of the information in the original dataset.
- Whale optimization algorithm (WOA) is used for feature selection.

The organization of this chapter is as follows, Section 20.2 represents the problem statement, Section 20.3 represents proposed work, Section 20.4 represents result, and Section 20.5 denotes the conclusion part.

20.2 PROBLEM STATEMENT

Every IoT device requires security in general. As smart cities link a wide range of devices to the internet, security becomes a major issue. Around 70% of IoT devices in a smart city (Khang & Rani et al., 2022) were vulnerable to attack owing to flaws such as insufficient authorization, poor software safeguards, and weakly encrypted communication protocols.

These flaws enable numerous risks and assaults, resulting in a number of security and privacy problems. Being small in size, IoT devices come with a variety of sensors. This presents a big problem in terms of battery life and cost since these gadgets need a continual supply of energy to function (Bhambri & Khang et al., 2022).

Connectivity among devices in a smart city that use multiple communication technologies relies heavily on this concept of interoperability. As stated by the World Economic Forum, the absence of common standards affects the IoT by preventing devices from various domains from communicating with one another (Khang & Gupta et al., 2023).

20.3 PROPOSED WORK

The cyber-attack dataset is first gathered and then preprocessed using the z-score and min-max normalization techniques. Data that has been preprocessed is separated into training and testing sets (Jebaraj & Khang et al., 2024).

To extract the pertinent features, we use PCA. The WOA technique is used to select the features as shown in Figure 20.2.

20.3.1 Dataset of Automated Guided Vehicles (AGV)

The "Canadian Institute for Cyber security (CIC) created the CICIDS2017 dataset in 2017". This collection contains the most current and harmless cyber-attacks, including "DoS, DDoS, PortScan, SQL injection, Infiltration, Brute Force, and Bot".

The "2,830,743 records in the CICIDS2017" are divided into 8 files, with 78 unique characteristics per record. In our study, we used "190,774 records, 148,777 of which are benign, and 41,997" of which include different forms of attacks. Table 20.1 indicates the proportion of attack and normal sample of CICIDS2017 datasets.

20.3.2 Data Preprocessing Using Normalization

Handling raw data in order to make it suitable for further data analysis procedures is known as data preprocessing. Before beginning the data analysis process, this has typically been a critical stage.

1. **Min-max normalization**

   Min-max normalization is a method of transforming linear data at the beginning of a range. The connection between distinct bits of data is retained using this strategy. Pre-defined boundaries with pre-defined boundaries are an important approach for accurately fitting information.

   The technique to normalization,

$$Q' = \left(\frac{Q - \text{min\_}value\, of\, Q}{maxvalue\, of\, Q - min\, value\, of\, Q} \right) * (T - K) + K, \tag{20.1}$$

   has min-max data, where one of the bounds is [K, T], and Q' is the actual data range, whereas Q represents the mapped data.

2. **Z-score**

   Z-score normalization, sometimes referred to as zero-mean normalization, normalizes each input feature vector by calculating the mean (M) and standard deviation (SD) of each feature across a training data and splitting it by the dataset's size. The SD and average for each variable are computed. As stated in the general formula, the transformation is required.

$$n' = \frac{(n - \mu)}{\sigma} \tag{20.2}$$

   The property n has an SD of and an M of. Before training can begin, each feature in the set of data is processed to z-score normalization. The mean and SD of each feature should be recorded once train data has been collected so that they may be used as algorithm weights.

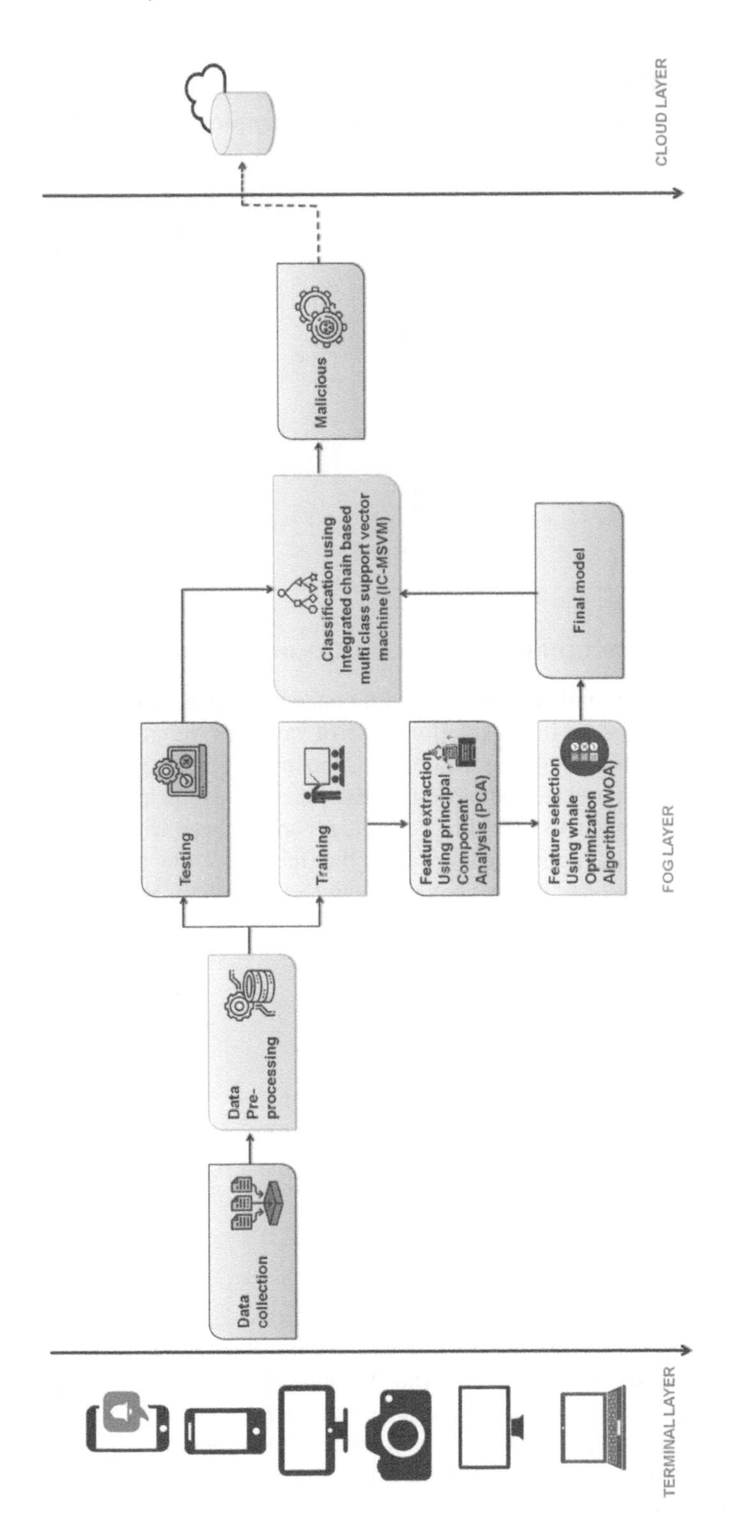

FIGURE 20.2 Framework of the proposed method.

TABLE 20.1
Proportion of Attack and Normal Sample of CICIDS2017 Datasets

Class	UNSW-NB15	CICIDS2017
Shell code	1133	–
Analysis	2000	1964
Generic	40,000	8001
Reconnaissance	10,491	2180
Worms	130	–
Dos	12,264	8000
Fuzzers	18,184	5897
Backdoors	1746	8000
Exploits	33,393	7938
Normal	56,000	148,777
Total	175,341	190,774

20.3.3 Feature Extraction Using Principal Component Analysis (PCA)

Principal component is a statistical approach for condensing information from a big collection of linked variables into a small number of variables (referred to as "principal components") without ignoring the dataset's intrinsic variability.

The variables in the dataset are combined linearly to create the principle components, with weights selected to make the principal components uncorrelated with one another.

The arrangement of the components ensures that the initial few account for the majority of the variability and that each component additionally provides fresh data about the set. Assume that population U is a random vector as:

$$V = (v1,\ v2,\ \dots\ vn)\ T \tag{20.3}$$

And that population's mean is represented by:

$$\mu_v = G\ (V) \tag{20.4}$$

The same dataset's covariance matrix is expressed in:

$$V = C\ (V - \mu_v) \tag{20.5}$$

By definition, the covariance matrix is always symmetric.

We may create an orthogonal basis by calculating the eigenvalues and eigenvectors of such a symmetric covariance matrix. An ordered orthogonal basis may be produced by arranging the eigenvectors in decreasing order of eigenvalues, with the first eigenvector designating the direction with the biggest variation in the data. This makes it possible for us to pinpoint the directions with the highest energy concentrations in the dataset.

Let B be a matrix with the row vectors being the eigenvectors of the covariance matrix. We get by transforming Equation (20.6) stated data vector U:

$$V = C^T A + \mu_v \tag{20.6}$$

This is a coordinate in the eigenvector-defined orthogonal system. The coordinates in the orthogonal basis may be thought of as components of A. From A, we may reassemble the initial data vector V in:

$$A = C_M \left(V - \mu_v\right) \tag{20.7}$$

As a result, the orthogonal basis' coordinate axes were projected onto the original vector V. The orthogonal basis vectors were then linearly combined to recreate the original vector.

We may opt to represent the information using just a few orthogonal basis vectors rather than all of the covariance matrix's eigenvectors. Similar transformations may be made if the matrix with the "M" initial eigenvectors is denoted as rows by C_M in:

$$V = C_M^T A + \mu v \tag{20.8}$$

$$E_v = G\{\left(V - \mu_v\right)\left(V - \mu_v\right)^T\} \tag{20.9}$$

PCA identifies characteristics that best characterize data. We're using PCA to extract features and reduce dimensionality. Based on our data, we may now test a variety of categorization or grouping approaches.

20.3.4 Feature Selection Using WOA

WOA is a member of the family of algorithms based on stochastic populations. It resembles humpback whales' bubble-net feeding. Using a net of bubbles, humpback whales hunt near the surface.

In a "6"-shaped path, they produce this net. The program simulates two stages: the exploitation stage, in which the prey is surrounded and attacked using a spiral bubble-net technique, and the exploration stage, in which the prey is sought out by random.

In exploration stage, Equations (20.10) and (20.11) are used to update the solution, since they provide a mathematical description of a whale's behavior as it approaches and consumes its prey.

$$T = \left|S.Y*(d) - Y(d)\right| \tag{20.10}$$

$$Y(d+1) = x*(d) - w.t \tag{20.11}$$

where d stands for the present iteration, Y* for the correct solution so far, Y for the current solution, | | for the absolute value, and • for a multiplication of each element by itself. The coefficient vectors W and S are determined by:

$$E = 2e.k - e \tag{20.12}$$

$$S = 2.k \tag{20.13}$$

where k is a random vector in the interval [0, 1], and a decrease linearly from 2 to 0. The solutions update their locations in accordance with the position of the best-known solution, according to Equation (20.11). The locations where a solution may be found nearby the optimum solution are controlled by changing the values of the W and S vectors.

The humpback whales approach their prey in a spiraling motion as their encircling mechanism becomes smaller. In WOA, the shrinking encircling behavior is reproduced by lowering the value in Equation (20.12) in accordance with Equation (20.13).

$$w = 2 - d\frac{2}{maxIter} \tag{20.14}$$

where d is the iteration number and MaxIter is the maximum allowable iterations. The spiral-shaped path is created by calculating the distance between Y and Y*. Then, a spiral equation is generated between the current and best (leading) solutions as:

$$Y(d+1) = T.a^{pf}.\cos(2\pi f) + Y^*(d) \tag{20.15}$$

where l is a random integer in the range [–1, 1], b specifies the spiral's form, and t represents the distance between a whale Y and its prey ($T = Y^*(d) - Y(d)|$).

In order to simulate the two mechanisms—the ascending spiral-shaped route and the diminishing encircling mechanism—a probability of 50% is used:

$$Y\ (d+1) = \begin{cases} Shrinking\ Encircling\ if\ (\mathrm{b} < 0.5) \\ Sprial - shapedpath\ if\ (\mathrm{b} \geq 0.5) \end{cases} \tag{20.16}$$

where b is a probability value between [0, 1].

In exploration phase (search for prey), WOA, a randomly picked solution, is used to update the location of the best solution identified so far to improve exploration. As a result, a vector A with random values bigger than 1 or less than 1 is employed to push a solution far from the most effective search agent. Mathematical models of this process may be found as:

$$T = S.Y_{\mathrm{rand}} - Y| \tag{20.17}$$

$$Y\ (\mathrm{d} + 1) = Y_{\mathrm{rand}} - E\ .\ T \tag{20.18}$$

where Y_{rand} is a randomly selected whale from the existing population.

20.3.5 Classification Using IC-MSVM

To identify attack and anomaly detection method based on an IC-MSVM. A group of classifiers that work together in some manner to categorize the test instances is known as an ICM of classifiers. It is well known that an ICM often performs significantly better than a single classifier. An ensemble performs more effectively in the following ways: consider a test set of data y and suppose there is an ensemble of N classifiers $\{l_1, l_2,\ldots.l_m\}$.

Each ICM component predicts test data Y with the same error if all the classifiers are equal, and as a result, the performance of an ICM and a single classifier are comparable. As a result, an ICM's strength depends on the fact that each member is unique and that their mistakes are unrelated to one another (Sharma & Khang, 2024).

When $l_j\,(y)$ prediction of y is inaccurate, the majority of the remaining classifiers may still be accurate, meaning the majority vote was accurate. More specifically, the probability B_A that the majority vote is erroneous is as follows if the error of each independent classifier is $b < 1/2$ as Equation (20.20).

$$B_A = \sum_{r=[n/2]}^{n} b^r (1-b)^{m-r} \tag{20.19}$$

As a result, as the number of classifiers n becomes larger, the probability *BA* shrinks. ICM-based classifiers may be generated using a variety of different techniques. They all have one goal in common: every single classifier should vary from one another as much as feasible. To satisfy this condition, we devised an ICM approach in the following manner.

First, we chose samples at random from the initial negative data. In order to develop a classifier, these negative samples were combined with the initial positive samples as Figure 20.3.

Assume that in the classification issue we want to create a model using our training data H to get the prediction f(y) at input y. First, N train models are created by randomly selecting replacement data from the original data, where $H^* = \{(y_1^*, x_1^*), (y_2^*, x_2^*), \ldots.., (y_N^*, x_N^*)\}$. Equation (20.20) displays how many models were used in the ensemble calculation.

$$N = N_{large} \,/\, N_{small} \tag{20.20}$$

where M_{large} is the proportion of larger samples, in this instance the negative sample, and M_{small} is the size of the tiny sample dataset, in this case the positive sample. It is necessary to aggregate the results of all developed separate classifiers.

To get these fundamental classifiers, methods like majority voting and simple averaging are often used. SVMs are capable of being extended to multiclass classification situations, which are frequent in remote sensing, while being created for binary classification.

For multiclass classification, SVM has recommended two primary strategies. Each method's fundamental premise was to break down the multiclass issue into a collection of binary problems, which made it possible to use a simple SVM strategy.

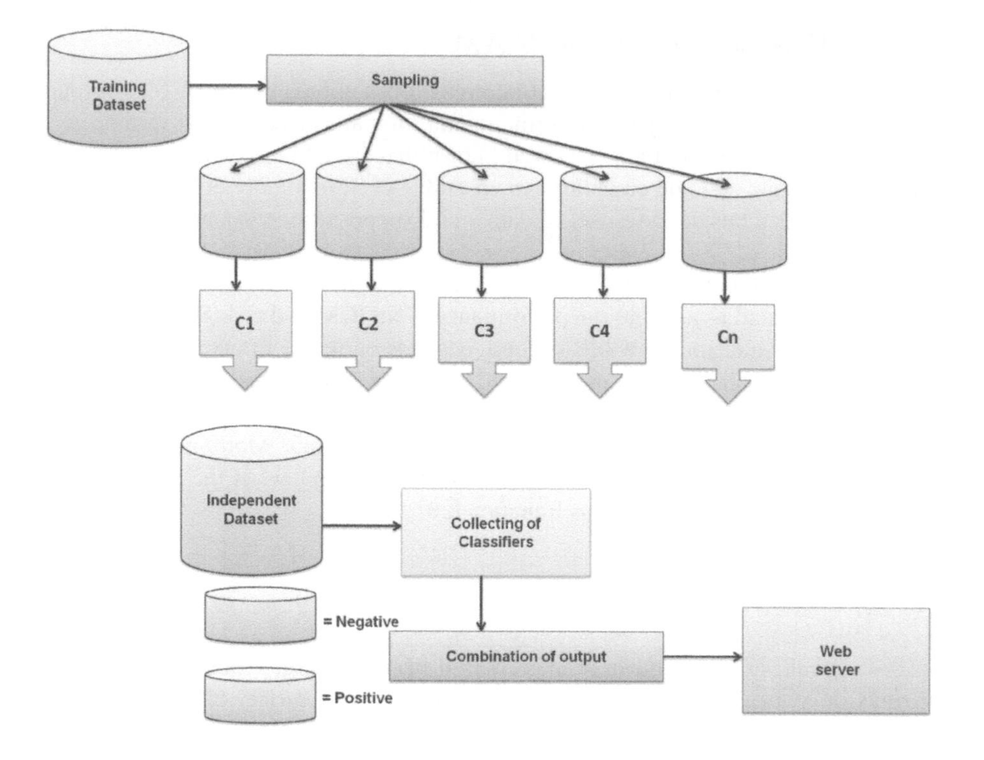

FIGURE 20.3 The IC-MSVM classification flowchart.

Although both approaches to breaking down the multiclass issue into a collection of binary classifications allow the SVM to be used, considering all classes at once results in a multiclass SVM and is a more suitable and computationally efficient technique. Let's specify the SVM's output T_i as $t_{j,i} \in \{-1,1\}, j = 1,\ldots,F$ *and* $i = 1,\ldots,S$ where S represents the sum of class labels and F represents the entire number of SVMs. As a result, the output of T_i is provided as an S-dimensional binary vector $[t_{j,1},\ldots,t_{j,S}]^D$.

If the final class is chosen by a_i, then $t_{j,i} = 1$; otherwise, $t_{j,i} = -1$. Therefore, if class a_i is available, the majority system will choose it and indicated by

$$\sum_{j=1}^{F} t_{j,r} = max_{i=1}^{S} \sum_{j=1}^{F} t_{j,i} \tag{20.21}$$

20.4 EXPERIMENTAL RESULT

The existing methods are RF [Gaber & El-Ghamry et al. (2022)], DT [Gaber & El-Ghamry et al. (2022)], and XGBoost [Elsisi and Tran (2021)], and DNN [Elsisi and Tran (2021)].

The ratio of accurate forecasts to all potential accurate predictions, including both benign and attack classes, is known as accuracy as

$$\text{Accuracy} = \frac{\text{TP} + \text{TN}}{\text{TP} + \text{TN} + \text{FP} + \text{FN}} \tag{20.22}$$

Figure 20.4 depicts the comparative evaluation of accuracy in suggested and traditional methods. When compared to the current techniques RF, DT, XGBoost, and DNN, which have accuracy of 85%, 87.25%, 89.5%, and 92.1%, respectively, the suggested IC-MSVM takes just 95.5%.

Therefore, when compared to current techniques like RF, DT, XGBoost, and DNN, the IC-MSVM enhances security.

Precision is defined as the percentage of the selected components in the retrieved information that are relevant as:

$$Precision = \frac{\text{TP}}{\text{TP} + \text{FP}} \tag{20.23}$$

When compared to the current techniques RF, DT, XGBoost, and DNN, which have precision of 86%, 86.8%, 87.2%, and 93.5%, respectively, the suggested IC-MSVM takes just 98%.

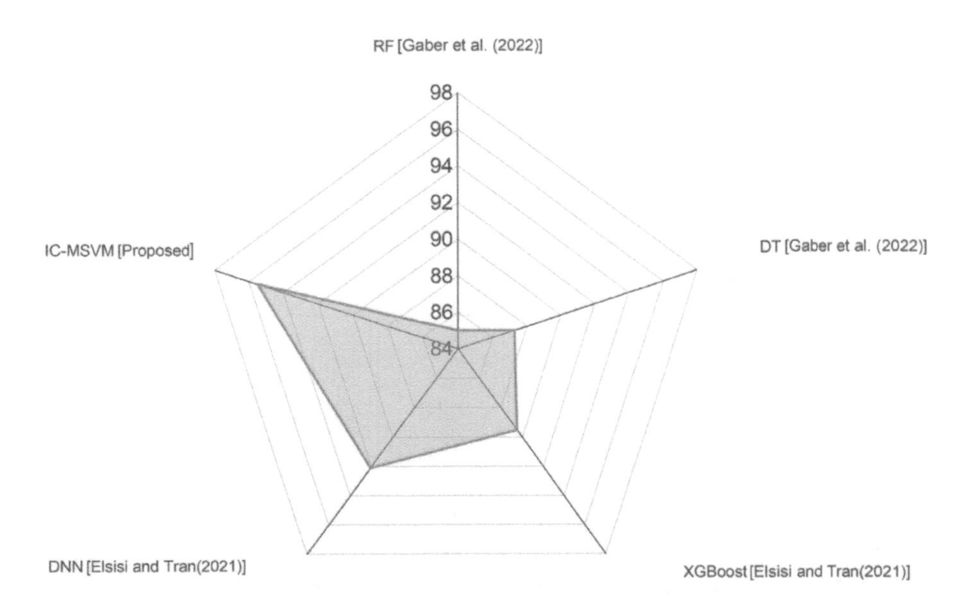

FIGURE 20.4 Comparative evaluation of accuracy in suggested and traditional methods.

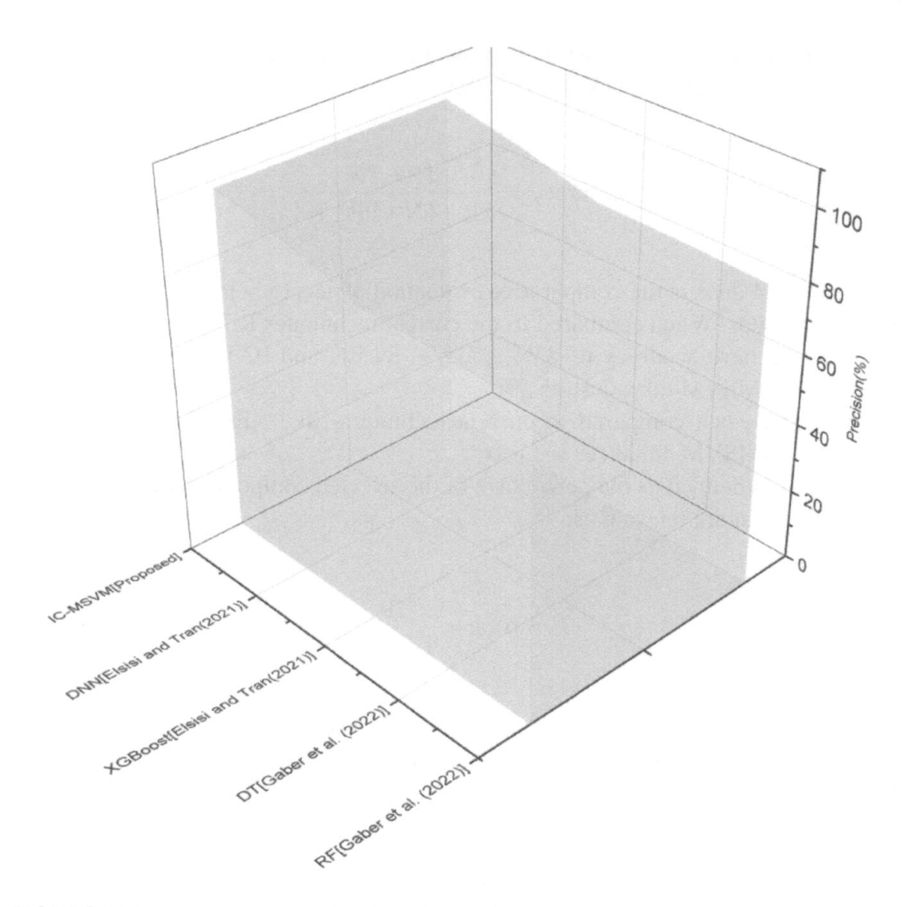

FIGURE 20.5 Comparative evaluation of precision in suggested and traditional methods.

Consequently, the IC-MSVM enhances performance as compared to existing methods like RF, DT, XGBoost, and DNN. Figure 20.5 depicts the comparative evaluation of precision in suggested and traditional methods.

Recall is defined as the number of relevant things that are selected from a total number of relevant elements as:

$$Recall = \frac{\text{TP}}{\text{TP} + \text{FP}} \tag{20.24}$$

Figure 20.6 depicts the comparative evaluation of recall in suggested and traditional methods. When compared to the current techniques RF, DT, XGBoost, and DNN, which have recall of 88.3%, 86.7%, 90.2%, and 94.6%, respectively, the

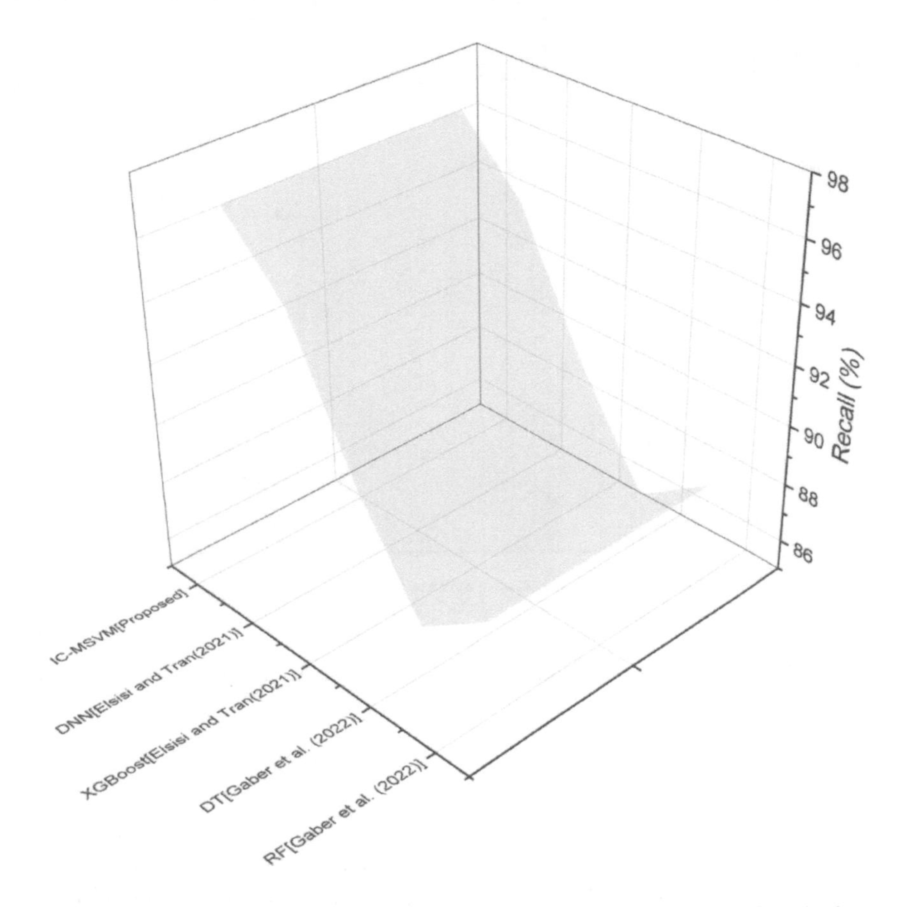

FIGURE 20.6 Comparative evaluation of recall in suggested and traditional methods.

suggested IC-MSVM takes just 96.8%. In conclusion, IC-MSVM is better than the RF, DT, XGBoost, and DNN.

Predicting F1 score requires calculating recall and precision as follows:

$$\text{F1 score} = 2 \times = \frac{\text{Precision} \times \text{Recall}}{\text{Precision} + \text{Recall}} \tag{20.25}$$

Figure 20.7 depicts the comparative evaluation of F1 score in suggested and traditional methods. When compared to the current techniques RF, DT, XGBoost, and DNN, which have F1 score of 89.5%, 88.2%, 91.6%, and 95%, respectively, the suggested IC-MSVM takes 97.3%. In summary, IC-MSVM is superior to RF, DT, XGBoost, and DNN.

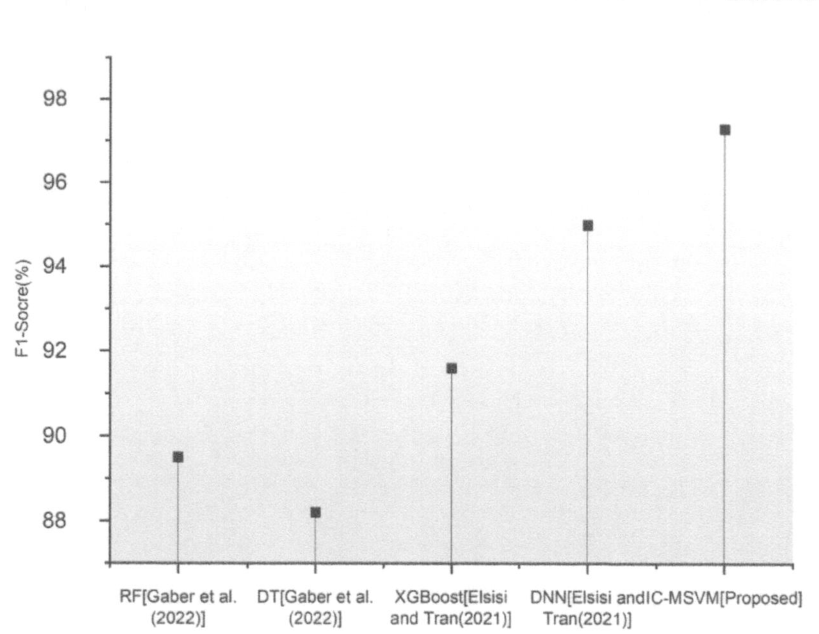

FIGURE 20.7 Comparative evaluation of F1 score in suggested and traditional methods.

20.5 CONCLUSION

In this study, we explore at a technique for detecting attacks and anomalies based on an IC-MSVM to prevent and reduce IoT cyber security risks in a smart city. The z-score and min-max normalization procedures are used to collect and preprocess the dataset (Rani & Khang et al., 2023).

Preprocessed data is split up into training and testing sets. To extract the pertinent features, we use PCA. The whale optimization technique is used to select the features (WOA). The values of performance metrics are accuracy (95.5%), precision (98%), recall (96.8%), and F1 score (97.3%).

WOA has several drawbacks, such as inconsistent outcomes in convergence and local optimal solution. To further improve the performance of IoT attack detection, our future study will investigate deep learning approaches (Khang & Vugar et al., 2024).

REFERENCES

Bhambri, P., Rani, S., Gupta, G., Khang, A. "*Cloud and Fog Computing Platforms for Internet of Things*" (2022). CRC Press. https://doi.org/10.1201/9781032101507

Chowdhury, A., Karmakar, G., Kamruzzaman, J. "The Co-Evolution of Cloud and IoT Applications: Recent and Future Trends", *Advances in Wireless Technologies & Telecommunication*, pp. 213–234 (2019). IGI Global. https://www.igi-global.com/chapter/the-co-evolution-of-cloud-and-iot-applications/225720

Elsisi, M., Tran, M.Q. "Development of an IoT architecture based on a deep neural network against cyber-attacks for automated guided vehicles", *Sensors (Basel)*, vol. 21, no. 24, p. 8467 (2021). https://www.mdpi.com/1424-8220/21/24/8467

Gaber, T., El-Ghamry, A., Hassanien, A.E. "Injection attack detection using machine learning for smart IoT applications", *Physical Communication*, vol. 52, p. 101685 (2022). https://www.sciencedirect.com/science/article/pii/S1874490722000490

Hahanov, V., Khang, A., Litvinova, E., Chumachenko, S., Hajimahmud, V.A., Alyar, A.V. "The Key Assistant of Smart City – Sensors and Tools", *AI-Centric Smart City Ecosystems: Technologies, Design and Implementation* (1st Ed.) (2022). CRC Press. https://doi.org/10.1201/9781003252542-17

Jebaraj, L., Khang, A., Vadivelraju, C., Antony, R.P., Kumar, S. "Smart City Concepts, Models, Technologies and Applications", *Smart Cities: IoT Technologies, Big Data Solutions, Cloud Platforms, and Cybersecurity Techniques* (1st Ed.) (2024). CRC Press. https://doi.org/10.1201/9781003376064-1

Khang, A. (2021). "Material4Studies", *Material of Computer Science, Artificial Intelligence, Data Science, IoT, Blockchain, Cloud, Metaverse, Cybersecurity for Studies.* https://www.researchgate.net/publication/370156102_Material4Studies

Khang, A., Gupta, S.K., Rani, S., Karras, D.A. *Smart Cities: IoT Technologies, Big Data Solutions, Cloud Platforms, and Cybersecurity Techniques* (2023). CRC Press. https://doi.org/10.1201/9781003376064

Khang, A., Hahanov, V., Abbas, G.L., Hajimahmud, V.A. "Cyber-Physical-Social System and Incident Management", *AI-Centric Smart City Ecosystems: Technologies, Design and Implementation* (1st Ed.) (2022). CRC Press. https://doi.org/10.1201/9781003252542-2

Khang, A., Rani, S., Sivaraman, A.K. *AI-Centric Smart City Ecosystems: Technologies, Design and Implementation* (1st Ed.) (2022). CRC Press. https://doi.org/10.1201/9781003252542

Khang, A., Abdullayev, V., Hahanov, V., Shah, V. "*Advanced IoT Technologies and Applications in the Industry 4.0 Digital Economy*" (1st Ed.) (2024). CRC Press. https://doi.org/10.1201/978-1-003-43426-9

Mihoub, A., Fredj, O.B., Cheikhrouhou, O., Derhab, A., Krichen, M. "Denial of service attack detection and mitigation for internet of things using looking-back-enabled machine learning techniques", *Computers and Electrical Engineering*, vol. 98, p. 107716 (2022).

Pahl, M.O., Aubet, F.X., Liebald, S. "Graph-based IoT microservice security", *Proceedings of the NOMS IEEE/IFIP Network Operations and Management Symposium* (2018). Taipei, Taiwan, 23–27 April 2018; 2018:1–3. https://ieeexplore.ieee.org/abstract/document/8406118/

Pu, C., Choo, K.R. "Lightweight sybil attack detection in IoT based on bloom filter and physical unclonable function", *Computers & Security*, vol. 113, p. 102541 (2022). https://www.sciencedirect.com/science/article/pii/S0167404821003655

Rani, S., Bhambri, P., Kataria, A., Khang, A., Sivaraman, A.K. "*Big Data, Cloud Computing and IoT: Tools and Applications*" (2023). Chapman and Hall/CRC. https://doi.org/10.1201/9781003298335

Rani, S., Chauhan, M., Kataria, A., Khang, A. "IoT Equipped Intelligent Distributed Framework for Smart Healthcare Systems", *Networking and Internet Architecture* (2021). CRC Press. https://doi.org/10.48550/arXiv.2110.04997

Restuccia, F., D'Oro, S., Melodia, T. *Securing the Internet of Things: New Perspectives and Research Challenges* (2018). Cryptography and Security. https://arxiv.org/abs/1803.05022

Sharma, A., Khang, A. "Exploratory Analysis on Current Application-Layer HTTP-Payload Based DDoS Attack on IoT Devices", *Smart Cities: IoT Technologies, Big Data Solutions, Cloud Platforms, and Cybersecurity Techniques.* (2024). CRC Press. https://doi.org/10.1201/9781003376064-21

Tyagi, V., Saraswat, A., Bansal, S., Khang, A. "An Analysis of Securing IoT Devices from Man-in-the-Middle (MIMA) and Denial of Service (DoS)", *Smart Cities: IoT Technologies, Big Data Solutions, Cloud Platforms, and Cybersecurity Techniques* (2024). CRC Press. https://doi.org/10.1201/9781003376064-20

21 Treatment Solution of the Environmental Water Pollution for Smart City

Gadirova Elmina Musrat, Abdullayev Vugar Hajimahmud, and Abuzarova Vusala Alyar

21.1 INTRODUCTION

The processes of ruralization and urbanization have recently been developing at an equal speed. In the very recent past, the acceleration of the urbanization process and the migration of people to cities have had many negative effects on the living environment of urban residents. Therefore, government of countries probably took this into account and began to pay more attention to the wastewater treatment systems for smart cities (Khang & Gupta et al., 2023).

Many important factors were manifested in the desire of people to move to cities. Many of them went to the cities out of necessity rather than desire. This, in turn, led to the increase and acceleration of the urbanization process. We can mention some of the factors influencing the acceleration of this process, as shown in Figure 21.1.

It is known that the birthplace of industrialization is cities, or in other words, the places where industry has set foot become urbanized later. Employment – most likely in the recent past before agricultural innovations – was greater in cities, which was the factor that attracted people most. Socialization was another key factor in accelerating urbanization and, also, other factors.

According to forecasts, at least 70% of the world population will live in cities in the near future around 2050. But the introduction of innovations – the introduction of smart village projects along with the application of smart technologies in agriculture – still makes people hopeful about the villages. However, there is still a stronger project than the smart village project, which is the smart city concept (Khanh & Khang, 2021).

The era of industrialization, the increase in human influence, and the further acceleration of urbanization cause many problems. So, along with industrialization, there has been a huge increase in the percentage of environmental pollution. Such pollution eventually leads to a decline in the quality of life of humans and other living beings and unfortunately continues to this day. However, continuous technological advances help to prevent pollution and improve the quality of life (Musrat & Hajimahmud et al., 2024).

The smart city model is a concept that combines the capabilities of ICT, that is, using its capabilities, aims to increase both public services and the quality of

DOI: 10.1201/9781003376064-21

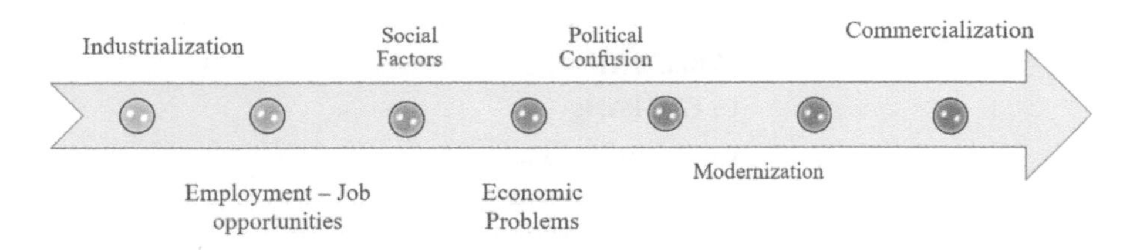

FIGURE 21.1 Some factors affecting the urbanization process.

life of people. From this point of view, various innovations are constantly presented in order to reduce environmental pollution within smart cities (Khang & Rani et al., 2022).

Given that, environmental pollution also causes cancer, one of the greatest scourges of our time; it is now the most global problem. With the help of technological possibilities applied in the smart city, it is estimated that in the near future, it will be possible to prevent environmental pollution – to a large extent (Rani & Khang et al., 2022).

In this chapter, the following directions are considered:

- Environmental pollution: impact of polycyclic aromatic hydrocarbon (PAH) on human health (Khang & Hajimahmud et al., 2023)
- Materials and methods
- Combating water pollution in smart cities

21.2 ENVIRONMENTAL POLLUTION: IMPACT OF PAH ON HUMAN HEALTH

PAHs are persistent organic pollutants (POPs) with two or more benzene rings in their chemical structure (Deelaman & Choochuay et al., 2023).

Organic compounds, such as PAHs, are a global environmental concern as they cause inflammation and skin cancer. As you know, there are two types of anthropogenic sources of hydrocarbons: petrogenic and pyrogenic.

Petrogenic sources include crude oil and petroleum-derived hydrocarbon compounds. Pyrogenic sources of hydrocarbon compounds are formed as a result of incomplete combustion of organic substances such as oil, wood, and coal.

About 6.1 million tons of oil products are thrown into the ocean annually most of which are of anthropogenic origin. It is known from the literature that according to the latest indicators, every year 6 million tons of oil and oil products are discharged to the world ocean.

Pyrogen and petrogen pollution of sea water contamination with ballast water and other factors cause the pollution of common water basins. These hydrosphere segments are a dynamic system which leads to a decrease and depletion of freshwater supplies over time (Hajiyeva & Gadirova, 2020).

TABLE 21.1
16 EPA PAHs

No	PAHs' Names
1	Acenaphthene
2	Acenaphthylene
3	Anthracene
4	Fluoranthene
5	Fluorene
6	Naphthalene
7	Phenanthrene
8	Pyrene
9	Benz[a]anthracene
10	Benzo[b]fluoranthene
11	Benzo[k]fluoranthene
12	Benzo[ghi]perylene
13	Benzo[a]pyrene
14	Chrysene
15	Dibenz[a,h]anthracene
16	Indeno[1,2,3-cd]pyrene

More than 100 PAHs can be found in nature. However, 16 of these are designated as specific pollutants by the United States Environmental Protection Agency. These are mentioned in Table 21.1.

These 16 EPA PAHs were identified in the 1970s (Hajiyeva & Gadirova, 2014). These compounds are among the priority pollutants as reasons (Alver & Demirci et al., 2012)

1. Having more information about these compounds than others.
2. More harmful PAHs are suspected to be harmful and demonstrate effects.
3. The risk of exposure to these PAHs is higher than others.
4. In America, these PAHs have the highest indicators on the list of national priorities (National Priorities List, NPL) to be considered (ATSDR, 1995).

Human exposure to PAHs can occur through inhalation, skin contact, or ingestion of food contaminated with PAHs. PAHs in air pollution are primarily associated with particulate matter. When PAHs are in the gas phase, their lifetime is less than a day.

Overall, current scientific evidence suggests that PAHs in ambient air are associated with increased cancer incidence in exposed populations. A positive association among environmental PAHs, breast cancer, uterine cancer, and lung cancer has been reported.

Epidemiological studies have shown that PAHs are associated with decreased lung function, exacerbation of asthma, obstructive pulmonary disease, and increased cardiovascular disease. Limited epidemiological evidence also suggests adverse effects on cognitive or behavioral function in children (PAH Report, 2021).

Here we will take a look at PAHs in water pollution – more precisely in the waters of the Caspian Sea. There are many ways to treat wastewater. Since these treatment methods are not completely effective, therefore, they become a serious threat to the flora and fauna of the Caspian Sea. We know that after treating wastewater generated from the oil refining industry, the polluted water is discharged into the Caspian Sea. Wastewater from the oil industry negatively affects the biota of the sea.

The Caspian Sea is playing a very sensitive role in ecosystem. Over the past decades under the influence of anthropogenic and biochemical factors, the state of ecosystems in general has deteriorated sharply and especially in the northeastern part of the sea (CASPECO, 2021).

Industrialization and urbanization in the Caspian region have developed rapidly over the past several decades and the associated increase in hydrocarbons is a concern in the region. Offshore production and accidental oil spills, industrial waste, wastewater, discharges flowing down from river water are considered the main source of anthropogenic hydrocarbons in the marine environment (Novikov, 2021).

Industry is believed to be the main source of oil that is polluting the Caspian Sea (Ostroumov, 2006). The total amount of industrial waste discharged into the Caspian Sea averages 2342.0 million m^3 per year. Such waters contain 122.5 thousand tons of oil, 1.1 thousand tons of phenols, and 9.9 thousand tons of organic chemistry products.

The total content of hydrocarbons in the northwestern part of the South Caspian was small – 32–54.2 μg/g. In this area, in the vicinity of oil fields, the concentration of phenol was 0.002–0.003 μg/g.

The major quantity of pollution (90% of the total) enters the Caspian Sea with river runoff. After purification, these wastewaters are discharged into the Caspian Sea and even in small quantities, these harmful substances are dangerous for the flora and fauna of the sea and the environment.

As is already known, PAHs are very dangerous for the environment for living organisms in aquatic ecosystems, and therefore, the identification of hazardous substances and the application of methods for their destruction are very important (Yunker & Macdonald et al., 2002).

21.3 MATERIALS AND METHODS

The quantitative analysis of PAHs as well as of phenol and its derivatives was carried out in a system, including an Agilent 6890N gas chromatograph which has an interface with an Agilent 5975 high-performance mass-selective detector manufactured by Agilent Technologies (USA).

The chromatograph was equipped with a split less injector and a ZB-5 capillary column (Phenomenex, USA). Column ZB-5 has the following specifications: 5%-biphenyl–95%-dimethylpolysiloxane with a copolymer length of 60 m, an inner diameter of 0.25 mm, and a film thickness of 0.25 μm. Helium (99.999% purity) with a flow rate of 1.5 ml/min was used as a carrier gas. The temperature rise was programmed from 40°C to 310°C.

The extracts were introduced using an automatic sampler in a volume of 1 μl (Dettmer & Engewald, 2014). Quantitative analysis was performed against a seven-point calibration against standard reference solutions. A mixture of deuterated PAHs containing naphthalene-d8 and phenanthrene-d10 (Cambridge Isotope Laboratories, Inc., Andover USA) was used as an internal standard for calculating the obtained results of chromatographic analysis (Khang & Gupta et al., 2023).

At first, PAHs were analyzed in water samples discharged from the oil industry into the Caspian Sea. The results of the analyses and chromatograms for each sample are given in Table 21.2.

TABLE 21.2
Determination of Polycyclic Aromatic Hydrocarbons in Water Samples Taken from the Oil Refinery

No	PAH, mg/l	Standard Indicator	Sample 1	Sample 2	Sample 3
1	Naphthalene	≥0.01	0.08	0.23	20
2	Acetylene	≥0.01	0.02	0.51	15
3	Acenaften	≥0.01	0.04	0.33	43
4	Fluorene	≥0.01	0.24	0.81	80
5	Phenanthrene	≥0.01	0.29	4.32	298
6	Anthracene	≥0.01	0.04	0.33	26
7	Fluoranthene	≥0.01	0.03	3.01	15
8	Piren	≥0.01	0.05	13	66
9	Benz(a)anthracene	≥0.01	0.01	3.5	17
10	Chrezen	≥0.01	0.02	15	41
11	Benz(b + j + k)fluoranthene	≥0.01	<0.01	2.2	5.0
12	Benz(a)pyrene	≥0.01	<0.01	1.3	5.0
13	Inden(1,2,3-cd)pyrene	≥0.01	<0.01	0.37	1.5
14	Benz(ghi)perilen	≥0.01	<0.01	0.63	1.7
15	Dibenz(ah)antracen	≥0.01	<0.01	0.57	1.8

It should be noted that excess PAHs are expected to be present in refinery wastewater. Wastewater of this type is biologically treated and then discharged into the Caspian Sea, as shown in Figures 21.2, 21.3, 21.4.

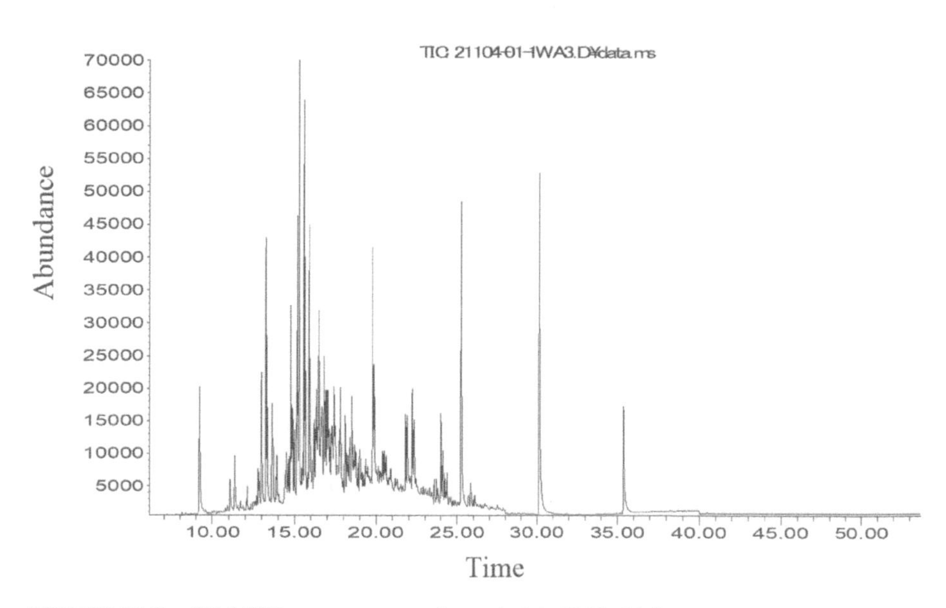

FIGURE 21.2 GC-MSD spectroscopy of sample 1 in Table 21.2.

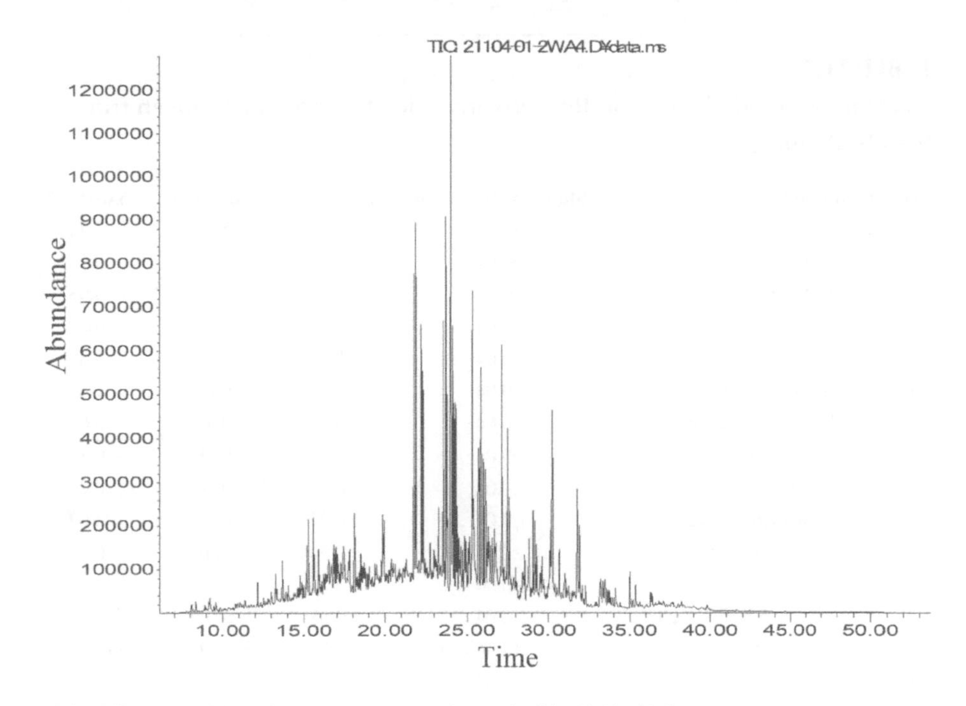

FIGURE 21.3 GC-MSD spectroscopy of sample 2 in Table 21.2.

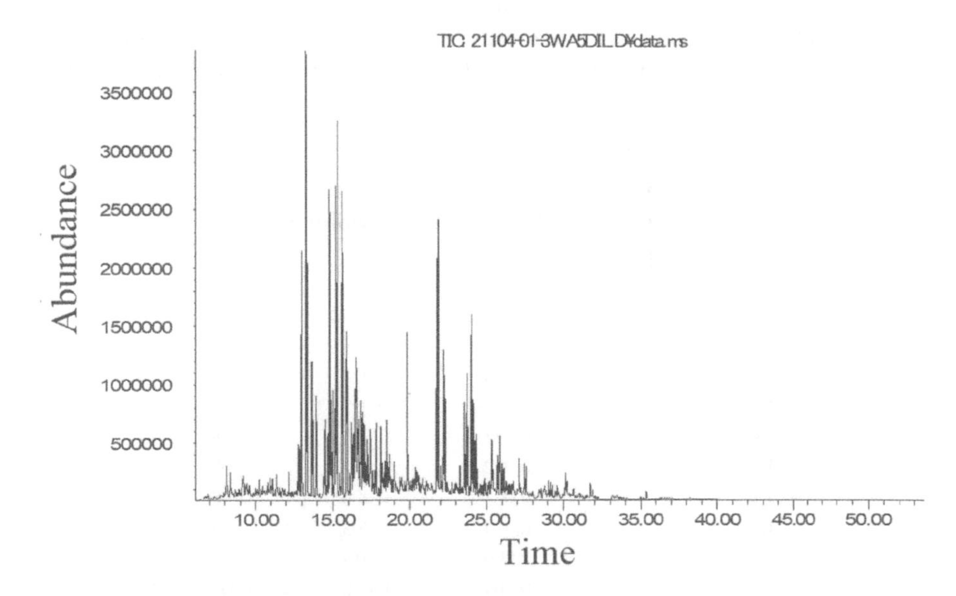

FIGURE 21.4 GC-MSD spectroscopy of sample 3 in Table 21.2.

TABLE 21.3
Determination of Phenol and Its Derivatives in Water Samples Taken from the Oil Refinery

No	Compounds (mg/l)	Standard Indicator	Sample 1	Sample 2	Sample 3
1	Phenol	≥0.04	5.34	5.65	1.44
2	o-Cresol	≥0.04	0.76	0.26	0.16
3	2-nitrophenol	≥0.04	0.59	0.48	0.08
4	2,4-dimethylphenol	≥0.04	31.99	8.71	0.20
5	m,p-cresol	≥0.04	21.59	3.81	0.85
6	2,6-dichlorphenol	≥0.04	1.14	0.67	1.80
7	4-chloro-3-methylphenol	≥0.04	0.63	0.43	0.69
8	2,4,5-TCP	≥0.04	0.48	0.28	0.18
9	2,4,6-TCP	≥0.04	0.14	0.04	0.25
10	2,3,4,6-tetrachlorophenol	≥0.04	0.27	0.18	0.002
11	2-methyl-4,6-dinitrophenol	≥0.04	<0.04	<0.04	<0.04
12	Pentachlorophenol	≥0.04	0.27	0.25	0.06
13	2-sec-Butyl-4.6-dinitrophenol	≥0.04	<0.08	<0.08	<0.08
14	Compounds (mg/l)	Standard indicator	Sample 1	Sample 2	Sample 3
15	Phenol	≥0.04	5.34	5.65	1.44

Phenol and its derivatives were also determined in the water samples of the oil industry and the results are given in Table 21.3.

The high amount of phenol in the two water samples is due to sampling from a water tank located closer to the production site. At the same time, heavy metal ions were identified in the water samples examined.

Heavy metals enter aquatic living organisms and the human body through the food chain. Determination of heavy metals was performed by ICP-OES, Optima 2100DV emission analysis method. The results are shown in Table 21.4.

TABLE 21.4
Number of Heavy Metals in Wastewater Samples of the Oil Refinery

Heavy metals, mg/l	Sample 1	Sample 2	Sample 3
Fe	0.048	0.047	0.042
Zn	0.143	0.078	0.023
Cu	0.023	0.028	0.020
Mn	0.021	0.105	0.038
Pb	0.022	0.017	0.009
Cd	0.007	0.006	0.008
Cr	0.004	0.003	0.004
Co	0.009	0.014	0.012
Ag	0.009	0.010	0.008

All these metals are dangerous if they exceed the MPC norm, but Cd is more dangerous. Norm for Cd is 0.001 mg/l (mg/dm^3). As can be seen from the table, the amount of Cd in the studied samples is higher than the norm.

21.4 COMBATING WATER POLLUTION IN SMART CITIES

We gave an analysis of PAHs in waters polluted by oil spills in the previous chapter.

Oil has become one of the main sources of energy since the 1950s, on the eve of the industrialization period. And it still remains so. Although the concept of smart cities is an "Environmentally friendly" project, there is still a need for oil-derived energy. However, the use of oil inevitably causes major environmental pollution. Most importantly, it causes ocean and sea water pollution.

We first analyzed PAHs and their effects on water pollution due to oil, as well as subsequent PAH levels in these waters.

Such environmental problems should be minimized through the smart city concept. Harmful sources of energy such as oil should be replaced with more efficient ones. However, although we cannot completely prevent water pollution in general, we can control it nowadays with a number of smart technologies.

We can look at this chapter in two directions:

- Prevent water pollution with simple efforts
- Steps to combat water pollution and depletion of freshwater resources through smart technologies

21.4.1 Prevent Pollution with Simple Efforts

The entire responsibility for preventing pollution cannot be placed on smart technologies. People can prevent pollution through simple efforts. By first protecting our own little world from pollution, we can significantly minimize environmental pollution as a whole.

Before the technological approach, people can implement a number of methods to prevent pollution.

21.4.1.1 Disposal of Toxic Chemicals

We use many chemical products for daily processes such as cleaning. Various bleaches and dyes cause serious problems. When such products are subsequently disposed of in the sewers, their harmful effects increase. The best way to avoid this is to apply recycling.

21.4.1.2 Proper Disposal of Medical Waste

Medicines should not be thrown into the sewer or into any water, such as a lake or the sea, because it affects the animals living there. It causes various diseases, and in the end, the water that people and cattle drink becomes polluted. This again poses a threat to a person's own health (Khang & Rana et al., 2023).

21.4.1.3 Do Not Pour Oil Down the Drain

Many experts repeat this many times. Products such as oil should not be thrown down into the drain. This again has a very heavy impact on water pollution.

21.4.1.4 Try to Avoid Using Plastic Containers

Plastic is one of the things that have a negative impact on nature. It continues to exist in nature for a long time. Similarly, thousands of plastics thrown into the seas have a negative effect on the creatures there. The main effort that can be made here is either not to use plastic items at all or to send plastic items to recycling.

21.4.1.5 Help Clean up Beaches and Rivers

Unfortunately, 75 out of 100 people make efforts to protect nature, while the other 25 do the opposite. In this regard, whenever possible, people should help to regularly clean the banks of rivers and seas from garbage thrown by other people. In fact, that garbage should not be thrown there.

On the other hand, it is the duty of every person to fight against the perpetrators, as well as to clean up. Because this nature is the abode of every living being.

21.4.1.6 Control Your Cars

People who use cars should regularly have their cars checked (monitored). Thus, the oils and various liquids leaking from the car flow into the sewers, from where they can eventually go to other water sources.

These are just some of the steps that need to be taken. And they are the simplest ones. People can use hundreds of different methods to prevent pollution. On the other hand, it is possible to prevent pollution through smart technologies (Khanh & Khang, 2021).

21.4.2 Steps to Combat Water Pollution and Depletion of Freshwater Resources through Smart Technologies

In fact, in a simple case, the main step here is the proper use of smart technologies. Due to water pollution, freshwater resources are in danger of diminishing. This eventually leads to drought. In order to prevent this, freshwater resources should be constantly monitored. This is a relatively difficult process without the help of technology.

Internet of Things based devices, especially applied in smart cities (Khang & Hrybiuk, et al., 2024), are the ideal smart tools that can help people in such processes. In general, the following smart technology projects have been proposed.

21.4.2.1 Data Collection Systems or Monitoring Systems

First, data collection systems are created for any system. Thus, changes in the amount and composition of water are constantly monitored, for which data is collected and analyzed. At the same time, various data analysis technologies are used for this. Big Data Analytics is also applicable (Bhambri & Khang et al., 2022).

21.4.2.2 IoT-Based Water Distribution Systems to Monitor Water Flow, Quantity, and Leakage

Smart water, Shayp technological innovations are among the main helpers in this direction. For irrigation systems, HydroPoint's Weather TRAK irrigation system applies.

21.4.2.3 ED Technology

Electrodialysis (ED) is notable for removing chromium and arsenic from water contaminated by sources such as textile dyeing, leather tanning, dye, and pigment industries (Abdallah & Samuel, et al., 2022). ED technology can recover wastewater and reclaim water through concentration, dilution, desalination, regeneration, and valorization.

21.4.2.4 Efficient Water Management Systems

Water management is important in terms of developing water-energy interconnection. For example, SCUBIC is an IoT smart energy management solution that uses data to improve the functionality of water utilities (Rani & Chauhan et al., 2021). By deploying sensors and collecting data, pumping schedules can be simplified to reduce costs and improve water safety (Deghles & Kurt, 2016).

Different countries adopt different acts to prevent water pollution. Accordingly, they use both the above-mentioned and a number of other technologies and build factories.

In Berlin, for instance, a phosphorus elimination plant was built to treat the effluent of the pharmaceutical industry in order to limit the quantity of phosphorus discharged into the rivers (Schimmelpfennig & Kirillin et al., 2012).

21.5 CONCLUSION

It is known that the wastewater of the oil industry is very dangerous for the aquatic ecosystem. Thus, it causes serious damage to flora and fauna of water ecosystem (Khang & Gupta et al., IoT, 2023).

For this purpose, three wastewater samples were taken from the oil industry and analyzed. The goal was to determine their composition. For this purpose, PAHs, phenolic organic compounds, and heavy metals were determined in water samples.

Chemical analysis of water samples was carried out on a GC-MSD gas chromatograph 6890N with a highly efficient mass-selective detecto-Agilent 5975. In water samples, heavy metals were analyzed on a PerkinElmer ICP/OES-2100DV.

A total of 15 PAHs, 13 phenolic, and heavy metals compounds were analyzed in water samples. In the analysis of PAHs, the most dangerous naphthalene was observed in the water samples 2 and 3. As can be seen from Table 21.2, the MPC limit for PAHs is 0.01 mg/l.

Naphthalene belongs to a class of high-risk substances and is considered the most hazardous among PAHs. Therefore, the main focus was on which areas of the water samples had the highest levels of naphthalene. The amount of phenol was higher in water samples 1 and 2.

In general, the permissible concentration of phenol for industrial water in the maximum case should be 0.1 mg/l.

As can be seen from Table 21.2, in 3, 1, and 2 samples, the amount of phenol gradually increases: 1.44–5.34–5.65 mg/l. As for heavy metals, most of them (mainly Cd) have exceeded the limit.

In general, we can briefly note the following results:

1. Environmental pollution is one of the most global problems of the time. One of the main reasons for this is the acceleration of the urbanization process. As people came to cities, industry became more developed, and of course, one of the reasons for pollution is the era of industrialization (Khang & Hrybiuk et al., 2024).
2. Environmental pollution is mainly presented in three forms: air, soil, and water pollution. In this chapter, we looked at water pollution and its contamination with PAHs. Various methods were applied and analysis was carried out here (Rani & Chauhan et al., 2021).
3. In order to prevent pollution, at least to reduce it to a minimum level, people need to apply various methods. A few of them were mentioned. On the other hand, several technological innovations applied in smart cities have been mentioned (Khang & Rani et al., 2022).

REFERENCES

Abdallah, C.K., Samuel, J.C., Mourad, K.A., Iddrisu, A., Ampofo, J.A., "Advances in sustainable strategies for water pollution control: a systematic review", *IntechOpen* (2022). https://www.intechopen.com/online-first/84270

Alver, E., Demirci, A., Özcimder, M., "Polisiklik Aromatik Hidrokarbonlar ve Sağlığa Etkileri", *Mehmet Akif Ersoy Üniversitesi Fen Bilimleri Enstitüsü Dergisi*, vol. 3, no. 1, pp. 45–52 (2012). https://dergipark.org.tr/en/pub/makufebed/issue/19422/206555

Bhambri, P., Rani, S., Gupta, G., Khang, A., *Cloud and Fog Computing Platforms for Internet of Things* (2022). CRC Press. https://doi.org/10.1201/9781032101507

CASPECO (2021). "Caspian Sea. State of the Environment." *Report of the Interim Secretariat of the Framework Convention for the Protection of the Marine Environment of the Caspian Sea and the Bureau for Management and Coordination of the CASPECO Project*, 2011, p.28. https://tadqiqot.uz/index.php/chemical/article/view/1125

Deelaman, W., Choochuay, C., Pongpiachan, S., Han, Y.. "Ecological and health risks of polycyclic aromatic hydrocarbons in the sediment core of Phayao Lake, Thailand", *Journal of Environmental Exposure Assessment*, vol. 2, no. 1, p. 3 (2023). https://www.oaepublish.com/jeea/article/view/5319

Deghles, A., Kurt, U., "Treatment of tannery wastewater by a hybrid electrocoagulation/electrodialysis process", *Chemical Engineering and Processing – Process Intensification*, vol. 104, pp. 43–50 (2016). doi:10.1016/J.CEP.2016.02.009

Dettmer, W., Engewald, K., *Werner Practical Gas Chromatography. Comprehensive Reference* (2014). ISBN 978-3-642-54640-2. http://doi.org/10.1007/978-3-642-54640-2

Hajiyeva, S.R., Gadirova, E.M., "Methods for cleaning water contaminated with oil", *Azerbaijan Chemistry Journal, Baku*, No. 1, p. 35–38 (2014). https://doi.org/10.32737/0005-2531

Hajiyeva, S.R., Gadirova, E.M. *Monitoring of the petro-genically polluted territories E3S Web of Conferences ICBTE* (2020), 01006, p. 212. https://doi.org/10.1051/e3sconf/202021201006

Khang, A., Gupta, S.K., Rani, S., Karras, D.A., *Smart Cities: IoT Technologies, Big Data Solutions, Cloud Platforms, and Cybersecurity Techniques* (1st Ed.) (2023). CRC Press. https://doi.org/10.1201/9781003376064

Khang, A., Gupta, S.K., Shah, V., Misra, A., *AI-Aided IoT Technologies and Applications in the Smart Business and Production* (2023). CRC Press. doi: 10.1201/9781003392224

Khang, A., Hrybiuk, O., Abdullayev, V., Shukla, A.K., *Computer Vision and AI-Integrated IoT Technologies in Medical Ecosystem* (1st Ed.) (2024). CRC Press. https://doi.org/10.1201/978-1-0034-2960-9

Khang, A., Rana, G., Tailor, R.K., Hajimahmud, V.A., *Data-Centric AI Solutions and Emerging Technologies in the Healthcare Ecosystem* (1st Ed.) (2023). CRC Press. https://doi.org/10.1201/9781003356189

Khang, A., Rani, S., Sivaraman, A.K., *AI-Centric Smart City Ecosystems: Technologies, Design and Implementation* (1st Ed.) (2022). CRC Press. https://doi.org/10.1201/9781003252542

Khanh, H.H., Khang, A., "The Role of Artificial Intelligence in Blockchain Applications", *Reinventing Manufacturing and Business Processes through Artificial Intelligence*, pp. 20–40 (2021). CRC Press. https://doi.org/10.1201/9781003145011-2

Musrat, G.E., Hajimahmud, A.V., Alyar, A.V., (Eds.). (2024). "Data-Centric Predictive Analytics for Solving Environmental Problems in the Building Stage of Smart City", *Smart Cities: IoT Technologies, Big Data Solutions, Cloud Platforms, and Cybersecurity Techniques* (1st Ed.). CRC Press. https://doi.org/10.1201/9781003376064-23

Novikov, Y.V., *Ecology of the environment and man: Moscow*, p. 347 (2021). https://link.springer.com/content/pdf/10.1007/3-540-28862-7.pdf#page=276

Ostroumov, S.A., *Problems of ecological safety of water supply sources*, No.5, p.17–21 (2006). https://www.e3s-conferences.org/articles/e3sconf/abs/2021/60/e3sconf_tpacee2021_02016/e3sconf_tpacee2021_02016.html

PAH Report, Human health effects of polycyclic aromatic hydrocarbons as ambient air pollutants – *Report of the Working Group on Polycyclic Aromatic Hydrocarbons of the Joint Task Force on the Health Aspects of Air Pollution*, 30 November 2021 | Report (2021). https://apps.who.int/iris/handle/10665/350636

Rani, S., Bhambri, P., Kataria, A., Khang, A., "Smart City Ecosystem: Concept, Sustainability, Design Principles and Technologies", *AI-Centric Smart City Ecosystems: Technologies, Design and Implementation* (1st Ed.) (2022). CRC Press. https://doi.org/10.1201/9781003252542-1

Rani, S., Chauhan, M., Kataria, A., Khang, A., "IoT Equipped Intelligent Distributed Framework for Smart Healthcare Systems", *Networking and Internet Architecture* (2021). CRC Press. https://doi.org/10.48550/arXiv.2110.04997

Schimmelpfennig, S., Kirillin, G., Engelhardt, C., Nützmann, G., Dünnbier, U., "Seeking a compromise between pharmaceutical pollution and phosphorus load: Management strategies for Lake Tegel, Berlin", *Water Research*, vol. 46, no. 13, pp. 4153–4163. DOI: 10.1016/j.watres.2012.05.024

Yunker, M.B., Macdonald, R.W., Vingarzan, R., et al. "PAHs in the Fraser River basin: a critical appraisal of PAH ratios as indicators of PAH source and composition", "*Organic Geochemistry*, vol. 33, pp. 489–515 (2002). ttps://www.sciencedirect.com/science/article/pii/S0146638002000025

Index

T

U

V

W

X

Z